SGP 2006

Symposium on Geometry Processing 2006

Fourth Eurographics Symposium on Geometry Processing

Cagliari, Sardinia, Italy
June 26 – 28, 2006

Paper Co-Chairs
Alla Sheffer, University of British Columbia
Konrad Polthier, Freie Universität Berlin

Proceedings Production Editors
Dieter Fellner (Graz University of Technology, Austria)
Stephen Spencer (The University of Washington, USA)

Co-sponsored by ACM SIGGRAPH and EUROGRAPHICS Association

Dieter W. Fellner, Werner Hansmann, Werner Purgathofer, François Sillion
Series Editors

Published by the Eurographics Association
-PO Box 16, CH-1288 Aire-la-Ville, Switzerland-
in cooperation with
Institute of Computer Graphics & Knowledge Visualization at Graz University of Technology
and
Institute of Scientific Computing at Technical University Brunswick.

Printed in Germany

Cover design by Stephen Spencer and Stefanie Behnke

ISBN 3-905673-36-3
ISSN 1727-8384

The electronic version of the proceedings is available from the Eurographics Digital Library at
http://diglib.eg.org

Table of Contents

Surface Modeling

Mesh Reconstruction

Representation and Coordinates

Table of Contents

Table of Contents

Preface

Geometry Processing is an emerging research field at the intersection of computer graphics, numerical computation and applied geometry. This book contains the research papers presented at the fourth Eurographics/ACM Symposium on Geometry Processing (SGP), held in Cagliari, Sardinia, June 26 to 28, 2006. SGP is the premier venue for disseminating cutting-edge research ideas and results in Geometry Processing. The research papers included in the book address diverse topics in Geometry Processing, including: surface reconstruction, model analysis and matching, computational geometry, surface fitting, remeshing, subdivision surfaces, and mesh editing.

This volume consists of 21 full research papers selected from a total of 79 submissions, plus a set of 5 short papers. All accepted submissions were reviewed by at least four reviewers, and at least two of them members of the program committee. Additionally, this year we introduced a virtual committee meeting with an online discussion among the program committee members.

In addition to the technical paper presentations the conference had two invited speakers, Markus Gross (ETH Zürich) and Ron Kimmel (Technion). We also held an industry panel discussion with representatives of major computer graphics companies, Richard Huddy (ATI), Olivier Paugam (mental images), Arnaud Ribadeau Dumas (Dassault Systèmes) and Rasmus Tamstorf (Walt Disney Feature Animation). This industry panel was a novel conference feature. It was introduced in response to participant feedback at SGP 2005 requesting a discussion platform on the most challenging research problems in industrial applications.

The success of the event relied on numerous helping hands of many friends and colleagues. First of all we wish to thank the wonderful management in Sardinia headed by the local organizers Riccardo Scateni and Roberto Scopigno. We are grateful to the many insights, tips, and helpful comments from the previous chairs, including Leif Kobbelt, Hugues Hoppe, Peter Schröder, Mathieu Desbrun, and Helmut Pottmann. Our special thanks go to Cici Koenig for the design of the conference logo and advertising material. We are indebted to Stephan Bischoff for his tremendous job running the online review management system and his invaluable support during the whole production of this volume. We also thank Stefanie Behnke and Dieter Fellner for production of these proceedings.

We greatly appreciate the contribution from the conference sponsors and would like to thank Microsoft Research, mental images, and IBM Research for their ongoing and continuous support of this symposium. We hope the research work and the novel results presented in this book will stimulate and initiate many more research activities. As paper chairs we had enjoyed collecting the state-of-the-art contributions of all authors into this volume, and we hope you as a reader will enjoy this volume too.

Konrad Polthier and Alla Sheffer

Sponsors

Eurographics Association

ACM**SIGGRAPH**

Microsoft
Research

mental images®

IBM Research

Keynotes

Are Points useful Modeling Primitives?

Markus Gross – Computer Graphics Laboratory, ETH Zürich
http://graphics.ethz.ch/~grossm/

Abstract

In recent years, point primitives have received a growing attention in graphics and modeling. There are two main reasons for this new interest in points: On one hand, we have witnessed a dramatic increase in the polygonal complexity of computer graphics models. The overhead of managing, processing, and manipulating very large polygonal meshes has led researchers to question the future utility of polygons as a fundamental modeling primitive. On the other hand, modern 3D digital photography and 3D scanning systems facilitate the ready acquisition of complex, real-world objects. These techniques generate huge volumes of point samples and create the need for advanced digital processing of points.

In this talk I will discuss the usefulness of sample-based representations for geometric and physically-based modeling. The first part of the talk is devoted to the general utility of points for geometric and graphics modeling. I will present an overview of the main research results in this area. Concepts for the representation of point sampled shapes will be addressed, as well as methods for the interactive modeling of point clouds. In addition, I will focus on data filtering algorithms as well as on digital geometry processing and compression of point data. In the second part of the talk I will address the more general issue of utilizing physically-based simulation in the interactive modeling process in order to create more intuitive modeling metaphors. I will use sample-based and meshless representations as an example and demonstrate their potential to simulate a wide range of real-world materials at interactive rates. The presented methods include real-time volumetric deformations as well as thin shells and approaches based on geometric shape matching. The talk will end with a critical discussion of the pros and cons of point sampled representations and the interplay of physical simulation and geometry processing.

Short Biography

Dr. Gross is a professor of computer science, head of the Institute of Computational Sciences, and director of the Computer Graphics Laboratory of the Swiss Federal Institute of Technology (ETH) in Zürich.

He received a Master of Science in electrical and computer engineering and a PhD in computer graphics, both from the University of Saarbrücken, Germany in 1986 and 1989. His research interests include scientific visualization and computer graphics, in particular point-based graphics, physics-based modeling, multiresolution analysis, and virtual reality.

He has widely published and lectured on these topics, and has taught courses at major conferences including ACM SIGGRAPH, IEEE Visualization, and Eurographics. Dr. Gross has served as a member of international program committees of many graphics and visualization conferences. He was a papers co-chair of the IEEE Visualization 1999 and 2002 conferences and of Eurographics 2000. This year, Dr. Gross has been chair of the papers committee of the ACM SIGGRAPH conference. He is a senior member of IEEE, a member of the IEEE Computer Society, a member of ACM and ACM Siggraph, and a member of the Eurographics Association. Dr. Gross is also a member of the advisory boards of various international research institutes and governmental agencies, and he has cofounded Cyfex and Novodex.

Keynotes

Biometrics, and Isometric Invariant Measures

Ron Kimmel – Computer Science, Technion, Haifa, Israel,
`www.cs.technion.ac.il/~ron`

Abstract

We will discuss the problem of matching metric spaces by embedding their intrinsic geometric structures in various domains. As for applications, we start with lip reading by flat embedding (joint with M. Aharon), and then continue to the 3DFace project (joint with the Bronstein brothers), and consider expression invariant face recognition. We review some theoretical problems and recent advances like the generalized multidimensional scaling (GMDS) that we use to embed the geometry of one surface into another.

As a side story, I will comment on the role of biometrics in biblical stories and fairy-tales.

Short Biography

Ron Kimmel received his B.Sc. (with honors) in computer engineering in 1986, the M.S. degree in 1993 in electrical engineering, and the D.Sc. degree in 1995 from the Technion – Israel Institute of Technology. During the years 1986-1991 he served as an R&D officer in the Israeli Air Force. During the years 1995-1998 he has been a postdoctoral fellow at the Computer Science Devision of Berkeley Labs, and the Mathematics Department, University of California, Berkeley. Since 1998, he has been a faculty member of the Computer Science Department at the Technion, Israel, where he is currently an associate professor.

He spent 2003-2004 as a visiting Professor at the Computer Science Department, Stanford University, and working with MediGuide Inc. His research interests are in computational methods and their applications in: Differential geometry, numerical analysis, image processing and analysis, and computer graphics.

Prof. Kimmel was awarded the Hershel Rich Technion innovation award (twice), the Henry Taub Prize for excellence in research, Alon Fellowship, the HTI Postdoctoral Fellowship, and the Wolf, Gutwirth, Ollendorff, and Jury fellowships.

He has been a consultant of HP research Labs in image processing and analysis during the years 1998-2000, and to Net2Wireless/Jigami research group during 2000-2001. He is on the advisory board of MediGuide (biomedical imaging 2002-2005), and has been on various program and organizing committees of conferences, workshops, and editorial boards of image processing and analysis journals, like International Journal of Computer Vision, and IEEE Trans. on Image Processing.

Prof. Kimmel is the author of "Numerical Geometry of Images" published by Springer, Nov. 2003.

Eurographics Symposium on Geometry Processing (2006)
Konrad Polthier, Alla Sheffer (Editors)

PriMo: Coupled Prisms for Intuitive Surface Modeling

| Mario Botsch | Mark Pauly | Markus Gross | Leif Kobbelt |
| ETH Zurich | ETH Zurich | ETH Zurich | RWTH Aachen |

Abstract

We present a new method for 3D shape modeling that achieves intuitive and robust deformations by emulating physically plausible surface behavior inspired by thin shells and plates. The surface mesh is embedded in a layer of volumetric prisms, which are coupled through non-linear, elastic forces. To deform the mesh, prisms are rigidly transformed to satisfy user constraints while minimizing the elastic energy. The rigidity of the prisms prevents degenerations even under extreme deformations, making the method numerically stable. For the underlying geometric optimization we employ both local and global shape matching techniques. Our modeling framework allows for the specification of various geometrically intuitive parameters that provide control over the physical surface behavior. While computationally more involved than previous methods, our approach significantly improves robustness and simplifies user interaction for large, complex deformations.

Categories and Subject Descriptors (according to ACM CCS): I.3.5 [Computer Graphics]: Computational Geometry and Object Modeling Geometric Transformations;

1. Introduction

In recent years, significant progress has been made in establishing triangle meshes as a representation for advanced geometric modeling. One of the most challenging geometry processing operations for meshes is high quality shape deformation. Ideally, mesh editing should be interactive and intuitive at the same time, but the large space of possible shape modifications often leaves the effects of user-controlled constraints hard to predict. With presently available methods achieving interactive or even real-time performance on large triangle meshes [BK04, SCOL*04, LSLCO05, BK05], the amount of "guidance" required from the designer remains a major bottleneck. Very often, the inherent limitations of the underlying deformation models force designers to split up complex deformations into a sequence of smaller ones.

Physically accurate surface deformations require the minimization of non-linear stretching and bending energies, resulting in the well known thin-plate and thin-shell functionals [TPBF87, CG91, WW92]. Linearization of curvatures with respect to a fixed reference mesh, an approximation often utilized by current mesh modeling approaches, leads to parametric distortions for large deformations and thus to a degradation of the surface. As a consequence, complex deformations have to be split up into multiple smaller ones, which complicates the overall modeling process.

We propose a new intuitive and robust shape modeling approach based on a *non-linear* surface deformation model. Our approach is inspired by the physical behavior of thin shells and computes intuitive deformations by emulating the natural material behavior of surfaces we experience in real-life. Rather than simulating accurate deformation physics we achieve *physically plausible* behavior while retaining interactive performance. Although our method is computationally more involved than previous approaches, it trades computational effort with the time the designer requires for guiding the modeling process.

Our underlying surface deformation model is based on a layer of *rigid* prisms which is enveloping the mesh faces. The prisms are coupled through non-linear elastic energies, which naturally resist stretching and bending and thus emulate the mechanical behavior of thin shells and plates. The rigidity of the prisms prevents them from degenerating even under extreme deformations, thus making our method numerically stable. The prisms' rigid motions are guided by a global shape matching procedure. We adapt and improve techniques developed for simultaneous registration of multiple objects [PLH02] for the efficient solution of the underlying geometric optimization. Its robustness and efficiency enable our surface model to be incorporated into an interactive and intuitive shape deformation application.

Figure 1: *The Goblin was posed by prescribing position and orientation for its head and right hand. For the left hand and foot positions were constrained only, thus enabling the automatic optimization of their orientations. The natural bending of joints was easily achieved by reducing the surface stiffness in these regions. The whole editing session took less than 5 minutes (see the accompanying video).*

Contributions. Our major contribution is a non-linear surface deformation model based on elastically connected rigid prisms, which features

- Robust and physically plausible large-scale deformations
- Intuitive preservation of surface details
- Hard constraints for positions and orientations
- Constraint-based and force-based deformations
- Intuitive geometric parameters for surface modeling

In addition, we present a robust and efficient numerical optimization method which combines local and global shape matching techniques. Based on these components, our surface deformation approach allows for intuitive and interactive shape deformations, as shown in Figure 1.

2. Related Work

Several shape editing approaches, like freeform deformation [SP86] and Wires [SF98], focus on shape *design* rather than on physically inspired shape *deformation*. While these methods typically provide more flexibility than physically-based ones, they in turn require more guidance if physically plausible deformations are actually desired.

The *design* of high quality surfaces is typically based on a constrained minimization of curvature energies [WW92, MS92], which results in thin-plate surfaces with planar rest states. Similarly, high quality physically-based *deformations* minimize stretching and bending (i.e., the change of curvature) under deformation constraints, which corresponds to thin-shell models of non-planar rest states [TPBF87, CG91].

Most recent physically inspired mesh deformation approaches can be categorized into two classes, one minimizing bending energies, the other one modifying differential coordinates. Figure 2 compares the behavior of a subset of the methods outlined below on simple synthetic examples.

Shape modeling based on a discretization of variational bending energy minimization (VARMIN in Figure 2) is mathematically well understood and yields smooth and tangent-continuous deformations [KCVS98, GSS99, BK04]. Similarly, [BK05] minimizes an analogous energy for space deformations (RBF in Figure 2). However, since for these approaches all computations and linearizations are performed w.r.t. a *fixed* reference mesh, large deformations might lead to shape distortions.

To correctly deform fine surface details, the above methods require a multi-scale decomposition, which splits a surface into a smooth base surface (low frequencies) and displacement vectors (high frequencies). Changing the smooth base surface and adding the details back onto it then yields the desired multi-scale deformation [KVS99]. The bottom row of Figure 2 shows the advantage of the *multi-scale* technique VARMIN over the *single-scale* RBF method, although even the multi-scale technique distorts the left-most bumps in a counter-intuitive manner.

Displacement volumes [BK03] encode the high frequencies by prism elements enclosed between the original and the base surface, which avoids detail distortion, but comes at the considerably higher cost of a non-linear detail reconstruction. Notice that displacement volumes are a multi-scale representation only, not a surface deformation technique, like the one presented in this paper. Although both representations (displacement vectors/volumes) can be combined with any underlying deformation technique, the required multi-scale decomposition can become quite involved for geometrically or topologically complex models.

To avoid the multi-scale decomposition, other methods modify differential surface properties instead of its spatial coordinates, and then solve a linear Poisson system for a deformed surface with the desired differential coordinates [LSCO*04, SCOL*04, YZX*04, ZRKS05, LSLCO05].

From this class, the methods of [YZX*04, ZRKS05] use gradients of affine deformations, i.e., their rotation and scale/shear components, for transforming surface gradients (GRAD), similar to [SP04, SZGP05]. As a consequence, these methods work well for rotations, but are insensitive to translations: Adding a translation to a given deformation does not change its gradient, and thus has no influence on the resulting surface gradients. But since even pure translations induce local rotations of tangent planes (Figure 2, bottom), these methods are counter-intuitive for modifications containing large translations. Although a special treatment of pure translations might be possible, deformations containing rotations and translations remain problematic.

In contrast, the shape editing approach of [SCOL*04] implicitly solves for local rotations of vertex neighborhoods, but due to linearizations their method has problems with large rotations, as was shown in their follow-up paper [LSLCO05]. In that paper, Lipman et al. minimize bending by preserving relative per-vertex orientations. They first

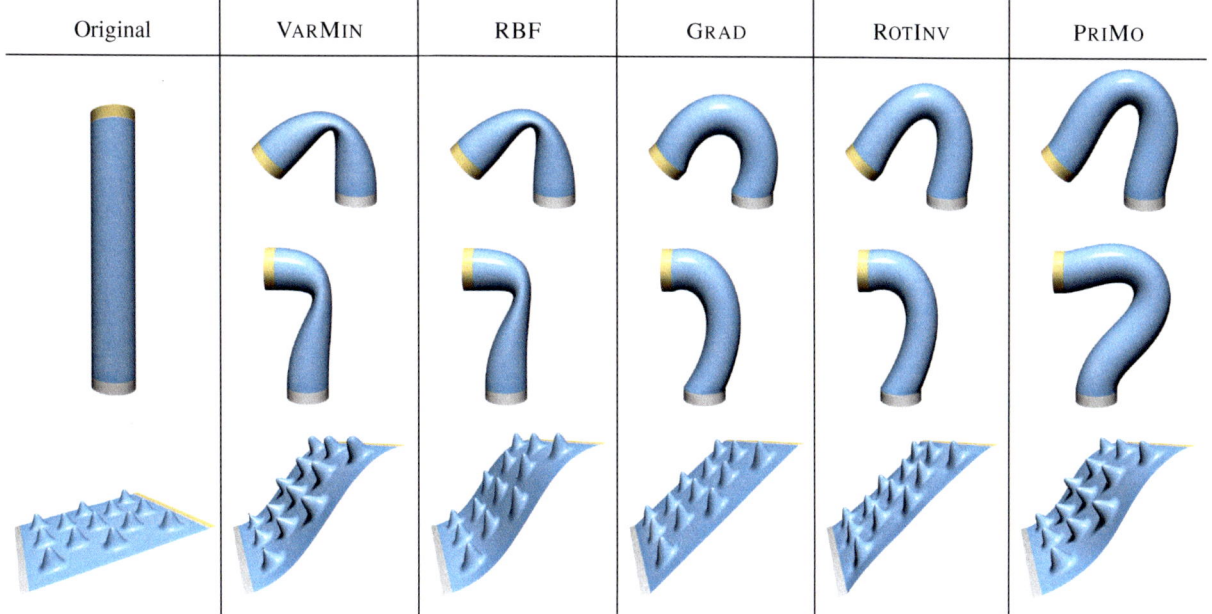

Original	VarMin	RBF	Grad	RotInv	PriMo

Figure 2: *Comparison of* VarMin *[BK04],* RBF *[BK05],* Grad *[ZRKS05],* RotInv *[LSLCO05], and the proposed* PriMo, *based on large, one-step deformations by rotations* (top), *rotations and translations* (center), *and pure translations* (bottom). *While* VarMin *and* RBF *work well for translations, they have problems with large rotations, whereas* Grad *and* RotInv *exhibit exactly the opposite behavior. Deformations containing large rotations as well as translations are therefore difficult for all of them. Moreover, the linear approaches lead to a noticeable loss of material for the two cylinder examples. In contrast, our non-linear deformable surface model* PriMo *successfully handles all cases.*

solve a linear system for per-vertex orientations, and from those reconstruct vertex positions in a second step. Since the first system does not consider position constraints, their technique also neglects the connection between translations and rotations. While their method works very well even for large rotations, it exhibits the same translation-insensitivity as gradient-based methods (RotInv in Figure 2).

The sketch-based deformation methods of [NSACO05, ZHS*05] provide more guidance to the system by deforming curves on the surface and propagating their local rotations over a region of interest. Since both methods are based on differential coordinates, they are in principle affected by translation-insensitivity, but the dense curve constraints avoid these problems in general. While a curve-based modeling metaphor would also be possible for our representation, we focus on sparse deformation constraints and a simple click & drag user interface.

Notice that all deformations of Figure 2 were done in a single large step, although the linear methods would perform better when splitting them into multiple smaller ones. But this requires to re-factorize the involved matrices in each step, which considerably increases computation costs. Moreover, since it changes the reference parametrization, a simple "undo" operation by moving the constrained vertices back to their original positions is disabled.

These inherent limitations of linear methods motivated us to investigate *non-linear* deformation techniques. In this context, Sheffer and Kraevoy [SK04] proposed non-linear, rigidly invariant pyramid coordinates. Their method corresponds to a non-linear extension of differential coordinates, and was shown to be capable of large deformations.

In contrast, we employ a constrained minimization of a non-linear bending energy. Since we aim at a qualitative emulation of the mechanical behavior of thin shells, we can provide more intuitive parameters for controlling the surface deformation. Notice that our method does not require a multi-scale decomposition, since our non-linear optimization correctly accounts for the coupling of translations and local frame rotations.

Although our approach is related to recent sophisticated shell simulations [GHDS03, WSG05], the latter usually follow different goals, since they are interested in the dynamic behavior of objects, including masses, inertia, and collisions. Their involved computations typically are not designed to robustly handle the arbitrarily extreme user constraints of interactive modeling applications. Instead of a *physically accurate* fully dynamic simulation we are explicitly targeting *physically plausible* shape deformations only, and thus can trade off physical accuracy for computational efficiency and numerical robustness.

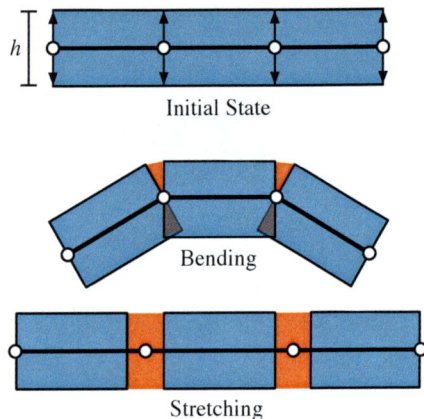

Figure 3: *After extruding prisms from the faces of the input mesh along vertex normals* (top), *the distortion of elastic joints between neighboring prisms is used to measure deformation energy, e.g., for bending or stretching.*

3. Prism Representation

Thin shells are volumetric objects of almost vanishing thickness and can therefore be modeled by a thin volumetric layer wrapped around a center surface [Bat95]. Following this intuition we consider the input mesh as the center surface, and the volume enclosed between two offset surfaces as the volumetric layer. This layer is built by an extrusion along vertex normals, which results in a (in general non-orthogonal) prism P_i for each mesh face F_i (cf. Figure 3, top).

A standard FEM formulation would be the straightforward way to generalize the shell's deformation energy to these prisms. However, most finite element techniques become numerically instable as soon as elements degenerate, since then neither volumes, areas, nor gradients can be computed robustly [ITF04]. Unfortunately, this kind of degeneracies is particularly likely to occur in interactive modeling applications, for instance due to extreme forces or constraints applied by the user.

In order to ensure numerical robustness even under extreme deformations, we prevent prisms from degenerating by keeping them rigid. We connect the rigid prisms along their common faces by *elastic joints*, which are stretched under deformations. The amount of stretching then yields the desired deformation energy (cf. Figure 3).

For the definition of the elastic joint energy we consider two neighboring prisms P_i and P_j. In the undeformed state the two prisms share a common face, but after a deformation these side faces might no longer coincide. The face of P_i neighboring P_j is a rectangular bi-linear patch $\mathbf{f}^{i \rightarrow j}(u,v)$, $(u,v) \in [0,1]^2$, which interpolates its four corner vertices $\{\mathbf{f}_{00}^{i \rightarrow j}, \mathbf{f}_{10}^{i \rightarrow j}, \mathbf{f}_{01}^{i \rightarrow j}, \mathbf{f}_{11}^{i \rightarrow j}\}$. Analogously we denote the opposite face by $\mathbf{f}^{j \rightarrow i}(u,v) \subset P_j$ (cf. Figure 4).

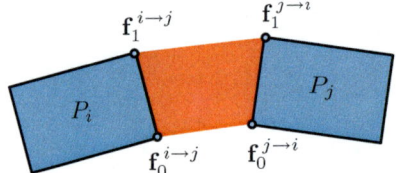

Figure 4: *Notation for prism elements.*

We define the energy between P_i and P_j as

$$E_{ij} := \int_{[0,1]^2} \left\| \mathbf{f}^{i \rightarrow j}(u,v) - \mathbf{f}^{j \rightarrow i}(u,v) \right\|^2 du\,dv , \quad (1)$$

which corresponds to an integral over infinitesimal elastic forces, that can be thought of as fibers of the elastic joint. As shown in the appendix, the above equation evaluates to a simple quadratic expression in the four difference vectors $(\mathbf{f}_{kl}^{j \rightarrow i} - \mathbf{f}_{kl}^{j \rightarrow i})$, $k, l \in \{0, 1\}$. The deformation energy of the whole mesh can now be defined as an accumulation of pairwise energies E_{ij}

$$E := \sum_{\{i,j\}} w_{ij} \cdot E_{ij} , \quad w_{ij} := \frac{\|\mathbf{e}_{ij}\|^2}{|F_i| + |F_j|} , \quad (2)$$

where the energy contribution of each pair P_i, P_j is weighted by the areas of the corresponding mesh faces F_i, F_j, and the squared length of their shared edge $\mathbf{e}_{ij}$ [GHDS03]. Notice that due to the zero rest length of the elastic joints the initial (undeformed) configuration of prisms is the unique global minimum of the energy, and any bending, shearing, twisting, or stretching increases it.

Our prism-based modeling metaphor works as follows: The user prescribes positions and/or orientations of an arbitrary subset of prisms. The optimization technique described in the next section then finds individual rigid motions for all unconstrained prisms, such that the global deformation energy (2) is minimized. The deformed surface mesh is finally derived from the resulting prisms by updating the position of each unconstrained vertex using the average transformation of its incident prisms.

As an additional benefit besides robustness, the prism formulation provides geometrically intuitive parameters for controlling the surface behavior: The extrusion amount h, i.e., the layer's thickness, determines the local surface stiffness, since for taller prisms the same bending angle induces a higher joint stretching (cf. Figure 5).

Another important property of our formulation is that it is not restricted to pure triangle meshes, but can also be applied to arbitrary polygonal meshes. Extruding a prism from an n-gon then simply generates an n-sided prism instead of a triangular one. Especially when dealing with regular quad meshes generated by CAD systems, inserting an arbitrary diagonal edge to split quads into triangles leads to asymme-

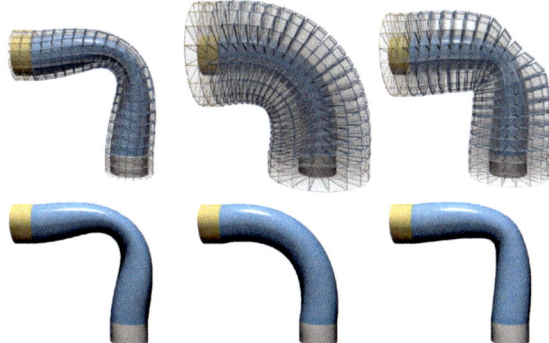

Figure 5: *The prisms' height intuitively controls the surface's resistance to bending. Local stiffness adjustment additionally allows to concentrate bending at desired joint locations, like the Goblin's shoulders and elbows in Figure 1.*

tries. In contrast, our method allows to directly process these meshes and thereby preserve their inherent symmetries.

Notice that the initial prism generation might lead to local self-intersections in the offset surfaces in regions of high curvature, which would lead to locally inverted prisms. However, as our energy only considers the elastic joints *between* prisms, it is not negatively affected by these configurations, which also holds for prisms interpenetrating during deformations, like those inside the cylinder in Figure 5.

4. Numerical Solution

In this section we propose a robust and efficient technique for the constrained minimization of the deformation energy defined in Equation (2). Our approach is based on generalized shape matching and adapts both local and global shape matching techniques in order to combine them to a hierarchical multigrid solver.

The user controls the surface deformation by constraining the position and/or orientation of certain prisms (respectively faces). The optimization then finds optimal rotations $\mathbf{R}_i$ and translations $\mathbf{t}_i$ for the unconstrained prisms P_i, such that the deformation energy (2) is minimized (we replace (u, v) by $\mathbf{u}$ for notational convenience):

$$\min_{\{\mathbf{R}_i, \mathbf{t}_i\}} \sum_{\{i,j\}} w_{ij} \int_{[0,1]^2} \left\| \mathbf{R}_i \mathbf{f}^{i \to j}(\mathbf{u}) + \mathbf{t}_i - \mathbf{R}_j \mathbf{f}^{j \to i}(\mathbf{u}) - \mathbf{t}_j \right\|^2 d\mathbf{u}.$$

(3)

This minimization actually corresponds to a generalized global shape matching problem: Discretizing the integrals by summations over sample points $\mathbf{f}^{i \to j}(u_k, v_k)$ would lead to a global alignment problem for multiple point sets, where rigid motions are to be found to minimize the sum of squared point distances.

As a consequence, the minimization (3) can be thought of as global alignment of prisms based on *continuous* face correspondences, instead of *discrete* point correspondences. This continuous formulation is mathematically more elegant compared to a sufficiently dense point-sampling of prism faces, and is also quite efficient, since the involved integrals evaluate to simple quadratic functions, as shown in the appendix.

For the global alignment of multiple point sets a large variety of techniques has been proposed, being based on either local pairwise alignment or simultaneous global registration. We will adapt both techniques to our problem in Section 4.1 and Section 4.2, and combine both them to an efficient multigrid solver in Section 4.3.

4.1. Local Shape Matching

A common approach to global registration is based on iterated pairwise alignment. In each iteration one prism P_i is randomly chosen and its position and orientation is optimized w.r.t. the remaining ones, which are kept fixed. Since each iteration minimizes the *local* shape matching error of P_i and does not change the other prisms, the *global* shape matching error decreases monotonically.

When picking a prism P_i for optimization, we have to match its faces $\mathbf{f}^{i \to j}$ to the corresponding faces $\mathbf{f}^{j \to i}$ of its neighbors, which we denote as P_j, $j \in \mathcal{N}_i$. Finding the best rigid motion $(\mathbf{R}_i, \mathbf{t}_i)$ yields a weighted pairwise shape matching problem

$$\min_{\mathbf{R}_i, \mathbf{t}_i} \sum_{j \in \mathcal{N}_i} w_{ij} \int_{[0,1]^2} \left\| \mathbf{R}_i \mathbf{f}^{i \to j}(\mathbf{u}) + \mathbf{t}_i - \mathbf{f}^{j \to i}(\mathbf{u}) \right\|^2 d\mathbf{u} \, ,$$

(4)

for which a simple closed-form solution can be computed by generalizing the method of [Hor87] to continuous face correspondences (see appendix).

This iterated pairwise matching is efficient and simple to implement, since each matching only requires an eigenvector decomposition of a 4×4 matrix (67k matches/sec on a 3.2GHz P4). Müller et al. [MHTG05] successfully used a similar discrete formulation for their deformation approach, but the number of clusters to be matched in their case is rather small compared to our number of prisms.

The main limitation of the local matching is that it corresponds to an error diffusion, and hence exhibits the typical behavior of iterative smoothers: for large systems the high frequencies of the error are rapidly attenuated, but the low frequencies — which correspond to the desired global deformations — take impractically long to converge.

4.2. Global Shape Matching

Instead of iterated pairwise registrations, several techniques for the simultaneous registration of multiple point sets have been proposed, see [KLMV05] and the references therein. Most of these methods factorize dense matrices, whose dimensions are proportional to the number of objects to be matched. While this is not critical when matching < 100 objects, in our setting a large number of prisms would lead to prohibitively complex matrices.

In contrast, Pottmann et al. [PLH02] propose an iterative simultaneous registration which involves solving *sparse* linear systems only, and hence can be adapted to our problem. Their technique corresponds to a Newton-type minimization of the registration error: In each iteration a linear system is solved for a descent direction, which corresponds to an *affine* motion per prism. A projection of those onto the manifold of rigid motions results in a *rigid* update for each prism. This process is iterated until convergence.

The descent direction of the Newton-type iteration requires first-order approximations A_i of rigid motions (R_i, t_i), which can be formulated in terms of linear and angular velocities v_i and ω_i:

$$R_i(\cdot) + t_i \approx (\cdot) + \omega_i \times (\cdot) + v_i =: A_i(\cdot) . \quad (5)$$

Reformulating the energy minimization (3) in terms of these first-order approximations yields

$$\min_{\{v_i, \omega_i\}} \sum_{\{i,j\}} w_{ij} \int_{[0,1]^2} \left\| A_i\left(f^{i\to j}(u)\right) - A_j\left(f^{j\to i}(u)\right) \right\|^2 du .$$
$$(6)$$

As all integrals can again by evaluated analytically, (6) represents a standard quadratic minimization in the linear and angular velocities, the optimal values for which can be found by solving a sparse linear system [PLH02].

The resulting optimal velocities (v_i, ω_i) correspond to the Newton descent direction and represent first-order approximations A_i. Since those are *affine* transformations, they have to be projected back onto the manifold of rigid motions before applying them to the prisms P_i. For this step, [PLH02] propose to choose (R_i, t_i) as the helical motion associated with (v_i, ω_i). However, this method turned out to be restricted to very small update steps, which is sufficient for registering pre-aligned point sets, but in our case leads to impractically slow convergence for large deformations.

We therefore propose to project A_i by finding the "closest" rigid motion (R_i, t_i), where we measure distances of transformations by comparing their effects on the prism P_i. We find the closest rigid motion by minimizing

$$\min_{R_i, t_i} \int_{[0,1]^2} \left\| R_i f^{i\to j}(u) + t_i - A_i\left(f^{i\to j}(u)\right) \right\|^2 du , \quad (7)$$

which yields another local shape matching problem, as depicted in the following figure.

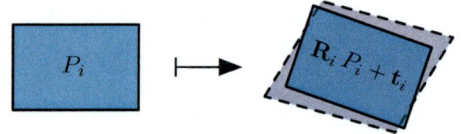

This geometrically intuitive projection operator allows for much larger update steps compared to the helical motions of [PLH02], which reduces the number of required Newton-type iterations by a factor of about 50 in all our examples. Although our projection is computationally more involved, its costs are still small compared to solving (6). Hence, the overall performance increases by roughly the same factor.

Finally, the Newton-like descent direction has to be scaled by a suitable step size λ. We thus derive the rigid motions (R_i, t_i) by projecting *scaled* velocities $(\lambda v_i, \lambda \omega_i)$ instead, where we simply start with $\lambda = 1$ and subsequently halve λ until the new rigid motions are found to decrease the energy (3). Although more elaborate methods exist [PHYH04], this simple technique turned out to be sufficient.

The computational complexity of the non-linear optimization is dominated by factorizing and solving the linear system corresponding to the minimization of (6) in each iteration. Since the matrix is sparse (about 16 non-zeros/row on average), symmetric, and positive definite, an efficient sparse Cholesky solver can be used [TCR03]. We can additionally exploit that the non-zero structure of the matrix stays fixed throughout all iterations. This allows us to precompute the symbolic part of the factorization [BBK05], which saves about 40% of the total time per iteration.

Combining the matching-based projection (7) with the symbolic pre-factorization reduces computation time by two orders of magnitude compared to [PLH02]. However, the optimization still achieves only 6500 prism updates per second on a 3.2GHz P4, which is not sufficient for interactive deformations of complex meshes. Given these limitations, neither the local nor the global shape matching yields a practically useful minimization technique by itself. But combining their respective strengths allows us to derive an efficient hierarchical method, as we will show in the next section.

4.3. Hierarchical Shape Matching

To maximize computational efficiency, we perform the shape matching on a multigrid hierarchy. For multigrid methods on irregular triangle meshes the successively coarser levels are built by mesh decimation [AKS05]. However, our framework does not require the hierarchy levels to represent consistent triangulations, since prisms can be generated from arbitrary polygons. This enables us to conveniently build the hierarchies levels (typically about 4) by successive clustering of neighboring faces and combining their corresponding prisms by considering them as one single rigid group.

A common practice for hierarchical multigrid solvers of linear systems [AKS05] is to use a *direct* solver on a coarse hierarchy level to obtain a low frequency approximation of the solution, which then is successively refined on higher levels using *iterative* techniques. Similarly, we start by applying the *global* shape matching on the coarsest hierarchy level in order to efficiently compute the low frequencies of the deformation. Since even for detailed surface meshes the shape deformations generally are smooth (low frequency) functions, this initial approximation typically is already very close to the exact solution. Since the *local* shape matching corresponds to an iterative error diffusion, we apply a few iterations (typically 2) on each finer hierarchy level, which rapidly smooths out the remaining high frequency errors.

Since we do not require consistent triangulations or sophisticated multigrid pre-conditioning, our hierarchical solver is considerably easier to implement compared to traditional multigrid techniques. The efficient combination of global and local shape matching yields a robust hierarchical non-linear optimization, which provides shape deformations of moderately complex models at interactive rates. Even our two most complex models, the 100k triangle Dragon of Figure 8 and the 180k triangle Goblin of Figure 1, can be edited interactively at one frame/sec (see the accompanying video).

4.4. Robustness

One of the main advantages of our method — and the main difference to existing shell-based techniques — is that during a deformation the shape quality of prisms will not degrade, since the individual prisms are kept rigid, which guarantees numerical robustness even for extreme deformations.

Even the initial shape quality of the prisms — which depends on the input mesh — only has a minor influence on the robustness of our method. The local and global shape matching techniques only fail for prisms that degenerate to a single line, which requires their corresponding triangles to degenerate to single points. However, the more likely cases of needle triangles (one extremely short edge) or caps triangles (one large angle) do not cause numerical problems, as long as stable normals can be computed for the prism extrusion. This allows us to process even meshes of low initial quality, which would be very likely to cause problems for classical FEM simulations.

A thorough convergence analysis of the Newton-like global shape matching can be found in [PHYH04]. In all our experiments the global matching converged robustly, with even extreme deformations requiring < 10 iterations. In theory, the minimization cannot be guaranteed to find the global minimum. Extreme user constraints that enforce the surface to form self-intersections might steer the iteration into a local minimum. However, as soon as the constraints are relaxed again, the optimization typically recovers, which is shown in Figure 6 and the accompanying video.

Figure 6: *The robust global optimization is able to fully recover the dragon model after fixing two prisms on the feet, collapsing all other prisms into one point, and randomly perturbing their orientations (left). The images show results after 1, 10, and 25 iterations of the global matching procedure.*

5. Results

In this section we show the flexibility of our prism-based modeling framework on a range of examples, including complex shape deformations and general surface processing.

In addition to robustness, our prism formulation also provides interesting, geometrically intuitive parameters for controlling the surface behavior. The rest state of the optimization can be adjusted by explicitly changing the prism shapes. Figure 5 already showed how surface stiffness can be specified in terms of prism heights. In addition to that, adjusting the prisms' widths allows to locally increase or decrease surface area. In the left image of Figure 7 the dragon model is T-Rex'ed by shrinking its arms and super-sizing its head.

Besides height and width, the prisms' deviation from orthogonality yields another interesting parameter. When prisms are generated by extrusion along vertex normals (as described in Section 3), the initial configuration is the rest state of the optimization. In contrast, extruding orthogonal prisms along face normals leads to a non-vanishing initial energy, which tries to achieve a locally planar state. Interpolating the extrusion directions between vertex normals and face normals therefore blends between a thin shell and thin plate behavior.

The latter tries to locally decrease curvature, which smooths the surface. However, since the size of prisms is kept fixed, the surface area is preserved, which avoids the typical shrinkage of Laplacian smoothing (cf. Figure 7, center). Moreover, extrapolating the face normals across vertex normals locally increases curvature and thus can be used for surface detail enhancement (cf. Figure 7, right).

Having these geometrically intuitive surface parameters at hand, the user can deform surfaces by simply selecting handle regions and moving them to their desired position. The respective transformations of the underlying prisms are automatically derived from these face constraints. Figure 8 shows a large-scale deformation of a complex dragon model

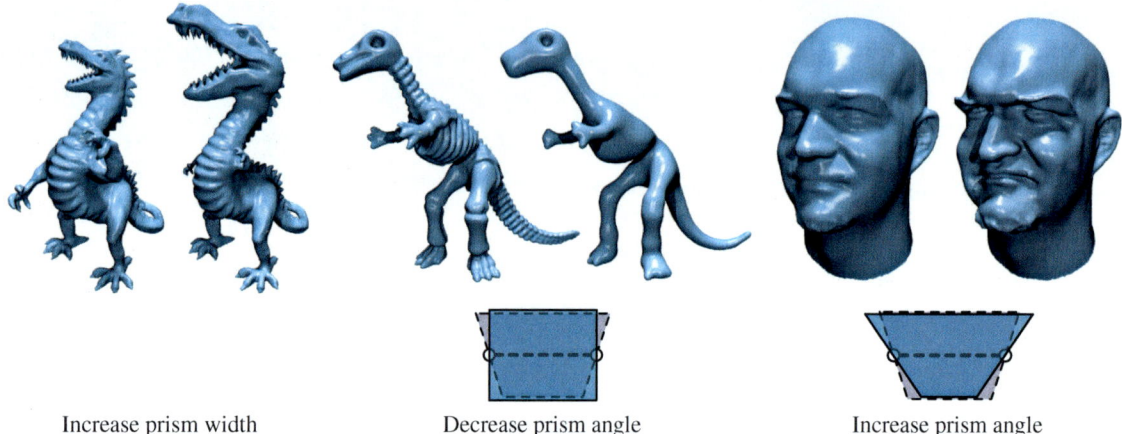

Increase prism width　　　　　　　　Decrease prism angle　　　　　　　　Increase prism angle

Figure 7: *Changing the prism shapes provides geometrically intuitive parameters for controlling the surface behavior. Adjusting the width of prisms can be used to locally shrink or enlarge surface area, which was done to convert the dragon to a T-Rex (left). Changing the prism's deviation from orthogonality blends between thin-shell and thin-plate behavior, which allows for non-shrinking smoothing (center). Increasing the prism angles instead amplifies surface curvature, and therefore enhances local surface details (right).*

and compares the result to the linear methods discussed in Section 2. While all linear methods fail to produce the desired result, our non-linear surface model deforms naturally, which can also be observed in the accompanying video.

Instead of fully constraining a prism's position and orientation, both the local and global shape matching formulations also allow to freeze either of them separately, such that the other term is free to be optimized. This enables a simple click & drag metaphor, where the user constrains the position of a dragged surface point, while its orientation is automatically optimized. This kind of interface would not be possible with methods based on differential coordinates, which require both rotation and translation constraints.

In addition to controlling surface deformations by enforcing hard constraints, our method also supports user-specified forces acting on the model. The squared point distances in Equation (3) can be interpreted as energies of zero-length springs, such that the shape matching solves for the steady-state of a (mass-less) spring system. User-defined spring forces can therefore be incorporated by adding surface points and corresponding target positions to the shape matching system. Depending on the application, this force-based modeling metaphor might provide physically more intuitive results, since enforcing hard constraints would correspond to extremely high forces.

The force-based metaphor, in combination with local stiffness control, was used to pose the Goblin model shown in Figure 1 in less than 5 minutes. Another example is shown in Figure 9, where a user-defined force pulls the Beetle's front upwards, performed for both a rather stiff and a more flexible surface material.

The limitation of our method is its computational performance, which restricts the global shape matching to about 10k prisms for interactive modeling. However, our hierarchical optimization provides interactive response rates even for complex meshes by performing the global optimization on a coarser level, for which 10k triangles are sufficient, since global shape deformations typically correspond to smooth functions.

6. Conclusion

We presented a non-linear surface deformation model based on elastically coupled rigid prisms, which allows for intuitive and physically plausible geometric modeling. In the past, non-linear techniques were rarely considered for interactive modeling applications because of their seemingly prohibitive computation costs, complicated implementation, and notorious numerical instabilities. In contrast, our new method combines ease of implementation and extreme robustness, while still achieving interactive rates for moderately complex models.

One promising direction for future work would be the application of our prism-based framework to physically inspired dynamic simulations, since in this context numerical robustness is also of major importance. The generalization of our global shape matching framework from thin shell surfaces to fully volumetric objects would also be an interesting extension.

Acknowledgments The authors are grateful to Jan Möbius for implementing [LSLCO05]. The Dinosaur and Head models (Figure 7) are courtesy of Cyberware, the Dragon model (Figure 8) is from the Stanford scanning repository.

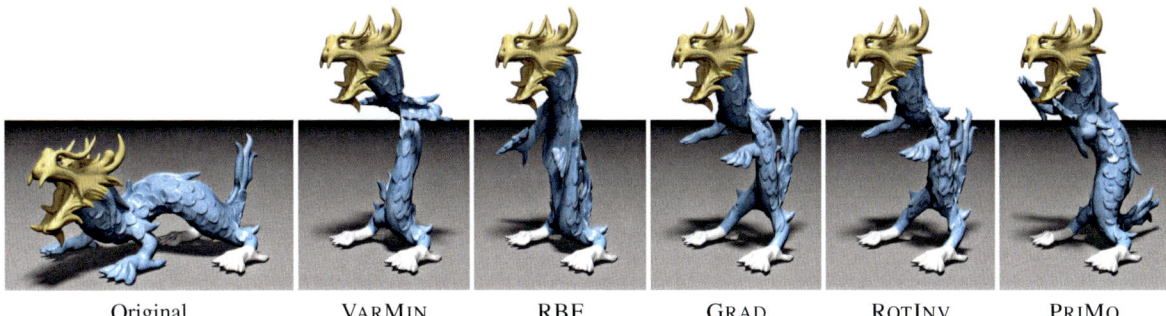

| Original | VARMIN | RBF | GRAD | ROTINV | PRIMO |

Figure 8: *The crouching dragon was lifted by fixing its hind feet and moving its head to the target position in a single step. Similar to Figure 2 the linear deformation methods yield counter-intuitive results, which even contain severe self-intersections. In contrast, our* PRIMO *technique leads to a very natural deformation.*

Figure 9: *In addition to hard constraints, our framework can also incorporate user-defined forces. In this example a force tries to lift the car's front, and center and right image show results for the same force on a rather stiff and a more flexible surface material, respectively.*

References

[AKS05] AKSOYLU B., KHODAKOVSKY A., SCHRÖDER P.: Multilevel Solvers for Unstructured Surface Meshes. *SIAM Journal on Scientific Computing 26*, 4 (2005).

[Bat95] BATHE K.-J.: *Finite Element Procedures.* Prentice Hall, 1995.

[BBK05] BOTSCH M., BOMMES D., KOBBELT L.: Efficient linear system solvers for geometry processing. In *11th IMA conference on the Mathematics of Surfaces* (2005).

[BK03] BOTSCH M., KOBBELT L.: Multiresolution surface representation based on displacement volumes. In *Proc. of Eurographics 03* (2003).

[BK04] BOTSCH M., KOBBELT L.: An intuitive framework for real-time freeform modeling. In *Proc. of ACM SIGGRAPH 04* (2004).

[BK05] BOTSCH M., KOBBELT L.: Real-time shape editing using radial basis functions. In *Proc. of Eurographics 05* (2005).

[CG91] CELNIKER G., GOSSARD D.: Deformable curve and surface finite-elements for free-form shape design. In *Proc. of ACM SIGGRAPH 91* (1991).

[GHDS03] GRINSPUN E., HIRANI A. N., DESBRUN M., SCHRÖDER P.: Discrete shells. In *Proc. of ACM SIGGRAPH/Eurographics symposium on Computer animation (SCA) '03* (2003).

[GSS99] GUSKOV I., SWELDENS W., SCHRÖDER P.: Multiresolution signal processing for meshes. In *Proc. of ACM SIGGRAPH 99* (1999).

[Hor87] HORN B. K. P.: Closed-form solution of absolute orientation using unit quaternions. *Journal of the Optical Society of America 4*, 4 (1987).

[ITF04] IRVING G., TERAN J., FEDKIW R.: Invertible finite elements for robust simulation of large deformation. In *Proc. of ACM SIGGRAPH/Eurographics symposium on Computer animation (SCA) '04* (2004).

[KCVS98] KOBBELT L., CAMPAGNA S., VORSATZ J., SEIDEL H.-P.: Interactive multi-resolution modeling on arbitrary meshes. In *Proc. of ACM SIGGRAPH 98* (1998).

[KLMV05] KRISHNAN S., LEE P. Y., MOORE J. B., VENKATA-SUBRAMANIAN S.: Global registration of multiple 3D point sets via optimization-on-a-manifold. In *Proc. of Eurographics symposium on Geometry Processing 05* (2005).

[KVS99] KOBBELT L., VORSATZ J., SEIDEL H.-P.: Multiresolution hierarchies on unstructured triangle meshes. *Comput. Geom. Theory Appl. 14*, 1-3 (1999).

[LSCO*04] LIPMAN Y., SORKINE O., COHEN-OR D., LEVIN D., RÖSSL C., SEIDEL H.-P.: Differential coordinates for interactive mesh editing. In *Proc. of Shape Modeling International 04* (2004).

[LSLCO05] LIPMAN Y., SORKINE O., LEVIN D., COHEN-OR D.: Linear rotation-invariant coordinates for meshes. In *Proc. of*

ACM SIGGRAPH 05 (2005).

[MHTG05] MÜLLER M., HEIDELBERGER B., TESCHNER M., GROSS M.: Meshless deformations based on shape matching. In *Proc. of ACM SIGGRAPH 05* (2005).

[MS92] MORETON H. P., SÉQUIN C. H.: Functional optimization for fair surface design. In *Proc. of ACM SIGGRAPH 92* (1992).

[NSACO05] NEALEN A., SORKINE O., ALEXA M., COHEN-OR D.: A sketch-based interface for detail-preserving mesh editing. In *Proc. of ACM SIGGRAPH 05* (2005).

[PHYH04] POTTMANN H., HUANG Q.-X., YANG Y.-L., HU S.-M.: *Geometry and convergence analysis of algorithms for registration of 3D shapes.* Tech. Rep. 117, Vienne University of Technology, 2004.

[PLH02] POTTMANN H., LEOPOLDSEDER S., HOFER M.: Simultaneous registration of multiple views of a 3D object. *Archives of the Photogrammetry, Remote Sensing and Spatial Information Sciences 34*, 3A (2002).

[SCOL*04] SORKINE O., COHEN-OR D., LIPMAN Y., ALEXA M., RÖSSL C., SEIDEL H.-P.: Laplacian surface editing. In *Proc. of Eurographics symposium on Geometry Processing 04* (2004).

[SF98] SINGH K., FIUME E.: Wires: A geometric deformation technique. In *Proc. of ACM SIGGRAPH 98* (1998).

[SK04] SHEFFER A., KRAEVOY V.: Pyramid coordinates for morphing and deformation. In *Proc. of Symp. on 3D Data Processing, Visualization and Transmission (3DPVT) '04* (2004).

[SP86] SEDERBERG T. W., PARRY S. R.: Free-form deformation of solid geometric models. In *Proc. of ACM SIGGRAPH 86* (1986).

[SP04] SUMNER R. W., POPOVIĆ J.: Deformation transfer for triangle meshes. In *Proc. of ACM SIGGRAPH 04* (2004).

[SZGP05] SUMNER R. W., ZWICKER M., GOTSMAN C., POPOVIĆ J.: Mesh-based inverse kinematics. In *Proc. of ACM SIGGRAPH 05* (2005).

[TCR03] TOLEDO S., CHEN D., ROTKIN V.: Taucs: A library of sparse linear solvers. http://www.tau.ac.il/~stoledo/taucs, 2003.

[TPBF87] TERZOPOULOS D., PLATT J., BARR A., FLEISCHER K.: Elastically deformable models. In *Proc. of ACM SIGGRAPH 87* (1987).

[WSG05] WICKE M., STEINEMANN D., GROSS M.: Efficient animation of point-sampled thin shells. In *Proc. of Eurographics 05* (2005).

[WW92] WELCH W., WITKIN A.: Variational surface modeling. In *Proc. of ACM SIGGRAPH 92* (1992).

[YZX*04] YU Y., ZHOU K., XU D., SHI X., BAO H., GUO B., SHUM H.-Y.: Mesh editing with Poisson-based gradient field manipulation. In *Proc. of ACM SIGGRAPH 04* (2004).

[ZHS*05] ZHOU K., HUANG J., SNYDER J., LIU X., BAO H., GUO B., SHUM H.-Y.: Large mesh deformation using the volumetric graph Laplacian. In *Proc. of ACM SIGGRAPH 05* (2005).

[ZRKS05] ZAYER R., RÖSSL C., KARNI Z., SEIDEL H.-P.: Harmonic guidance for surface deformation. In *Proc. of Eurographics 05* (2005).

Appendix A: Continuous Face-Based Shape Matching

We show how to extend the local and global shape matching approaches of [Hor87] and [PLH02] from discrete point correspondences to continuous face correspondences.

Suppose we are given two functions $a(\mathbf{u})$ and $b(\mathbf{u})$ defined by bi-linear interpolation of four values $\{a_{00},a_{10},a_{01},a_{11}\}$ and $\{b_{00},b_{10},b_{01},b_{11}\}$, respectively. Then their L_2 inner product simplifies to a weighted sum of 16 combinations of corner values:

$$\int_{[0,1]^2} a(\mathbf{u})\cdot b(\mathbf{u})\,d\mathbf{u} = \frac{1}{9}\sum_{i,j,k,l=0}^{1} a_{ij}\cdot b_{kl}\cdot 2^{(-|i-k|-|j-l|)}$$
$$=: \langle a,b\rangle_2 \ .$$

Using this, the continuous pairwise energy of Equation (1) evaluates to

$$E_{ij} = \left\langle \mathbf{f}^{i\to j}-\mathbf{f}^{j\to i}, \mathbf{f}^{i\to j}-\mathbf{f}^{j\to i}\right\rangle_2 \ .$$

In order to generalize the local shape matching of [Hor87], we first compute the weighted centroids $\mathbf{c}^i$ and $\mathbf{c}^*$ of the two face sets to be aligned, which leads to

$$\left.\begin{array}{c}\mathbf{c}^i\\\mathbf{c}^*\end{array}\right\} = \frac{1}{\sum_{j\in\mathcal{N}_i}w_{ij}}\sum_{j\in\mathcal{N}_i}\frac{w_{ij}}{4}\sum_{k,l=0}^{1}\left\{\begin{array}{c}\mathbf{f}^{i\to j}_{k,l}\\\mathbf{f}^{j\to i}_{k,l}\end{array}\right. .$$

To derive the optimal rotation $\mathbf{R}_i$ according to [Hor87], we build the matrix $\mathbf{N}=$

$$\begin{pmatrix}S_{xx}+S_{yy}+S_{zz} & S_{yz}-S_{zy} & S_{zx}-S_{xz} & S_{xy}-S_{yx}\\ S_{yz}-S_{zy} & S_{xx}-S_{yy}-S_{zz} & S_{xy}+S_{yx} & S_{zx}+S_{xz}\\ S_{zx}-S_{xz} & S_{xy}+S_{yx} & -S_{xx}+S_{yy}-S_{zz} & S_{yz}+S_{zy}\\ S_{xy}-S_{yx} & S_{zx}+S_{xz} & S_{yz}+S_{zy} & -S_{xx}-S_{yy}+S_{zz}\end{pmatrix}$$

from the component-wise L_2 inner products

$$S_{xx} = \sum_{j\in\mathcal{N}_i}w_{ij}\left\langle\left(\mathbf{f}^{i\to j}-\mathbf{c}^i\right)_x,\left(\mathbf{f}^{j\to i}-\mathbf{c}^*\right)_x\right\rangle_2 \ ,$$
$$S_{xy} = \sum_{j\in\mathcal{N}_i}w_{ij}\left\langle\left(\mathbf{f}^{i\to j}-\mathbf{c}^i\right)_x,\left(\mathbf{f}^{j\to i}-\mathbf{c}^*\right)_y\right\rangle_2 \ ,$$

and analogously for the other components. The eigenvector corresponding to the largest eigenvalue of $\mathbf{N}$ gives the optimal rotation $\mathbf{R}_i$ when interpreted as a unit quaternion. The optimal translation finally is $\mathbf{t}_i = \mathbf{c}^* - \mathbf{R}_i\,\mathbf{c}^i$.

For the generalization of the global shape matching approach of [PLH02] we have to adjust the linear system corresponding to the minimization of Equation (6). Assume two corresponding points $\mathbf{p}^{i\to j}$ and $\mathbf{p}^{j\to i}$ sampled from neighboring faces $\mathbf{f}^{i\to j}(\mathbf{u})$ and $\mathbf{f}^{j\to i}(\mathbf{u})$. Their contribution to the global energy is

$$w_{ij}\left\|\left(\mathbf{p}^{i\to j}+\omega_i\times\mathbf{p}^{i\to j}+\mathbf{v}_i\right)-\left(\mathbf{p}^{j\to i}+\omega_j\times\mathbf{p}^{j\to i}+\mathbf{v}_j\right)\right\|^2 \ ,$$

which gives four 6×6 matrix blocks. From those the global block structure of the matrix and the numeric values are easily derived. The continuous formulation then only requires to replace the involved products of the form $(\mathbf{p}^{i\to j})_x\cdot(\mathbf{p}^{j\to i})_y$ by the inner products $\left\langle\left(\mathbf{f}^{i\to j}\right)_x,\left(\mathbf{f}^{j\to i}\right)_y\right\rangle_2$ in the respective matrix entries.

Eurographics Symposium on Geometry Processing (2006)
Konrad Polthier, Alla Sheffer (Editors)

Hierarchical Error-Driven Approximation
of Implicit Surfaces from Polygonal Meshes

Takashi Kanai[1]   Yutaka Ohtake[2]   Kiwamu Kase[2]

[1] The University of Tokyo, Graduate School of Arts and Sciences, Japan
[2] RIKEN, VCAD Modeling Team, Japan

Abstract

This paper describes an efficient method for the hierarchical approximation of implicit surfaces from polygonal meshes. A novel error function between a polygonal mesh and an implicit surface is proposed. This error function is defined so as to be scale-independent from its global behavior as well as to be area-sensitive on local regions. An implicit surface tightly-fitted to polygons can be computed by the least-squares fitting method. Furthermore, this function can be represented as the quadric form, which realizes a compact representation of such an error metric. Our novel algorithm rapidly constructs a SLIM (Sparse Low-degree IMplicit) surface which is a recently developed non-conforming hierarchical implicit surface representation. Users can quickly obtain a set of implicit surfaces with arbitrary resolution according to errors from a SLIM surface.

Categories and Subject Descriptors (according to ACM CCS): I.3.5 [Computer Graphics]: Curve, Surface, Solid and Object Representations, G.1.2 [Numerical Analysis]: Approximation of Surfaces and Contours

1. Introduction

Polygonal mesh is nowadays recognized as a *de facto* standard geometric representation in the area of Computer Graphics (CG). A large variety of applications to generate or edit polygonal meshes have been developed. However, these polygonal meshes often contain defects such as gaps, T-junctions, self-intersections, and non-manifold structure. These problems must be handled with care for many purposes. Geometrically unfavorable conditions such as bad aspect ratios also render them unsuitable for other purposes such as numerical simulations.

On the other hand, an implicit surface is a well-known surface representation. Geometric details of an object can be represented using less surface primitives than meshes. Since the combination of multiple implicit surfaces can define a solid model strictly, issues for meshes described above do not occur. Implicit surfaces are convenient even for geometry processing because we do not need to take into consideration the connectivity of neighbor surfaces.

This paper provides a tool for rapidly converting polygonal meshes into implicit surfaces. We specifically deal with SLIM (*Sparse Low-degree IMplicit*) surfaces [OBA05]. Our

algorithm can approximate a SLIM surface which has the hierarchical structure of implicit surfaces. Each node of a SLIM surface contains an error between parts of a polygonal mesh and an implicit surface which is calculated in our algorithm. It is then possible to quickly extract different resolutions of SLIM nodes according to the error specified by the user.

The technical contributions of our approach include:

Surface Fitting using Polygon-Implicit Error Metrics.
We propose a novel error function of an implicit surface for polygons. It combines the algebraic distance and the normal error distance and is a natural extension of error functions for points. The function is also represented as a quadric form. This provides several advantages, for example, a compact storage and efficient approximation. Coefficients of an implicit function can be calculated by the least-squares fitting method. This requires only solving a small linear system.

Hierarchical Implicit Surface Approximation. Our scheme for the approximation to a SLIM surface from a polygonal mesh is originated from the mesh simplification method. The algorithm itself is quite simple, fast, and robust for creating a hierarchical tree structure of

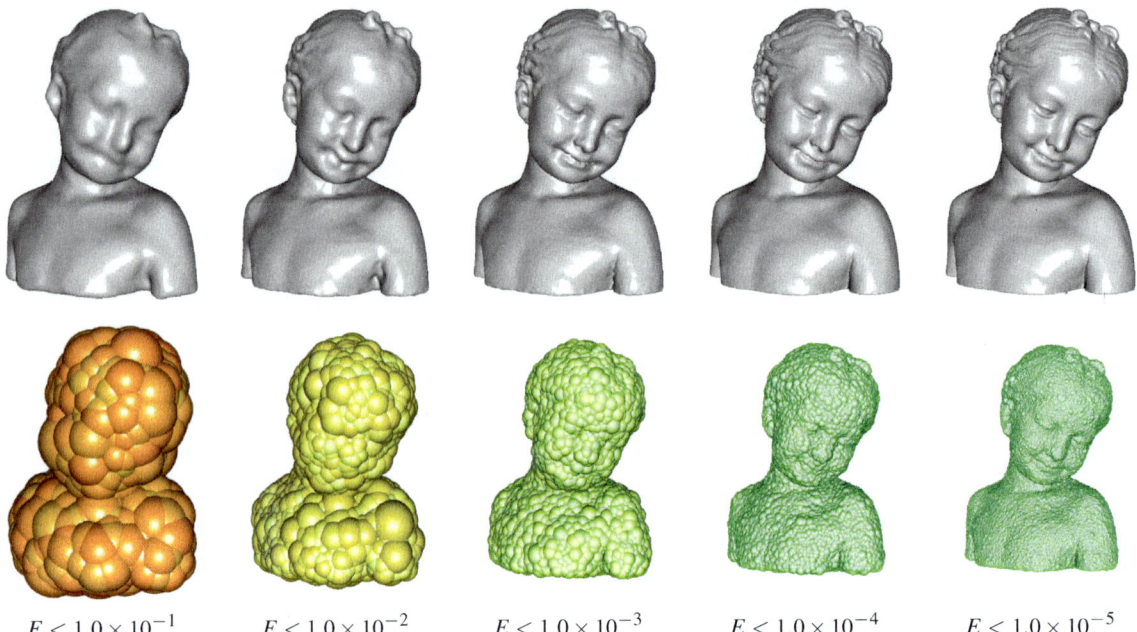

$$E < 1.0 \times 10^{-1} \qquad E < 1.0 \times 10^{-2} \qquad E < 1.0 \times 10^{-3} \qquad E < 1.0 \times 10^{-4} \qquad E < 1.0 \times 10^{-5}$$

Figure 1: *Error-driven approximation of SLIM surfaces for "bimba" mesh (1.2M polygons). Different specifications of error thresholds enable us to quickly generate various resolutions of SLIM surfaces. The number of nodes is: 721, 2,258, 7,076, 22,317, and 70,273 (from left to right). Color balls shown in the bottom denote a set of supports for the corresponded SLIM surfaces. Colors are assigned according to the error E which decreases from red to green.*

implicit surfaces. Since our algorithm computes a set of implicit surfaces in the order of increasing errors, it is easy to build such a hierarchy based on errors. Geometric features such as creases or corners can be preserved in the process of our algorithm.

Partition of Unity Evaluation with Sharp Features. To polygonize and visualize implicit surfaces with sharp features, we propose a new Partition of Unity (PU) evaluation method which can exactly estimate crease edges. This new PU method is valid for the generic representation of SLIM surfaces and is well-fitted to our approximation algorithm.

Fig. 1 clearly demonstrates the characteristics of our approach. We assign an error value calculated in the approximation process to each node of a SLIM surface. Using the hierarchical structure of a SLIM surface, we can then quickly extract a set of implicit surfaces within an error threshold specified by the user. Such the run-time extraction is useful for LOD rendering of SLIM surfaces. Also, the idea of such an error-driven extraction is a natural approach to control fitting errors for the use of the applications such as CAD.

1.1. Related Work

The most relevant research to ours can be seen in [SOS04]. They have proposed a construction method of implicit surfaces which approximates or interpolates polygonal meshes. Their method uses Moving Least Squares (MLS) surfaces

imposed with additional constraints such as points, normals, or integrals over polygons. The difference between their approach and ours is the algorithm of surface construction. Their algorithm first collects neighbor primitives (e.g. points, polygons) within an error threshold for each primitive. For such neighbor primitives, they fit a plane to generate a MLS surface. In contrast, our algorithm is performed hierarchically. A set of implicit surfaces with different resolutions can be generated by executing the algorithm only once.

Many approaches on implicit surface reconstruction from a set of input points have been proposed. In [SPOK95, CBC*01, TO02, YT02] the function is represented by globally-supported radial splines. This class of functions has a favorable property related to a solution's global behavior. One disadvantage of these approaches is that it takes considerable time to obtain the functions. [YT02] has proposed a method to roughly match polygons by using these splines. However, the resulting implicit surfaces still deviate substantially from the input.

Other types of tools for reconstruction use locally-supported implicit functions [Mur91, MYC*01, OBA*03, OBA05]. These approaches have the advantage that the position on a surface or its gradient can be rapidly evaluated due to their local property. Hence the local fitting scheme can be used to represent even a dense object by a large set of implicit surfaces. Our approach presented in this paper also uses this type of function.

More recently, the implicit surface fitting scheme is utilized to obtain surface objects in some applications. [IH03] fits implicit surfaces to generate smooth 3D objects from sketch information. Basic primitives such as spheres or cylinders are locally fitted to a mesh to extract characteristic features in [WK05].

1.2. SLIM Surfaces

Our approach mainly uses SLIM (*Sparse Low-degree IMplicit*) surfaces [OBA05]. It is composed of hierarchical tree-structured surfel nodes, each of which has low-degree implicit polynomial functions. Each surfel node is a local approximation of an object and an implicit function of a node is a rough approximation of those of its all children. A position or its gradient on a SLIM surface is represented as a blend of several wrapped neighbor surfels. MPU (*Multi-level Partition of Unity*) implicit surface [OBA*03] is quite a similar representation to the SLIM surface. The only difference is that a hierarchical structure of a MPU surface is created by the spatial partitioning of its object.

Although the original SLIM surface [OBA05] supports up to cubic degree polynomials, we use here only quadratic implicit functions for convenience. However, our approach can be applied to cubic or higher-degree polynomials. In these cases, formulae discussed in later sections are relatively complicated.

2. Implicit Quadratic Surface Fitting for Polygonal Meshes

In this section, we propose a novel method to fit an implicit surface to a polygonal mesh. We indicate here a special function to measure the distance between a polygon and an implicit surface.

The implicit curve or surface fitting problem has a history of over twenty years in the area of computer vision (See [Pra87] and [Tau91] for summary). One well-known function is the so called *algebraic distance*. It uses an absolute value of a function as an approximate distance. Although using an exact distance between a point and surface achieves better fit, it requires high computational costs because non-linear equations need to be solved [KP90].

The other function solved by linear equations is the *3L algorithm* proposed in [BLCC00]. Additional points are sampled in the shrunken and expanded regions and their functional values are computed. A combined error function based on these function values is minimized by solving a linear system. [TTC00] proposed the *gradient one algorithm* which adds the gradient constraint of implicit curves or surfaces. [HBM04] extended an approach of [TTC00] to fit an implicit curve robustly.

The difference between these error functions and ours is that our function evaluates over a polygon itself, whereas the others evaluate at a point set. We will show the comparison of results later in this section.

In the computer graphics community, similar algebraic functions to ours are proposed in [GWH01, CSAD04]. In their functions the distance between two polygons is measured. On the contrary, our functions described in the following subsections are formulated so as to efficiently compute the distance between an implicit surface to a polygon.

2.1. Polygon-Implicit Error Metrics

A quadratic implicit function $f(\mathbf{x})$ and its gradient $\nabla f(\mathbf{x})$ ($\mathbf{x} = (x, y, z)$) are represented by:

$$
\begin{aligned}
f(\mathbf{x}) &= f(x, \ y, \ z) \\
&= a_1 x^2 + a_2 y^2 + a_3 z^2 + a_4 xy + a_5 yz \\
&\quad + a_6 zx + a_7 x + a_8 y + a_9 z + a_{10} = 0, \quad (1) \\
\nabla f(\mathbf{x}) &= (2a_1 x + a_4 y + a_6 z + a_7, \\
&\quad\quad 2a_2 y + a_4 x + a_5 z + a_8, \\
&\quad\quad 2a_3 z + a_5 y + a_6 x + a_9). \quad\quad (2)
\end{aligned}
$$

On the other hand, a point $\mathbf{x}$ on a triangle $T \equiv \{\mathbf{v}_0, \mathbf{v}_1, \mathbf{v}_2\}$ is represented using barycentric coordinates (s, t):

$$
\mathbf{x} = s\mathbf{v}_0 + t\mathbf{v}_1 + (1 - s - t)\mathbf{v}_2, \quad (3)
$$
$$
0 \le s \le 1 - t, \quad 0 \le t \le 1.
$$

As noted above, we know that the exact distance between $\mathbf{x}$ and $f(\mathbf{x})$ can be calculated. However, it is non-linear and requires high computational cost [KP90]. Instead, we use the algebraic distance $|f(\mathbf{x})|$ adopted in previous approaches on implicit surface fitting as an approximate distance. We define a *distance error function* ε^{dis} as the squared algebraic distance integrated over a triangle:

$$
\varepsilon^{dis}(T) \equiv A \int_0^1 \left(\int_0^{1-t} |f(\mathbf{x})|^2 ds \right) dt, \quad (4)
$$

where A denotes the area of a triangle.

In addition, we define a *gradient error function* ε^{nrm} as an error between a normal vector $\mathbf{n}$ of a triangle and a gradient $\nabla f(\mathbf{x})$ integrated over a triangle:

$$
\varepsilon^{nrm}(T) \equiv A \int_0^1 \left(\int_0^{1-t} |\mathbf{n} - \nabla f(\mathbf{x})|^2 ds \right) dt, \quad (5)
$$
$$
\mathbf{n} = \frac{(\mathbf{v}_1 - \mathbf{v}_0) \times (\mathbf{v}_2 - \mathbf{v}_0)}{|(\mathbf{v}_1 - \mathbf{v}_0) \times (\mathbf{v}_2 - \mathbf{v}_0)|}.
$$

Note that $|\mathbf{n} - \nabla f(\mathbf{x})|$ is a more strict representation than a function $|\mathbf{n}\nabla f(\mathbf{x}) - 1|$ used in the gradient one algorithm [TTC00].

Fig. 2 illustrates the geometric meanings of our error functions. ε^{dis} (Fig. 2 left) is the integral sum of squared function values. It is then regarded as a volume surrounded at a triangle (hatched region in Fig. 2). Since such a volume is often not changed when we slide a surface to the horizontal direction of a triangle, ε^{dis} has a high degree of freedom in this direction. In contrast, ε^{nrm} (Fig. 2 right) prescribes in order to approach the average direction of gradients to the direction of a normal vector. In this case, ε^{nrm} has a high degree of freedom in the vertical direction of a triangle. Consequently, we expect that the above two functions restrain the movements of each other.

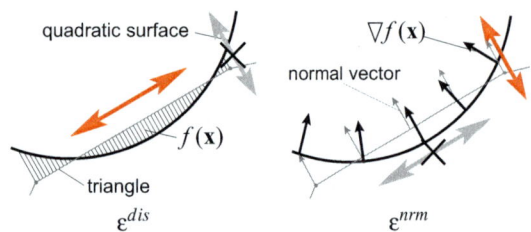

Figure 2: *Geometric meanings of error metrics.*

ε^{dis} and ε^{nrm} have a good property for the optimization of implicit functions. Let a coefficient vector of an implicit function $f(\mathbf{x})$ in (1) be $\mathbf{p} = (a_1, a_2, \ldots, a_{10})$. Both ε^{dis} and ε^{nrm} are then represented by the *quadric forms* as follows:

$$\varepsilon^{dis}(T) = \mathbf{p}\mathbf{A}^{dis}\mathbf{p}^T, \qquad (6)$$

$$\varepsilon^{nrm}(T) = \mathbf{p}\mathbf{A}^{nrm}\mathbf{p}^T - 2\mathbf{b}^{nrm}\mathbf{p}^T + c^{nrm}. \qquad (7)$$

The derivation of formulae from (4), (5) to (6), (7) is described in Appendix. Both $\mathbf{A}^{dis}$ and $\mathbf{A}^{nrm}$ are a 10×10 symmetric matrix respectively. Each matrix is composed of 55 floating point elements. $\mathbf{b}^{nrm}$ is a 10-dimensional vector (10 floating points) and c^{nrm} is a scalar (a floating point). We can then store 55 and 66 floating points for the above two functions. A total set of elements $\{\mathbf{A}^{dis}, \mathbf{A}^{nrm}, \mathbf{b}^{nrm}, c^{nrm}\}$ represents an error metric between a triangular polygon and implicit surface. We call it the quadratic *Polygon-Implicit Error Metric* (PIEM) here.

PIEM is an analogy of QEM (Quadric Error Metric) proposed in [GH97]. It inherits good properties from QEM: An addition of error functions for two triangles T_1, T_2 can be written by:

$$\varepsilon^{dis}(T_1) + \varepsilon^{dis}(T_2) = \mathbf{p}\left(\mathbf{A}_1^{dis} + \mathbf{A}_2^{dis}\right)\mathbf{p}^T,$$

$$\varepsilon^{nrm}(T_1) + \varepsilon^{nrm}(T_2) = \mathbf{p}\left(\mathbf{A}_1^{nrm} + \mathbf{A}_2^{nrm}\right)\mathbf{p}^T$$
$$- 2\left(\mathbf{b}_1^{nrm} + \mathbf{b}_2^{nrm}\right)\mathbf{p}^T$$
$$+ \left(c_1^{nrm} + c_2^{nrm}\right).$$

This is because the function is independently defined for a triangle.

2.2. Implicit Surface Fitting Using PIEMs

We discuss here that a quadratic implicit function $f(\mathbf{x})$ is fitted to a mesh composed of a set of triangles $M = \{T_i;$

$i = 1 \ldots n\}$. We define an error function $E(M)$ as follows:

$$E(M) \equiv E^{dis}(M) + \lambda E^{nrm}(M)$$
$$= \sum_{i=1}^{n}\left(\varepsilon^{dis}(T_i)\right) + \lambda\sum_{i=1}^{n}\left(\varepsilon^{nrm}(T_i)\right), \qquad (8)$$

where λ denotes a parameter for adjusting the dimensional scale between $E^{dis}(M)$ and $E^{nrm}(M)$. ε^{dis} is a two-dimensional quantity defined as the squared distance, whereas ε^{nrm} is a dimensionless quantity. If we do not set λ appropriately, fitting results would be changed according to the scale of an object. For this scale parameter λ, we set here the sum of triangle areas $\lambda = \sum_{i}^{n} A_i$. This setting works very well in all our experiments.

It should be noted that from (8) an error function for two neighbor meshes M_1, M_2 is given by:

$$E(M_1 + M_2) = E^{dis}(M_1) + E^{dis}(M_2)$$
$$+ (\lambda_1 + \lambda_2)\left(E^{nrm}(M_1)\right.$$
$$+ \left. E^{nrm}(M_2)\right), \qquad (9)$$

which does not satisfy the linearity described above. For this, we put λ in (8) to a PIEM as the 122nd element. An extended PIEM is then $\{\mathbf{A}^{dis}, \mathbf{A}^{nrm}, \mathbf{b}^{nrm}, c^{nrm}, \lambda\}$. Each element of a PIEM including λ is independently added in (9).

From (6) and (7), (8) is finally represented as the following quadric form:

$$E(M) = \mathbf{p}\mathbf{A}\mathbf{p}^T - 2\mathbf{b}\mathbf{p}^T + c. \qquad (10)$$

A coefficient vector $\mathbf{p}$ minimizing E can be computed by solving the following linear system:

$$\mathbf{A}\mathbf{p} = \mathbf{b}. \qquad (11)$$

We use SVD [PFTV92] to compute the inverse matrix of $\mathbf{A}$ because $\mathbf{A}$ becomes singular in rare occasions.

Evaluation of our error metrics. We investigate here the property of our error metric by comparing with the error function for a point set. We evaluate the fitting of curves in 2D to visually understand the results. A 2D error function is essentially the same as that in 3D. In the case of 2D, the length of a line segment of a poly-line is used as weight instead of the area of a polygon. We use the gradient one algorithm [TTC00] as a reference for the fitting method for a point set. However, we alternate a gradient error function to the 2D version of our function instead of the original function in [TTC00]. We measure the arithmetic average $\tilde{d}$ of the following approximate distances from a sampled point set proposed in [Tau94]:

$$\tilde{d} = \frac{1}{N}\sum_{i}^{N}\left(\frac{|f(\mathbf{x}_i)|}{|\nabla f(\mathbf{x}_i)|}\right). \qquad (12)$$

Fig. 3 shows the fitting results to an implicit curve from a poly-line consisting of six vertices or from its sampled point set. It can be seen from the results that our function in

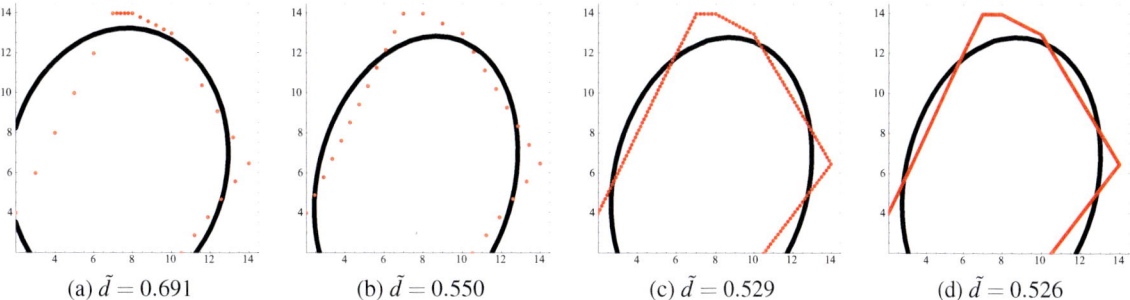

| (a) $\tilde{d} = 0.691$ | (b) $\tilde{d} = 0.550$ | (c) $\tilde{d} = 0.529$ | (d) $\tilde{d} = 0.526$ |

Figure 3: *2D implicit curve fitting results. A point or a line colored as red shows the geometry to be fitted. A black-colored bold curve denotes a contour curve of an implicit function $f(\boldsymbol{x}) = 0$. (a) Fitted to a sparse point set with constant sampling for each line segment (26 points). (b) Fitted to a sparse point set with spatially uniform sampling (27 points). (c) Fitted to a dense point set with spatially uniform sampling (138 points). (d) Fitted to a poly-line (6 vertices) by our approach. $\tilde{d}$ represents the average of approximate distances to an implicit curve from points.*

Fig. 3(d) has a minimum average distance among four examples and thus achieves the best-fit. In Fig. 3(a)-(c) three types of sampling methods are used. The same number of points (5 points) is sampled for each line segment of a poly-line in (a). In this case, a curve is attracted to a high-density region due to spatially irregular sampling. In (b) and (c), points are generated by spatially-uniform sampling. The sampling of (b) is sparser than that of (c). In these cases, the fitting result with more densely sampled points approaches our result (d).

Consequently, it can be thought that our error metric yields the same effects as fitting with a highly dense point set. This means that our approach is effective compared to approaches for a point set. In case of the fitting using a point set instead of a polygon, it is difficult to determine how dense we should sample points from a polygon to achieve better approximation. In contrast, our approach does not need to take into consideration such sampling rate issues. Moreover, our approach has the additional advantage that it is fast compared to the fitting computation for highly dense point sets.

3. SLIM Surface Approximation

In this section, we describe a novel algorithm to approximate a polygonal mesh to a SLIM surface based on PIEMs for the accurate restoration of surface geometry. Since a well-known mesh simplification scheme adopts the fine-to-coarse approach, we can obtain the hierarchical structure of a SLIM surface once the algorithm is performed.

3.1. Implicitization of Polygonal Meshes

The first step of our algorithm is the *implicitization* of a polygonal mesh i.e. to convert each triangle of a polygonal mesh to an implicit surface. A plane of a triangle is also defined as a linear degree implicit polynomial: That is, $a_1, a_2, \ldots, a_6$ in the coefficients of (1) are zero. We set these polynomials to a SLIM surface as leaf nodes with $E = 0$. We also define the support center $\mathbf{c}$ as a barycenter of a triangle.

The support radius r is calculated by the distance between the farthest point (one of three vertices) of a triangle and $\mathbf{c}$.

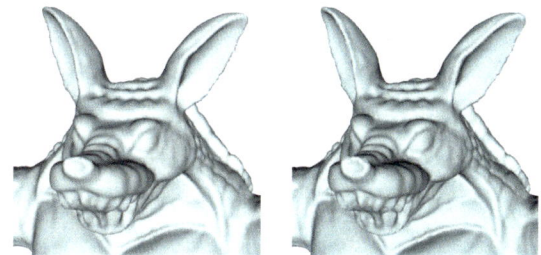

Figure 4: *Visual comparison between a polygonal mesh (left) and SLIM surface with leaf nodes (right).*

It can be seen from Fig. 4 that there is almost no visual difference between the rendering result of a polygonal mesh and that of leaf nodes in its converted SLIM surface.

3.2. Hierarchical Surface Approximation Using Face Clustering

Our simple scheme for approximation to a SLIM surface is based on the hierarchical face clustering [GWH01, She01, SSGH01] proposed as the approach for mesh simplification.

The algorithm begins to construct the *dual graph* from an input mesh. In the dual graph, the *node* corresponds to a face of a mesh, and the *edge* corresponds to the connectivity between two neighbor faces. For each edge of a dual graph, we next compute a minimum value E_{min} of an error function as well as approximating to an implicit surface in case two end nodes are combined together (we call the *edge-collapse* operation hereafter). This is done by minimizing a combined error function of two end nodes described in (9). We then push such an edge to a priority queue setting E_{min} as a key.

In the approximation algorithm, we pull out an edge of a minimum key and perform an edge-collapse operation using two end nodes (See Fig. 5). An implicit surface by solving

(11) is then added as a node of a SLIM surface. A combined node of a dual graph is a parent of two end nodes. Neighbor edges of each node are copied to such a newly-created node (duplicated edges are deleted here). Each E_{min} of neighbor edges in a combined node is re-computed and a priority queue is updated. We repeat the above processes until the priority queue becomes empty. When the algorithm terminates, a SLIM surface which has twice the number of hierarchical implicit surfaces as faces of a mesh is created.

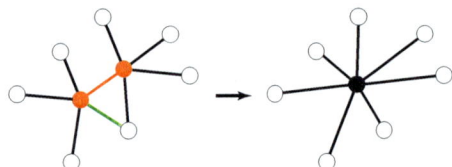

Figure 5: *Edge-collapse operation. One of two nodes shown in red-filled circles is collapsed to create a combined node shown in a black-filled circle. A green-colored line shows a neighbor edge and is also collapsed due to the duplication of two end nodes at a combined node.*

Dual graph construction. If a mesh face has the connectivity of neighbor faces, the construction of a dual graph is quite easy. We simply create an edge for each pair of neighbor faces. Even if a mesh face does not have the connectivity of neighbor faces (called as *polygon soup*), we can construct such a connectivity by using, for example, a simplified version of the approach proposed in [BDK98]. We first store edges of mesh faces to a spatial data structure such as an octree or a kD-tree. We next find a pair of neighbor faces by applying an adjacency processing method described in [BDK98]. An edge of a dual graph is created if a pair of faces is judged as adjacent to each other.

Approximation by PIEMs vs. on-the-fly approximation. There are two types of methods to manage error functions during the approximation. One is to use PIEMs similarly as QEMs described in [GH97]. In this case, we always store 122 floating points of a PIEM $\{\mathbf{A}^{dis}, \mathbf{A}^{nrm}, \mathbf{b}^{nrm}, c^{nrm}, \lambda\}$ in each node of a dual graph. In an edge-collapse operation, we simply sum up two PIEMs by using (9) and an implicit surface is approximated by using (11). Since the preparation for the approximation is only the addition of two PIEMs, the computational cost is dramatically reduced. However, quite a large memory space is required because we have to store 122 floating points for each node. In our implementation, the construction of a SLIM surface from 0.2M polygons requires 325MB memory space.

The other is the so called *on-the-fly* approximation by storing a list of mesh faces for each node. In a leaf node, only one face for creating an implicit surface is stored. In an edge-collapse operation, we combine face lists of two end nodes and compute $\mathbf{A}$ and $\mathbf{b}$ in (11) directly using a combined face list. We then free these elements as soon as the computation

is finished. In this case, the memory space can be reduced. However, the computational cost considerably increases especially before the end of the algorithm due to the need to query a face list in each edge-collapse operation.

The best choice seems to use a *hybrid* scheme of the above two methods especially for more than 1M polygons. In our current implementation, the algorithm starts with the on-the-fly approximation scheme. As the algorithm processes, we change to the management scheme to the approximation by using PIEMs when the number of the rest nodes in a priority queue becomes less than one-tenth of the number of mesh faces. Since the above two computations are exactly the same, such a change of the management schemes makes no difference in the final result.

Support center and radius for a new node. We compute a support center and radius in a combined node by the method introduced in [JP04]. [JP04] shows two approaches for these computations. One is the *wrapped hierarchy* which a parent sphere tightly bounds the geometry of a set of triangles in two child nodes. The other is the *layered hierarchy* which a parent sphere bounds two child spheres. We adopt here a layered hierarchy which can be computed by using only a support center and radius. This is especially advantageous for both two management schemes described above.

3.3. PU Evaluation and Sharp Features

A SLIM surface created by our proposed method is defined by the Partition of Unity (PU) evaluation method [OBA*03] as follows:

$$\tilde{f}(\mathbf{x}) = \frac{\sum_i w_i(\mathbf{x}) f_i(\mathbf{x})}{\sum_i w_i(\mathbf{x})}, \tag{13}$$

where $w(\mathbf{x})$ denotes an average weight function. As in [OBA*03], we use a one-dimensional quadric B-spline basis function whose parameter is the distance between a point $\mathbf{x}$ and a support center.

Preserving sharp features such as creases or corners is quite important especially for mechanical objects. Given a SLIM surface, we can obtain a globally smooth surface by using (13). However, if we perform a PU evaluation method as it is, sharp features are rounded as shown in Fig. 7 (a). One approach to address this issue is to define two or more implicit functions with Boolean operations for a node around a crease. Sharp features for such nodes can be evaluated by using the PU method as proposed in [OBA*03]. However in this case, our approximation algorithm is more complicated including a special process for such a node.

Instead, we introduce a different approach for evaluating sharp features. Based on an alternative approach, we can preserve such features by simple extension of our approximation scheme. In a new PU evaluation method, sharp features can be exactly evaluated with considering the discontinuities

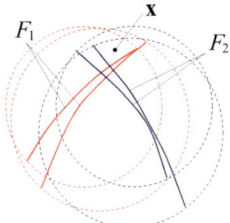

Figure 6: *Separating implicit functions into two clusters at a point x to be evaluated. Red and blue bold lines indicate separated sets of implicit functions F_1, F_2 respectively. Dot lines show the support spheres.*

of surface normals, even if only an implicit function is assigned to each node.

A new PU evaluation method is based on the *on-the-fly* estimation of the discontinuities of surface normals, which is composed of the following four steps:

1. Implicit functions $F = \{f_i | w_i(\mathbf{x}) > 0\}$ where a point $\mathbf{x}$ is included in their support spheres are collected.
2. A normalized gradient $\mathbf{g}_i(\mathbf{x}) = \nabla f_i(\mathbf{x})/|\nabla f_i(\mathbf{x})|$ is evaluated for each function.
3. F is clustered according to $\{\mathbf{g}_i(\mathbf{x})\}$ by using the method proposed in [OBA*03] (See Fig. 6 which is in the case of two clusters).
4. For each cluster, a function (13) is evaluated. A max/min operation is applied according to the convex/concave judgment at $\mathbf{x}$.

In Step 4, the convex/concave judgment for a pair of clusters F_1 and F_2 is performed as follows: For each cluster, we first compute a PU of normalized gradients $\mathbf{g}_1$ ($\mathbf{g}_2$) and an average position of support centers $\mathbf{c}_1$ ($\mathbf{c}_2$). If $(\mathbf{g}_2 - \mathbf{g}_1) \cdot (\mathbf{c}_2 - \mathbf{c}_1)$ is positive, a point is on a concave region because both the variation of normals and that of their surface positions are in the same direction (here we assume that normals are oriented to the inside of an object). In case of three clusters, we perform the convex/concave judgment for two pairs of clusters (F_1, F_2), (F_2, F_3) and then apply Boolean operations to the resulting two judgments. In our current implementation, we can handle up to three clusters.

Extension of the approximation algorithm for sharp features. For polygonal meshes with sharp features, we additionally extend our approximation algorithm described in Section 3.2. We first judge an edge as a crease if a dihedral angle of neighbor faces is more than a threshold when constructing edges of a dual graph. In the optimization process, we simply multiply a big number (e.g. 1,000) to an error function value in (9) when collapsing such a crease edge. This avoids the binding of neighbor faces around a crease.

Moreover, we add a "crease edge" flag to each node of an implicit function including a crease edge. When simplifying the nodes, we inherit such a flag to a newly-created node if at least one of two nodes has a flag. A new PU evaluation

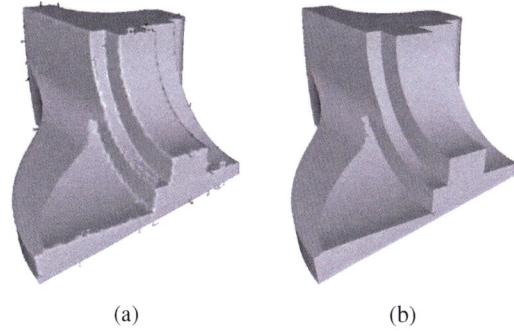

(a)   (b)

Figure 7: *Fitting results of a "fandisk" mesh with sharp features. (a) A SLIM surface (5K nodes) without preserving sharp features. (b) A SLIM surface (5K nodes) with preserving sharp features.*

method described above is applied only if a collected set of implicit functions in Step 1 has more than two flags. This achieves more robust evaluations in the vicinity of sharp features.

Nodes around crease edges are preserved wherever possible by applying the above extended algorithm. Each of such nodes has an implicit function created from a face neighboring a crease edge. This is a suitable state to use a new PU evaluation method. It should be noted, however, that an extended algorithm works well if support spheres of nodes around crease edges are sufficiently small (only a crease edge is covered with a sphere). To do so, the simplification of nodes around such crease edges needs to be restrained. This is a limitation of our new PU evaluation method.

Fig. 7 (b) shows the result of our approximation considered with sharp features. In Fig. 7 (b), two clusters around a crease edge and three clusters around a corner are created on the new PU evaluation method. Since we can use the tree structure for searching the local functions which include a point $\mathbf{x}$, the time-complexity per each evaluation of $\tilde{f}(\mathbf{x})$ is $\log(N)$, where N is the number of local functions. Thus, we can achieve reasonable speed for polygonizing our implicit surfaces.

4. Results and Discussion

To display implicit surfaces, we first polygonize the contour defined by such surfaces and then render created polygons. For polygonizing our implicit surfaces $\tilde{f}(\mathbf{x}) = 0$, we firstly sample $\tilde{f}(\mathbf{x})$ at each grid point on a uniformly sized grid, then polygons are generated by using Dual Contouring proposed by Ju et al. [JLSW02]. Since our implicit function $\tilde{f}(\mathbf{x})$ is not defined if the point $\mathbf{x}$ is not included in any supports, we sampled $\tilde{f}(\mathbf{x})$ only near implicit functions like in [Blo94]. Roughly 30K evaluations per second can be achieved by the new PU evaluation method. The polygonization of a model of 1M polygons requires approximately two minutes.

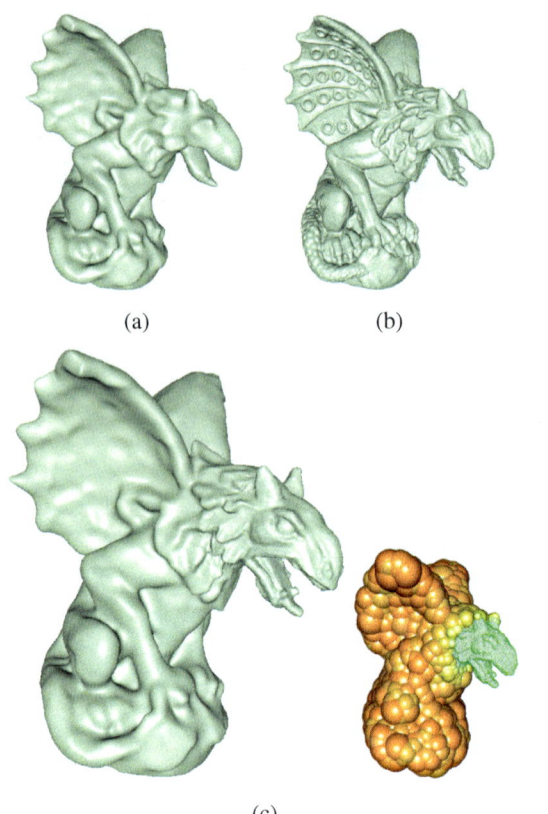

(a)   (b)

(c)

Figure 8: *Fitting results of a "gargoyle" mesh (1.7M faces). (a) A low-res. SLIM surface (1K nodes). (b) A high-res. SLIM surface (90K nodes). (c) Adaptive refinement of a SLIM surface (9.4K nodes). Only a part (head) of a gargoyle is composed of high-res. nodes. Color balls shown in the right denote a set of support spheres.*

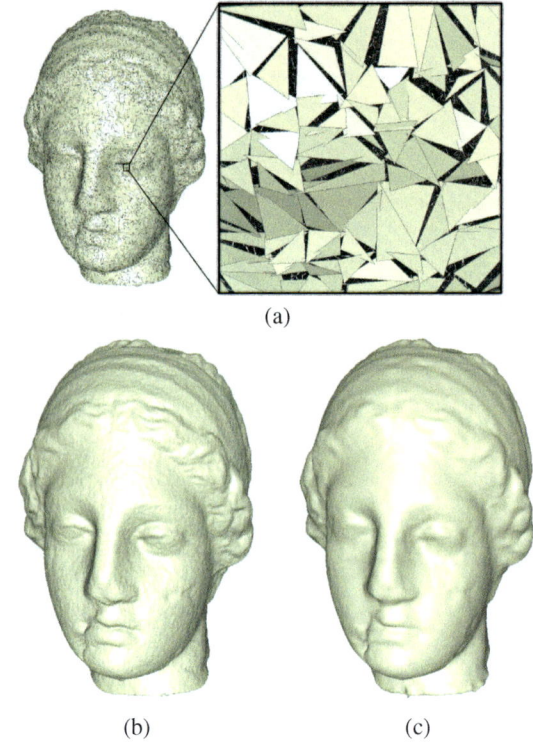

(a)

(b)   (c)

Figure 9: *Fitting results of a "venus" polygon soup (67K faces). (a) Random noise is added to each vertex. (b) A SLIM surface (67K nodes) by the implicitization of a polygon soup. (c) A SLIM surface (2.5K nodes).*

Fig. 1 and Fig. 8 show the results of our hierarchical approximation scheme of implicit surfaces. By just one execution of our approximation algorithm, we obtain a hierarchy of a SLIM surface. A set of implicit surfaces with any resolution is quickly extracted by traversing such a hierarchy. We can also extract a set of implicit surfaces where a part of an object has different resolution as shown in Fig. 8 (c). Such an extraction is especially useful for view-dependant LOD rendering of a SLIM surface described in [OBA05].

Fig. 9 shows the approximation results of a *polygon soup* which does not have the connectivity between neighbor faces. In Fig. 9, random noise is added to each vertex. Our approximation scheme is also applicable for such a polygon soup which contains noises. Fig. 9 (b) shows the rendering result of leaf nodes by applying implicitization. As shown in this figure, our approach can be performed as a simple method for repairing a mesh: We first implicitize a polygon soup, and then polygonize such constructed implicit surfaces. Meshes with different resolutions can be easily obtained by our approximation approach.

Our approximation algorithm of implicit surfaces is simple but very robust. Even if an edge construction method on a dual graph is not as *perfect* in the example as shown in Fig. 9 (a), our algorithm could never fail. We then prescribe the conditions *loosely* for the construction of a dual graph, in other words, a larger threshold for the judgment of neighbor faces can be set. Such loose conditions enable generation of redundant edges. However, such edges are deleted as duplicated edges in the approximation algorithm process.

Tab. 1 shows the statistical summary of all examples in this paper. In our experiments we used an Athlon 64 3500+ PC with 2GB RAM. In the example of an "Armadillo" mesh, three types of management schemes for error metrics are tested. In Tab. 1, it can be seen that both the on-the-fly scheme and hybrid scheme use the same amount of peak memory space, whereas the PIEM scheme needs a larger space. On the other hand, a PIEM scheme has lowest computational cost among the three schemes. The advantage of a hybrid scheme can be confirmed by the examples of a "bimba" mesh. Although the computational time is slightly longer, memory space is dramatically reduced. Note that the approximation for a "gargoyle" mesh via a PIEM scheme failed due to the lack of memory space in our implementation.

Roughly 70-80% of the computation time for each exam-

	#faces	type	mem. (MB)	time (sec.)
fandisk (Fig. 7)	12,946	PIEM	25	3.8
venus (Fig. 9)	67,178	PIEM	112	22.5
Armadillo (Fig. 4)	345,944	otf.	388	160.5
Armadillo (Fig. 4)	345,944	PIEM	563	107.1
Armadillo (Fig. 4)	345,944	hyb.	388	114.5
bimba (Fig. 1)	1,005,382	PIEM	1,568	313.5
bimba (Fig. 1)	1,005,382	hyb.	983	314.4
gargoyle (Fig. 8)	1,726,420	hyb.	1,643	586.7

Table 1: *Statistical summary. From left to right: the number of faces of an input mesh, the type of management schemes for error metrics, a memory space the algorithm used, and the computation time. PIEM, otf. and hyb. denote the approximation by PIEMS, on-the-fly approximation, and a hybrid approach respectively.*

ple is spent to solve linear equations by SVD. Our approach is fast compared to the approach in [SOS04]. Moreover, we can obtain a hierarchy of implicit surfaces including all resolutions in an execution of our algorithm, whereas the algorithm in [SOS04] creates a single resolution of implicit surfaces by a fixed error parameter.

Limitations. Since our algorithm requires considerable memory space, it cannot be applied to a large mesh with more than millions of faces. This is partially due to our implementation: Most of the memory space is spent for the storage of created SLIM nodes. If we improve this part by using approaches such as an out-of-core strategy, the required memory space can be reduced.

In implicit surface fitting, unexpected surfaces, e.g. a hyperboloid, may be generated especially just before the end of the approximation algorithm. It tends to give rise to visually undesirable results. This is because a part of the polygonal mesh to be fitted has a rather complicated shape and so it is unreasonable to fit it to a quadratic polynomial implicit surface. Several improved approaches need to be considered for this problem.

5. Conclusion and Future Work

In this paper, we have provided an effective method to convert a polygonal mesh to a set of implicit surfaces. A hierarchy of a SLIM surface can be obtained by just one execution of our algorithm. Thus, the resolution control can be achieved by only traversing a hierarchy of a SLIM surface, without re-calculating the algorithm. Our novel error metric can be defined as the quadric form. This provides compact storage and allows several efficient approximation schemes. Our algorithm can preserve sharp features such as creases or corners. Moreover, our approach can be applied for polygon soups and can act as a simple method for mesh repair.

A SLIM surface obtained from our method seems to be

useful for alternative surface representation from a polygonal mesh. In future work, we will try to develop modeling methods for such SLIM surfaces as well as to use them for CG/CAD applications.

Acknowledgements. The models used in this paper are courtesy of VCG-ISTI via the AIM@SHAPE Shape Repository (bimba, gargoyle), the Stanford 3D Scanning Repository (armadillo), and Cyberware Inc. (venus). We are grateful to anonymous reviewers for their useful comments and suggestions.

Appendix. Error Function as Quadric Form

Both a quadratic implicit function $f(\mathbf{x})$ and its gradient $\nabla f(\mathbf{x}) = (f_x(\mathbf{x}), f_y(\mathbf{x}), f_z(\mathbf{x}))$ are represented by an inner product of two 10-dimensional vectors:

$$f(\mathbf{x}) = \mathbf{f}\,\mathbf{p}^T, \tag{14}$$

$$f_x(\mathbf{x}) = \mathbf{f}_x\,\mathbf{p}^T, f_y(\mathbf{x}) = \mathbf{f}_y\,\mathbf{p}^T, f_z(\mathbf{x}) = \mathbf{f}_z\,\mathbf{p}^T, \tag{15}$$

$$\mathbf{p} \equiv (a_1, a_2, a_3, a_4, a_5, a_6, a_7, a_8, a_9, a_{10}),$$

$$\mathbf{f} \equiv (x^2, y^2, z^2, xy, yz, zx, x, y, z, 1),$$

$$\mathbf{f}_x \equiv (2x, 0, 0, y, 0, z, 1, 0, 0, 0),$$

$$\mathbf{f}_y \equiv (0, 2y, 0, x, z, 0, 0, 1, 0, 0),$$

$$\mathbf{f}_z \equiv (0, 0, 2z, 0, y, x, 0, 0, 1, 0).$$

From (14), a distance error function ε^{dis} is written by:

$$\begin{aligned}
\varepsilon^{dis}(T) &= A\int_0^1\int_0^{1-t}(\mathbf{f}\,\mathbf{p}^T)^2 dt ds \\
&= A\int_0^1\int_0^{1-t}(\mathbf{f}\,\mathbf{p}^T)^T\mathbf{f}\,\mathbf{p}^T dt ds \\
&= \mathbf{p}\left(A\int_0^1\int_0^{1-t}\mathbf{f}^T\mathbf{f} dt ds\right)\mathbf{p}^T \\
&= \mathbf{p}\,\mathbf{A}^{dis}\,\mathbf{p}^T. \tag{16}
\end{aligned}$$

Note that $\mathbf{A}^{dis}$ can be evaluated directly using a closed form since it is possible to integrate the polynomial $\mathbf{f}^T\mathbf{f}$ in analytical way.

From (15), a part of a gradient error function $\varepsilon^{nrm}|_x$ for x component of a gradient f_x is also represented by:

$$\begin{aligned}
\varepsilon^{nrm}|_x(T) &= A\int_0^1\int_0^{1-t}\left(\mathbf{f}_x\,\mathbf{p}^T - n_x\right)^2 dt ds \\
&= A\int_0^1\int_0^{1-t}\left((\mathbf{f}_x\,\mathbf{p}^T)^T\mathbf{f}_x\,\mathbf{p}^T\right. \\
&\quad \left. -2n_x\mathbf{f}_x\,\mathbf{p}^T + (n_x)^2\right) dt ds \\
&= \mathbf{p}\left(A\int_0^1\int_0^{1-t}\mathbf{f}_x^T\mathbf{f}_x dt ds\right)\mathbf{p}^T \\
&\quad -2\left(A\int_0^1\int_0^{1-t}n_x\mathbf{f}_x dt ds\right)\mathbf{p}^T + A(n_x)^2 \\
&= \mathbf{p}\,\mathbf{A}^{nrm}|_x\,\mathbf{p}^T - 2\mathbf{b}^{nrm}|_x + c^{nrm}|_x.
\end{aligned}$$

$\varepsilon^{nrm}|_y, \varepsilon^{nrm}|_z$ are also represented in the same manner. Consequently, we can put them together into the following quadric form:

$$\varepsilon^{nrm}(T) = \mathbf{p}\,\mathbf{A}^{nrm}\,\mathbf{p}^T - 2\mathbf{b}^{nrm}\mathbf{p}^T + c^{nrm}. \qquad (17)$$

References

[BDK98] BAREQUET G., DUNCAN C. A., KUMAR S.: RSVP: A geometric toolkit for controlled repair of solid models. *IEEE Transactions on Visualization and Computer Graphics 4*, 2 (1998), 162–177.

[BLCC00] BLANE M. M., LEI Z., CIVI H., COOPER D. B.: The 3L algorithm for fitting implicit polynomial curves and surfaces to data. *IEEE Transactions on Pattern Analysis and Machine Intelligence 22*, 3 (2000), 298–313.

[Blo94] BLOOMENTHAL J.: An implicit surface polygonizer. In *Graphics Gems IV*. AK Peters, 1994, pp. 324–349.

[CBC*01] CARR J. C., BEATSON R. K., CHERRIE J. B., MITCHELL T. J., FRIGHT W. R., MCCALLUM B. C., EVANS T. R.: Reconstruction and representation of 3D objects with radial basis functions. In *Computer Graphics (Proc. SIGGRAPH 2001)* (2001), ACM Press, New York, pp. 67–76.

[CSAD04] COHEN-STEINER D., ALLIEZ P., DESBRUN M.: Variational shape approximation. *ACM Transactions on Graphics (Proc. SIGGRAPH 2004) 23*, 3 (2004), 905–914.

[GH97] GARLAND M., HECKBERT P. S.: Surface simplification using quadric error metrics. In *Computer Graphics (Proc. SIGGRAPH '97)* (1997), ACM Press, New York, pp. 209–216.

[GWH01] GARLAND M., WILLMOTT A., HECKBERT P. S.: Hierarchical face clustering on polygonal surfaces. In *Proc. ACM Symposium on Interactive 3D graphics 2001* (2001), ACM Press, New York, pp. 49–58.

[HBM04] HELZER A., BARZOHAR M., MALAH D.: Stable fitting of 2D curves and 3D surfaces by implicit polynomials. *IEEE Transactions on Pattern Analysis and Machine Intelligence 26*, 10 (2004), 1283–1294.

[IH03] IGARASHI T., HUGHES J. F.: Smooth meshes for sketch-based freeform modeling. In *Proc. ACM Symposium on Interactive 3D Graphics 2003* (2003), ACM Press, New York, pp. 139–142.

[JLSW02] JU T., LOSASSO F., SCHAEFER S., WARREN J.: Dual contouring of hermite data. In *Computer Graphics (Proc. SIGGRAPH 2002)* (New York, NY, USA, 2002), ACM Press, pp. 339–346.

[JP04] JAMES D. L., PAI D. K.: BD-tree: Output-sensitive collision detection for reduced deformable models. *ACM Transactions on Graphics (Proc. SIGGRAPH 2004) 23*, 3 (2004), 393–398.

[KP90] KRIEGMAN D. J., PONCE J.: On recognizing and positioning curved 3D objects from image contours. *IEEE Transactions on Pattern Analysis and Machine Intelligence 12*, 12 (1990), 1127–1137.

[Mur91] MURAKI S.: Volumetric shape description of range data using blobby model. In *Computer Graphics (Proc. SIGGRAPH '91)* (1991), ACM Press, New York, pp. 227–235.

[MYC*01] MORSE B. S., YOO T. S., CHEN D. T., RHEINGANS P., SUBRAMANIAN K. R.: Interpolating implicit surfaces from scattered surface data using compactly supported radial basis functions. In *Proc. 3rd International Conference on Shape Modeling and Applications* (2001), IEEE CS Press, Los Alamitos, CA, pp. 89–98.

[OBA*03] OHTAKE Y., BELYAEV A., ALEXA M., TURK G., SEIDEL H.-P.: Multi-level partition of unity implicits. *ACM Transactions on Graphics (Proc. SIGGRAPH 2003) 22*, 3 (2003), 463–470.

[OBA05] OHTAKE Y., BELYAEV A. G., ALEXA M.: Sparse low-degree implicits with applications to high quality rendering, feature extraction, and smoothing. In *Proc. 3rd Eurographics Symposium on Geometry Processing* (2005), Eurographics Association, Aire-la-Ville, Switzerland, pp. 149–158.

[PFTV92] PRESS W. H., FLANNERY B. P., TEUKOLSKY S. A., VETTERLING W. T.: *Numerical Recipes: The Art of Scientific Computing*, 2nd ed. Cambridge University Press, Cambridge (UK) and New York, 1992.

[Pra87] PRATT V.: Direct least-squares fitting of algebraic surfaces. In *Computer Graphics (Proc. SIGGRAPH '87)* (1987), ACM Press, New York, pp. 145–152.

[She01] SHEFFER A.: Model simplification for meshing using face clustering. *Computer Aided Design 33*, 13 (2001), 925–934.

[SOS04] SHEN C., O'BRIEN J. F., SHEWCHUK J. R.: Interpolating and approximating implicit surfaces from polygon soup. *ACM Transactions on Graphics (Proc. SIGGRAPH 2004) 23*, 3 (2004), 896–904.

[SPOK95] SAVCHENKO V. V., PASKO A. A., OKUNEV O. G., KUNII T. L.: Function representation of solids reconstructed from scattered surface points and contours. *Computer Graphics Forum 14*, 4 (1995), 181–188.

[SSGH01] SANDER P. V., SNYDER J., GORTLER S. J., HOPPE H.: Texture mapping progressive meshes. In *Computer Graphics (Proc. SIGGRAPH 2001)* (2001), ACM Press, New York, pp. 409–416.

[Tau91] TAUBIN G.: Estimation of planar curves, surfaces, and nonplanar space curves defined by implicit equations with applications to edge and range image segmentation. *IEEE Transactions on Pattern Analysis and Machine Intelligence 13*, 11 (1991), 1115–1138.

[Tau94] TAUBIN G.: Distance approximations for rasterizing implicit curves. *ACM Transactions on Graphics 13*, 1 (1994), 3–42.

[TO02] TURK G., O'BRIEN J. F.: Modelling with implicit surfaces that interpolate. *ACM Transactions on Graphics 21*, 4 (2002), 855–873.

[TTC00] TASDIZEN T., TAREL J.-P., COOPER D.: Improving the stability of algebraic curves for applications. *IEEE Transactions on Image Processing 9*, 3 (2000), 405–416.

[WK05] WU J., KOBBELT L. P.: Structure recovery via hybrid variational surface approximation. *Computer Graphics Forum (Proc. Eurographics 2005) 24*, 3 (2005), 277–284.

[YT02] YNGVE G., TURK G.: Robust creation of implicit surfaces from polygonal meshes. *IEEE Transactions on Visualization and Computer Graphics 8*, 4 (2002), 346–359.

Eurographics Symposium on Geometry Processing (2006)
Konrad Polthier, Alla Sheffer (Editors)

Constructing Curvature-continuous Surfaces by Blending

Denis Zorin

New York University

Abstract

In this paper we describe an approach to the construction of curvature-continuous surfaces with arbitrary control meshes using subdivision. Using a simple modification of the widely used Loop subdivision algorithm we obtain perturbed surfaces which retain the overall shape and appearance of Loop subdivision surfaces but no longer have flat spots or curvature singularities at extraordinary vertices. Our method is computationally efficient and can be easily added to any existing subdivision code.

1. Introduction

Subdivision surfaces are well-established as a practical representation for geometric modeling with many useful properties. However, classical subdivision schemes like Loop and Catmull-Clark suffer from a number of problems: probably the best-known is the lack of C^2-continuity at the extraordinary vertices, i.e. vertices of the control mesh of valence different from 6 (Loop surfaces) and 4 (Catmull-Clark surfaces).

Several relatively simple solutions were proposed to this problem (e.g. [PU98]). However, ensuring formal C^2-continuity is not sufficient to solve all problems associated with absence of C^2-continuity. In particular, all simple approaches to making Loop or Catmull-Clark surfaces C^2-continuous at extraordinary vertices result in surfaces with *flat spots*: at surface points associated with an extraordinary vertex, the curvatures are forced to be zero. Careful rule tuning may make this artifact difficult to notice visually in most circumstances, but it will still exhibit itself for certain geometric configurations and certain types of lighting (e.g. reflection lines).

Even more importantly, C^2-continuity and absence of flat spots are needed for several types of numerical computation on surfaces. Examples include computation of curvature lines, which have singularities at flat spots and curvature singularities, computation of fairness functionals which require second derivatives and surface-surface intersection computations (e.g. one can construct examples of C^1 curves and surfaces intersecting in infinitely many isolated points).

The absence of flat spots for C^2 surfaces is more precisely

described as *surface 2-flexibility*. Following Reif [Rei96], we say that a C^2-surface representation is 2-flexible, if for some C^2 parameterization any desired first and second derivatives can be obtained at a given point by a suitable choice of positions of the control points (see Section 4 for a precise definition). Surfaces with flat spots, or parametric points where the Gaussian curvature is always positive, are not flexible. On the other hand, if a surface is 2-flexible, the user is able to make the surface locally a paraboloid or a saddle with arbitrary orientation at any point. Flexibility is also related to surface approximation quality. If a surface has a flat spot, it cannot approximate C^2-surface in C^2-norm: the error remains constant.

In this paper, we introduce a new method for the construction of curvature-continuous flexible surfaces on arbitrary meshes, based on the idea of blending subdivision surfaces with locally defined surface patches. Our approach is a simple extension of common subdivision algorithms and can be easily implemented on the top of an existing subdivision framework. The appearance of the resulting surfaces is similar to the appearance of standard subdivision surfaces and have insignificantly higher computational cost. We were able to verify that resulting surfaces are flexible everywhere under certain assumptions are imposed on the control mesh.

Compared to a existing constructions of C^2 subdivision surfaces (Section 2), the distinguishing feature of our specific construction is that it remains very close to standard subdivision, while eliminating curvature discontinuity and flat-spot related problems and maintaining 2-flexibility away from extraordinary vertices. While we describe our construction for Loop surfaces, and our proofs are restricted to this case, the extension to Catmull-Clark surfaces is straightforward and similar techniques can be used for analysis.

2. Previous Work

A large number of C^2 constructions for arbitrary meshes of various types were proposed over years. We mention some representative work. Hagen and Pottmann [HP89] C^2 interpolants of boundary position, tangent and curvature data are constructed. Gregory and Hahn [GH89] describe a C^2 hole-filling algorithm; Bohl and Reif [BR97] describe C^2- conditions on degenerate patches and how N patches can be joined at a point. C^2 spline surfaces on arbitrary meshes were constructed by Peters [Pet96] and higher order spline surfaces are described by Prautzsch in [Pra97]. More recently, various types of constructions based on polynomial patches were proposed in [Pet02], [Loo04] and [KP05].

C^2 subdivision algorithms based on standard schemes and with zero curvature at extraordinary vertices were proposed by Umlauf [PU98] and Biermann *et al.* [BLZ00].

The idea of obtaining smooth surfaces for arbitrary meshes using blending and appropriate local parameterizations, while known in geometric modeling for a long time (e.g. [GH89]), is used in more general form in the work on manifold-based surfaces [GH95, NG00, YZ04].

A closely related technique for subdivision surfaces was independently developed by Levin [Lev06].

The flexibility of resulting surfaces at arbitrary points is rarely addressed explicitly but, for many spline constructions, can be relatively easily inferred from the surface construction. For representations based on blending, a complete analysis is far more complex.

Despite the broad variety of options proposed in the research literature, the practice is dominated by non-C^2-continuous algorithms.

The difficulty of constructing a practical C^2-continuous surfaces appears to be in achieving the right tradeoff between mathematical properties (C^2-continuity and flexibility), visual quality, which can be captured by fairness measures, computational expense and the difficulty of implementation.

Compared to previous work, our main contribution is to propose a simple algorithm which can be added to an existing implementation of Loop subdivision with minimal effort and in most cases, yields surfaces closely approximating the standard Loop surfaces, yet curvature-continuous and 2-flexible everywhere.

The crucial ideas we build on are: obtaining smooth surfaces by blending in appropriate parameterization and using characteristic maps [Rei95] to obtain such parameterizations.

3. Overview

The basic idea of our construction is to blend a subdivision surface with parametric quadratic patches near extraordinary vertices. The quadratic patches are constructed from the control points in such a way that flexibility is guaranteed at the vertices.

For a given extraordinary vertex v, we use inverse of the *the characteristic map* to obtain local parameterization of the surface, which is C^2 away from v. The quadratic patch is defined as a function on to domain of the parameterization, i.e. the characteristic map image.

The blending basis function for the domain is taken to be the subdivision basis function corresponding to the extraordinary vertex, computed using a flat-spot modification of the Loop scheme.

Near the extraordinary vertex, the surface is blended with the quadratic patch using the blending function, so that the weight of the surface at v is zero. As we discuss below, this leads to C^2 surfaces flexible at vertices.

The distinguishing feature of the proposed construction is that three components of the construction (the surface itself, the characteristic map, and the blending function) can be computed using the same subdivision code, and the remaining component (the quadratic patch) is easy to evaluate.

4. Notation and terminology

To describe our construction and its properties in detail we briefly review the necessary notation and terms. We use boldface letters to denote 3D or 2D vectors.

Flexibility.

Definition 1 Let F be a parametric family of functions with values in $\mathbf{R}^n$ defined on a domain $D \subset \mathbf{R}^2$. We say that this family is *parametrically r-flexible* at a point $x \in D$, if for any given set of prescribed values of all partial derivatives $d_{ij}, i = 0 \ldots r, j = 0 \ldots i$ up to order r, there is a function $f(x_1, x_2) \in F$ with this set of derivatives at point $x \in D$:

$$\frac{\partial^i f}{\partial^j x_1 \partial^{i-j} x_2}(x) = d_{ij}$$

In particular, a function family is 2-flexible at x, if there is a function in the family with any prescribed values and prescribed first and second derivatives at x, which implies that it is possible to obtain arbitrary prescribed curvatures and curvature directions.

As explained below, subdivision defines a family (or more precisely a linear space) of functions $\sum \mathbf{p}_v B_v$ on a mesh, parameterized by the control points. For this family, r-flexibility means that we can choose the positions of the control points in such a way that at any fixed point x of the mesh we have prescribed partial derivative values up to order r, with derivatives computed with respect to certain local parameterizations (characteristic map parameterizations).

Subdivision surfaces as functions on meshes. It is well known [Rei95] that common subdivision surfaces such as Loop and Catmull-Clark can be thought of as infinite collections of polynomial patches. The domains for these patches can be taken to be subtriangles or subquads associated with faces of the initial mesh. In particular, one can regard a patch of the subdivision surface corresponding to a ring of triangles adjacent to an extraordinary vertex to be a function on a regular k-gon U_k centered at $(0,0)$. While in the interior of each triangle, this function is C^2 for subdivision schemes extending C^2-continous splines. In general, one can only expect C^0 continuity between triangles. A different parameterization that we described below is needed to obtain C^2 on edges. Such parameterization is provided by characteristic maps.

Subdivision matrix and characteristic maps. A characteristic map is defined using the eigenstructure of the *subdivision matrix*, introduced in [DS78]. Consider a vertex v, and let p be the vector of control points in a neighborhood of the vertex; For the Loop subdivision scheme, we use all control points in a double ring of triangles around the vertex. From now on, we call the double ring of triangles around a vertex v the *2-neighborhood* of v; we call the single ring of triangles the *1-neighborhood* of v.

Let S be the $N \times N$ matrix of subdivision coefficients relating the vector of control points p^j on subdivision level j to the vector of control points p^{j+1} in a similar neighborhood on the next subdivision level. Many properties of the subdivision scheme can be deduced from the eigenstructure of the matrix. This is seen by decomposing the vector of control points p with respect to the eigenbasis $\{x^i\}$ of S, $i = 0..N-1$, assuming it exists:

$$p = \mathbf{a}_0 x^0 + \mathbf{a}_1 x^1 + \mathbf{a}_2 x^2 + \ldots \quad (1)$$

where the multiplication of three-dimensional coefficients $\mathbf{a}_i$ with N-dimensional vectors x_i is understood in tensor-product sense and yields a $N \times 3$ matrix (the vector of 3d control points). We assume that the eigenvectors x^i are arranged in the order of non-increasing eigenvalues, and the first eigenvalue λ_0 is 1, which is required for convergence of subdivision. Furthermore, we assume that the corresponding eigenvector x^0 is the vector of ones necessary for affine invariance.

For control points in general position, the limit position of the center control point is $\mathbf{a}_0$ and the tangent directions at this position are $\mathbf{a}_1$ and $\mathbf{a}_2$.

Two subdominant eigenvectors x^1 and x^2 are used to construct the characteristic map for valence k. In the cases of interest to us, the corresponding eigenvalues are equal $\lambda_1 = \lambda_2 = \lambda$. The characteristic map Φ_k is defined as the

limit function of a subdivision for a 2D mesh which is constructed as follows. The mesh consists of two rings of triangles around a vertex of valence k. The coordinates of the vertices of the initial mesh are taken to be the components of eigenvectors x^1 and x^2 respectively, $\mathbf{p}_j = (x_j^1, x_j^2)$, $j = 0 \ldots N-1$. The 1-neighborhood of the central vertex is typically a regular k-gon, or can be mapped to one by a piecewise-linear mapping. The characteristic map Φ_k is the limit function with values in $\mathbf{R}^2$ generated by subdivision from this initial mesh, and restricted to the regular k-gon (Figure 1).

The following property is most important for us: the characteristic maps for all valences are injective. Moreover, F is the parameterization of subdivision surface over a regular k-gon described above; the composition $F \circ \Phi_k^{-1}$ is C^2-continuous if the scheme is C^2-continuous in the regular case.

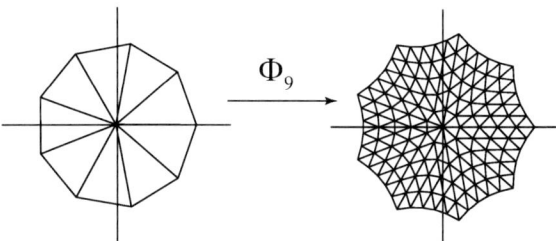

Figure 1: *Characteristic map of the Loop subdivision scheme for valence $k = 9$. On the left, a piecewise linear approximation to the image of the map is shown.*

More generally, the subdivision surface can be represented as a linear combination of *eigenbasis functions* f_i i.e. functions obtained from the eigenvectors x_i by subdivision (this amounts to a change of basis in (1)). An eigenbasis f_i satisfies $f_i(t/2) = \lambda_i f(t)$ on the regular k-gon and its characteristic map reparameterization satisfies $(f_i \circ \Phi_k^{-1})(\lambda x) = \lambda_i(f_i \circ \Phi_k^{-1})(x)$. It easily follows from this formula [Zor00] that the function f_i changes as $|x|^\alpha$ near $x = (0,0)$ with $\alpha = \log \lambda_i / \log \lambda$.

Subdivision scheme. We use two versions of Loop subdivision. The first is a commonly used version of Loop subdivision with the vertex rule

$$p_0^{j+1} = (5/8)p_0^j + (3/8k) \sum_{i=1}^{k} p_i^j \quad (2)$$

Note that a special rule is normally necessary for $k = 3$ as in this case the resulting surface is not C^1-continuous for the rule (2).We use the same rule for $k = 3$ as the smoothness of subdivision surface at the extraordinary vertex does not affect the smoothness of the blended surface.

A modified version of this scheme is used to compute the

blending function used in our construction. We use a special case of the scheme from [BLZ00]. On the first subdivision step, values for vertices immediately adjacent to an extraordinary vertex are computed using

$$p^1 = \mathbf{a}_0 + \lambda \mathbf{a}_1 x_1 + \mathbf{a}_2 \lambda x_2 \qquad (3)$$

Note that this forces the control values adjacent to the extraordinary vertex to be in the same plane, which turns out to be sufficient to ensure that the limit function is C^2 with zero second derivatives in the characteristic map parameterization. After the first step, the standard rules are used.

Also note that eigenvectors of the subdivision matrix are not affected by the modification. The only change to the eigenvalues is that for a subset of eigenbasis functions the coefficients are set to zero. These eigenbasis functions include all functions which are obtained as linear combinations of basis functions corresponding to projected control points in the one-ring, excluding f_0, f_1 and f_2.

5. Blending local patches with subdivision surfaces

First, we describe how to blend a quadratic patch $\mathbf{Q}(t)$ associated with a vertex v with the subdivision surface produced by the Loop scheme. We assume that the patch $\mathbf{Q}(t)$ is defined as a smooth function from the plane into $\mathbf{R}^3$, so t is a point on the plane. A specific approach to computing local quadratic patches is described in Section 5.1.

We start by applying one step of refinement to the subdivision surface, to obtain a new control mesh M^1. The 1-neighborhoods of the vertices of M in M^1 share only isolated points (Figure 2).

The blending is restricted to the 1-neighborhoods of extraordinary vertices in M^1 i.e. at every point of the surface at most one quadratic patch is blended with the surface.

As we have discussed, a subdivision surface is a function defined on the control mesh, $\mathbf{F}(x) = \sum_v \mathbf{p}_v B_v(x)$, where $\mathbf{p}_v$ are the control points, $B_v(x)$ are the basis functions of subdivision.

Figure 2: *1-neighborhoods of the vertices of the initial mesh M in the once-refined mesh M^1.*

We construct the blending function B_k^2 as follows. Take a

mesh R_k with a single central extraordinary vertex v of valence k, and containing a double ring of vertices around v. Subdivide this mesh once to obtain R_k^1. Assign the value 1 to v and zeros to all other vertices and apply subdivision to these values. This yields a scalar basis function defined on a 2-neighborhood of v in R_k^1, i.e. on 1-neighborhood of v in R_k, which we identify with the regular k-gon. We rescale function values so that the value at the extraordinary vertex is 1. The resulting function $B_k^2 : U_k \to \mathbf{R}$ is the blending function we use.

Let Φ_k be the characteristic map as defined in Section 4.

We use the following formula for blending the patch with the surface on the regular k-gon U_k:

$$\mathbf{F}^{\text{blended}}(x) = \left(1 - B_k^2(x)\right)\mathbf{F}(x) + B_k^2(x)\mathbf{Q}(\Phi_k(x)) \quad (4)$$

Our new surface is a blend of the old surface $\mathbf{F}(x)$ and the new locally defined patch $\mathbf{Q}(t)$ with the contribution of $\mathbf{Q}(t)$ reducing to zero at the boundary of 1-neighborhood of v in M^1, i.e. half-way to the adjacent vertex in the original mesh M. As we will see, all three components required to compute $\mathbf{F}^{\text{blended}}$ can be easily evaluated.

The proof of C^2-continuity at extraordinary vertices. As it was discussed in Section 4, if Φ_k is the characteristic map, $\mathbf{F} \circ \Phi_k^{-1}$ is differentiable.

We reparameterize the blended surface $\mathbf{F}^{\text{blended}}(x)$ over a domain in the plane (the image of the characteristic map),

$$\mathbf{F}^{\text{blended}}(\Phi_k^{-1}(t)) = (1 - B_k^2(\Phi_k^{-1}(t)))\mathbf{F}(\Phi_k^{-1}(t)) + \\ B_k^2(\Phi_k^{-1}(t))\mathbf{Q}(t) \qquad (5)$$

For standard Loop subdivision near extraordinary vertices the second derivatives of $\mathbf{F} \circ \Phi_k^{-1}$ grow no faster than $t^{\log \lambda_2 / \log \lambda - 2}$, where λ_2 is the next largest eigenvalue after the subdominant eigenvalue λ [Zor00]. Specific flat-spot subdivision rules described in Section 4 ensure that the only eigenbasis functions that contribute to the blending function derivative behavior at zero are the eigenbasis functions corresponding to control points outside the one-ring of the vertex. The eigenvalues of these functions are known to be less than $1/8$, so the decay rate of these functions near the extraordinary point is faster than $|x|^{\alpha_1}$ for $\alpha_1 = \log(1/8)/\log \lambda$ in the characteristic map parameterization. For the surface itself the decay rate can be estimated to be no slower than $\alpha_2 = \log \lambda_2 / \log \lambda$. As a consequence the increase rate of second derivatives does not exceed $|x|^{\alpha}$ for $\alpha = \alpha_2 - 2$, if $\lambda_2 > \lambda^2$.

Using these observations and the product rule for second derivatives one concludes that the rate of change of second derivatives of $(1 - B_k^2(x))\mathbf{F}(x)$ in the characteristic map reparametrization is $\alpha_1 + \alpha_2 - 2$, which can be verified to be positive if $\lambda_2 < 8\lambda^2$, which can be shown to hold for any valence for Loop subdivision.

The blending function $B_k^2 \circ \Phi_k^{-1}$ is C^2-continuous by construction.

We conclude that the characteristic map parametrization of the blended surface is C^2-continuous because all included functions are C^2-continuous. Moreover, all first and second derivatives of the first part at $(0,0)$ are zero by construction, and the derivatives of the second part coincide with the derivatives of $Q(t)$; thus we have complete control over surface flexibility through the choice of $Q(t)$.

5.1. Local quadratic patches

The patch $Q(t)$ for an extraordinary vertex v is constructed as a function in the plane, with values in $\mathbf{R}^3$. The idea of the construction is to find a quadratic patch that follows the local shape of the mesh; at the same time, we try to make our surfaces as similar as possible to the surfaces generated using subdivision.

Thus we use the formulas for the limit positions of Loop subdivision for $Q(0)$. Remarkably, the rest of the coefficients can be found using least-squares fit, and the resulting formulas for the coefficients for the first derivatives coincide with the standard formulas for tangents to the Loop subdivision surfaces.

The coefficients can be obtained as follows. Let (r, φ) be the polar coordinates in the plane. Then the second-order approximation to the surface can be written as

$$Q(r, \varphi) = \mathbf{b}_0 + (\mathbf{b}_{11} \cos \varphi + \mathbf{b}_{12} \sin \varphi) r +$$
$$(\mathbf{b}_{20} + \mathbf{b}_{21} \cos 2\varphi + \mathbf{b}_{22} \sin 2\varphi) r^2$$

We assume that $\mathbf{b}_0$ is computed using the formula for the limit positions of a control point for Loop subdivision, that is, $\mathbf{b}_0 = \mathbf{p}_0/2 + (1/2k) \sum_i \mathbf{p}_i$, for $k \neq 3$, $a_0 = 2\mathbf{p}_0/5 + \sum_i \mathbf{p}_i/5$. We determine the other five coefficients by fitting a quadric with $\mathbf{b}_0 = 0$ to the shifted control points $\mathbf{p}_i - \mathbf{p}_0$, $i = 0 \ldots k - 1$.

A simple calculation shows that the least squares fit to k points of the 1-neighborhood $\mathbf{p}_1 - \mathbf{p}_0 \ldots \mathbf{p}_k - \mathbf{p}_0$ assumed to be values at $(\cos(2\pi i/k), \sin(2\pi i/k))$, $i = 0..k - 1$, leads to

$$\mathbf{b}_{11} = \frac{2}{k} \sum_i \mathbf{p}_i \cos \frac{2\pi i}{k}; \quad \mathbf{b}_{12} = \frac{2}{k} \sum_i \mathbf{p}_i \sin \frac{2\pi i}{k}$$

$$\mathbf{b}_{20} = -\mathbf{p}_0 + \frac{1}{k} \sum_i \mathbf{p}_i$$

$$\mathbf{b}_{21} = \frac{2}{k} \sum_i \mathbf{p}_i \cos \frac{4\pi i}{k}; \quad \mathbf{b}_{22} = \frac{2}{k} \sum_i \mathbf{p}_i \sin \frac{4\pi i}{k}$$

Note that the formulas for $\mathbf{b}_{11}$ and $\mathbf{b}_{21}$ coincide with the standard formulas for the tangents to the Loop subdivision surface, and $\mathbf{b}_{20}, \mathbf{b}_{21}, \mathbf{b}_{22}$, with appropriate variable changes, produce second derivatives in the regular case. As a result, we obtain a set of simple rules for computing the coefficients of an approximating quadratic surface which can be used as function $Q(t)$ in (4). A similar construction can be used for the boundary, but we do not consider it here. Example set of quadratic patches is shown in Figure 3.

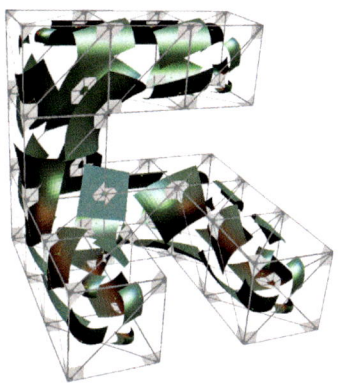

Figure 3: *Local quadratic patches are used to approximate a surface near extraordinary points.*

Flexibility. It is easy to show that for valences $k \geq 5$, the patches are parametrically 2-flexible: for any specified first and second derivatives, we can solve for patch coefficients $\mathbf{b}_i$ yielding these derivatives. For $k = 3, 4$, the patches are not 2-flexible, as the number of control points is less than the total number of the derivatives of order ≤ 2.

To obtain surfaces that are parametrically flexible for all valences, we need to use special rules for $k = 3, 4$. Such rules can be obtained in a similar manner, but require using triangles outside the 1-neighborhood of the vertex. We use the standard masks for $\mathbf{b}_0$ (the limit position) and $\mathbf{b}_{11}, \mathbf{b}_{12}$ (tangents). In both cases $\mathbf{b}_{20}$ can be computed using a single ring of vertices and expressions specified above. In the case of $k = 4$, the mask for $\mathbf{b}_{21}$ does not require changes either. In the remaining cases (both "saddle" coefficients $\mathbf{b}_{21}$ and $\mathbf{b}_{22}$ for $k = 3$ and one of "saddle" coefficients $\mathbf{b}_{22}$ for $k = 4$) additional vertices need to be used. We obtain these coefficients using a least squares fit to the single ring of control points around the vertex augmented by k triangles outside the ring. The resulting masks are shown in Figure 4.

Even if these special masks for valences 3 and 4 are used, the surfaces may still be not parametrically 2-flexible for some meshes. There are four small closed meshes for which the

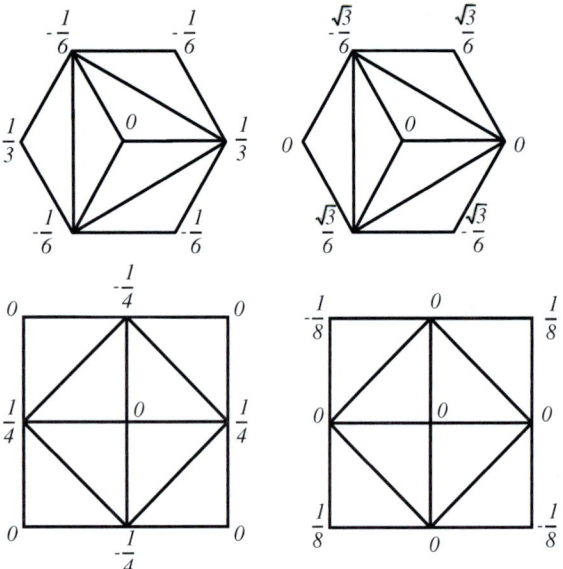

Figure 4: *Masks for computing coefficients $\mathbf{b}_{21}$ and $\mathbf{b}_{22}$ for valences $k = 3$ and $k = 4$.*

surface is not parametrically 2-flexible, including the octahedron and tetrahedron (Figure 5, *b,c,f,g*), due to insufficient total number of degrees of freedom in the double ring.

In addition, the surface is not parametrically flexible for meshes containing the three configurations *a,d,e* shown in Figure 5. While configuration *e* is somewhat unusual, the configuration *d* occurs, for example, if a pyramid is joined with its reflected image at the base. There does not appear to be a simple solution to this problem in our framework. While it is unlikely to be an issue for tetrahedron and octahedron, it is likely that a larger mesh would contain a neighborhood shown in Figure 5. It should be noted that the degree of inflexibility is relatively small in this case: a mixed second derivative is constrained to be zero.

5.2. Algorithm

Finally we summarize the basic algorithm for computing our surfaces. If a direct evaluation routine of the type described in [Sta98] is available, than the implementation amounts to defining a set of control meshes for the characteristic map and blending functions, as described in more detail below, and calling this routine to evaluate (4).

The output of the algorithm is a mesh approximating the surface described by (4) after n refinement steps. All terms in (4) are computed using subdivision modified Loop subdivision with zero curvature.

One refinement step is performed first; the refined mesh is M^1. Each triangle of the refined mesh either has a single ex-

traordinary vertex (valence not equal to 6) or all its vertices are regular. All control points that are inserted in triangles with regular vertices are computed using standard subdivision rules.

For each triangle T which has an extraordinary vertex, we can characterize any vertex obtained by refining this triangle by its barycentric coordinates (u, v, w). The barycentric coordinates have the form $(i/2^n, j/2^n, 1 - (i+j)/2^n)$, because all new vertices are inserted using midpoint subdivision. We always choose the coordinates in such a way that the weight w corresponds to the single extraordinary vertex in the triangle; the last coordinate w can be dropped, as $u + v + w = 1$. Finally, assuming n fixed, each vertex is identified by the pair of indices (i, j). For most common representations of the subdivision meshes the indices (i, j) or a some modified form of these indices is readily available. When evaluating (4) we are interested only in values at the vertices obtained after n subdivision steps, which can be enumerated using the indices $(i, j), i, j = 0 \ldots 2^{n-1}, i + j < 2^n$, as above, with $(0, 0)$ corresponding to the extraordinary vertex.

To compute an approximation to the surface after n subdivision steps, the algorithm proceeds as follows.

Precomputation.

1. For each extraordinary vertex, we precompute and store the coefficients of the quadric associated with the extraordinary vertex applying the coefficients of Section 5.1 to a ring of vertices around the extraordinary vertices.

2. For each valence k present in the mesh, we precompute the limit values of the characteristic map Φ_k and $B_k^2(x)$ at vertices (i, j) at the subdivision level n for one triangular sector (all other values can be obtained by applying an appropriate rotation). For each valence k, we create two small meshes, both consisting of 2 rings of triangles around an extraordinary vertex of valence k. To the vertices of the mesh used to compute B_k^2 we assign scalar initial values. To obtain B_k^2, we refine twice first, then set all values to zero except the value in the center, which is set to 2, so that the limit value in the center is 1. The second mesh, used to compute Φ_k, is initialized to values from $\mathbf{R}^2$. The two components of each value are the components of the subdominant eigenvectors for valence k. As we care only about the values of the functions on the first triangle T_0 of the k-gon, we need to refine only this triangle. After refining n times and using the standard limit rules, we obtain the values $\Phi_k(i, j)$ and $B_k^2(i, j)$ at all vertices of n times refined triangle T_0.

Evaluation. We first subdivide the mesh n times using Loop rules, and evaluate the limit position for each vertex. Fix a triangle $T = (v_0, v_1, v_2)$ of M^1 with an extraordinary vertex v_0.

For each vertex inserted into a triangle of M^1 adja-

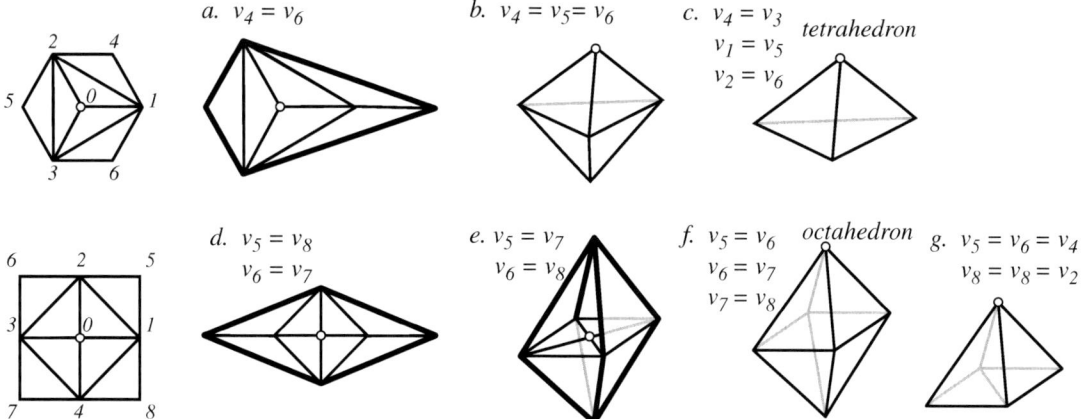

Figure 5: *Inflexible meshes. Upper row: meshes inflexible at a vertex of valence 3. Lower row: meshes inflexible at a vertex of valence 4. For each mesh, the circle indicates the vertex where the mesh is not 2-flexible. Naming convention for vertices is shown on the left. Each inflexible mesh is obtained by identifying some vertices, as indicated by the equations. Only three meshes are open, that is, have boundary edges (shown as thick) and can be submeshes of larger meshes. The remaining meshes are closed. Invisible edges are shown in gray.*

cent to an extraordinary vertex, we determine the indices (i, j) and compute the final value $(1 - B_k^2(i,j))\mathbf{F}(i,j) + \mathbf{Q}(\Phi_k(i,j))B_k^2(i,j)$.

Note that computation of B_k^2 and Φ_k have to be performed only once per valence, that is, they have little impact on the performance of the algorithm; the only additional expense for each vertex is the lookup of the values $\Phi_k(i,j)$ and $B_k^2(i,j)$, and computation the linear combination (4).

A simple implementation of the algorithm required less than 500 lines of code on the top of an existing subdivision library.

5.3. Surface Quality

As it can be seen in Figure 7, for common meshes the difference between surface appearance is difficult to see. However, in some cases the difference can be significant (Figure 6). We have chosen to compare surface appearance for standard Loop surfaces and surfaces obtained by blending using a saddle-shaped control mesh. The reason for this is that for valences higher than six surfaces corresponding to such control meshes have in some sense least curvature continuity. As it is discussed in detail in [Zor00] the lack of curvature continuity is due to the fact that certain eigenvalues of the subdivision matrix have magnitude larger than λ^2 where λ is the subdominant eigenvalue. For the Loop scheme, the largest among these eigenvalues is the eigenvalue $3/8 + (1/4)\cos 4\pi/k$; the corresponding eigenvector is v_{21}. A saddle-like arrangement of control points has the largest magnitude of the corresponding coefficient $\mathbf{b}_{12}$ in the eigenbasis decomposition, which results in pronounced vi-

olation of curvature continuity especially for high valences. For example in Figure 6 for valence 20 one can see that the surface develops a visible kink. At the same time for valences close to 6 no artifacts are visible. However, the curvature still diverges for valences higher than 6, which causes numerical problems for algorithms such as surface-surface intersection and geodesic tracing. In contrast, the overall shape of blended surfaces is not much affected by valence and the curvatures converge smoothly to a limit value (Figure 6). It appears that the surfaces are slightly bumpier away from the extraordinary points but on the models that we have considered the effect was not very pronounced. The fact that the Gaussian curvature for the saddle becomes positive is also undesirable and is likely to be correlated with bumpiness. The visible ripple artifacts for valence 20 are due to the artifacts present in the Loop basis function used for blending. While these artifacts are absent when standard Loop is applied to the saddle eigenvector configuration, they immediately appear once other eigenvectors are present as it is the case for any real surface. One can hope that a blending function with no ripple artifacts would produce even better results, but it is not clear how to construct such functions with k-gonal support or if it is possible to use different support shapes without introducing a different type of artifact (in [Lev06] a different blending function is explored.)

6. Analysis of Flexibility

Parametric 2-flexibility at extraordinary vertices with respect to the characteristic map parameterization was easily established by construction.

However our construction does not *a priori* guarantee that

valence 7   *valence 10*   *valence 20*

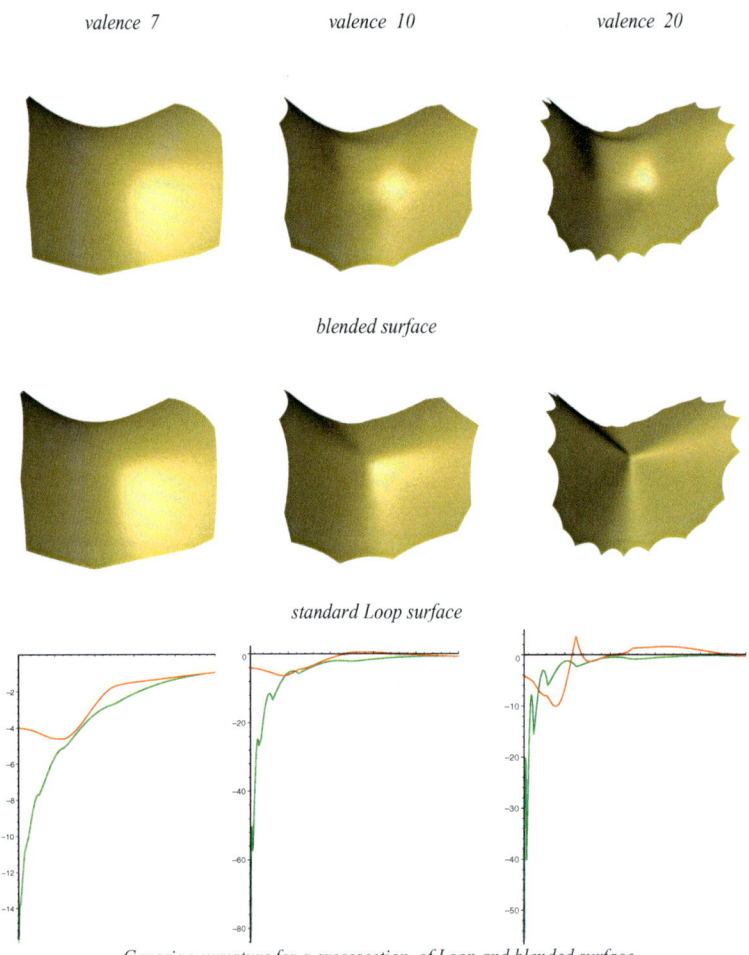

blended surface

standard Loop surface

Gaussian curvature for a crosssection of Loop and blended surface

Figure 6: *Behavior of the standard Loop surface and the blended surface for valences 7,10 and 20. The range for curvature plots is 0.01 to 0.5 measured barycentric coordinates along a line passing through the origin. The red curve is the curvature for the blended surface and the green curve is the curvature for the red surface.*

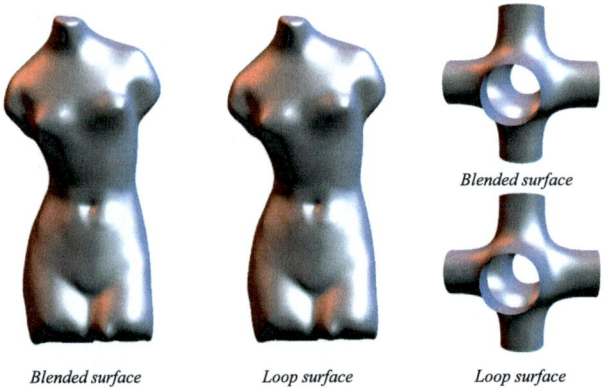

Blended surface   *Loop surface*   *Loop surface*

Figure 7: *Left: A slight difference exists but hard to identify visually for this model. Right: the difference in the shape of highlights for simple models is more apparent. We reiterate that our goal not as much improve surface fairness but retain it and obtain surfaces with better mathematical properties in a simple way.*

the resulting surfaces are flexible away from the vertices of the top-level mesh. While not very probable, it is possible that as a result of blending we introduce inflexible points at locations away from the extraordinary vertex. Thus, to establish that the proposed approach is guaranteed to result in 2-flexible everywhere for a sufficiently broad class of meshes further analysis is required. The analysis of flexibility of blended at points away from extraordinary points, while relatively straightforward conceptually proved to be difficult technically. Here we outline the idea of the proof which extensively uses computer algebra. More details, and relevant computer algebra code is presented as a separate report. *A note to the reviewers: the report is included in the paper submission.*

We are able to show that the surfaces produced by the scheme are flexible on closed control meshes with two additional constraints imposed:

C1: Extraordinary vertex separation. No two extraordinary vertices are adjacent.

C2: Simple neighborhood topology. All $k + 6$ vertices in the 1-neighborhood of any triangle which has a vertex of valence $k \neq 6$ are distinct.

The argument can be also easily extended to the case when any two adjacent extraordinary vertices both have valence higher than six.

The case of meshes with adjacent low-valence extraordinary vertices presents considerable difficulty. In some cases, the resulting surfaces cannot be flexible as the number of basis functions with support overlapping some of the points of the surface can be less than six, the minimal number required for flexibility. There are few configurations like this however, and in most other case it is highly likely that the resulting surfaces are flexible. A brute-force analysis would require analyzing a large number of local configurations, computing the explicit piecewise polynomial representation of the limit surface for each. We leave such analysis for our scheme and other C^2 constructions which are potentially 2-flexible as future work.

As we analyze parametric 2-flexibility, analysis can be performed for each coordinate separately; therefore it is sufficient to consider scalar functions defined by subdivision.

Theorem 1 For a closed mesh M satisfying constraints C1 and C2, the blended surface defined by (4) is parametrically 2-flexible at any point of M.

Outline of the proof. For the vertices of the control mesh, flexibility immediately follows from the surface definition. We need to prove 2-flexibility for all other points of the surface. If any two extraordinary vertices are separated by regular vertices, we only need to consider triangles which have a single extraordinary vertex.

Next, consider the part of the surface that is defined on the

1-neighborhood of an extraordinary vertex. This part of the surface vertex depends on a double ring of control points around this vertex. Rather than proving that one can choose the positions of all of these control points to obtain the desired result, we use a set of 6 coefficients in the eigenbasis decompositions as degrees of freedom, i.e. we prove that for some choice of these coefficients we can obtain any desired first and second derivatives at any given point with barycentric coordinates (u, v, w) in a triangle adjacent to the extraordinary vertex (one can easily show that any set of these coefficients can be obtained using a suitable combination of control points).

As we have already mentioned, the surfaces we are interested in can be evaluated using Stam's algorithm at any point (u, v, w), with $u + v + w = 1$. Recall that using this algorithm, for a point (u, v, w), we evaluate the surface as a value of a quartic box spline patch with control points expressed as functions of control points of the subdivision surface and subdivision level i depending on (u, v, w). More specifically, the control points of this patch are linear combinations of control points with coefficients of the form $C\lambda_j^i$, where λ_j are eigenvalues of the subdivision matrix. It turns out that in the specific case that we consider a variable change simplifies the expressions for the surface at an arbitrary location to a polynomial $F(u, v, w, \varepsilon, c)$ in 5 variables u, v, w, ε, c, where $c = \cos(\pi/k)$, k is the valence, $\varepsilon = 2^{-i}$ is the subdivision level, with coefficients linearly depending on a_{ij}, $0 < j < i \leq 2$, and independent of k (the dependence on valence is completely captured by c).

Differentiating this polynomial with respect to u and v to obtain 6 derivatives of order ≤ 2; and prescribing the values of these derivatives yields a system of 6 equations in 6 variables a_{ij}, with coefficients which are polynomials in u, v, w, ε and c. The system has a solution whenever its determinant is not zero. As all coefficients are polynomials, the determinant is also a polynomial in the same variables; the ranges of the variables are $0 \leq u \leq 1$, $0 \leq v \leq 1$, $0 \leq u + v \leq 1$, $0 < c < 1$ (for $k \geq 5$), $0 < \varepsilon \leq 1$. To prove flexibility it is sufficient to show that this polynomial is greater than a constant C on the domain defined by listed inequalities. We achieve this by converting it to Bezier form and verifying that all coefficients are positive. The case $k = 3$ is considered in the same way but separately, with one less variable.

7. Conclusions and Future Work

The method that we have described easily extends to other types (e.g. Catmull-Clark) or higher-smoothness subdivision surfaces, e.g. if the subdivision rule is C^k in the regular case, the technique can yield C^k surfaces. However, flexibility away from extraordinary vertices needs to be verified separately in each case.

The choice of subdivision basis functions as blending functions in this paper was primarily motivated by considerations of simplicity and efficiency, as well as the possibility of proving 2-flexibility away from extraordinary points. In general, the construction of blending functions need not be the same as this construction. Moreover, one can use blending functions to combine local surface patches of different types.

While proposed surfaces are C^2 and flexible, they still do not allow exact reproduction of certain important simple shapes. For example, it would be useful to be able to reproduce a sphere without seams, a modeling task, which, to the best of our knowledge, cannot be achieved by any existing parametric surface representation (a NURBS sphere has a seam), except trigonometric schemes proposed in [MWW01].

7.0.0.1. Acknowledgements. This work was partially supported by NSF award CCR-0093390 and a Sloan Foundation Fellowship. The implementation was based on the code developed in collaboration with H. Biermann.

References

[BLZ00] BIERMANN H., LEVIN A., ZORIN D.: Piecewise smooth subdivision surfaces with normal control. In *SIGGRAPH 2000 Conference Proceedings* (2000), pp. 113–120.

[BR97] BOHL H., REIF U.: Degenerate bézier patches with continuous curvature. *Comput. Aided Geom. Des. 14*, 8 (1997), 749–761.

[DS78] DOO D., SABIN M.: Analysis of the behaviour of recursive division surfaces near extraordinary points. *Computer Aided Design 10*, 6 (1978), 356–360.

[GH89] GREGORY J. A., HAHN J. M.: A C^2 polygonal surface patch. *Comput. Aided Geom. Design 6*, 1 (1989), 69–75.

[GH95] GRIMM C. M., HUGHES J. F.: Modeling surfaces of arbitrary topology using manifolds. *Proceedings · of SIGGRAPH 95* (August 1995), 359–368. ISBN 0-201-84776-0. Held in Los Angeles, California.

[HP89] HAGEN H., POTTMANN H.: Curvature continuous triangular interpolants. In *Mathematical methods in computer aided geometric design (Oslo, 1988)*. Academic Press, Boston, MA, 1989, pp. 373–384.

[KP05] KARČIAUSKAS K., PETERS J.: Polynomial C^2 spline surfaces guided by rational multisided patches. In *Computational methods for algebraic spline surfaces*. Springer, Berlin, 2005, pp. 119–134.

[Lev06] LEVIN A.: Modified subdivision surfaces with continuous curvature. In *Proceedings of SIGGRAPH 2006* (2006). to appear.

[Loo04] LOOP C.: Second order smoothness over extraordinary vertices. In *SGP '04: Proceedings of the 2004*

Eurographics/ACM SIGGRAPH symposium on Geometry processing (New York, NY, USA, 2004), ACM Press, pp. 165–174.

[MWW01] MORIN G., WARREN J., WEIMER H.: A subdivision scheme for surfaces of revolution. *Computer Aided Geometric Design 18*, 5 (June 2001), 483–502. ISSN 0167-8396.

[NG00] NAVAU J. C., GARCIA N. P.: Modeling surfaces from meshes of arbitrary topology. *Computer Aided Geometric Design 17*, 7 (2000), 643–671.

[Pet96] PETERS J.: Curvature continuous spline surfaces over irregular meshes. *Computer-Aided Geometric Design 13*, 2 (Feb 1996), 101–131.

[Pet02] PETERS J.: C^2 free-form surfaces of degree $(3,5)$. *Comput. Aided Geom. Design 19*, 2 (2002), 113–126.

[Pra97] PRAUTZSCH H.: Freeform splines. *Comput. Aided Geom. Design 14*, 3 (1997), 201–206.

[PU98] PRAUTZSCH H., UMLAUF G.: A G^2-subdivision algorithm. In *Geometric Modelling, Computing Suppl. 13*, Farin G., Bieri H., Brunnet G.,, DeRose T., (Eds.). Springer-Verlag, 1998, pp. 217–224.

[Rei95] REIF U.: A unified approach to subdivision algorithms near extraordinary points. *Computer-Aided Geometric Design 12* (1995), 153–174.

[Rei96] REIF U.: A degree estimate for polynomial subdivision surfaces of higher regularity. *Proc. Amer. Math. Soc. 124* (1996), 2167–2174.

[Sta98] STAM J.: Exact evaluation of catmull-clark subdivision surfaces at arbitrary parameter values. In *SIGGRAPH '98: Proceedings of the 25th annual conference on Computer graphics and interactive techniques* (New York, NY, USA, 1998), ACM Press, pp. 395–404.

[YZ04] YING L., ZORIN D.: A simple manifold-based construction of surfaces of arbitrary smoothness. *ACM Trans. Graph. 23*, 3 (2004), 271–275.

[Zor00] ZORIN D.: Smoothness of subdivision on irregular meshes. *Constructive Approximation vol. 16*, 3 (2000), 359–397.

Eurographics Symposium on Geometry Processing (2006)
Konrad Polthier, Alla Sheffer (Editors)

Robust Reconstruction of Watertight 3D Models from Non-uniformly Sampled Point Clouds Without Normal Information

Alexander Hornung[†] and Leif Kobbelt[‡]

Computer Graphics Group, RWTH Aachen University

Abstract

We present a new volumetric method for reconstructing watertight triangle meshes from arbitrary, unoriented point clouds. While previous techniques usually reconstruct surfaces as the zero level-set of a signed distance function, our method uses an unsigned distance function and hence does not require any information about the local surface orientation. Our algorithm estimates local surface confidence values within a dilated crust around the input samples. The surface which maximizes the global confidence is then extracted by computing the minimum cut of a weighted spatial graph structure. We present an algorithm, which efficiently converts this cut into a closed, manifold triangle mesh with a minimal number of vertices. The use of an unsigned distance function avoids the topological noise artifacts caused by misalignment of 3D scans, which are common to most volumetric reconstruction techniques. Due to a hierarchical approach our method efficiently produces solid models of low genus even for noisy and highly irregular data containing large holes, without loosing fine details in densely sampled regions. We show several examples for different application settings such as model generation from raw laser-scanned data, image-based 3D reconstruction, and mesh repair.

Categories and Subject Descriptors (according to ACM CCS): I.3.5 [Computer Graphics]: Computational Geometry and Object Modeling

1. Introduction

The high quality reconstruction of a proper, watertight surface mesh from scattered point samples remains a difficult problem in many areas of computer graphics, including laser-scanning or image-based surface reconstruction techniques as well as repairing non-manifold or topologically noisy meshes.

Most previous work can be classified into computational geometry approaches based on Voronoi diagrams and global volumetric reconstruction techniques based on signed distance functions. Voronoi based approaches reconstruct a mesh directly from the input samples, with the particular strength of reconstructing even fine surface details. However, it is generally difficult to guarantee the reconstruction of a smooth and manifold surface, especially in the presence of noise and for varying sampling density.

Volumetric methods on the other hand attempt to reconstruct a signed distance function to the point cloud samples, and then reconstruct the zero level-set using, e.g., Marching Cubes [LC87]. The generation of the signed distance function however requires that the unstructured cloud of input points comes with consistently oriented normal information. This, however, is known to be one of the most critical steps in the reconstruction pipeline. The derivation of consistent normals from a point cloud poses a number of significant conceptual and computationally intensive problems, especially in the presence of noise, non-uniform sampling, or thin features. As a consequence, methods based on signed distance functions generally cannot guarantee that the resulting model is of the lowest possible genus. For instance, misaligned or noisy 3D scans are known to lead to severe topological noise artifacts (cf. Fig. 1 and 14).

[†] hornung@cs.rwth-aachen.de
[‡] kobbelt@cs.rwth-aachen.de

These artifacts are a fundamental problem of methods based on extracting the zero level-set of a signed distance function. An unnecessarily high genus prevents the possibility of immediate post-processing such as mesh decimation, hence these techniques generally require subsequent post-processing [ESV97, GW01, NT03, WHDS04] for artifact removal or general mesh repair [BNK02, Ju04, BPK05].

In this paper we present a robust algorithm to overcome these drawbacks. In contrast to the above mentioned approaches, our method reconstructs the surface from a volumetric *unsigned* distance function, which represents the probability that the surface passes through a given voxel. Since the unsigned distance function does not carry information about the local surface orientation we are able to process input data consisting solely of 3D sample positions without any normal information. Moreover, since the surface extraction does not depend on a sign-change of the implicit representation anymore, our method is immune to noisy and non-uniformly distributed samples (cf. Fig. 1). As an important consequence our method produces meshes of low genus without the small-scale topological artifacts.

The particular contribution of this paper is a method to compute an unsigned distance function from pure point cloud data, from which a closed surface can be extracted via graph-cut based energy minimization. We show how this algorithm can be embedded into a hierarchical framework allowing for efficient processing of highly non-uniformly sampled input data with large gaps, without loosing fine details in densely sampled regions. Finally we present a new algorithm to convert the graph-cut representation of the surface into a smooth and guaranteed watertight triangle mesh.

2. Related Work

Previous work on surface extraction from point clouds can be roughly classified into the following approaches.

Voronoi based techniques such as [ABK98, ACK01, BC02, DG03] have the advantage of computing output meshes with a complexity in the order of the input data, and produce good results for data sets with known sampling density. Wrapping approaches such as [BMR*99] provide a good local feature preservation. For non-uniformly sampled or noisy input data containing outliers, however, both types of approaches often cannot guarantee the reconstruction of a globally optimal, watertight surface. Improvements in these fields concerning noisy input data and outliers have been achieved recently in [MAVdF05, SFS05].

Methods based on deformable models for point cloud reconstruction have been presented in [EBV05, SLS*06]. They solve the problem of computing a watertight surface by incrementally deforming an initial mesh along an energy field induced by the point cloud. Although they guarantee watertight reconstructions, they have the potential problem of cre-

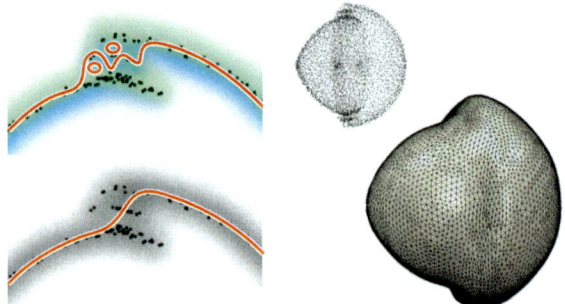

Figure 1: *The fundamental problem of surface reconstruction methods based on a signed distance function is the fact that local inconsistencies of the input data caused, e.g., by unreliable normal estimations, lead to frequent sign changes. This generally results in reconstructions of undesirably high genus, with significant topological artifacts (upper left). In contrast, our method is based on an unsigned distance function, which gracefully handles such inconsistencies (lower left). The right images show a corresponding point cloud and our genus 0 reconstruction.*

ating overly smoothed surfaces since it is often difficult to find appropriate surface tension parameters.

Most related to our work are approaches such as [HDD*92, CL96, CBC*01, ABCO*01, DMGL02, OBA*03, OBS04] which reconstruct the unknown surface as the zero level-set of a signed distance function. These methods, however, often rely on accurate normal orientation and fairly uniform sampling densities, which are both requirements generally not met by real world data sets. Furthermore they can be quite sensitive to noise or outliers, e.g., for badly aligned scan patches they tend to introduce topological artifacts such as handles or bridges due to spurious zero crossings of the signed distance function (cf. Fig. 1). This often leads to reconstructions with significantly increased genus. Recently, these issues have been addressed in [FCOS05, Kaz05].

To summarize, most of the above methods have in common that it is generally difficult or even impossible to generate proper meshes from highly non-uniformly sampled point cloud data without reliable normal information. We explicitly address these problems in our work.

Recently research on combinatorial energy minimization has shown that globally optimal solutions to discrete volumetric segmentation problems can be found efficiently by reformulating them into a maximum flow / minimum cut problem of a specific spatial graph structure [BK03, KB05]. Applications using graph cuts have been presented for problems such as image segmentation [LSGX05] or 3D stereo reconstruction [HK06, VTC05]. We will show in this work, how our method presented in [HK06] can be extended to the problem of point cloud reconstruction.

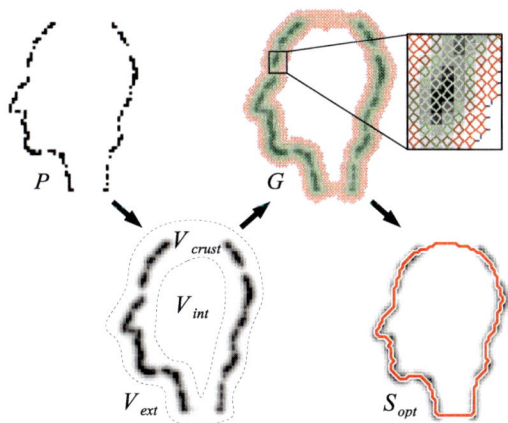

Figure 2: *This figure illustrates the point cloud reconstruction process in 2D. From left to right: Based on an input point cloud P we first compute an unsigned distance function by volumetric diffusion. The unknown surface is supposed to lie in the voxel crust V_{crust} between the outer boundary V_{ext} and the inner boundary V_{int}. We embed a spatial graph structure G within the voxel grid, with small edge weights for high confidence voxels and vice versa. The boundaries are connected a sink and a source node, respectively. Computing the min-cut of this graph yields the surface S_{opt}.*

3. Overview

In this section we outline the four central ideas and corresponding processing steps of our algorithm to achieve our goals. The different phases are then described in detail in the following sections.

The input to our algorithm is a point cloud, i.e., a set of scattered 3D position samples $p \in P$ of a surface S. Our basic idea is to first derive a *confidence map* in the vicinity of these point samples (similar to Pauly et al. [PMG04]), representing the probability that the unknown surface passes through a certain part of 3D space. We compute these confidence values as an unsigned distance function $\phi : v \to c \in [0,1]$ over the voxels $v \in V$ in a volumetric grid , where c can be interpreted as the pseudo-distance of a voxel to the closest point sample p (cf. Fig. 2). Since this representation, unlike signed distance functions, does not imply any local orientation properties of the unknown surface, noise or non-uniform sampling of the input samples do not significantly influence the quality of the reconstruction. Sect. 4 describes the steps for computing ϕ in detail.

Given such a grid of confidence weighted voxels we want to extract a minimal subset $S_{opt} \subseteq V$, representing a closed surface with maximum confidence, i.e., for a faithful approximation of the unknown surface S the sum of unsigned distance values has to be minimized $\sum_{v \in S_{opt}} \phi(v) \to min$. Meth-

ods for iso-surface extraction are not suitable to reconstruct a surface represented by such a probability distribution. Previous work [BK03] has shown that similar types of combinatorial optimization problems involving the minimization of certain energy functionals can be efficiently solved by transforming them into a max-flow / min-cut problem of an embedded spatial graph G. Our specific problem formulation for surface reconstruction from a set of confidence weighted voxels is highly related to our previous work on image based stereo reconstruction [HK06]. We will show in Sect. 5 how this graph based algorithm can be adapted to the setting of this work.

For non-uniformly sampled point clouds it is generally difficult to estimate an optimal volumetric grid resolution such that holes in sparsely sampled areas can be efficiently detected and closed without loosing details in densely sampled regions. On the other hand, simply computing the above mentioned unsigned distance function and surface extraction on a high resolution grid would result in a significant computational overhead. In Sect. 6 we show how to integrate the confidence estimation and graph-based surface extraction into a hierarchical framework such that the above mentioned problems are effectively resolved.

Once the desired target resolution is reached the voxel based representation of the surface S_{opt} has to be converted into a triangle mesh to be usable for further geometric processing steps. Sect. 7 describes a new algorithm to generate a smooth and manifold mesh derived from S_{opt} and the min-cut edges of G.

4. Surface Confidence Estimation

Initially we insert each 3D sample $p \in P$ into a volumetric grid V, resulting in a sparse set of occupied voxels v (cf. Fig. 2, upper left). As mentioned above the probability or confidence that a voxel v is part of the unknown surface can be approximated by an unsigned distance function ϕ over V. To compute ϕ in the vicinity of S we first apply several steps of a morphological dilation operator to the 6-neighborhood of occupied voxels, generating an extended crust of voxels V_{crust}. The distance function $\phi(v)$ for each voxel $v \in V_{crust}$ is then computed by volumetric diffusion (cf. Fig. 2, lower left).

For the graph-based surface computation we have to ensure that the computed crust is watertight (i.e., 6-connected) and has two interfaces V_{ext} and V_{int}, to an outer and inner volumetric component, respectively (cf. Fig. 3 b). In most cases the number of necessary dilation steps for computing this crust can be computed robustly with a simple heuristic. By flood-filling unoccupied voxels from the outer boundaries of V we can easily determine the current number of different volumetric components separated by V_{crust}. Initially we generally have only one (outer) component. This number increases during the dilation process as the crust grows,

(*a*)   (*b*)

(*c*)   (*d*)

Figure 3: *An example in 3D for point cloud dilation and confidence estimation for the dragon model. We increased the crust size for visualization purposes. Image (a) shows the initial set of occupied voxels which contain a sample of P. In (b) the dilated voxels generate a watertight crust V_{crust} in which the unknown surface is contained. Note that the outer boundary is genus 0, while the inner component is split into several components. (c) shows a section of the unsigned distance function in the head region after the diffusion process. Darker values indicate higher surface confidence. The set S_{opt} of surface intersected voxels after the graph-cut computation is shown in (d). The resulting model has the correct genus 1.*

and eventually drops down again to one component when the "interior" of the point cloud P is full of occupied voxels. V_{int} is then simply defined by the voxels conquered during the last dilation steps (e.g., 3 in our experiments). Please note that this repeated flood-filling and dilation process is computationally irrelevant in our overall hierarchical setting (cf. Sect. 6), since we generally start at low volumetric resolutions of 64^3 or 128^3.

For point clouds covering only a part of the surface of an object (cf. Fig. 13), or objects with relatively thin, elongated features and non-uniform sampling density (cf. Fig. 12) it is sometimes not possible to compute a proper interior component V_{int}. In these cases our algorithm computes a fast approximation to the medial axis of the dilated crust by estimating normal orientations on V_{ext}, and propagating them inwards through V_{crust} by an averaging filter. For each voxel we then estimate a normal cone by collecting the normals of all 26 neighboring voxels and label each voxel as V_{int} if the opening angle of this cone lies above a threshold of $\theta = \pi/2$. The actual choice of this threshold however does not have a significant influence on the results, since we basically just want to find discontinuities in the normal field.

Concerning the diffusion process to compute a smooth distance function we first assign distance values $\phi(v) = 0$ to voxels containing surface samples p, and $\phi(v) = 1$ for the

remaining voxels in V_{crust}. The diffusion is then simply performed by iterative averaging over the 6-neighborhood $N(v)$ (in V_{crust}) of a voxel

$$\phi(v) = \frac{1}{|N(v)|+1} \left(\phi(v) + \sum_{u \in N(v)} \phi(u) \right) , \quad (1)$$

while keeping $\phi(v) = 0$ fixed for voxels containing surface samples (cf. Fig. 3 c). The overall algorithm is not very sensitive to the number of diffusion steps. In fact, a valid surface can already be computed after the initialization of ϕ without any diffusion. However, the surface becomes slightly smoother with more diffusion steps, and we additionally show in Sect. 7 how the unsigned distance values allow for confidence weighted mesh smoothing of the extracted mesh. In our results presented in Sect. 8 we simply use three diffusion steps for all reconstructed models.

We also experimented with initial confidence values computed from the sample density within a voxel instead of setting all occupied voxels to $\phi(v) = 0$. However, the current approach has the advantage of handling strongly non-uniformly sampled regions in a more uniform manner. Similarly, keeping $\phi(v) = 0$ fixed for voxels containing surface samples instead of including them in the diffusion process preserves fine details more faithfully.

5. Graph-based Surface Extraction

Given a function ϕ of surface confidence values in a volumetric region V_{crust}, we have presented a method [HK06] to compute a closed 2-manifold surface S embedded in V_{crust} which minimizes an energy functional

$$E(S) = \int_S \phi(x) \, dx + \int_S a \, dS , \quad (2)$$

with a being a regularizing parameter of the surface tension. Our algorithm efficiently computes a set of surface voxels $S_{opt} \subseteq V_{crust}$, which minimizes the discretization of $E(S)$ by computing the minimum cut of a weighted graph G embedded into the volumetric grid. For completeness we briefly describe the graph construction and edge weight computation.

In G, a graph node is associated with each voxel face, and a weighted graph edge is created for each voxel edge, such that each voxel contains an octahedral subgraph (cf. Fig. 4 a). An edge weight $w(v)$,

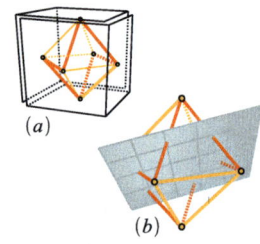

(*a*)

(*b*)

Figure 4: *Because of the duality of the voxel and the octahedron, a cut through the octahedral subgraph corresponds to a split of the voxel faces into an exterior and an interior component (a). The splitting surface is visualized in (b).*

depending on the voxel's unsigned distance value $\phi(v)$ and the constant a, is assigned to all edges of the corresponding subgraph

$$w(v) = \phi(v)^s + a \ . \qquad (3)$$

The exponent s can be used to tune the unsigned distance function to some extend, such that the maxima of the confidence values are emphasized more or less strongly. However, in all our experiments we set the parameters for the edge weight computation (cf. Eq. 3) as proposed in [HK06] to $s = 4$ and $a = 10^{-5}$. The outer and inner boundaries nodes (voxel faces) exposed at the interface to V_{ext} and V_{int} are connected to a sink and a source node, respectively.

The minimum cut of G then yields a set of cut-edges C which form a watertight, manifold separation between the sink and the source node (and hence V_{ext} and V_{int}) and are the globally optimal solution in terms of the surface energy functional (2) (cf. Fig. 2 and 3 d), and hence can be considered as a faithful approximation to the surface S. The corresponding set of surface voxels S_{opt} is defined by those voxels containing at least one cut edge.

6. Hierarchical Hole Filling and Detail Preservation

For high volumetric resolutions the computational complexity for generating the unsigned distance function and the graph cut computation become impractical, especially for strongly non-uniform data containing sparsely sampled regions as well as fine details. Hence we employ an iterative hierarchical framework on an adaptive volumetric grid (e.g., using an Octree), and use the surface approximation obtained on a lower volumetric resolution to constrain the crust and surface computation on the respective higher level. Starting at a low volumetric resolution allows for an efficient generation of a proper initial crust even for highly non-uniformly sampled point clouds with large gaps.

Previous work (e.g., [LSGX05]) has shown however that a simple hierarchical refinement of voxels within a fixed distance to the cut surface S_{opt} potentially leads to a loss of fine details if the corresponding input data samples are not contained inside this fixed distance crust. In our application setting however, where explicit data samples are available, we can derive an efficient hierarchical algorithm which effectively avoids the above mentioned problems.

On a given volumetric refinement level l we compute a surface approximation S_{opt}^l within a crust V_{crust}^l as described in the previous sections. Then, we compute a new, thinned crust V_{crust}^{l+1} on the next higher resolution level by refining the surface voxels $v \in S_{opt}^l$, and applying a number of morphological dilation steps. This effectively constrains the volumetric region for surface extraction based on our current surface proxy S_{opt}^l. To preserve fine details represented by point samples of P outside of this crust, we re-insert the corresponding input samples as occupied cells into the volumetric grid at the new resolution $l + 1$, and dilate these cells

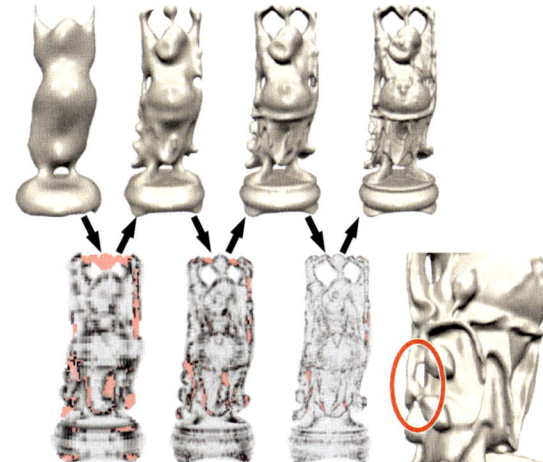

Figure 5: *Starting at a volumetric resolution of 64^3 this image sequence shows a hierarchical reconstruction of the Buddha model from the Stanford 3D Scanning Repository up to 512^3. On each level the images show the reconstructed mesh and the crust of consistency values in alternating order. Re-inserted detail samples are shown in red. Note that reconstructed thin features and the genus of the model do not depend on the respective approximation on the previous levels, but automatically adapt to the current resolution. Our algorithm reduces the original genus of the input model from >100 to 10. A few additional holes are introduced in regions, where the point samples of opposite surface sheets lie very close compared to the volumetric resolution, e.g., as shown in the lower right close-up.*

until they merge with V_{crust}^{l+1} (cf. Fig. 5). In our experiments we simply used a fixed number of 3 dilation steps. The cut computation then automatically includes these voxels for the surface extraction, and fine details are preserved.

The number of dilation steps to compute the thinned crust V_{crust}^{l+1} basically depends on the amount of noise of the input data P. Without noise two dilation step at level $l + 1$ would be sufficient assuming that the low frequency parts of the surface have been reconstructed up to voxel accuracy at level l. In all our experiments with real data four dilation steps proved to be a good choice. However, since all original data points outside of the crust are re-inserted and dilated in the higher resolution grid anyway, the choice of this parameter is mostly of importance for the computational performance. An unnecessarily thick crust would include too many voxels which do not contribute to the surface extraction at all, while a too thin crust would lead to a higher number of samples outside of the crust, such that a higher number of additional dilation steps is necessary.

To summarize, our hierarchical point cloud reconstruction consists of the following three iterative phases. Starting at a volumetric resolution l:

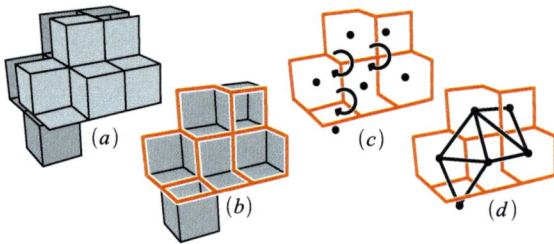

Figure 6: *The computed min-cut of G splits the voxels in S_{opt} into exterior and interior faces (a). The set of cut-edges C defines a loop of split-edges for each voxel, corresponding to a non-planar polygonal face (b). In the dual mesh, the non-planar polygonal faces are defined by cut-corners which are shared by at least 3 surface voxels. We extract the corresponding mesh by placing a single mesh vertex at the center of each voxel, and visiting the voxels associated with a cut-corner by cycling over shared cut-edges (c). Triangulating the corresponding polygons with a triangle fan yields the final mesh (d).*

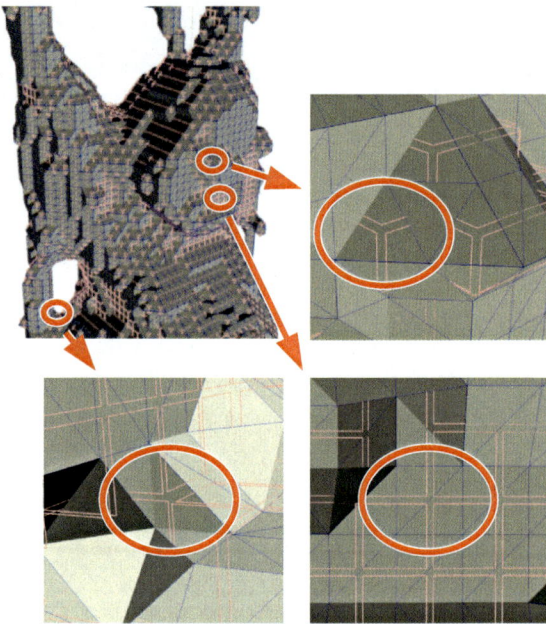

Figure 7: *Different cut-corner configurations for a low resolution mesh of the buddha model with 3, 4, and 5 incident cut-edges (shown in red), respectively.*

1. Surface confidence estimation

   - Insert points $p \in P$ as occupied voxels v into the volumetric grid V^l.
   - Dilate these voxels to a crust V^l_{crust}.
   - Compute the unsigned distance function $\phi(v)$ for all $v \in V^l_{crust}$ by volumetric diffusion (cf. Eq. 1).

2. Graph-based surface extraction

   - Generate graph G consisting of octahedral subgraphs for each voxel $v \in V^l_{crust}$ with edge weights according to Eq. 3. Nodes at the boundaries of V^l_{crust} are connected to a corresponding terminal node of G.
   - Compute the min-cut of G (e.g., using [BK04]), resulting in a set of surface intersected voxels S^l_{opt} and cut edges C.
   - If l is the target resolution then terminate and extract the final surface mesh (cf. Sect. 7).

3. Volumetric refinement

   - Refine surface voxels S^l_{opt} to the next higher resolution and set $l = l + 1$.
   - Compute the new crust V^l_{crust} by dilation and proceed with step 1.

Please note that the computation of the unsigned distance function and the graph cut in steps 1 and 2 are not affected by the hierarchical approach, since all voxels in the current crust V^l_{crust} are at the same refinement level.

This algorithm computes a closed surface representation even from strongly non-uniformly sampled point clouds. Large gaps are effectively closed with a reasonable surface due to the surface tension in Eq. 2, and fine details are preserved due to the iterative point insertion and dilation.

7. Mesh Extraction

In the final step of the surface reconstruction algorithm we have to extract a polygonal representation from the set of voxels S_{opt}. Due to the duality of the octahedron and the cube, we can interpret the cut *through* the octahedron edges in G as a cut *along* the cube edges ("cut-edges", see Fig. 6 a,b). By this, the global graph cut through S_{opt} defines a polygonal mesh $\mathcal{M}$ with non-planar faces, which corresponds to a closed manifold surface. The vertices of $\mathcal{M}$ lie on the voxel corners and the mesh edges coincide with voxel cut-edges (cf. Fig. 6 b). In [HK06] an algorithm is described to extract a triangle mesh by generating a triangle fan for each polygon/voxel.

However, since confidence values are estimated per voxel and also in order to reduce the output mesh complexity, it seems more natural to extract a polygon mesh $\mathcal{M}'$ which is dual to $\mathcal{M}$, i.e. vertices are lying in the voxel centers and the non-planar polygonal faces correspond to voxel corners (cf. Fig. 8).

The polygonal mesh $\mathcal{M}'$ can easily be generated by running through all $2 \times 2 \times 2$ blocks of voxels. For each block B the center voxel corner corresponds to a polygon face of $\mathcal{M}'$ if the block contains at least three voxels from S_{opt} (cf. Fig. 6 c and 7). The edges for a polygonal face of $\mathcal{M}'$ are enumerated by cycling through the voxels of the $2 \times 2 \times 2$ block. Since every voxel in S_{opt} has exactly two cut-edges incident to the corner in the center of B, the ordering is given

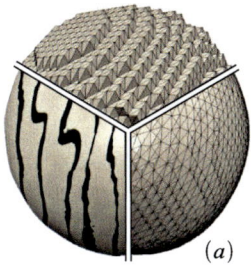 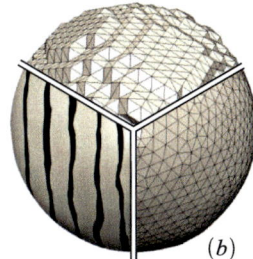

(a) *(b)*

Figure 8: *This figure compares the meshing algorithm presented in [HK06] (a) to our new mesh extraction (b). Both meshes have been generated from the same graph cut of a sphere data set. The upper sectors show the mesh extracted directly from the cut edges with voxel grid discretization artifacts. The lower sectors show the vertex distribution and reflection lines after an equal number of Laplacian smoothing steps. Due to the oversampling of grid artifacts in (a) the surface smoothing for eliminating discretization artifacts converges much slower and more non-uniformly. Since our new meshing algorithm creates exactly one vertex per surface voxel we achieve a significantly better vertex distribution with a more regular mesh topology (i.e., valence 6 vertices), resulting in significantly improved smoothing convergence and lower mesh complexity.*

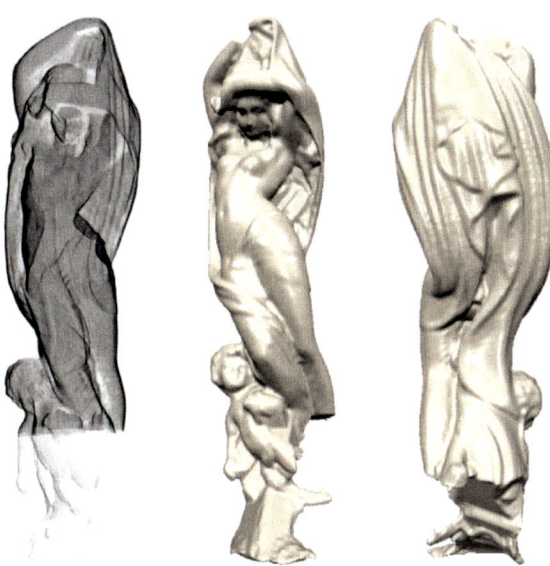

Figure 9: *Solid genus 0 reconstruction of a statue from non-uniformly sampled 3D points from raw laser scanned data, with significant outliers and holes. The backside of the upper arm and the lower part of the model are only partially sampled from the front, without any samples at the back of the object.*

by cut-edges that two neighboring voxels have in common. The following pseudo-code describes the procedure:

```
for each 2×2×2 block B with at least
      three voxels in S_opt
   pick a starting voxel v from S_opt ∩ B;
   pick a cut−edge e in v adjacent to
         the center voxel corner c of B;
   do
      find the second cut−edge f in v
            adjacent to c;
      find the neighbor voxel w from S_opt ∩ B
            sharing the cut−edge f with v;
      generate a polygon edge from v to w;
      v ← w, e ← f;
   until the first voxel is reached again;
```

This code works correctly because each voxel from $S_{opt} \cap B$ has exactly two cut edges adjacent to the center corner and no more than two voxels share a common cut-edge. A consistent orientation of the faces can be propagated by mesh traversal. The polygonal faces may be converted into triangle fans afterwards and decimated if required.

The resulting mesh is watertight but shows grid artifacts due to the fact that we initially placed the mesh vertices at the voxel centers. However our surface confidence map ϕ computed for each voxel can be applied to the mesh vertices accordingly. We can exploit this information for a confidence weighted smoothing algorithm, which allows for error bounded surface smoothing in confident surface areas, such

that only grid artifacts are removed, while less confident or noisy parts of the mesh can be smoothed stronger. We implement this algorithm by applying an iterative bi-Laplacian smoothing operator [DMSB99] for each vertex v:

$$v \leftarrow v - \frac{1}{d} \triangle^2 v , \quad d = 1 + \frac{1}{n_v} \sum_j n_{v,j} \qquad (4)$$

with n_v and $n_{v,j}$ being the valences of vertex v and its j-th one-ring neighbor. The surface confidence values $\phi(v)$ prescribe how much every vertex is allowed to deviate from its original position during smoothing. We stop the movement of a vertex if $\delta p < \delta v (\phi(v) + 1)^s$ is violated. δp is the difference between the original and the smoothed vertex position, δv represents the voxel size, and s allows for emphasized smoothing in inconfident regions. For all of our presented results, however, we simply set $s = 1$.

This algorithm computes smooth meshes while preserving the original surface approximation quality of the computed cut surface.

8. Results

In this section we present the results of our method applied to a variety of different data sets such as point clouds acquired from laser scans and stereo vision based point reconstructions as well as model repair. All reconstructions are based purely on 3D sample positions without any normal in-

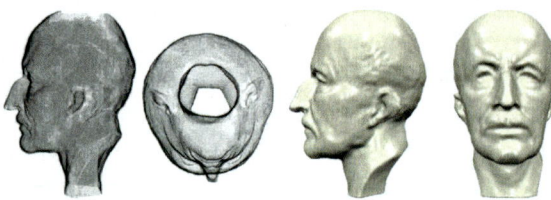

Figure 10: *The point cloud of the Max-Planck consists of a set of circularly acquired laser scans. The second image shows that top and bottom of the bust as well as some smaller areas around the ears do not contain any samples. Our method closes these holes and produces a genus 0 mesh.*

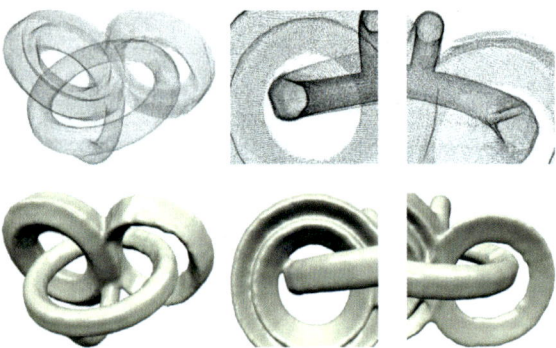

Figure 11: *Solid reconstruction of a genus 3 object from a noisy scan. Due to significant noise, however, two of the rings are merged, resulting in a genus 4 reconstruction. On the other hand, even significantly misaligned parts as shown in the right images are easily handled without producing topological artifacts.*

Figure 12: *The Leo point cloud was generated with a "Voxel Coloring" algorithm for image based 3D reconstruction [SD97]. It is quite noisy and highly non-uniform in terms of sample density and surface "thickness". Nevertheless our method succeeds in reconstructing a proper genus 1 model.*

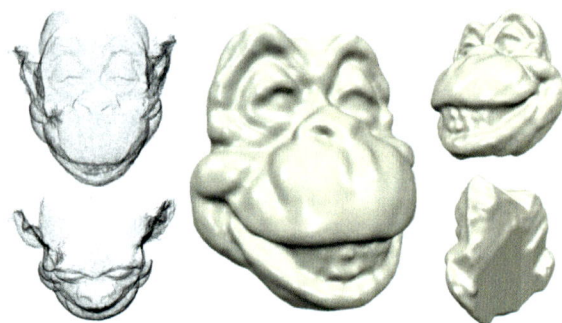

Figure 13: *The point cloud for the Monkey model was created by an image based stereo reconstruction algorithm. Despite the fact that only samples for the front of the model are available, our algorithm is capable of computing a closed mesh. We computed the interior component with the medial axis approximation described in Sect. 4. The ears however get cut away since they would include too many inconfident voxels in particular on the back of the head, and the resulting energy (cf. Sect. 5) would be higher than the given result.*

formation. We also provide quantitative evaluations in terms of the computation performance and the resulting meshes in Table 1. All experiments were performed on a 3.2 GHz Pentium P4 with 2 Gb of main memory.

The statue shown in Fig.9 is reconstructed from raw laser-scanning data at a volumetric resolution of 1024^3. The input point cloud contains significant noise, outliers and large gaps, e.g., at the bottom part or at the backside of the upper arm. Nevertheless our algorithm reconstructs a proper, watertight genus 0 model. This model is particularly difficult to reconstruct because of large regions with completely missing samples on the backside of the statue.

Further reconstructions from raw laser scanned point clouds are shown in Figures 10 and 11, respectively. While the Max-Planck example has a highly non-uniform sample distribution with large holes, especially at the top and the bottom, the Rings example contains significant noise and alignment artifacts.

The point cloud for the Leo (Fig. 12) as well as for the Monkey (Fig. 13) model have been acquired by image based

3D stereo reconstruction methods. The Leo has a particularly non-uniform and noisy sample distribution, with large clusters of points inside the model, and larger gaps at the tail and the legs. The Monkey model has a more uniform sampling but consists only of samples for the front side of the face. Creating watertight models from such models without at least approximate surface normals has been very challenging for previous methods.

As examples for mesh repair we applied our algorithm to the VRIP-reconstructions [CL96] of the Buddha (Fig. 5)

Model		Resolution	Timings	Genus	Vertices
Rings	(Fig. 11)	256^3	45 s	4	91 K
Leo	(Fig. 12)	256^3	48 s	1	47 K
Monkey	(Fig. 13)	256^3	82 s	0	72 K
Buddha	(Fig. 5)	512^3	112 s	10	264 K
Dragon	(Fig. 14)	512^3	150 s	1	318 K
Max-Planck	(Fig. 10)	512^3	199 s	0	320 K
Statue	(Fig. 9)	1024^3	269 s	0	448 K

Table 1: *The time and space complexity for all presented reconstructions. The timings include all processing steps, from confidence estimation to mesh smoothing. However, the most significant processing time is currently used for creating the actual graph structure. As discussed in Sect. 9, we expect a much better performance by computing the cut directly on the voxel grid.*

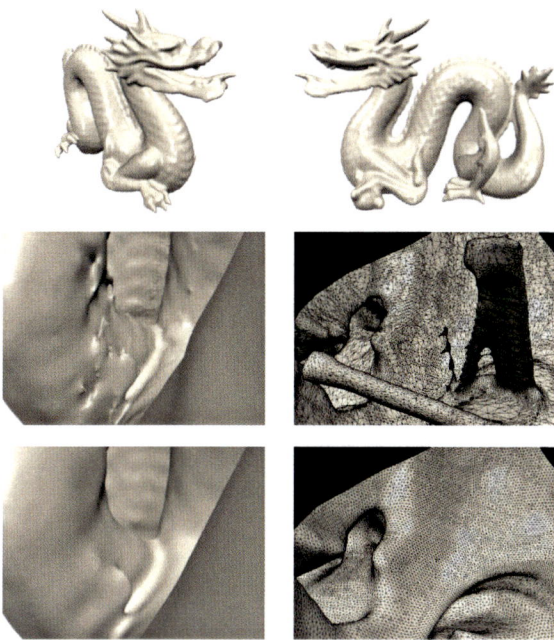

Figure 14: *The original Dragon model in the Stanford 3D Scanning Repository contains numerous small holes and topological artifacts such as bridges, resulting in a very high genus (>400). When using the 3D point samples of this mesh as input to our algorithm, all topological artifacts are removed and the resulting watertight mesh has genus 1. The close-ups compare the original backside of one of the legs and a view from inside the model to our result.*

and the Dragon (Fig. 14), available at the Stanford 3D Scanning Repository. Both models have a very high genus due to topological artifacts. Our method faithfully reconstructs watertight models with low genus. For the Dragon model we also show the triangle quality generated by our meshing algorithm.

9. Conclusions

In this paper we presented a robust algorithm to reconstruct a watertight triangle mesh from point clouds without requiring normal information. It generates surfaces of low genus without the topological artifacts produced by many other techniques due to the use of an unsigned distance function and graph cut minimization for surface extraction. Our method supports highly non-uniform sample densities without loosing details due to an efficient hierarchical scheme. Finally we showed how to extract a proper triangle mesh from the cut representation with just one vertex per voxel.

The resolution of our output models is currently restricted to 1024^3 because we explicitly generate the spatial graph structure G using a graph cut library [BK04], resulting in a noticeable memory overhead. However, due to the duality of the voxels and the embedded octahedral subgraphs it should be possible to alleviate this limitation by computing the cut directly on the voxel grid.

Since the surface is reconstructed at voxel accuracy, flat surface areas with a slight slope with respect to the main axes of the volumetric grid can lead to staircase artifacts, which converge only slowly to a smooth surface during our smoothing process. Although this fact does not influence the accuracy of the reconstructed model, we would like to find a solution for visually improved results, e.g., by integrating direct solvers.

References

[ABCO*01] ALEXA M., BEHR J., COHEN-OR D., FLEISHMAN S., LEVIN D., SILVA C. T.: Point set surfaces. In *IEEE Visualization* (2001).

[ABK98] AMENTA N., BERN M. W., KAMVYSSELIS M.: A new voronoi-based surface reconstruction algorithm. In *SIGGRAPH* (1998), pp. 415–421.

[ACK01] AMENTA N., CHOI S., KOLLURI R. K.: The power crust. In *Symposium on Solid Modeling and Applications* (2001), pp. 249–266.

[BC02] BOISSONNAT J.-D., CAZALS F.: Smooth surface reconstruction via natural neighbour interpolation of distance functions. *Comput. Geom. 22*, 1-3 (2002), 185–203.

[BK03] BOYKOV Y., KOLMOGOROV V.: Computing geodesics and minimal surfaces via graph cuts. In *ICCV* (2003), pp. 26–33.

[BK04] BOYKOV Y., KOLMOGOROV V.: An experimental comparison of min-cut/max-flow algorithms for energy minimization in vision. *IEEE Trans. Pattern Anal. Mach. Intell. 26*, 9 (2004), 1124–1137.

[BMR*99] BERNARDINI F., MITTLEMAN J., RUSH-

MEIER H. E., SILVA C. T., TAUBIN G.: The ball-pivoting algorithm for surface reconstruction. *IEEE Trans. Vis. Comput. Graph. 5*, 4 (1999), 349–359.

[BNK02] BORODIN P., NOVOTNI M., KLEIN R.: Progressive gap closing for mesh repairing. In *Advances in Modelling, Animation and Rendering*, Vince J., Earnshaw R., (Eds.). Springer Verlag, July 2002, pp. 201–213.

[BPK05] BISCHOFF S., PAVIC D., KOBBELT L.: Automatic restoration of polygon models. *ACM Trans. Graph. 24*, 4 (2005), 1332–1352.

[CBC*01] CARR J. C., BEATSON R. K., CHERRIE J. B., MITCHELL T. J., FRIGHT W. R., MCCALLUM B. C., EVANS T. R.: Reconstruction and representation of 3d objects with radial basis functions. In *SIGGRAPH* (2001), pp. 67–76.

[CL96] CURLESS B., LEVOY M.: A volumetric method for building complex models from range images. In *SIGGRAPH* (1996), pp. 303–312.

[DG03] DEY T. K., GOSWAMI S.: Tight cocone: a watertight surface reconstructor. In *Symposium on Solid Modeling and Applications* (2003), pp. 127–134.

[DMGL02] DAVIS J., MARSCHNER S. R., GARR M., LEVOY M.: Filling holes in complex surfaces using volumetric diffusion. In *3DPVT* (2002), pp. 428–438.

[DMSB99] DESBRUN M., MEYER M., SCHRÖDER P., BARR A. H.: Implicit fairing of irregular meshes using diffusion and curvature flow. In *SIGGRAPH* (1999), pp. 317–324.

[EBV05] ESTEVE J., BRUNET P., VINACUA A.: Approximation of a variable density cloud of points by shrinking a discrete membrane. *Comput. Graph. Forum 24*, 2 (2005), 791–807.

[ESV97] EL-SANA J., VARSHNEY A.: Controlled simplification of genus for polygonal models. In *IEEE Visualization* (1997), pp. 403–412.

[FCOS05] FLEISHMAN S., COHEN-OR D., SILVA C. T.: Robust moving least-squares fitting with sharp features. *ACM Trans. Graph. 24*, 3 (2005), 544–552.

[GW01] GUSKOV I., WOOD Z. J.: Topological noise removal. In *Graphics Interface* (2001), pp. 19–26.

[HDD*92] HOPPE H., DEROSE T., DUCHAMP T., MCDONALD J. A., STUETZLE W.: Surface reconstruction from unorganized points. In *SIGGRAPH* (1992), pp. 71–78.

[HK06] HORNUNG A., KOBBELT L.: Hierarchical volumetric multi-view stereo reconstruction of manifold surfaces based on dual graph embedding. In *CVPR* (2006).

[Ju04] JU T.: Robust repair of polygonal models. *ACM Trans. Graph. 23*, 3 (2004), 888–895.

[Kaz05] KAZHDAN M. M.: Reconstruction of solid models from oriented point sets. In *Symposium on Geometry Processing* (2005), pp. 73–82.

[KB05] KOLMOGOROV V., BOYKOV Y.: What metrics can be approximated by geo-cuts, or global optimization of length/area and flux. In *ICCV* (2005), pp. 564–571.

[LC87] LORENSEN W., CLINE H.: Marching cubes: A high resolution 3d surface reconstruction algorithm. In *SIGGRAPH* (1987), pp. 163–169.

[LSGX05] LOMBAERT H., SUN Y., GRADY L., XU C.: A multilevel banded graph cuts method for fast image segmentation. In *ICCV* (2005), pp. 259–265.

[MAVdF05] MEDEROS B., AMENTA N., VELHO L., DE FIGUEIREDO L. H.: Surface reconstruction for noisy point clouds. In *Symposium on Geometry Processing* (2005), pp. 53–62.

[NT03] NOORUDDIN F. S., TURK G.: Simplification and repair of polygonal models using volumetric techniques. *IEEE Trans. Vis. Comput. Graph. 9*, 2 (2003), 191–205.

[OBA*03] OHTAKE Y., BELYAEV A. G., ALEXA M., TURK G., SEIDEL H.-P.: Multi-level partition of unity implicits. *ACM Trans. Graph. 22*, 3 (2003), 463–470.

[OBS04] OHTAKE Y., BELYAEV A. G., SEIDEL H.-P.: 3d scattered data approximation with adaptive compactly supported radial basis functions. In *SMI* (2004), pp. 31–39.

[PMG04] PAULY M., MITRA N., GUIBAS L.: Uncertainty and variability in point cloud surface data. In *Eurographics Symposium on Point-Based Graphics* (2004).

[SD97] SEITZ S. M., DYER C. R.: Photorealistic scene reconstruction by voxel coloring. In *CVPR* (1997), pp. 1067–1073.

[SFS05] SCHEIDEGGER C. E., FLEISHMAN S., SILVA C. T.: Triangulating point set surfaces with bounded error. In *Symposium on Geometry Processing* (2005), pp. 63–72.

[SLS*06] SHARF A., LEWINER T., SHAMIR A., KOBBELT L., COHEN-OR D.: Competing fronts for coarse-to-fine surface reconstruction. In *Eurographics* (2006), p. to appear.

[VTC05] VOGIATZIS G., TORR P., CIPOLLA R.: Multiview stereo via volumetric graph-cuts. In *CVPR* (2005), pp. 391–398.

[WHDS04] WOOD Z. J., HOPPE H., DESBRUN M., SCHRÖDER P.: Removing excess topology from isosurfaces. *ACM Trans. Graph. 23*, 2 (2004), 190–208.

Eurographics Symposium on Geometry Processing (2006)
Konrad Polthier, Alla Sheffer (Editors)

Reconstruction with Voronoi Centered
Radial Basis Functions

M. Samozino[1], M. Alexa[2], P. Alliez[1] and M. Yvinec[1]

[1]INRIA Sophia-Antipolis, France
[2]TU Berlin

Abstract

We consider the problem of reconstructing a surface from scattered points sampled on a physical shape. The sampled shape is approximated as the zero level set of a function. This function is defined as a linear combination of compactly supported radial basis functions. We depart from previous work by using as centers of basis functions a set of points located on an estimate of the medial axis, instead of the input data points. Those centers are selected among the vertices of the Voronoi diagram of the sample data points. Being a Voronoi vertex, each center is associated with a maximal empty ball. We use the radius of this ball to adapt the support of each radial basis function. Our method can fit a user-defined budget of centers: The selected subset of Voronoi vertices is filtered using the notion of lambda medial axis, then clustered to fit the allocated budget.

Categories and Subject Descriptors (according to ACM CCS): I.3.5 [Computer Graphics]: Surface Reconstruction from Scattered Data, Radial Basis Functions.

1. Introduction

Recent improvements in automated shape acquisition has stimulated a profusion of surface reconstruction techniques over the past few years for computer graphics and reverse engineering applications. Data collected from scanning processes of physical objects are often provided as large point sets.

Reconstruction methods can be roughly classified as Voronoi-based and mesh-free. Voronoi based reconstruction algorithms compute the Delaunay triangulation of the sample points, the dual to the Voronoi diagram. A subcomplex interpolating the sampled surface is then extracted from the Delaunay triangulation [AGJ02, AB98, AS00, CSD04, DGH01, DG04, ACK01, KSO04]. Detailed surveys are presented in [CG04, Dey04]. In the mesh-free approaches, the surface is approximated or interpolated using explicit methods such as deformable models [DQ01, Set99], parametric methods such as NURBS, B-Spline [Far02] or implicit methods such as RBF or MLS (see [TO02] for a survey). Among the many techniques developed for surface reconstruction with implicit methods, the radial basis functions (RBF) approach has shown successful at reconstructing surfaces from point sets scattered on surfaces of arbitrary topology [Buh03, Duc77, Isk04, Wen04].

Radial Basis Functions (RBF) were introduced by Broomhead and Lowe in the neural network community [BL88]. Techniques based on radial basis functions are now common tools for geometric data analysis [FN80, LF99], pattern recognition [Kir01] and statistical learning [HTF01]. The radial basis functions approach is volumetric in the sense that it approximates the input surface as the zero level-set of a scalar 3D function. This function is expressed as a weighted sum of radial functions, whose centers commonly coincide with the input data points. The function is constrained to be zero on the input data points and to be non-zero on other points in order to avoid the trivial constant solution. Given a set of centers, a set of constraint points and a type of radial basis function, reconstructing the sampled surface amounts to finding the set of weights which minimize a least-squares error to fit the constraints.

Although Voronoi-based reconstruction has long been criticized for its computational burden, recent developments in the implementation of fast algorithms have alleviated this issue. As an example, computing the Delaunay of 50K points takes 1s using the CGAL library [FGK*00]. Such methods still depend on the quality of the sampling and on the differential and topological properties of the surface. In particular, sparsity, redundancy, noisiness of the sampling, or non-smoothness and boundaries of the surface makes the surface reconstruction a challenging problem. Besides, Voronoi-based reconstruction methods often fail to produce watertight surfaces.

Radial basis functions, on the other hand, still have issues with picking the right non-zero constraints (to avoid disconnected components), and with efficiently computing the weights. Functions with unbounded support give the best reconstruction results, but also lead to dense matrices. The only viable solution to this problem so far is the multipole expansion for polyharmonic functions developed by Carr et al. [CFB97]. Unfortunately this approach is notoriously intricate and difficult to reproduce. Compactly supported functions lead to sparse matrices [Wen95]. However, finding a proper support size for the functions in case of irregularly sampled surfaces is difficult. A recent trend is to perform a set of local reconstructions, which may be mixed with quadric or higher-order jet fitting, and to blend them using the partition of unity [TI04, OBS04]. Although a great deal of effort has been put into the elaboration of multi-level techniques with local reconstructions to deal with large data sets, less effort has been spent at improving the compactness of the representation by center selection and optimization [CFB97, TI04, OBS04]. Besides, when the basis functions is compactly supported, the computed function is only defined in the vicinity of the input data points.

1.1. Contributions

Our approach combines both worlds and eliminates some of the aforementioned shortcomings. The sampled surface S is still reconstructed as the zero-level of a function f expressed as a linear combination of radial basis functions. The main advance in our method is to use radial basis functions centered at vertices of the Voronoi diagram of the data points. More precisely, centers of radial basis functions are chosen among a subset of those Voronoi vertices, which are called poles. Under certain sampling conditions, the poles are known to be closed to the medial axis of the sampled surface S [AB98]. Furthermore, each pole is the center of a Delaunay ball hereafter called polar ball. A polar ball is a maximal ball empty of sampled points. Such a ball is close to a maximal ball in $\mathbb{R}^3 \setminus S$. Considering that any smooth surface S can be viewed as the envelope of the maximal balls in $\mathbb{R}^3 \setminus S$, using poles as centers for radial basis functions is a rather natural idea. Furthermore, in our reconstruction process, we use the radius of each polar ball as a guidance for choosing the support of the corresponding basis functions. Hence, the support of each basis functions is locally adapted to the geometry and topology of the sampled shape. Also, because the radius of each polar ball is a good estimate of the distance between the pole and the sampled surface, we use this radius to set, as additional constraints, the value of the function at the poles. This leads to a reconstruction technique with the following features:

- The surface is represented as the zero-level set of a signed

function, which is a good approximation of the signed distance field to the surface.
- The function is defined as a weighted combination of locally supported radial functions; The number of functions is independent from the number of input points and typically significantly smaller. The function can thus be evaluated faster than when using traditional (even compactly supported) RBF.
- While the computation of the weights potentially incorporates all data points as constraints, the size of the system matrix only depends on the number of centers, not on the number of constraints.
- A filtering of the poles based on the notion of λ-medial axis allows the surface to degrade gracefully with noise.

In comparison with Voronoi-based reconstruction, the most important advantages of our technique are the resilience to noise and the construction of a smooth watertight surface that approximates all data points. In comparison to the common compactly supported RBF, fewer centers are used for the same accuracy. This leads to faster computation of the weights and faster evaluation of the functions. Using poles associated with their Voronoi ball radius as additional constraints leads to a better approximation of the distance field to the surface, and to fewer topological issues such as superfluous connected components away from the input points.

1.2. Overview

Our algorithm proceeds as follows: given a 3D point set scattered on a surface, we first compute its Delaunay triangulation and the dual Voronoi diagram. Our algorithm then repeatedly refines a subset of the Voronoi vertices. In the first stage, poles are extracted from the Voronoi vertices and are classified as inside or outside. In the second stage, we select a user-defined number of centers, m, among the set of poles. The selection is performed by filtering, then clustering the set of poles. Poles are filtered in order to adjust the level of detail to the budget of centers and clustered in order to achieve a center distribution nicely spread on the medial axis. We choose as radial basis function a Gaussian-like function with a compact support [Wen95], where the support size is locally adapted. As constraints, we impose the function f to be zero at the data points and to be non zero at the center points. A value set at a center point approximates the signed distance from this point to the sampled surface. The weights are obtained by computing the best least squares approximation of the function f with respect to the constraint points.

For completeness we list some key notions behind the radial basis functions in Section 2. Section 3 details the main steps of our algorithm. We show several experimental results in Section 4. Some work in progress and perspective directions are discussed in Section 5.

2. Background

Definition 1 The approximation problem is formulated as follows. Given $\{p_i\}_{i=1\ldots n}$ a set of n points and n scalar numbers $F = \{f_i\}_{i=1\ldots n}$, find a function $f : \mathbb{R}^3 \to \mathbb{R}$ satisfying the approximation condition:

$$f^* = \text{argmin}_f E(f), \qquad (1)$$

where E is the least squares error :

$$E(f) = \sum_{i=1}^{n} (f_i - f(p_i))^2. \qquad (2)$$

In the RBF approach, the function f is defined from a class of basis functions $\Phi_j : \mathbb{R}^3 \times \mathbb{R}^3 \to \mathbb{R}$, as a linear combination

$$f(x) = \sum_{j=1}^{m} \alpha_j \Phi_j(x, c_j), \qquad (3)$$

where $\{c_j\}_{j=1\ldots m}$ denotes a set of m center points and $\{\alpha_j\}_{j=1\ldots m}$ denotes a set of unknown weights to be solved for.

The reconstruction problem boils down to determine the vector $\alpha = \{\alpha_1, \ldots, \alpha_n\}$, by solving a linear system of equations resulting from the minimization of E (Eq.2) :

$$\alpha = \left[G_{P,\Phi}^t G_{P,\Phi} \right]^{-1} G_{P,\Phi}^t F, \qquad (4)$$

where matrix $G_{P,\Phi} = [\Phi(p_i, c_j)]_{i=1..n, j=1..m}$ and $F = [f_i]_{i=1..n}$.

In the following, the set of points, P, where the function value is specified a priori are called *constraints*. The set P includes the data points where the function f should vanish by definition, i.e. where all the f_i should be zero. To avoid the trivial solution $\alpha = \vec{0}$, during the minimization of E in (2), several interior and exterior constraints are added where the function f does not vanish. For each additional constraint point p_k, we assign to f a signed value f_k. This value is commonly the approximated signed distance between p_k and the sampled surface. The N constraints $\{p_i\}_{i=1..N}$ are now composed of the n input points and of the additional off-surface constraints where the function f is specified.

Most approaches locate centers both at the input data points and at the off-surface constraints, therefore the number of centers is such that $m = N$ and the minimization process (1) reduces to solving a $N \times N$ linear system which requires $O(N^3)$ machine operations and $0(N^2)$ bits for storage. Then, each evaluation of $f(x)$ requires $O(N)$ operations. This approach is therefore not suitable to a number of constraints greater than several thousands. To reduce the computational complexity, one first idea is to reduce the number of constraints. Notice that since most algorithms use the same points as constraints and as centers, this also leads to center reduction. This approach is commonly called

center reduction in the literature.

Center reduction consists of optimizing the trade-off between fitting accuracy and number of centers. A greedy algorithm is proposed in [CBC*01]: centers are iteratively added at locations where the fitting error is maximum until a satisfactory accuracy is reached. Another idea to further reduce the number of centers while maintaining decent fitting accuracy is to relax the one-to-one correspondence between the centers and the constraints. This approach, which we follow in this paper, is called *Generalized Radial Basis Functions* (GRBF) in the neural networks community [PG89]. Let m be a user-defined number of centers, possibly located anywhere in space, and N the number of constraints, such that $m << N$. The size of the matrix to be inverted and stored is now $m \times m$, independently of the number of constraints. $O(m)$ operations are now required for a single point-wise evaluation. Each term of the matrix $G^t G$ being a sum of contributions arising from each constraint, the number of constraints conditions the cost for assembling the matrix. This paper investigates one of the most important degrees of freedom offered by the RBF method: the location of centers and constraints to obtained a satisfactory trade-off between number of centers and fitting accuracy.

3. Algorithm

The input data for our algorithm is a point set $P = \{p_i\}_{i=1..n} \subset \mathbb{R}^3$. All the input data points are supposed to lie on the surface so the function value f is set to zero a these points:

$$f_i = f(p_i) = 0, \qquad \forall i = 1 \ldots n. \qquad (5)$$

We structure this section by the main components of the reconstruction algorithm, namely the choices made for the centers, for the constraints and for the radial basis functions.

3.1. Centers

Centers for RBFs are selected from the vertices of the Voronoi diagram of the input points. Selection is performed by refining a set of candidates in three steps.

Pole Extraction Let $\mathcal{O}$ be a shape with a closed continuous boundary $S = \partial \mathcal{O}$. A ball $\mathcal{B}$, included in $\mathbb{R}^3/S$, is said to be a maximal ball if there exists no other ball included in $\mathbb{R}^3/S$ and containing $\mathcal{B}$.

Definition 2 Medial axis:
The medial axis M of S is the topological closure of the set of points of $\mathbb{R}^3$ that have at least two nearest neighbors on S. Every point in M is the center of a maximal ball.

Definition 3 Voronoi diagram:
The Voronoi diagram of a point set P is a partition of the space in regions with the same closest point in P. Every

Voronoi cell corresponds to exactly one point p_i and contains all points in the space that are closer to p_i than to any other points in P.

$$V(p_i, P) = \{x \in \Omega : \forall q \in P \ \|x - p_i\| \leq \|x - q\|\}. \quad (6)$$

For the problem of reconstructing surfaces from point sets, we assume that all points are sampled on the surface. In 2D, it has been shown that if the sample is dense enough, all Voronoi vertices are closed to the medial axis. However, a similar result does not hold in 3D, where some Voronoi vertices may be located close to the surface and thus far from the medial axis, even when the sample density goes to infinity. The notion of a *pole* was previously introduced to handle this problem.

Definition 4 Pole:
A vertex of the Voronoi cell, $V(p_i, P)$, of a sample point $p_i \in P$ is called a pole if :

- either it is the vertex v_i of $V(p_i)$ that is the farthest from p_i
- or it is the vertex w_i of $V(p_i, P)$ that is the farthest from p_i in the halfspace H_i^-, set of points x such that $(v_i - p_i) \cdot (x - p_i) \leq 0$.

As a pole is a Voronoi vertex, there exists a maximal ball centered at each pole. This ball is called a *polar ball*. Amenta et al. [ACK01] and Boissonnat and Cazals [BC00] show that under some conditions the poles are close to the medial axis of the sampled shape. The conditions are that the surface is smooth and the sampling is dense enough. More precisely, the sample has to be an ε-sample. This means that for any point, x, on the surface, the distance from x to the sample is not greater than ε times the distance from x to the medial axis. The poles are shown to exhibit interesting properties:

- if v_i is a pole of the cell $V(p_i, P)$, the direction $\overline{v_i p_i}$ is a good approximation of the normal at p_i;
- the radius of the Delaunay ball centered at v_i is a good approximation of the distance from v_i to the sampled surface.

Let m be the user-defined budget of centers. Generally, the number of poles is greater than m, and we must select m relevant poles as centers. If m is small, there is no hope to reconstruct very small details and thus we need to remove the poles which correspond to the smallest details (which are not distinguishable from noise). This task is performed by filtering the poles based on the notion of the λ-medial axis. Notice that this filtering stage is different from the clustering stage, which is designed to distribute the final budget of centers on the λ-medial axis with a proper sampling density.

Pole Filtering A major problem arises when trying to approximate the medial axis of a sampled shape from the Voronoi vertices of the data points: The medial axis is known to be highly unstable with respect to small details of the shape. This means that even if two objects are very close

with respect to the Hausdorff distance, they may have very different medial axis (Fig.1). Thus, the set of poles extracted from the Voronoi diagram of a sampled surface is very unstable with respect to noise as well. Several approaches have been proposed to tackle this problem [AM96, DZ03]. In this paper we follow the recent work of Chazal and Lieutier [CL05], which defines the notion of λ-medial axis.

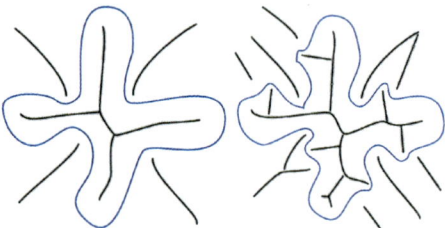

Figure 1: *Instability of the medial axis. Left: a smooth shape and its medial axis (black). Right: the same shape with some bumps added and its (unstable) medial axis.*

For any point p, we denote by $\Gamma(p)$ the set points of the boundary $\partial \mathcal{O}$ that are closest to p.

$$\Gamma(p) = \{y \in \partial \mathcal{O}, d(x, y) = d(x, \partial \mathcal{O})\}. \quad (7)$$

The medial axis M of $\mathcal{O}$ can be viewed as the set of points $x \in \mathcal{O}$ such that $|\Gamma(x)| > 2$. For each point p, there is a smallest ball enclosing $\Gamma(p)$. We define the real-valued function $\gamma(p)$ as the radius of the smallest ball enclosing $\Gamma(p)$. The λ-medial axis M_λ is defined as :

$$M_\lambda = \{p \in \mathcal{O} | \gamma(p) > \lambda\}. \quad (8)$$

M_λ is a closed subset of the medial axis. Moreover, the medial axis is obtained for $\lambda = 0$. Chazal and Lieutier have shown that for any value for λ which is not a singular value of the map $\lambda \longmapsto M_\lambda$, the λ-medial axis of a surface is stable under small perturbations and can be estimated from a dense sampling. Roughly speaking, restricting the λ-medial axis with increasing value of λ, smooths out both small features and noise. We use this idea of medial axis filtering to smooth noise and adapt the level of detail of the reconstruction to the allocated budget of centers. More precisely, this means that we determine the value λ suitable to the sampled shape and to the budget of centers, and filter out the poles which are not close to the λ-medial axis. To estimate if a pole v is close to the λ-medial axis, we compute the radius $\gamma(v)$ of the smallest ball enclosing the set $\Gamma(v)$ of sample points closest to v. Poles with radius $\gamma(v)$ smaller than λ are discarded.

Pole Clustering The filtered set of poles now forms a set of *possible* centers, PC. Generally, the size of PC remains larger than m, the user-defined budget of centers. In order to select m centers from PC, we perform a k-means clustering over the set of possible centers [Mac67]. The goal is to obtain a sampling of the λ-medial axis with a local sizing field at a pole v_i proportional to the radius of the polar ball $r(v_i)$.

Figure 2: *Medial axis filtering on a 2D shape (blue). The λ-medial axis is depicted in black. Top left: all extracted poles. Top right: pole filtering with parameter $\lambda = 0.01$. Bottom left: $\lambda = 0.03$. Bottom right: $\lambda = 0.05$. To get a better sense of these parameters: the diagonal length of the bounding box of the input point set is 1.4.*

Therefore, we compute the centroid, c, of a clustering cell $\mathscr{C}$ as

$$c = \sum_{v_i \in \mathscr{C}} \omega_i v_i, \qquad (9)$$

using for each pole, v_i, a weight, ω_i

$$\omega_i = \frac{d(v_i)}{\rho(v_i)}, \qquad (10)$$

where $d(v_i)$ denotes a quadrature term taking into account the actual pole density, and $\rho(v_i)$ denotes the desired local density. More precisely, and owing to the energy equi-distribution property [DFG99], we know that the density function $\rho(v_i)$ must be proportional to $\frac{1}{r(v_i)^{d+2}}$ to obtain a cluster density matching the field $r(v_i)$ in a underlying space of dimension d. In our case $d = 2$, because the filtered poles approximate the medial axis, which is a generically a two-dimensional manifold. As for the quadrature term $d(v_i)$, we take it proportional to $\frac{V(v_i)}{r(v_i)}$, where $V(v_i)$ is the volume of the cell of v_i in the Voronoi diagram of the filtered poles, and $r(v_i)$ is the polar ball radius since each filtered pole v_i roughly represents the area $\frac{V(v_i)}{r(v_i)}$ of the λ-medial axis.

After convergence of the clustering procedure, the centroid of each cluster is replaced by the closest pole within its cluster, so that the final centers are guaranteed to be located near the medial axis of the sampled surface.

3.2. Constraints

We take as constraints both the input points where the function f is specified to be zero, and a set of additional constraints where f is specified to be non-zero. Recall that our

Figure 3: *Pole clustering on the Bimba model (100K input data points). 200K poles have been extracted and clustered to 15K poles. Left: All poles (100K inside poles depicted in orange, 100K outside poles depicted in green). Right: After clustering to 15K poles (8K inside depicted in red, 7K outside depicted in green).*

goal is to consider as an approximation of the shape the zero-level set of f. Therefore, we wish to define a signed function f which is positive outside the shape, negative inside and with a non-zero gradient close to the sampled surface. A good candidate is a function approximating the signed distance function to the sampled shape where the distance is positive for points outside the shape and negative inside (Fig.4). At each pole the radius of the polar ball corresponds

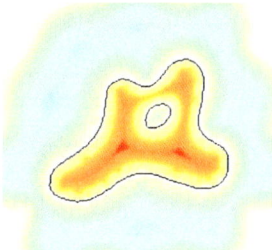

Figure 4: *2D shape (black) and the computed function. Colors range from cold color tones for positive distance values to hot color tones for negative distance values.*

to an approximation of its distance to the input point set. Thus, poles can be used as a constraints in order to approximate a distance function to the sampled surface. It remains however to determine the sign of this value, and therefore to *classify* the poles as inside or outside.

Pole Classification Pole classification is the process of labeling the poles as inside or outside the surface. Common approaches use an algorithm to propagate the pole labels through the graph built from adjacency relationships between the poles. In our implementation, we classify the poles using a variant of the algorithm proposed by Amenta [ACK01]. This variant, due to F.Cazals (internal communication), is more efficient and more robust against to noise. During the classification process, a location tag (inside, outside and undetermined) and a confidence value are attributed to each pole. If the confidence of a pole is lower than a certain threshold, the pole will not be taken into account as a constraint.

3.3. basis functions

The reconstructed surface is required to be independent of Euclidean transformation. The function Φ is thus restricted to the set of radial functions :

$$\Phi(x, c_i) = \phi(\|x - c_i\|) \qquad (11)$$

where $\|.\|$ denotes the Euclidean distance and $\phi : \mathbb{R}^+ \to \mathbb{R}$.

When the ϕ function has a unbounded support, the corresponding constraint equations lead to a dense linear system. Recovering a solution is therefore tractable only for small data sets. In order to obtain a sparse interpolation matrix, compactly supported RBFs have been introduced by Wendland in [Wen95]. Other compactly supported RBFs (CSRBF) can be used for reconstruction as proposed in [Sch95,Wu95]. As centers are poles, each center c_i, has a corresponding to a scalar value, r_i, the radius of its polar ball. Our function of choice ϕ is compactly supported, and the support size s_i for the function centered on c_i is computed using to r_i. The ϕ function (11) centered on c_i is scaled according to the local support s_i:

$$\phi_i(\|x - c_i\|) = \phi\left(\frac{\|x - c_i\|}{s_i}\right) * s_i. \qquad (12)$$

The basis functions chosen in our implementation is

$$\phi(r) = (1 - r)_+^4 (1 + 4r) \qquad (13)$$

where the symbol $+$ means $(x)_+ = x$ if $x > 0$ and $(x)_+ = 0$ otherwise.

3.4. Solver

The centers are the set $\{c_j\}_{j=1...m}$ of m points in $\mathbb{R}^3$. The constraints are the set $\{p_i\}_{i=1...N}$ of N points where the value of f is known.

Let G be the matrix $[\phi_j(\|p_i - c_j\|)]_{i=1..N, j=1..m}$ and F be the vector $[f_i]_{i=1..N}$. The constraints points $\{p_i\}_{i=1..N}$ include both the n input points and the additional off surface points where we specify the function f value.

$$\mathbf{G} = \begin{pmatrix} \phi_1(\|p_1 - c_1\|) & \cdots & \phi_m(\|p_1 - c_m\|) \\ \vdots & \vdots & \ddots & \vdots \\ \phi_1(\|p_N - c_1\|) & \cdots & \phi_m(\|p_N - c_m\|) \end{pmatrix} \qquad (14)$$

An approximation using the least squares method implies

solving the system (4). With the new notations, the system is

$$G^t G \cdot \alpha = G^t F. \qquad (15)$$

The size of the matrix is $m \times m$, where m is the number of centers. The use of compactly supported functions ϕ_i leads to a sparse matrix with about 90% zero elements.

Assembling of the matrix Each term a_{ij} of the matrix $G^t G$ is computed as a sum:

$$a_{i,j} = \sum_{k=1}^{N} \phi_i(\|p_k - c_i\|)\phi_j(\|p_k - c_j\|). \qquad (16)$$

For each constraint p, we need to find the list l_p of centers which contain p in their support. To avoid searching exhaustively, we use a 3D Delaunay triangulation of the centers. The constraint p is first located in the triangulation, then our algorithm search outwards from p in the triangulation until all centers containing p in their support are found. For each pair of centers (c_i, c_j) contained in the list l_p, we add a term for p to a_{ij}.

4. Results

We have implemented our algorithm in C++. The Voronoi diagram and Delaunay triangulation are computed using the CGAL library [FGK*00]. The linear system is solved using the TAUCS library [Tol01]. We use an implementation of the *marching cube* algorithm [Blo94] to extract the zero-level set of the reconstructed function . To evaluate the fitting accuracy, we use the *Taubin distance* [Tau94] from the input points (17)

$$Err(f) = \frac{1}{N}\sum_{i=1}^{N}\left(\frac{f_i - f(p_i)}{\|\nabla f(p_i\|}\right)^2. \qquad (17)$$

This distance is a first order approximation of the Euclidean distance between the input points and the zero level set of the function f. Since the gradient can vanish or go to infinity with compactly supported basis functions, we need to use a threshold S_1 such that :

$$Err_t(f) = \frac{1}{N}\sum_{i=1}^{N}\left(\frac{f_i - f(p_i)}{\Gamma(\|\nabla f(p_i)\|)}\right)^2, \qquad (18)$$

where :

$$\Gamma(\|\nabla f(p_i)\|) = \begin{cases} S_1 & if \ g < S_1 \\ \|\nabla f(p_i)\| & if \ S_1 < g \end{cases}$$

Figure 6 summarizes all steps of our algorithm on a 2D shape.

As a typical example for our algorithm, we detail the timings of each reconstruction step for the *omotondo* model (80K points) (Fig.7).

1. point insertion in the Delaunay triangulation: 6.3s;

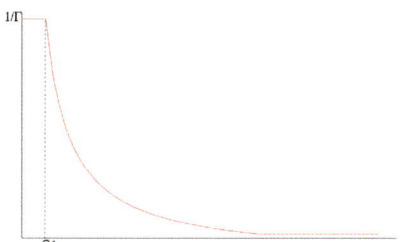

Figure 5: *Error function. 1/Γ function.*

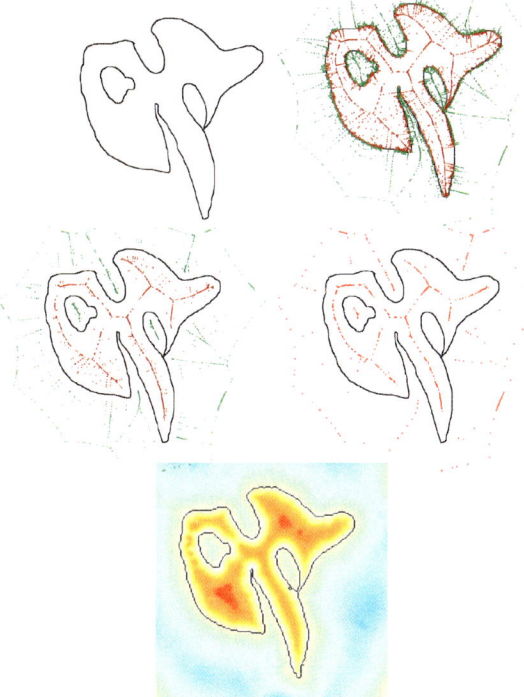

Figure 6: *The main steps of our algorithm on a 2D shape. From left to right: input data points (black), all poles are extracted and classified from the Voronoi diagram (red inside, green outside), poles are filtered, poles are clustered into centers, the 2D scalar function is computed and the zero-level set is extracted (black).*

2. extraction of 16K poles: 2.75s;
3. classification (8K inside poles and 8K outside poles): 20s;
4. filtering and clustering to got 13k centers : 230s;
5. assembling the linear system: 674s;
6. solving the linear system: 78s.

In our current implementation, most of the time is spent assembling the linear system, specifically finding all pairs of centers whose supports intersect a constraint. Although the

use of a 3D Delaunay triangulation avoids the naive exhaustive search, this part could be further optimized.

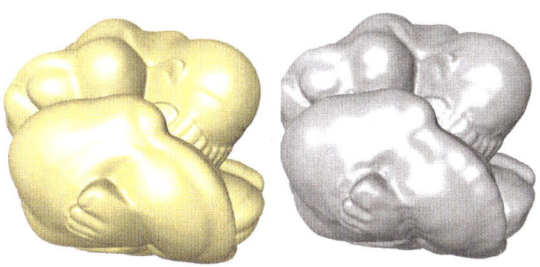

Figure 7: *Reconstruction of the Omotondo model (80K points) with 13K centers. Fitting accuracy: 2.8×10^{-6}. Left : the original model (gold); Right : the reconstructed surface (silver).*

The importance of our choice for the centers is shown graphically by Figure 8. We plot the error against the number of centers for our method and for the common method where constraints and centers coincide. In the common method, the set of data points is subsampled and the off constraints are taken along the normals estimated at the subsampled points.

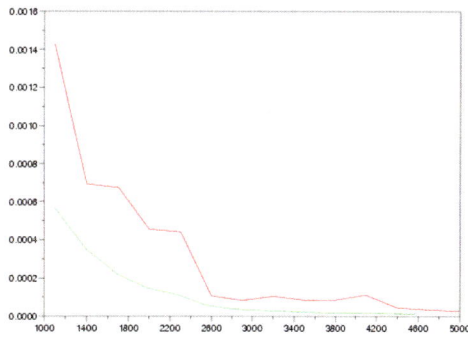

Figure 8: *Plot of the error against the number of centers (from 1K and 5K). The red curve corresponds to the common method. The green curve corresponds to our method.*

Figure 9 illustrates several reconstructions of the Dinosaur with increasing number of centers.

As Fig. 10 depicts, our function is defined all over the space around the sampled shape. In contrast, when compactly supported radial basis functions are centered at the input data points, the function is only defined in a tubular neighborhood of the sampled surface.

The clustering step redistributes the centers among the set of poles as shown in figure 11.

Figure 9: *Reconstruction sequence of the Dinosaur with increasing number of centers. From left to right: original, then reconstruction with 1K, 3K, 4K and 5K centers.*

Figure 10: *Reconstructed function. The colors represent the function values (cold tones for positive, hot tones for negative values and white for the zero values). Left: the reconstructed function for the common approach; The function does not vanish only in a tubular neighborhood of the point set. Right: the reconstructed function for our method.*

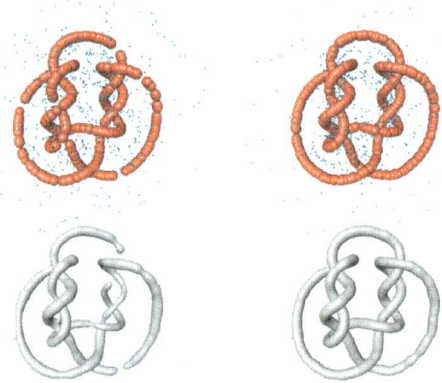

Figure 11: *Knot model (6K input points). Left: centers and reconstruction without clustering (inside centers with their polar balls are depicted in red, outside centers are depicted in green). Right: centers and reconstruction with clustering (12K poles are clustered into 1K centers).*

The pole filtering step of our algorithm is useful to adapt the level of detail to the user-defined number of centers (Fig. 12), as well as to improve robustness against noise (Fig. 13). It also shows the effect of filtering when the allocated budget of centers is low.

Figure 13 illustrates an extreme example with a substantial amount of noise due to the misregistration of three range maps. Moreover, the sampling is highly non isotropic and non uniform due to the acquisition system. Figure 13 depicts the main stages of our algorithm applied to a noisy point set sampled on a hand. Although noise in the input data points leads to misclassified poles, the λ-medial axis is stable under such perturbations, and theses misclassified poles are filtered.

5. Conclusion

We have presented a new approach for reconstructing surfaces from scattered points, combining generalized radial basis functions and Voronoi-based surface reconstruction. In contrast to the Voronoi-based approaches, our method creates a smooth and watertight surface, similarly to the RBF approaches. The resulting function is an approximation of the signed distance to the sampled surface defined all around the sampled shape, instead of being defined only in a small neighborhood as in previous work. Our approach relies on a theoretically sound framework for pole extraction and λ-medial axis filtering. This framework provides us with reliable estimates of the normal at each data point, with an approximation of the distance to the sampled surface at each pole, as well as with a filtering method based on the stable

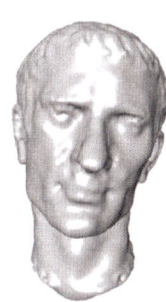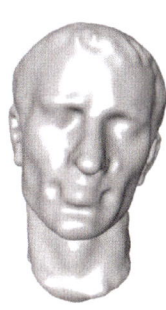

Figure 12: *Effect of the filtering step on the Julius model (80K input points). The number of centers is $m = 5K$. Left: without filtering; Middle: poles filtered with $\lambda = 0.01$; Right: poles filtered with $\lambda = 0.02$ (to get a better sense of these parameters, the diagonal length of the bounding box of the input point set is 1.47).*

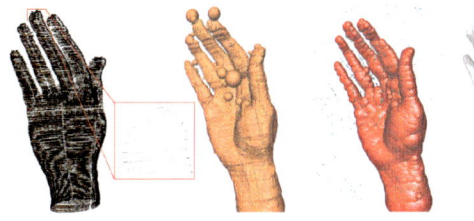

Figure 13: *Noisy hand reconstruction. Left: noisy hand model (90K input points). The input points result from registering three range maps. Middle left: inside poles with their polar balls (88K poles, some of them being misclassified); Middle right: 2K centers after filtering and clustering (inside and outside centers with their polar balls (resp. red and green); Right: reconstructed hand.*

λ-medial axis. As a result we can reduce the number of parameters for our algorithm to two: the number of centers, and λ, used to filter the medial axis.

In terms of efficiency, the only stage which impairs scalability is the assembling of the final matrix. We are expect to greatly improve this aspect by an optimized implementation or using new geometric data structure. In our study the medial axis filtering stage allows us to adapt the level of details to a user-defined budget of centers, the value for λ being fixed experimentally. In a future work, we will investigate how to automatically adjust this parameter to accommodate for the allocated budget of centers.

Acknowledgements Our sponsors include the EU Network of Excellence AIM@SHAPE (IST NoE No 506766). The dinosaur model is courtesy of Cyberware, other models being courtesy of the AIM@SHAPE shape repository.
Work partially supported by the IST Programme of the EU as a Shared-cost RTD (FET Open) Project under Contract No IST-006413 (ACS - Algorithms for Complex Shapes) and by the European Network of Excellence AIM@shape (FP6 IST NoE 506766).

References

[AB98] AMENTA N., BERN M.: Surface reconstruction by voronoi filtering. In *SCG '98: Proceedings of the fourteenth annual symposium on Computational geometry* (New York, NY, USA, 1998), ACM Press, pp. 39–48.

[ACK01] AMENTA N., CHOI S., KOLLURI R. K.: The power crust. In *Proceedings of the sixth ACM symposium on Solid modeling and applications* (2001), ACM Press, pp. 249–266.

[AGJ02] ADAMY U., GIESEN J., JOHN M.: Surface reconstruction using umbrella filters. *Comput. Geom. Theory Appl. 21*, 1 (2002), 63–86.

[AM96] ATTALI D., MONTANVERT A.: Modeling noise for a better simplification of skeletons, 1996.

[AS00] ATTENE M., SPAGNUOLO M.: Automatic surface reconstruction from point sets in space. *Computer Graphics Forum 19*, 3 (2000), 457–465.

[BC00] BOISSONNAT J.-D., CAZALS F.: Smooth surface reconstruction via natural neighbour interpolation of distance functions. In *SCG '00: Proceedings of the sixteenth annual symposium on Computational geometry* (New York, NY, USA, 2000), ACM Press, pp. 223–232.

[BL88] BROOMHEAD D. S., LOWE D.: Multivariable functional interpolation and adaptive networks. *ComSys 2* (1988), 321–355.

[Blo94] BLOOMENTHAL J.: An implicit surface polygonizer. 324–349.

[Buh03] BUHMAN M.: *Radial basis functions : theory and implementations*, cambridge monographs on applied and computational mathematics ed., vol. 12. 2003.

[CBC*01] CARR J. C., BEATSON R. K., CHERRIE J. B., MITCHELL T. J., FRIGHT W. R., MCCALLUM B. C., EVANS T. R.: Reconstruction and representation of 3D objects with radial basis functions. In *SIGGRAPH 2001, Computer Graphics Proceedings* (2001), Fiume E., (Ed.), ACM Press / ACM SIGGRAPH, pp. 67–76.

[CFB97] CARR J., FRIGHT W., BEATSON R.: Surface interpolation with radial basis functions for medical imaging, 1997.

[CG04] CAZALS F., GIESEN J.: *Delaunay Triangulation Based Surface Reconstruction: Ideas and Algorithms*. Research Report RR-5393, INRIA, BP93, 06902 Sophia-Antipolis, France, 2004.

[CL05] CHAZAL F., LIEUTIER A.: The "λ-medial axis". *Graph. Models 67*, 4 (2005), 304–331.

[CSD04] COHEN-STEINER D., DA F.: A greedy delaunay-based surface reconstruction algorithm. *Vis. Comput. 20*, 1 (2004), 4–16.

[Dey04] DEY T. K.: Curve and surface reconstruction. In *Handbook of Discrete and Computational Geometry* (2004), Goodman and O' Rourke eds. , CRC press, 2nd edition.

[DFG99] DU Q., FABER V., GUNZBURGER M.: Centroidal Voronoi Tessellations: Applications and Algorithms. *SIAM Review 41*, 4 (1999), 637–676.

[DG04] DEY T. K., GOSWAMI S.: Provable surface reconstruction from noisy samples. In *SCG '04: Proceedings of the twentieth annual symposium on Computational geometry* (New York, NY, USA, 2004), ACM Press, pp. 330–339.

[DGH01] DEY T. K., GIESEN J., HUDSON J.: Delaunay based shape reconstruction from large data. In *PVG '01: Proceedings of the IEEE 2001 symposium on parallel and large-data visualization and graphics* (Piscataway, NJ, USA, 2001), IEEE Press, pp. 19–27.

[DQ01] DUAN Y., QIN H.: Intelligent balloon: a subdivision-based deformable model for surface reconstruction of arbitrary topology. In *SMA '01: Proceedings of the sixth ACM symposium on Solid modeling and applications* (New York, NY, USA, 2001), ACM Press, pp. 47–58.

[Duc77] DUCHON J.: Spline minimizing rotation-invariant semi-norms in sobolev spaces. In *Constructive Theory of Functions of Several Variables* (1977), Schempp W., Zeller K., (Eds.), vol. 571 of *Lecture Notes in Mathematics*, pp. 85–100.

[DZ03] DEY T. K., ZHAO W.: Approximating the medial axis from the voronoi diagram with a convergence guarantee. *Algorithmica 38*, 1 (2003), 179–200.

[Far02] FARIN G.: *Curves and surfaces for CAGD: a practical guide*. Morgan Kaufmann Publishers Inc., San Francisco, CA, USA, 2002.

[FGK*00] FABRI A., GIEZEMAN G.-J., KETTNER L., SCHIRRA S., SCHÖNHERR S.: On the Design of CGAL, a Computational Geometry Algorithms Library. *Softw. – Pract. Exp. 30*, 11 (2000), 1167–1202.

[FN80] FRANKE R., NIELSON G.: Smooth interpolation of large sets of scattered data. *Internat. J. Numer. Methods Engrg. 15*, 11 (1980), 1691–1704.

[HTF01] HASTIE T., TIBSHIRANI R., FRIEDMAN J. H.: *The elements of statistical learning: data mining, inference, and prediction: with 200 full-color illustrations*. New York: Springer-Verlag, 2001.

[Isk04] ISKE A.: *Multiresolution Methods in Scattered Data Modelling*, vol. 37. Springer-Verlag, 2004.

[Kir01] KIRBY M.: *Geometric Data Analysis*. John Wiley & Sons, 2001.

[KSO04] KOLLURI R., SHEWCHUK J. R., O'BRIEN J. F.: Spectral surface reconstruction from noisy point clouds. In *Symposium on Geometry Processing* (July 2004), ACM Press, pp. 11–21.

[LF99] LODHA S. K., FRANKE R.: Scattered data techniques for surfaces. In *Dagstuhl '97, Scientific Visualization* (Washington, DC, USA, 1999), IEEE Computer Society, pp. 181–222.

[Mac67] MACQUEEN J. B.: Some methods for classification and analysis of multivariate observations. vol. 1, University of Clafifornia Press, Berkeley, Calif., pp. 281–297.

[OBS04] OHTAKE Y., BELYAEV A., SEIDEL H.-P.: 3d scattered data approximation with adaptive compactly supported radial basis functions. In *SMI* (2004), pp. 31–39.

[PG89] POGGIO T., GIROSI F.: A theory of networks for approximation and learning, 1989.

[Sch95] SCHABACK R.: Creating surfaces from scattered data using radial basis functions, 1995.

[Set99] SETHIAN J. A.: *Level Set Methods and Fast Marching Methods*. Cambridge Monograph on Applied and Computational Mathematics. Cambridge University Press, 1999.

[Tau94] TAUBIN G.: Distance approximations for rasterizing implicit curves. *ACM Trans. Graph. 13*, 1 (1994), 3–42.

[TI04] TOBOR I. REUTER P. S. C.: Efficient reconstruction of large scattered geometric datasets using the partition of unity and radial basis functions. vol. 12 of *Journal of WSCG 2004*, pp. 467–474.

[TO02] TURK G., O'BRIEN J. F.: Modelling with implicit surfaces that interpolate. *ACM Trans. Graph. 21*, 4 (2002), 855–873.

[Tol01] TOLEDO S.: *TAUCS Version 2.0*, November 2001.

[Wen95] WENDLAND H.: Piecewise polynomial, positive definite and compactly supported radial functions of minimal degree. *Advances in Computational Mathematics 4* (1995), 389–396.

[Wen04] WENDLAND H.: *Scattered Data Approximation*. Cambridge University Press, 2004.

[Wu95] WU Z.: Compactly supported positive definite radial functions. *Advances in Computational Mathematics 4* (1995), 283.292.

Eurographics Symposium on Geometry Processing (2006)
Konrad Polthier, Alla Sheffer (Editors)

Poisson Surface Reconstruction

Michael Kazhdan[1], Matthew Bolitho[1] and Hugues Hoppe[2]

[1]Johns Hopkins University, Baltimore MD, USA
[2]Microsoft Research, Redmond WA, USA

Abstract

We show that surface reconstruction from oriented points can be cast as a spatial Poisson problem. This Poisson formulation considers all the points at once, without resorting to heuristic spatial partitioning or blending, and is therefore highly resilient to data noise. Unlike radial basis function schemes, our Poisson approach allows a hierarchy of locally supported basis functions, and therefore the solution reduces to a well conditioned sparse linear system. We describe a spatially adaptive multiscale algorithm whose time and space complexities are proportional to the size of the reconstructed model. Experimenting with publicly available scan data, we demonstrate reconstruction of surfaces with greater detail than previously achievable.

1. Introduction

Reconstructing 3D surfaces from point samples is a well studied problem in computer graphics. It allows fitting of scanned data, filling of surface holes, and remeshing of existing models. We provide a novel approach that expresses surface reconstruction as the solution to a Poisson equation.

Like much previous work (Section 2), we approach the problem of surface reconstruction using an implicit function framework. Specifically, like [Kaz05] we compute a 3D *indicator function* χ (defined as 1 at points inside the model, and 0 at points outside), and then obtain the reconstructed surface by extracting an appropriate isosurface.

Our key insight is that there is an integral relationship between oriented points sampled from the surface of a model and the indicator function of the model. Specifically, the gradient of the indicator function is a vector field that is zero almost everywhere (since the indicator function is constant almost everywhere), except at points near the surface, where it is equal to the inward surface normal. Thus, the oriented point samples can be viewed as samples of the gradient of the model's indicator function (Figure 1).

The problem of computing the indicator function thus reduces to inverting the gradient operator, i.e. finding the scalar function χ whose gradient best approximates a vector field $\vec{V}$ defined by the samples, i.e. $\min_\chi \|\nabla\chi - \vec{V}\|$. If we apply the divergence operator, this variational problem transforms into a standard Poisson problem: compute the scalar func-

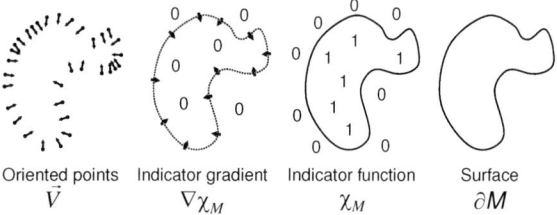

Oriented points	Indicator gradient	Indicator function	Surface
$\vec{V}$	$\nabla\chi_M$	χ_M	∂M

Figure 1: *Intuitive illustration of Poisson reconstruction in 2D.*

tion χ whose Laplacian (divergence of gradient) equals the divergence of the vector field $\vec{V}$,

$$\Delta\chi \equiv \nabla \cdot \nabla\chi = \nabla \cdot \vec{V}.$$

We will make these definitions precise in Sections 3 and 4.

Formulating surface reconstruction as a Poisson problem offers a number of advantages. Many implicit surface fitting methods segment the data into regions for local fitting, and further combine these local approximations using blending functions. In contrast, Poisson reconstruction is a global solution that considers all the data at once, without resorting to heuristic partitioning or blending. Thus, like radial basis function (RBF) approaches, Poisson reconstruction creates very smooth surfaces that robustly approximate noisy data. But, whereas ideal RBFs are globally supported and non-decaying, the Poisson problem admits a hierarchy of *locally supported* functions, and therefore its solution reduces to a well-conditioned sparse linear system.

Moreover, in many implicit fitting schemes, the value of the implicit function is constrained only near the sample points, and consequently the reconstruction may contain spurious surface sheets away from these samples. Typically this problem is attenuated by introducing auxiliary "off-surface" points (e.g. [CBC*01, OBA*03]). With Poisson surface reconstruction, such surface sheets seldom arise because the gradient of the implicit function is constrained at *all* spatial points. In particular it is constrained to zero away from the samples.

Poisson systems are well known for their resilience in the presence of imperfect data. For instance, "gradient domain" manipulation algorithms (e.g. [FLW02]) intentionally modify the gradient data such that it no longer corresponds to any real potential field, and rely on a Poisson system to recover the globally best-fitting model.

There has been broad interdisciplinary research on solving Poisson problems and many efficient and robust methods have been developed. One particular aspect of our problem instance is that an accurate solution to the Poisson equation is only necessary near the reconstructed surface. This allows us to leverage adaptive Poisson solvers to develop a reconstruction algorithm whose spatial and temporal complexities are proportional to the size of the reconstructed surface.

2. Related Work

Surface reconstruction The reconstruction of surfaces from oriented points has a number of difficulties in practice. The point sampling is often nonuniform. The positions and normals are generally noisy due to sampling inaccuracy and scan misregistration. And, accessibility constraints during scanning may leave some surface regions devoid of data. Given these challenges, reconstruction methods attempt to infer the topology of the unknown surface, accurately fit (but not overfit) the noisy data, and fill holes reasonably.

Several approaches are based on combinatorial structures, such as Delaunay triangulations (e.g. [Boi84, KSO04]), alpha shapes [EM94, BBX95, BMR*99]), or Voronoi diagrams [ABK98, ACK01]. These schemes typically create a triangle mesh that interpolates all or a most of the points. In the presence of noisy data, the resulting surface is often jagged, and is therefore smoothed (e.g. [KSO04]) or refit to the points (e.g. [BBX95]) in subsequent processing.

Other schemes directly reconstruct an approximating surface, typically represented in implicit form. We can broadly classify these as either global or local approaches.

Global fitting methods commonly define the implicit function as the sum of radial basis functions (RBFs) centered at the points (e.g. [Mur91, CBC*01, TO02]). However, the ideal RBFs (polyharmonics) are globally supported and non-decaying, so the solution matrix is dense and ill-conditioned. Practical solutions on large datasets involve adaptive RBF reduction and the fast multipole method [CBC*01].

Local fitting methods consider subsets of nearby points at a time. A simple scheme is to estimate tangent planes and define the implicit function as the signed distance to the tangent plane of the closest point [HDD*92]. Signed distance can also be accumulated into a volumetric grid [CL96]. For function continuity, the influence of several nearby points can be blended together, for instance using moving least squares [ABCO*01, SOS04]. A different approach is to form point neighborhoods by adaptively subdividing space, for example with an adaptive octree. Blending is possible over an octree structure using a multilevel partition of unity, and the type of local implicit patch within each octree node can be selected heuristically [OBA*03].

Our Poisson reconstruction combines benefits of both global and local fitting schemes. It is global and therefore does not involve heuristic decisions for forming local neighborhoods, selecting surface patch types, and choosing blend weights. Yet, the basis functions are associated with the ambient space rather than the data points, are locally supported, and have a simple hierarchical structure that results in a sparse, well-conditioned system.

Our approach of solving an indicator function is similar to the Fourier-based reconstruction scheme of Kazhdan [Kaz05]. In fact, we show in Appendix A that our basic Poisson formulation is mathematically equivalent. Indeed, the Fast Fourier Transform (FFT) is a common technique for solving *dense, periodic* Poisson systems. However, the FFT requires $O(r^3 \log r)$ time and $O(r^3)$ space where r is the 3D grid resolution, quickly becoming prohibitive for fine resolutions. In contrast, the Poisson system allows adaptive discretization, and thus yields a scalable solution.

Poisson problems The Poisson equation arises in numerous applications areas. For instance, in computer graphics it is used for tone mapping of high dynamic range images [FLW02], seamless editing of image regions [PGB03], fluid mechanics [LGF04], and mesh editing [YZX*04]. Multigrid Poisson solutions have even been adapted for efficient GPU computation [BFGS03, GWL*03].

The Poisson equation is also used in heat transfer and diffusion problems. Interestingly, Davis et al [DMGL02] use diffusion to fill holes in reconstructed surfaces. Given boundary conditions in the form of a clamped signed distance function d, their diffusion approach essentially solves the homogeneous Poisson equation $\Delta d = 0$ to create an implicit surface spanning the boundaries. They use a local iterative solution rather than a global multiscale Poisson system.

Nehab et al [NRDR05] use a Poisson system to fit a 2.5D height field to sampled positions and normals. Their approach fits a given parametric surface and is well-suited to the reconstruction of surfaces from individual scans. However, in the case that the samples are obtained from the union of multiple scans, their approach cannot be directly applied to obtain a connected, watertight surface.

3. Our Poisson reconstruction approach

The input data S is a set of samples $s \in S$, each consisting of a point $s.p$ and an inward-facing normal $s.\vec{N}$, assumed to lie on or near the surface ∂M of an unknown model M. Our goal is to reconstruct a watertight, triangulated approximation to the surface by approximating the indicator function of the model and extracting the isosurface, as illustrated in Figure 2.

The key challenge is to accurately compute the indicator function from the samples. In this section, we derive a relationship between the gradient of the indicator function and an integral of the surface normal field. We then approximate this surface integral by a summation over the given oriented point samples. Finally, we reconstruct the indicator function from this gradient field as a Poisson problem.

Defining the gradient field Because the indicator function is a piecewise constant function, explicit computation of its gradient field would result in a vector field with unbounded values at the surface boundary. To avoid this, we convolve the indicator function with a smoothing filter and consider the gradient field of the smoothed function. The following lemma formalizes the relationship between the gradient of the smoothed indicator function and the surface normal field.

Lemma: Given a solid M with boundary ∂M, let χ_M denote the indicator function of M, $\vec{N}_{\partial M}(p)$ be the inward surface normal at $p \in \partial M$, $\tilde{F}(q)$ be a smoothing filter, and $\tilde{F}_p(q) = \tilde{F}(q-p)$ its translation to the point p. The gradient of the smoothed indicator function is equal to the vector field obtained by smoothing the surface normal field:

$$\nabla \left(\chi_M * \tilde{F} \right)(q_0) = \int_{\partial M} \tilde{F}_p(q_0) \vec{N}_{\partial M}(p) dp. \tag{1}$$

Proof: To prove this, we show equality for each of the components of the vector field. Computing the partial derivative of the smoothed indicator function with respect to x, we get:

$$
\begin{aligned}
\left. \frac{\partial}{\partial x} \right|_{q_0} \left(\chi_M * \tilde{F} \right) &= \left. \frac{\partial}{\partial x} \right|_{q=q_0} \int_M \tilde{F}(q-p) dp \\
&= \int_M \left(-\frac{\partial}{\partial x} \tilde{F}(q_0-p) \right) dp \\
&= -\int_M \nabla \cdot \left(\tilde{F}(q_0-p), 0, 0 \right) dp \\
&= \int_{\partial M} \left\langle \left(\tilde{F}_p(q_0), 0, 0 \right), \vec{N}_{\partial M}(p) \right\rangle dp.
\end{aligned}
$$

(The first equality follows from the fact that χ_M is equal to zero outside of M and one inside. The second follows from the fact that $(\partial/\partial q)\tilde{F}(q-p) = -(\partial/\partial p)\tilde{F}(q-p)$. The last follows from the Divergence Theorem.)

A similar argument shows that the y-, and z-components of the two sides are equal, thereby completing the proof. $\square$

Approximating the gradient field Of course, we cannot evaluate the surface integral since we do not yet know the

Figure 2: *Points from scans of the "Armadillo Man" model (left), our Poisson surface reconstruction (right), and a visualization of the indicator function (middle) along a plane through the 3D volume.*

surface geometry. However, the input set of oriented points provides precisely enough information to approximate the integral with a discrete summation. Specifically, using the point set S to partition ∂M into distinct patches $\mathscr{P}_s \subset \partial M$, we can approximate the integral over a patch $\mathscr{P}_s$ by the value at point sample $s.p$, scaled by the area of the patch:

$$
\begin{aligned}
\nabla(\chi_M * \tilde{F})(q) &= \sum_{s \in S} \int_{\mathscr{P}_s} \tilde{F}_p(q) \vec{N}_{\partial M}(p) dp \\
&\approx \sum_{s \in S} |\mathscr{P}_s| \, \tilde{F}_{s.p}(q) \, s.\vec{N} \equiv \vec{V}(q).
\end{aligned} \tag{2}
$$

It should be noted that though Equation 1 is true for any smoothing filter $\tilde{F}$, in practice, care must be taken in choosing the filter. In particular, we would like the filter to satisfy two conditions. On the one hand, it should be sufficiently narrow so that we do not over-smooth the data. And on the other hand, it should be wide enough so that the integral over $\mathscr{P}_s$ is well approximated by the value at $s.p$ scaled by the patch area. A good choice of filter that balances these two requirements is a Gaussian whose variance is on the order of the sampling resolution.

Solving the Poisson problem Having formed a vector field $\vec{V}$, we want to solve for the function $\tilde{\chi}$ such that $\nabla \tilde{\chi} = \vec{V}$. However, $\vec{V}$ is generally not integrable (i.e. it is not curl-free), so an exact solution does not generally exist. To find the best least-squares approximate solution, we apply the divergence operator to form the Poisson equation

$$\Delta \tilde{\chi} = \nabla \cdot \vec{V}.$$

In the next section, we describe our implementation of these steps in more detail.

4. Implementation

We first present our reconstruction algorithm under the assumption that the point samples are uniformly distributed over the model surface. We define a space of functions with high resolution near the surface of the model and coarser resolution away from it, express the vector field $\vec{V}$ as a linear sum of functions in this space, set up and solve the Poisson equation, and extract an isosurface of the resulting indicator function. We then extend our algorithm to address the case of non-uniformly sampled points.

4.1. Problem Discretization

First, we must choose the space of functions in which to discretize the problem. The most straightforward approach is to start with a regular 3D grid [Kaz05], but such a uniform structure becomes impractical for fine-detail reconstruction, since the dimension of the space is cubic in the resolution while the number of surface triangles grows quadratically.

Fortunately, an accurate representation of the implicit function is only necessary near the reconstructed surface. This motivates the use of an adaptive octree both to represent the implicit function and to solve the Poisson system (e.g. [GKS02,LGF04]). Specifically, we use the positions of the sample points to define an octree $\mathcal{O}$ and associate a function F_o to each node $o \in \mathcal{O}$ of the tree, choosing the tree and the functions so that the following conditions are satisfied:

1. The vector field $\vec{V}$ can be precisely and efficiently represented as the linear sum of the F_o.
2. The matrix representation of the Poisson equation, expressed in terms of the F_o can be solved efficiently.
3. A representation of the indicator function as the sum of the F_o can be precisely and efficiently evaluated near the surface of the model.

Defining the function space   Given a set of point samples S and a maximum tree depth D, we define the octree $\mathcal{O}$ to be the minimal octree with the property that every point sample falls into a leaf node at depth D.

Next, we define a space of functions obtained as the span of translates and scales of a fixed, unit-integral, base function $F : \mathbb{R}^3 \to \mathbb{R}$. For every node $o \in \mathcal{O}$, we set F_o to be the unit-integral "node function" centered about the node o and stretched by the size of o:

$$F_o(q) \equiv F\left(\frac{q - o.c}{o.w}\right) \frac{1}{o.w^3}.$$

where $o.c$ and $o.w$ are the center and width of node o.

This space of functions $\mathscr{F}_{\mathcal{O},F} \equiv \mathrm{Span}\{F_o\}$ has a multiresolution structure similar to that of traditional wavelet representations. Finer nodes are associated with higher-frequency functions, and the function representation becomes more precise as we near the surface.

Selecting a base function   In selecting a base function F, our goal is to choose a function so that the vector field $\vec{V}$, defined in Equation 2, can be precisely and efficiently represented as the linear sum of the node functions $\{F_o\}$.

If we were to replace the position of each sample with the center of the leaf node containing it, the vector field $\vec{V}$ could be efficiently expressed as the linear sum of $\{F_o\}$ by setting:

$$F(q) = \tilde{F}\left(\frac{q}{2^D}\right).$$

This way, each sample would contribute a single term (the normal vector) to the coefficient corresponding to its leaf's node function. Since the sampling width is 2^{-D} and the samples all fall into leaf nodes of depth D, the error arising from the clamping can never be too big (at most, on the order of half the sampling width). In the next section, we show how the error can be further reduced by using trilinear interpolation to allow for sub-node precision.

Finally, since a maximum tree depth of D corresponds to a sampling width of 2^{-D}, the smoothing filter should approximate a Gaussian with variance on the order of 2^{-D}. Thus, F should approximate a Gaussian with unit-variance.

For efficiency, we approximate the unit-variance Gaussian by a compactly supported function so that (1) the resulting Divergence and Laplacian operators are sparse and (2) the evaluation of a function expressed as the linear sum of F_o at some point q only requires summing over the nodes $o \in \mathcal{O}$ that are close to q. Thus, we set F to be the n-th convolution of a box filter with itself resulting in the base function F:

$$F(x,y,z) \equiv (B(x)B(y)B(z))^{*n} \quad \text{with} \quad B(t) = \begin{cases} 1 & |t| < 0.5 \\ 0 & \text{otherwise} \end{cases}$$

Note that as n is increased, F more closely approximates a Gaussian and its support grows larger; in our implementation we use a piecewise quadratic approximation with $n = 3$. Therefore, the function F is supported on the domain $[-1.5, 1.5]^3$ and, for the basis function of any octree node, there are at most $5^3 - 1 = 124$ other nodes at the same depth whose functions overlap with it.

4.2. Vector Field Definition

To allow for sub-node precision, we avoid clamping a sample's position to the center of the containing leaf node and instead use trilinear interpolation to distribute the sample across the eight nearest nodes. Thus, we define our approximation to the gradient field of the indicator function as:

$$\vec{V}(q) \equiv \sum_{s \in S} \sum_{o \in \mathrm{Ngbr}_D(s)} \alpha_{o,s} F_o(q) s.\vec{N} \tag{3}$$

where $\mathrm{Ngbr}_D(s)$ are the eight depth-D nodes closest to $s.p$ and $\{\alpha_{o,s}\}$ are the trilinear interpolation weights. (If the neighbors are not in the tree, we refine it to include them.)

Since the samples are uniform, we can assume that the area of a patch $\mathscr{P}_s$ is constant and $\vec{V}$ is a good approximation, up to a multiplicative constant, of the gradient of the smoothed indicator function. We will show that the choice of multiplicative constant does not affect the reconstruction.

4.3. Poisson Solution

Having defined the vector field $\vec{V}$, we would like to solve for the function $\tilde{\chi} \in \mathscr{F}_{\mathcal{O},F}$ such that the gradient of $\tilde{\chi}$ is closest to $\vec{V}$, i.e. a solution to the Poisson equation $\Delta\tilde{\chi} = \nabla \cdot \vec{V}$.

One challenge of solving for $\tilde{\chi}$ is that though $\tilde{\chi}$ and the coordinate functions of $\vec{V}$ are in the space $\mathscr{F}_{\mathscr{O},F}$ it is not necessarily the case that the functions $\Delta\tilde{\chi}$ and $\nabla\cdot\vec{V}$ are.

To address this issue, we need to solve for the function $\tilde{\chi}$ such that the projection of $\Delta\tilde{\chi}$ onto the space $\mathscr{F}_{\mathscr{O},F}$ is closest to the projection of $\nabla\cdot\vec{V}$. Since, in general, the functions F_o do not form an orthonormal basis, solving this problem directly is expensive. However, we can simplify the problem by solving for the function $\tilde{\chi}$ minimizing:

$$\sum_{o\in\mathscr{O}}\left\|\langle\Delta\tilde{\chi}-\nabla\cdot\vec{V},F_o\rangle\right\|^2 = \sum_{o\in\mathscr{O}}\left\|\langle\Delta\tilde{\chi},F_o\rangle-\langle\nabla\cdot\vec{V},F_o\rangle\right\|^2.$$

Thus given the $|\mathscr{O}|$-dimensional vector v whose o-th coordinate is $v_o=\langle\nabla\cdot\vec{V},F_o\rangle$, the goal is to solve for the function $\tilde{\chi}$ such that the vector obtained by projecting the Laplacian of $\tilde{\chi}$ onto each of the F_o is as close to v as possible.

To express this in matrix form, let $\tilde{\chi}=\sum_o x_o F_o$, so that we are solving for the vector $x\in\mathbb{R}^{|\mathscr{O}|}$. Then, let us define the $|\mathscr{O}|\times|\mathscr{O}|$ matrix L such that Lx returns the dot product of the Laplacian with each of the F_o. Specifically, for all $o,o'\in\mathscr{O}$, the (o,o')-th entry of L is set to:

$$L_{o,o'}\equiv\left\langle\frac{\partial^2 F_o}{\partial x^2},F_{o'}\right\rangle+\left\langle\frac{\partial^2 F_o}{\partial y^2},F_{o'}\right\rangle+\left\langle\frac{\partial^2 F_o}{\partial z^2},F_{o'}\right\rangle.$$

Thus, solving for $\tilde{\chi}$ amounts to finding

$$\min_{x\in\mathbb{R}^{|\mathscr{O}|}}\|Lx-v\|^2.$$

Note that the matrix L is sparse and symmetric. (Sparse because the F_o are compactly supported, and symmetric because $\int f''g = -\int f'g'$.) Furthermore, there is an inherent multiresolution structure on $\mathscr{F}_{\mathscr{O},F}$, so we use an approach similar to the multigrid approach in [GKS02], solving the restriction L_d of L to the space spanned by the depth d functions (using a conjugate gradient solver) and projecting the fixed-depth solution back onto $\mathscr{F}_{\mathscr{O},F}$ to update the residual.

Addressing memory concerns In practice, as the depth increases, the matrix L_d becomes larger and it may not be practical to store it in memory. Although the number of entries in a column of L_d is bounded by a constant, the constant value can be large. For example, even using a piecewise quadratic base function F, we end up with as many as 125 non-zero entries in a column, resulting in a memory requirement that is 125 times larger than the size of the octree.

To address this issue, we augment our solver with a block Gauss-Seidel solver. That is, we decompose the d-th dimensional space into overlapping regions and solve the restriction of L_d to these different regions, projecting the local solutions back into the d-dimensional space and updating the residuals. By choosing the number of regions to be a function of the depth d, we ensure that the size of the matrix used by the solver never exceeds a desired memory threshold.

4.4. Isosurface Extraction

In order to obtain a reconstructed surface $\partial\tilde{M}$, it is necessary to first select an isovalue and then extract the corresponding isosurface from the computed indicator function.

We choose the isovalue so that the extracted surface closely approximates the positions of the input samples. We do this by evaluating $\tilde{\chi}$ at the sample positions and use the average of the values for isosurface extraction:

$$\partial\tilde{M}\equiv\{q\in\mathbb{R}^3\mid\tilde{\chi}(q)=\gamma\}\quad\text{with}\quad\gamma=\frac{1}{|S|}\sum_{s\in S}\tilde{\chi}(s.p).$$

This choice of isovalue has the property that scaling $\tilde{\chi}$ does not change the isosurface. Thus, knowing the vector field $\vec{V}$ up to a multiplicative constant provides sufficient information for reconstructing the surface.

To extract the isosurface from the indicator function, we use a method similar to previous adaptations of the Marching Cubes [LC87] to octree representations (e.g. [WG92, SFYC96, WKE99]). However, due to the nonconforming properties of our tree, we modify the reconstruction approach slightly, defining the positions of zero-crossings along an edge in terms of the zero-crossings computed by the finest level nodes adjacent to the edge. In the case that an edge of a leaf node has more than one zero-crossing associated to it, the node is subdivided. As in previous approaches, we avoid cracks arising when coarser nodes share a face with finer ones by projecting the isocurve segments from the faces of finer nodes onto the face of the coarser one.

4.5. Non-uniform Samples

We now extend our method to the case of non-uniformly distributed point samples. As in [Kaz05], our approach is to estimate the local sampling density, and scale the contribution of each point accordingly. However, rather than simply scaling the *magnitude* of a *fixed*-width kernel associated with each point, we additionally adapt the kernel width. This results in a reconstruction that maintains sharp features in areas of dense sampling and provides a smooth fit in sparsely sampled regions.

Estimating local sampling density Following the approach of [Kaz05], we implement the density computation using a kernel density estimator [Par62]. The approach is to estimate the number of points in a neighborhood of a sample by "splatting" the samples into a 3D grid, convolving the "splatting" function with a smoothing filter, and evaluating the convolution at each of the sample points.

We implement the convolution in a manner similar to Equation 3. Given a depth $\hat{D}\le D$ we set the density estimator to be the sum of node functions at depth $\hat{D}$:

$$W_{\hat{D}}(q)\equiv\sum_{s\in S}\sum_{o\in\mathrm{Ngbr}_{\hat{D}}(s)}\alpha_{o,s}F_o(q).$$

Since octree nodes at lower resolution are associated with functions that approximate Gaussians of larger width, the parameter $\hat{D}$ provides away for specifying the locality of the density estimation, with smaller values of $\hat{D}$ giving sampling density estimates over larger regions.

Computing the vector field Using the density estimator, we modify the summation in Equation 3 so that each sample's contribution is proportional to its associated area on the surface. Specifically, using the fact that the area is inversely proportional to sampling density, we set:

$$\vec{V}(q) \equiv \sum_{s \in S} \frac{1}{W_{\hat{D}}(s.p)} \sum_{o \in \mathrm{Ngbr}_D(s)} \alpha_{o,s} F_o(q).$$

However, adapting only the magnitudes of the sample contributions results in poor noise filtering in sparsely sampled regions as demonstrated later in Figure 7. Therefore, we additionally adapt the width of the smoothing filter $\tilde{F}$ to the local sampling density. Adapting the filter width lets us retain fine detail in regions of dense sampling, while smoothing out noise in regions of sparse sampling.

Using the fact that node functions at smaller depths correspond to wider smoothing filters, we define

$$\vec{V}(q) \equiv \sum_{s \in S} \frac{1}{W_{\hat{D}}(s.p)} \sum_{o \in \mathrm{Ngbr}_{\mathrm{Depth}(s.p)}(s)} \alpha_{o,s} F_o(q).$$

In this definition, $\mathrm{Depth}(s.p)$ represents the desired depth of a sample point $s \in S$. It is defined by computing the average sampling density W over all of the samples and setting:

$$\mathrm{Depth}(s.p) \equiv \min\left(D, D + \log_4(W_{\hat{D}}(s.p)/W)\right)$$

so that the width of the smoothing filter with which s contributes to $\vec{V}$ is proportional to the radius of its associated surface patch P_s.

Selecting an isovalue Finally, we modify the surface extraction step by selecting an isovalue which is the weighted average of the values of $\tilde{\chi}$ at the sample positions:

$$\partial \tilde{M} \equiv \{q \in \mathbb{R}^3 \mid \tilde{\chi}(q) = \gamma\} \quad \text{with} \quad \gamma = \frac{\sum \frac{1}{W_{\hat{D}}(s.p)} \tilde{\chi}(s.p)}{\sum \frac{1}{W_{\hat{D}}(s.p)}}.$$

5. Results

To evaluate our method we conducted a series of experiments. Our goal was to address three separate questions: How well does the algorithm reconstruct surfaces? How does it compare to other reconstruction methods? And, what are its performance characteristics?

Much practical motivation for surface reconstruction derives from 3D scanning, so we have focused our experiments on the reconstruction of 3D models from real-world data.

5.1. Resolution

We first consider the effects of the maximum octree depth on the reconstructed surface.

Figure 3 shows our reconstruction results for the "dragon" model at octree depths 6, 8, and 10. (In the context of reconstruction on a regular grid, this would correspond to resolutions of 64^3, 256^3, and 1024^3, respectively.) As the tree depth is increased, higher-resolution functions are used to fit the indicator function, and consequently the reconstructions capture finer detail. For example, the scales of the dragon, which are too fine to be captured at the coarsest resolution begin appearing and become more sharply pronounced as the octree depth is increased.

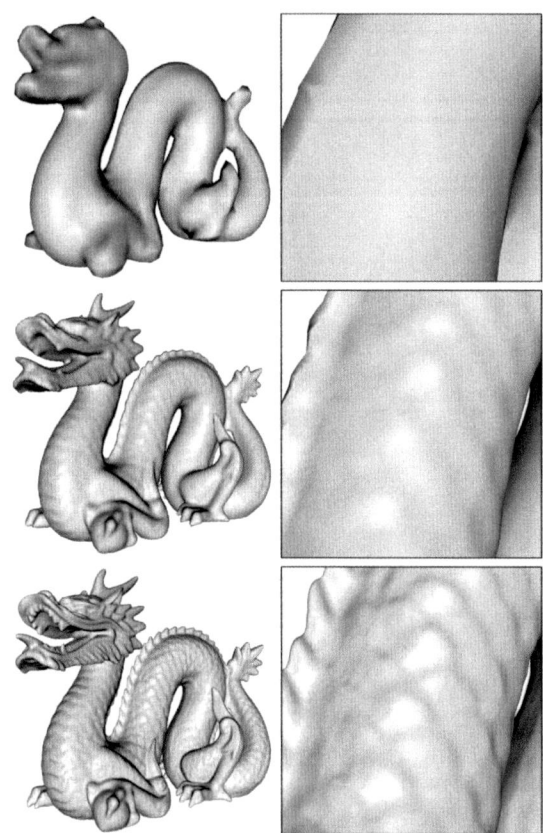

Figure 3: *Reconstructions of the dragon model at octree depths 6 (top), 8 (middle), and 10 (bottom).*

5.2. Comparison to Previous Work

We compare the results of our reconstruction algorithm to the results obtained using Power Crust [ACK01], Robust Cocone [DG04], Fast Radial Basis Functions (FastRBF) [CBC*01], Multi-Level Partition of Unity Implicits (MPU) [OBA*03], Surface Reconstruction from Unorganized Points [HDD*92], Volumetric Range Image Processing (VRIP) [CL96], and the FFT-based method of [Kaz05].

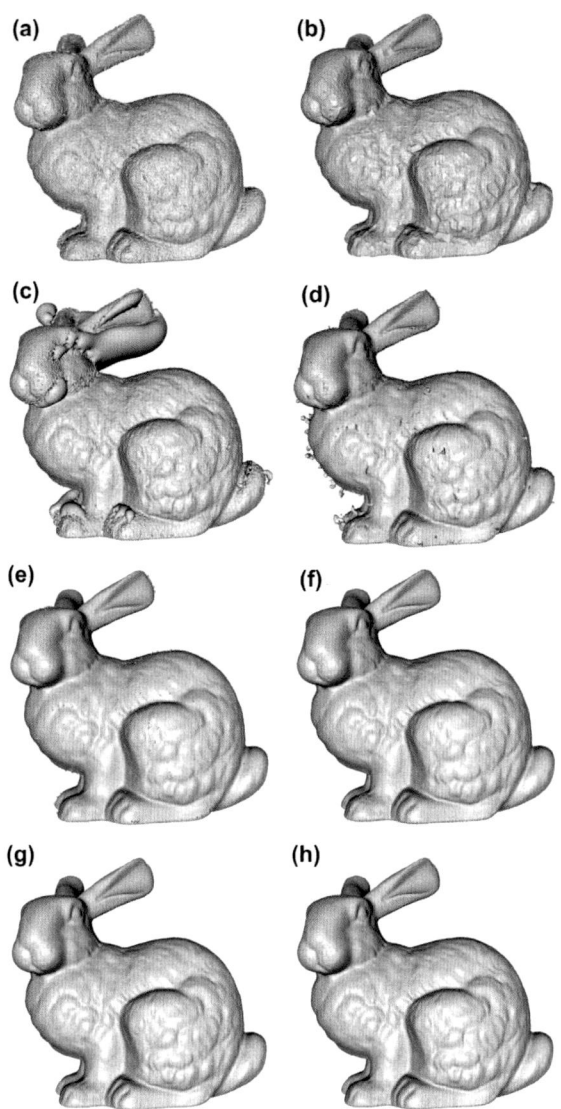

Figure 4: *Reconstructions of the Stanford bunny using Power Crust (a), Robust Cocone (b), Fast RBF (c), MPU (d), Hoppe et al.'s reconstruction (e), VRIP (f), FFT-based reconstruction (g), and our Poisson reconstruction (h).*

Our initial test case is the Stanford "bunny" raw dataset of 362,000 points assembled from ten range images. The data was processed to fit the input format of each algorithm. For example, when running our method, we estimated a sample's normal from the positions of the neighbors; Running VRIP, we used the registered scans as input, maintaining the regularity of the sampling, and providing the confidence values.

Figure 4 compares the different reconstructions. Since the scanned data contains noise, interpolatory methods such as Power Crust (a) and Robust Cocone (b) generate surfaces that are themselves noisy. Methods such as Fast RBF (c) and MPU (d), which only constrain the implicit function near

the sample points, result in reconstructions with spurious surface sheets. Non-interpolatory methods, such as the approach of [HDD*92] (e), can smooth out the noise, although often at the cost of model detail. VRIP (f), the FFT-based approach (g), and the Poisson approach (h) all accurately reconstruct the surface of the bunny, even in the presence of noise, and we compare these three methods in more detail.

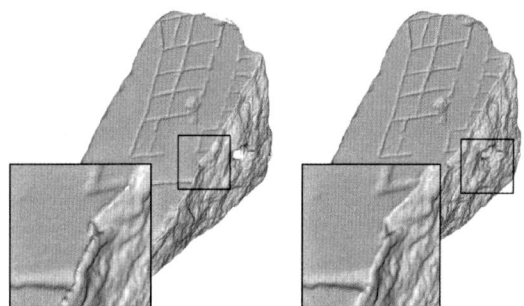

Figure 5: *Reconstructions of a fragment of the Forma Urbis Romae tablet using VRIP (left) and the Poisson solution (right).*

Comparison to VRIP A challenge in surface reconstruction is the recovery of sharp features. We compared our method to VRIP by evaluating the reconstruction of sample points obtained from fragment 661a of the Forma Urbis Romae (30 scans, 2,470,000 points) and the "Happy Buddha" model (48 scans, 2,468,000 points), shown in Figures 5 and 6. In both cases, we find that VRIP exhibits a "lipping" phenomenon at sharp creases. This is due to the fact that VRIP's distance function is grown perpendicular to the view direction, not the surface normal. In contrast, our Poisson reconstruction, which is independent of view direction, accurately reconstructs the corner of the fragment and the sharp creases in the Buddha's cloak.

Comparison to the FFT-based approach As Figure 4 demonstrates, our Poisson reconstruction (h) closely matches the one obtained with the FFT-based method (g). Since our method provides an adaptive solution to the same problem, the similarity is a confirmation that in adapting the octree to the data, our method does not discard salient, high-frequency information. We have also confirmed that our Poisson method maintains the high noise resilience already demonstrated in the results of [Kaz05].

Though theoretically equivalent in the context of uniformly sampled data, our use of adaptive-width filters (Section 4.5) gives better reconstructions than the FFT-based method on the non-uniform data commonly encountered in 3D scanning. For example, let us consider the region around the left eye of the "David" model, shown in Figure 7(a). The area above the eyelid (highlighted in red) is sparsely sampled due to the fact that it is in a concave region and is seen only by a few scans. Furthermore, the scans that do sample

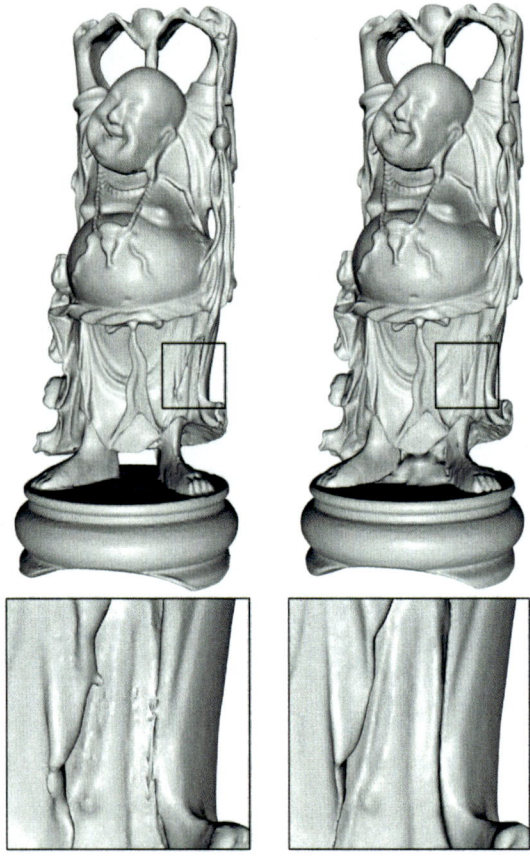

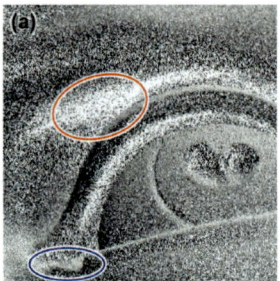

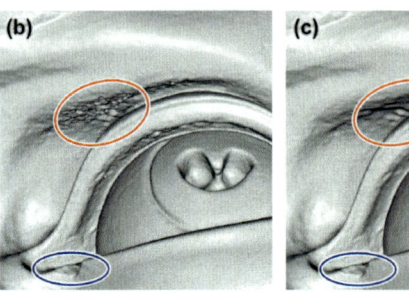
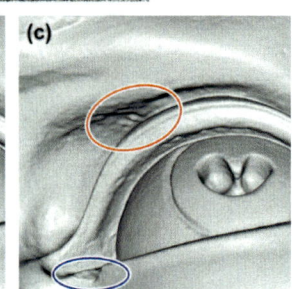

Figure 7: *Reconstruction of samples from the region around the left eye of the David model (a), using the fixed-resolution FFT approach (b), and Poisson reconstruction (c).*

Figure 6: *Reconstructions of the "Happy Buddha" model using VRIP (left) and Poisson reconstruction (right).*

the region tend to sample at near-grazing angles resulting in noisy position and normal estimates. Consequently, fixed-resolution reconstruction schemes such as the FFT-based approach (b) introduce high-frequency noise in these regions. In contrast, our method (c), which adapts both the scale and the variance of the samples' contributions, fits a smoother reconstruction to these regions, without sacrificing fidelity in areas of dense sampling (e.g. the region highlighted in blue).

Limitation of our approach A limitation of our method is that it does not incorporate information associated with the acquisition modality. Figure 6 shows an example of this in the reconstruction at the base of the Buddha. Since there are no samples between the two feet, our method (right) connects the two regions. In contrast, the ability to use secondary information such as line of sight allows VRIP (left) to perform the space carving necessary to disconnect the two feet, resulting in a more accurate reconstruction.

5.3. Performance and Scalability

Table 1 summarizes the temporal and spatial efficiency of our algorithm on the "dragon" model, and indicates that the

memory and time requirements of our algorithm are roughly quadratic in the resolution. Thus, as we increase the octree depth by one, we find that the running time, the memory overhead, and the number of output triangles increases roughly by a factor of four.

Tree Depth	Time	Peak Memory	# of Tris.
7	6	19	21,000
8	26	75	90,244
9	126	155	374,868
10	633	699	1,516,806

Table 1: *The running time (in seconds), the peak memory usage (in megabytes), and the number of triangles in the reconstructed model for the different depth reconstructions of the dragon model. A kernel depth of 6 was used for density estimation.*

The running time and memory performance of our method in reconstructing the Stanford Bunny at a depth of 9 is compared to the performance of related methods in Table 2. Although in this experiment, our method is neither fastest nor most memory efficient, its quadratic nature makes it scalable to higher resolution reconstructions. As an example, Figure 8 shows a reconstruction of the head of Michelangelo's David at a depth of 11 from a set of 215,613,477 samples. The reconstruction was computed in 1.9 hours and 5.2GB of RAM, generating a 16,328,329 triangle model. Trying to compute an equivalent reconstruction with methods such as the FFT approach would require constructing two voxel grids at a resolution of 2048^3 and would require in excess of 100GB of memory.

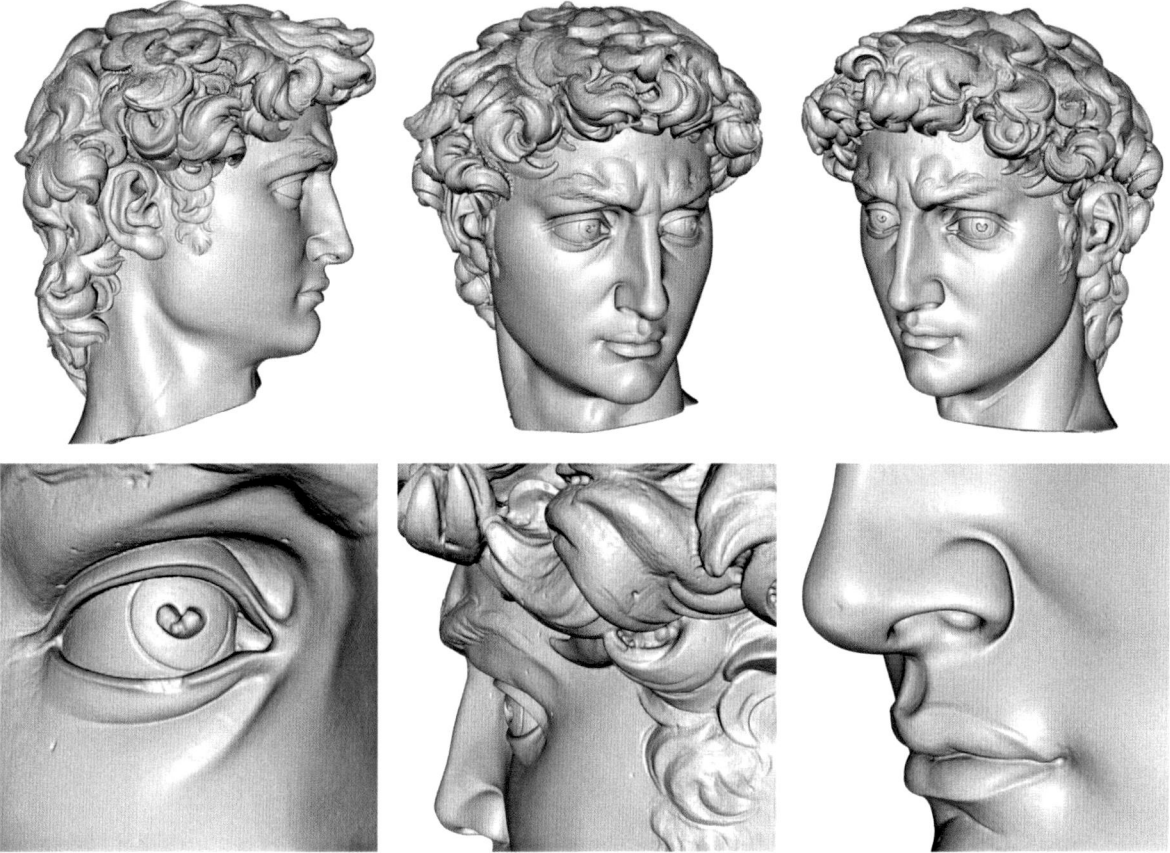

Figure 8: *Several images of the reconstruction of the head of Michelangelo's David, obtained running our algorithm with a maximum tree depth of 11. The ability to reconstruct the head at such a high resolution allows us to make out the fine features in the model such as the inset iris, the drill marks in the hair, the chip on the eyelid, and the creases around the nose and mouth.*

Method	Time	Peak Memory	# of Tris.
Power Crust	380	2653	554,332
Robust Cocone	892	544	272,662
FastRBF	4919	796	1,798,154
MPU	28	260	925,240
Hoppe et al 1992	70	330	950,562
VRIP	86	186	1,038,055
FFT	125	1684	910,320
Poisson	263	310	911,390

Table 2: *The running time (in seconds), the peak memory usage (in megabytes), and the number of triangles in the reconstructed surface of the Stanford Bunny generated by the different methods.*

6. Conclusion

We have shown that surface reconstruction can be expressed as a Poisson problem, which seeks the indicator function that best agrees with a set of noisy, non-uniform observations, and we have demonstrated that this approach can robustly recover fine detail from noisy real-world scans.

There are several avenues for future work:
- Extend the approach to exploit sample confidence values.
- Incorporate line-of-sight information from the scanning process into the solution process.
- Extend the system to allow out-of-core processing for huge datasets.

Acknowledgements

The authors would like to express their thanks to the Stanford 3D Scanning Repository for their generosity in distributing their 3D models. The authors would also like to express particular gratitude to Szymon Rusinkiewicz and Benedict Brown for sharing valuable experiences and ideas, and for providing non-rigid body aligned David data.

References

[ABCO*01] ALEXA M., BEHR J., COHEN-OR D., FLEISHMAN S., LEVIN D., SILVA C.: Point set surfaces. In *Proc. of the Conference on Visualization '01* (2001), 21–28.

[ABK98] AMENTA N., BERN M., KAMVYSSELIS M.: A new Voronoi-based surface reconstruction algorithm. *Computer Graphics (SIGGRAPH '98)* (1998), 415–21.

[ACK01] AMENTA N., CHOI S., KOLLURI R.: The power crust, unions of balls, and the medial axis transform. *Computational Geometry: Theory and Applications 19* (2001), 127–153.

[BBX95] BAJAJ C., BERNARDINI F., XU G.: Automatic reconstruction of surfaces and scalar fields from 3d scans. In *SIGGRAPH* (1995), 109–18.

[BFGS03] BOLZ J., FARMER I., GRINSPUN E., SCHRÖDER P.: Sparse matrix solvers on the GPU: Conjugate gradients and multigrid. *TOG 22* (2003), 917–924.

[BMR*99] BERNARDINI F., MITTLEMAN J., RUSHMEIER H., SILVA C., TAUBIN G.: The ball-pivoting algorithm for surface reconstruction. *IEEE TVCG 5* (1999), 349–359.

[Boi84] BOISSONNAT J.: Geometric structures for three dimensional shape representation. *TOG* (1984), 266–286.

[CBC*01] CARR J., BEATSON R., CHERRIE H., MITCHEL T., FRIGHT W., MCCALLUM B., EVANS T.: Reconstruction and representation of 3D objects with radial basis functions. *SIGGRAPH* (2001), 67–76.

[CL96] CURLESS B., LEVOY M.: A volumetric method for building complex models from range images. *Computer Graphics (SIGGRAPH '96)* (1996), 303–312.

[DG04] DEY T., GOSWAMI S.: Provable surface reconstruction from noisy samples. In *Proc. of the Ann. Symp. Comp. Geom.* (2004), 428–438.

[DMGL02] DAVIS J., MARSCHNER S., GARR M., LEVOY M.: Filling holes in complex surfaces using volumetric diffusion. In *Int. Symp. 3DPVT* (2002), 428–438.

[EM94] EDELSBRUNNER H., MÜCKE E.: Three-dimensional alpha shapes. *TOG* (1994), 43–72.

[FLW02] FATTAL R., LISCHINKSI D., WERMAN M.: Gradient domain high dynamic range compression. In *SIGGRAPH* (2002), 249–256.

[GKS02] GRINSPUN E., KRYSL P., SCHRÖDER P.: Charms: a simple framework for adaptive simulation. In *SIGGRAPH* (2002), 281–290.

[GWL*03] GOODNIGHT N., WOOLLEY C., LEWIN G., LUEBKE D., HUMPHREYS G.: A multigrid solver for boundary value problems using programmable graphics hardware. In *Graphics Hardware* (2003), 102–111.

[HDD*92] HOPPE H., DEROSE T., DUCHAMP T., MCDONALD J., STUETZLE W.: Surface reconstruction from unorganized points. *Computer Graphics 26* (1992), 71–78.

[Kaz05] KAZHDAN M.: Reconstruction of solid models from oriented point sets. *SGP* (2005), 73–82.

[KSO04] KOLLURI R., SHEWCHUK J., O'BRIEN J.: Spectral surface reconstruction from noisy point clouds. In *SGP* (2004), 11–21.

[LC87] LORENSEN W., CLINE H.: Marching cubes: A high resolution 3d surface reconstruction algorithm. *SIGGRAPH* (1987), 163–169.

[LGF04] LOSASSO F., GIBOU F., FEDKIW R.: Simulating water and smoke with an octree data structure. *TOG (SIGGRAPH '04) 23* (2004), 457–462.

[Mur91] MURAKI S.: Volumetric shape description of range data using "blobby model". *Computer Graphics 25* (1991), 227–235.

[NRDR05] NEHAB D., RUSINKIEWICZ S., DAVIS J., RAMAMOORTHI R.: Efficiently combining positions and normals for precise 3D geometry. *TOG (SIGGRAPH '05) 24* (2005).

[OBA*03] OHTAKE Y., BELYAEV A., ALEXA M., TURK G., SEIDEL H.: Multi-level partition of unity implicits. *TOG* (2003), 463–470.

[Par62] PARZEN E.: On estimation of a probability density function and mode. *Ann. Math Stat. 33* (1962), 1065–1076.

[PGB03] PÉREZ P., GANGNET M., BLAKE A.: Poisson image editing. *TOG (SIGGRAPH '03) 22* (2003), 313–318.

[SFYC96] SHEKHAR R., FAYYAD E., YAGEL R., CORNHILL J.: Octree-based decimation of marching cubes surfaces. In *IEEE Visualization* (1996), 335–342.

[SOS04] SHEN C., O'BRIEN J., SHEWCHUK J.: Interpolating and approximating implicit surfaces from polygon soup. *TOG (SIGGRAPH '04) 23* (2004), 896–904.

[TO02] TURK G., O'BRIEN J.: Modelling with implicit surfaces that interpolate. In *TOG* (2002), 855–873.

[WG92] WILHELMS J., GELDER A. V.: Octrees for faster isosurface generation. *TOG 11* (1992), 201–227.

[WKE99] WESTERMANN R., KOBBELT L., ERTL T.: Real-time exploration of regular volume data by adaptive reconstruction of iso-surfaces. *The Visual Computer 15* (1999), 100–111.

[YZX*04] YU Y., ZHOU K., XU D., SHI X., BAO H., GUO B., SHUM H.: Mesh editing with Poisson-based gradient field manipulation. *TOG (SIGGRAPH '04) 23* (2004), 641–648.

Appendix A:

The solution to surface reconstruction described in this paper approaches the problem in a manner similar to the solution of [Kaz05] in that the reconstructed surface is obtained by first computing the indicator function and then extracting the appropriate isosurface.

While the two methods seem to approach the problem of computing the indicator function in different manners ([Kaz05] uses Stokes' Theorem to define the Fourier coefficients of the indicator function while we use the Poisson equation), the two methods are in fact equivalent.

To show this, we use the fact that the Poisson equation $\Delta u = f$ where f is periodic can be solved using the Fourier transform. The Fourier series expansion is $-|\zeta|^2 \hat{u}(\zeta) = \hat{f}(\zeta)$, or equivalently $\hat{u}(\zeta) = \frac{-1}{|\zeta|^2} \hat{f}(\zeta)$.

Thus, our Poisson equation $\Delta\chi = \nabla \cdot \vec{V}$ can be solved using $\hat{\chi} = \frac{-1}{|\zeta|^2} \widehat{\nabla \cdot \vec{V}}$. With the well known identity $\hat{f}' = -i\zeta\hat{f}$ and its generalization $\widehat{\nabla \cdot \vec{V}} = -i\zeta \cdot \hat{\vec{V}}$, we get $\hat{\chi} = \frac{i}{|\zeta|^2} \zeta \cdot \hat{\vec{V}}$, which is *identical* to [Kaz05].

Eurographics Symposium on Geometry Processing (2006)
Konrad Polthier, Alla Sheffer (Editors)

Error Bounds and Optimal Neighborhoods for MLS Approximation

Yaron Lipman   Daniel Cohen-Or   David Levin

Tel-Aviv University

Abstract

In recent years, the moving least-square (MLS) method has been extensively studied for approximation and reconstruction of surfaces. The MLS method involves local weighted least-squares polynomial approximations, using a fast decaying weight function. The local approximating polynomial may be used for approximating the underlying function or its derivatives. In this paper we consider locally supported weight functions, and we address the problem of the optimal choice of the support size. We introduce an error formula for the MLS approximation process which leads us to developing two tools: One is a tight error bound independent of the data. The second is a data dependent approximation to the error function of the MLS approximation. Furthermore, we provide a generalization to the above in the presence of noise. Based on the above bounds, we develop an algorithm to select an optimal support size of the weight function for the MLS procedure. Several applications such as differential quantities estimation and up-sampling of point clouds are presented. We demonstrate by experiments that our approach outperforms the heuristic choice of support size in approximation quality and stability.

Categories and Subject Descriptors (according to ACM CCS): I.3.3 [Computer Graphics]: Surface approximation, Point clouds, Meshes, Differential quantities estimation

1. Introduction

A fundamental problem in surface processing is the reconstruction of a surface or estimating its differential quantities from scattered (sometimes noisy) point data [HDD*92]. A common approximation approach is fitting local polynomials, explicitly [ABCO*01], or implicitly [OBA*03], to approximate the surface locally. This approach can be realized by the moving least-squares (MLS) method, where, for each point x, a polynomial is fitted, in the least-squares sense, using neighboring points x_i. This technique works well assuming that the surface is smooth enough [Lev98, Wen01]. The local polynomial fitting enables up and down sampling of the surface [ABCO*01], estimating differential quantities such as normal or curvature data [CP05], and performing other surface processing operations [PKKG03].

In recent years, the MLS technique has gained much popularity, and the method is now well studied. However, proper choice of neighboring points x_i to be used in the approximation still remains an important open problem. Apparently, there is a large degree of freedom in choosing the

points participating in the approximation since the number of data points is usually very large, while the degree of polynomial is usually very small. Naturally, one would like to make use of these large degrees of freedom to achieve the "best" approximating polynomial. Several researchers [ABCO*01, PGK02, PKKG03] have used different heuristic approaches, such as using a neighborhood proportional to the local sampling density measured via the radius of the ball containing the K nearest neighbors, or using Voronoi triangulation [FR01] to choose the neighboring points. In this paper, we compare a heuristic method in the spirit of these approaches with a new approach based on error analysis.

Since the problem of choosing the points to be used in the approximation is closely related to multivariate interpolation, it is known that the choice of the points depends on the geometry of the points, and not only their number, as in the sampling density based approaches. However, the geometric configuration of points which admits a stable interpolation/approximation problem is a hard problem in the field of multivariate polynomial approximation [Bos91, SX95, GS00]. Loosely speaking, a 'stable' points'

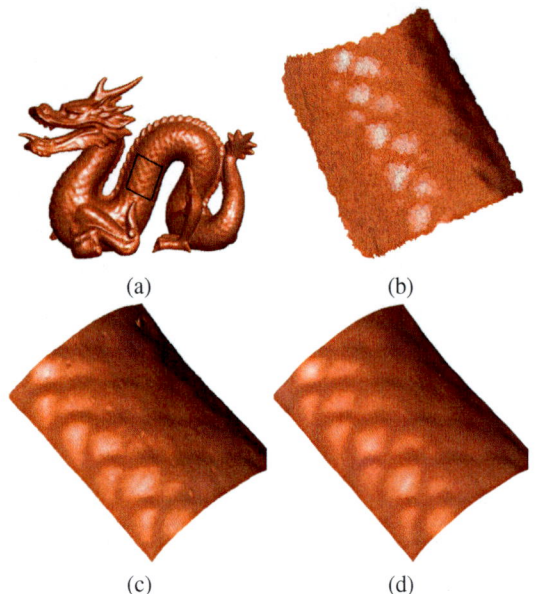

(a)   (b)

(c)   (d)

Figure 1: *Resampling of a noisy surface. A patch of the dragon model (a) with additional white noise (b) is resampled with two methods: Using the local density (heuristic) to determine the local neighborhood yields some noticeable artifacts (c). The proposed method, assuming the maximal value of the noise is known, faithfully reconstruct the surface.*

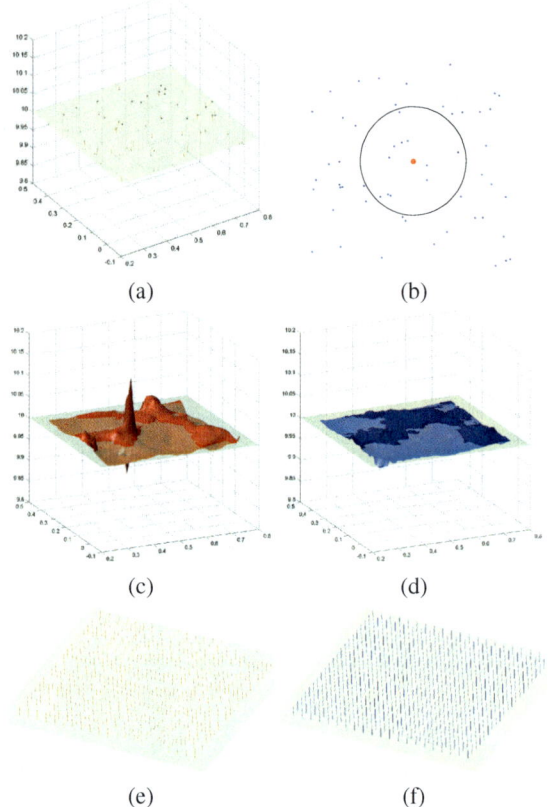

(a)   (b)

(c)   (d)

(e)   (f)

Figure 2: *An example of the amplifying effect of noise in the data. In (a) the 'true surface', i.e., a plane, and the sampled points with white noise errors of maximal magnitude 0.015. In (b), the sample data (blue) and a red point where an approximation is sought. The black circle indicates the points used by the heuristic method. In (c) the surface reconstructed by this heuristic, note that the high peak is created at the place indicated by the red point in (b). In (d) the proposed algorithm for neighborhood was used. In (e) surface (plane) normals were estimated using the heuristic algorithm and in (f) the normals were estimated using the new proposed algorithm. We used approximation by quadratics, i.e., $N = 2$.*

configuration is such that is 'far' from a degenerate point configuration, where a degenerate point configuration lies on an algebraic curve of the dimension of the interpolation space.

Preliminary example. Consider the points $X_h = \{h(\cos j2\pi/6, \sin j2\pi/6), j = 0, 1, .., 5\}$. X forms a degenerate point configuration for bi-variate quadratic interpolation. Although this point set seems nicely distributed around $(0,0)$, quadratic interpolation cannot be used for approximation at $(0,0)$, no matter how small we take h. This can be seen by taking, for example, the values associated with $x_i \in X_h$ to be zeros and noting that both $f_1(x) \equiv 0$ and $f_2(x) = (h^2 - x_1^2 - x_2^2)/h^3$ solve the interpolation problem, but $|f_1(0) - f_2(0)| \sim 1/h$. Another phenomenon is shown in Figure 1 where a noisy part (b) of the dragon model (a) was re-sampled using a density based heuristic (c), which caused undesired artifacts, and in (d) the new proposed method was used. In Figure 2, another example of this kind is shown. As elaborated in Section 3.5, this configuration of points causes amplification of the noise level in the data by a factor of ~ 10 when using MLS to approximate the surface value at the red point in (b), and a factor of ~ 100 when this point configuration is used to approximate $\partial_x f$ at that point. The resulting approximations, i.e., point evaluation (c), and normal approximation (e), are useless at this point. Using the new approach presented here yields bounded errors which

don't exceed much the initial noise level in the data, see (d) and (f).

Recognizing that there are no simple nor intuitive rules which distinguish 'bad' point sets from 'good' point sets for approximation at a given point, we take a different route to decide which neighbors should be used in the approximation process: We introduce an error formula which provides means to understand and to evaluate the approximation quality of MLS approximation. From it, we derive two tools. The first, is a tight bound independent of the data, assuming only that the local corresponding derivatives of the function are bounded. The second is a data dependent approximation to the error function of the MLS interpolant. We examine the practical usage of these tools and compare them to a carefully chosen heuristic method.

Based on the above bounds, we develop an algorithm to select an optimal radial neighborhood for the MLS procedure. Loosely speaking, since the underlying surface from which the sampled points are taken is unknown, the optimality of the chosen neighborhoods is in the sense of the approximation error having the lowest error bound.

MLS approximations are used in a variety of cases, but the problem is always reduced to the functional case, by defining some parameter domain [Lev03, ABCO*01]. Thus, for the error analysis, it is enough to consider the functional case.

We develop the various theoretical error terms and bounds in Section 3, and based on these results, in Section 4, we introduce an algorithm for selecting the optimal neighborhood. In Section 5 we derive a heuristic rule which we use for comparison. In the following section, we briefly describe the MLS technique, and define the terms and notation to be used in the paper. In Section 6 we present some numerical experiments, and in Section 7 we conclude.

2. Background

Surfaces or 2-manifolds embedded in $\mathbb{R}^3$ are most commonly represented by a set of spacial points, with neighboring relations (meshes) or without (point clouds). A common way to estimate the value of the surface in a new point, or to estimate differential quantities of the surface at any point, is by fitting a local polynomial and extracting it's value or derivatives.

In order to reduce the problem to the functional case a local parameter plane is constructed, and the local polynomial is defined over this parameter space. Eventually, one ends up with the problem of fitting a polynomial $p \in \Pi$, where Π is some polynomial subspace, given data points $(x_i, f(x_i)) \in \Omega \times \mathbb{R}$, $i = 1, .., I$, where Ω is a domain in $\mathbb{R}^d$. The goal is to approximate a functional L^x at a point $x \in \Omega$, where L^x can be a function evaluation at x or a derivative evaluation, e.g., $L^x(f) = f(x)$ or $L^x(f) = (\partial_{xy} f)(x)$. A common way to do it is by fitting the polynomial locally in the least-squares sense:

$$\min \left\{ \sum_{i=1}^{I} (f(x_i) - p(x_i))^2 w(\|x_i - x\|) \;\;,\;\; p \in \Pi \right\}, \quad (1)$$

where $w(r)$ is a radial weight function. When the minimizer polynomial p is achieved, the approximation functional L^x is applied to it to form the approximation

$$L^x(f) \approx L^x(p). \quad (2)$$

As showed in [Lev98],

$$L^x(p) = \sum_{i=1}^{I} a_i f(x_i),$$

where a_i are the solution to the constrained quadratic minimization problem

$$\left\{ \min \sum_{i=1}^{I} w(\|x - x_i\|)^{-1} |a_i|^2 \; s.t. \; \sum_{i=1}^{I} a_i p(x_i) = p(x), \; \forall p \in \Pi \right\}.$$

The weight function w is usually chosen to ensure fast decay of the magnitude of the a_i for points distant from the evaluation point x. The decay rate is heuristically chosen to be as fast as possible while keeping enough points in the significant weights area to keep the problem well-posed. Furthermore, a smooth weight function implies smooth approximation. In this paper we have chosen to use the weight function of finite support [Lev98], $w(r) = w_h(r)$, where

$$w_h(r) = e^{-\frac{r^2}{(h-r)^2}} \chi_{[0,h)}(r). \quad (3)$$

The main objective of this paper is to present an algorithm for choosing the support size h which best assures a minimal approximation error using the procedure (1)-(2). This is accomplished in two independent ways: First by minimizing a novel, tight, local error bound formula. This procedure also supplies a bound on the error which is achieved in the approximation process, given that a bound on local corresponding derivatives of f is known. Second, a novel approximation of the error in the MLS approximation is constructed, and the best support size h is chosen as before. The latter generally performs better than the former.

3. Error analysis
3.1. Settings

The settings of the problem consists of a data set $(x_i, f(x_i))$, $X = \{x_i\}_{i=1}^{I} \subset \Omega \subset \mathbb{R}^d$, sampled from a smooth function $f \in C^k(\Omega)$, and another point x where an approximation $L^x f = D^\alpha f(x)$, where $D^\alpha = \partial_{x^{(1)}}^{\alpha_1} \cdot ... \cdot \partial_{x^{(d)}}^{\alpha_d}$, is sought. Denote by N the degree of the polynomials used as the approximation space. $J = \binom{N+d}{d}$ is the dimension of the space Π_N of d-variate polynomial of degree N. Also define $p_1, p_2, ..., p_J \in \Pi$ to be the standard basis of Π_N shifted to x, that is, $\{(\cdot - x)^\alpha\}_{|\alpha| \leq N}$, where we use the multi-index notation $\alpha = (\alpha_1, ..., \alpha_d)$, $\alpha! = \alpha_1! \cdot ... \cdot \alpha_d!$, $|\alpha| = \alpha_1 + ... + \alpha_d$, and for $x = (x^{(1)}, ..., x^{(d)})$ $x^\alpha = (x^{(1)})^{\alpha_1} ... (x^{(d)})^{\alpha_d}$. We also define the generalized Vandermonde matrix E by $E_{i,\beta} = p_\beta(x_i)$, $i = 1, .., I$, $|\beta| \leq N$.

Denote the subset $X_h = X \cap B_h(x)$, where $B_h(x)$ denotes a ball of radius h with center x, and let $I = |X_h|$, the number of data points in $B_h(x)$. Then, for a fixed h, we define the approximation $D^\alpha f(x) \approx D^\alpha p(x)$, where $p \in \Pi_N$, is defined by (1), and $w = w_h$ defined by (3).

Let us introduce some test functions. The first one is taken from [Fra82],

$$F_1 = \frac{30}{4} e^{-\frac{(9x+5)^2 + (9y+5)^2}{16}} + \frac{30}{4} e^{-\frac{(9x+11)^2}{196} - \frac{9y+11}{20}} \quad (4)$$
$$+ \frac{10}{2} e^{-\frac{(9x-5)^2 + (9y+3)^2}{16}} - \frac{10}{5} e^{-\frac{(9x+1)^2 + (9y-5)^2}{4}}.$$

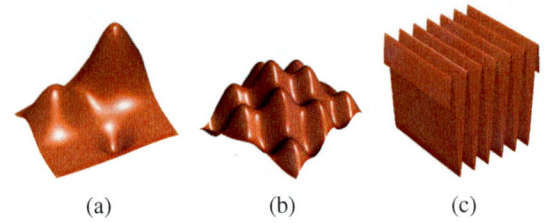

(a)   (b)   (c)

Figure 3: *The test function used in the paper. (a) is the graph of F_1, (b) of F_2 and (c) of F_3.*

$$F_2 = 0.3\cos(8x)\sin(6y) + e^{-x^2 - y^2}. \qquad (5)$$
$$F_3 = \cos(20x). \qquad (6)$$

In Figure 3, we plotted the graphs of these test functions. These functions were selected since they seem to represent well several smooth surface types: F_1 is a standard test function and has been used in numerous papers. F_2 has interesting 'details', and F_3 is an anisotropic surface with very high derivatives.

3.2. Pointwise error in the MLS approximation

In this section we lay out the formula for the error in the MLS approximation which forms the basis for all latter developments in the paper:

Theorem 3.1 Denote by p the fitted polynomial defined by (1) to the data $(X_h, f(X_h)) \subset \Omega \times \mathbb{R}$, sampled from a smooth function $f \in C^{N+1}(\Omega)$, then for $x \in \Omega$,

$$R(x) = D^{\alpha}p(x) - D^{\alpha}f(x) = \qquad (7)$$

$$\frac{\alpha!}{(N+1)!}\sum_{i,v}D^v f(\eta_i(x_i - x) + x)(x_i - x)^v \frac{\det\left(E^t W E_{\alpha \leftarrow e_i}\right)}{\det\left(E^t W E\right)}$$

where $\sum_{i,v}$ stands for $\sum_{|v|=N+1}\sum_{i=1}^{I}$, $0 \le \eta_i \le 1$, and E is the Vandermonde matrix $E_{i,\beta} = p_{\beta}(x_i)$. $E_{\alpha \leftarrow e_i}$ denotes the matrix E where the α column is replaced by the standard basis vector $e_i = \delta_{j,i}$. The weight matrix W is defined by $W = diag(w_h(\|x_1 - x\|), ..., w_h(\|x_I - x\|))$.

Proof. The proof relies on the polynomial reproduction property of the Least-Squares method and is based upon local Taylor expansions as approximations of f.

First, w.l.o.g, we may assume $x = 0$. Denote by $p_1, ..., p_J$ the standard basis of the multivariate polynomials of degree $\le N$, that is $p_{\alpha}(x) = x^{\alpha}$, $|\alpha| \le N$. Next, writing $p = \sum_{\beta}c_{\beta}p_{\beta}$, leads to

$$D^{\alpha}p(0) = \sum_{\beta}c_{\beta}D^{\alpha}p_{\beta}(0) = \sum_{\beta}c_{\beta}\alpha!\delta_{\alpha,\beta} = \alpha!c_{\alpha}.$$

The multivariate Vandermonde matrix E is ordered by the multi-index β, i.e., $E_{i,\beta} = p_{\beta}(x_i)$, as we do also for the vector $c = \{c_{\beta}\}_{|\beta|\le N}$ of the unknown coefficients. The fitted polynomial p is then defined as the solution in the least-squares

sense. That is, c satisfies the normal equations:

$$E^t W E c = E^t W F. \qquad (8)$$

Using Taylor expansion,

$$f(x_i) = \sum_{|v|\le N}\frac{D^v f(0)}{v!}x_i^v + \frac{1}{(N+1)!}\sum_{|v|=N+1}D^v f(\eta_i x_i)x_i^v,$$

where $1 \le \eta_i \le v$. Hence, the vector F can be written as

$$F = \sum_{|v|\le N}\frac{D^v f(0)}{v!}E_v + \frac{1}{(N+1)!}\sum_{|v|=N+1}Q_v E_v,$$

where $Q_v = diag(D^v f(\eta_1 x_1), ..., D^v f(\eta_I x_I))$, and E_v denotes the v column vector of matrix E. Then, For the solution of (8) we have,

$$c_v = \frac{D^v f(0)}{v!} + \frac{1}{(N+1)!}\left(\sum_{|v|=N+1}(E^t W E)^{-1}E^t W Q_v E_v\right)_v.$$

Next, $E^t W Q_v E_v = \sum_{i=1}^{I}(E^t W)_i D^v f(\eta_i x_i)p_v(x_i)$, with $p_v(x_i) = x_i^v$, and by Cramer's rule and the linearity of the determinant $\left((E^t W E)^{-1}E^t W Q_v E_v\right)_v =$

$$\sum_{i=1}^{I}D^v f(\eta_i x_i)x_i^v \frac{\det(E^t W E_{v \leftarrow e_i})}{\det(E^t W E)}.$$

Finally we get for $v = \alpha$, $\alpha!c_{\alpha} - D^{\alpha}f(0) =$

$$\frac{\alpha!}{(N+1)!}\sum_{|v|=N+1}\sum_{i=1}^{I}D^v f(\eta_i x_i)x_i^v \frac{\det(E^t W E_{v \leftarrow e_i})}{\det(E^t W E)},$$

where $|v| = N+1$ and $i = 1, .., n$.

Corollary 3.1 Denote by p the fitted polynomial defined by (1) to the data $(X_h, f(X_h)) \in \Omega \subset \mathbb{R}^d \times \mathbb{R}$, sampled from a smooth function $f \in C^{N+1}(\Omega)$, then the following is a *tight* error bound,

$$|R(x)| \le \frac{\alpha!C}{(N+1)!}\sum_{i,v}|x_i - x|^v \left|\frac{\det\left(E^t W(E)_{\alpha \leftarrow e_i}\right)}{\det\left(E^t W E\right)}\right|,$$

where C is the bound: $\max_{|v|=N+1, x\in\Omega}|D^v f(x)| \le C$.

3.3. Data Independent Bound

For a given support size h, an error bound for the polynomial fitting procedure based on the data X_h can be calculated via the tight bound in Corollary (3.1). We define the bounding function B_{α} by

$$B_{\alpha} = B_{\alpha}(x, X_h) = \frac{\alpha!}{(N+1)!}\sum_{\beta,i}|x_i - x|^{\beta}\frac{|\det(E^t W E_{\alpha \leftarrow e_i})|}{|\det(E^t W E)|}, \qquad (9)$$

where, as before, $\sum_{i,\beta}$ is a short notation for $\sum_{|v|=N+1}\sum_{i=1}^{I}$.

From the computational point of view, in order to compute (9), we note that $\frac{\det(E^t W E_{\alpha \leftarrow e_i})}{\det(E^t W E)}$ is the α coordinate of the solution to the linear system:

$$E^t W E c = (E^t W)_i,$$

where $(E^t W)_i$ denotes the i-th column of matrix $E^t W$. Therefore, in the calculation of (9), one should calculate the solution V to $E^T W E V = E^t W$, and then set

$$V_{\alpha,i} = \frac{\det(E^t W E_{\alpha \leftarrow e_i})}{\det(E^t W E)}. \tag{10}$$

Then formula (9) reduces to,

$$B_\alpha(x, X_h) = \frac{\alpha!}{(N+1)!} \sum_{\beta,i} |x_i - x|^\beta |V_{\alpha,i}|. \tag{11}$$

3.4. Data dependent error approximation

In this section we construct a data dependent approximation to the error function in the MLS approximation for $f \in C^{N+2}(\Omega)$. This error approximation uses the known values at the points X_h in order to better approximate the error in the approximation (7). In particular we note that $D^\nu f(\eta_i(x_i - x) + x) = D^\nu f(x) + O(h)$, for $|\nu| = N+1$, where h is the support size used. Therefore, Eq. (7) can be written as:

$$R(x) = \frac{\alpha!}{(N+1)!} \sum_{i,\nu} D^\nu f(x)(x_i - x)^\nu V_{\alpha,i} + O(h^{N+2-|\alpha|}).$$

The idea is to improve the error estimate by approximating the unknown values $f_\nu = D^\nu f(x)$, $|\nu| = N+1$. Such approximations can be derived by using the error formula at points x_k near x: We have

$$p(x_k) - f(x_k) = R(x_k) = \frac{\alpha!}{(N+1)!} \sum_\nu f_\nu \left(\sum_i (x_i - x_k)^\nu V_{\alpha,i}^k \right), \tag{12}$$

where $V_{\alpha,i}^k$ are defined similar to $V_{\alpha,i}$ in (10), using the shifted basis $\{(\cdot - x_k)^\alpha\}_{|\alpha| \le N}$. The points $\{x_i\}$ are taken from a ball of radius $h = 3h_J$ centered at x, where h_J denotes the radius of the ball which contains the J nearest points to x. The points $\{x_k\}$ are taken as the $2J$ nearest points to x. The system (12), of $2J$ equations and $N+2$ unknowns f_ν is solved in the least-squares sense.

Plugging the resulting estimated values f_ν into the error bound (7) we get an approximation of the error term:

$$\tilde{R}_\alpha = \tilde{R}_\alpha(x, X_h) = \frac{\alpha!}{(N+1)!} \sum_{i,\nu} f_\nu (x_i - x)^\nu V_{\alpha,i}.$$

This error approximation incorporates the given data values, and as shown in Section 6, in practice it approximate the actual error better than the tight bound described in Section 3.3. In Figure 4, we demonstrate the high similarity of the approximated error function $\tilde{R}_\alpha$ to the true error function R. We used in this example the test functions introduced in Eq. (4)-(6).

A drawback of this approach that it is not a bound, but

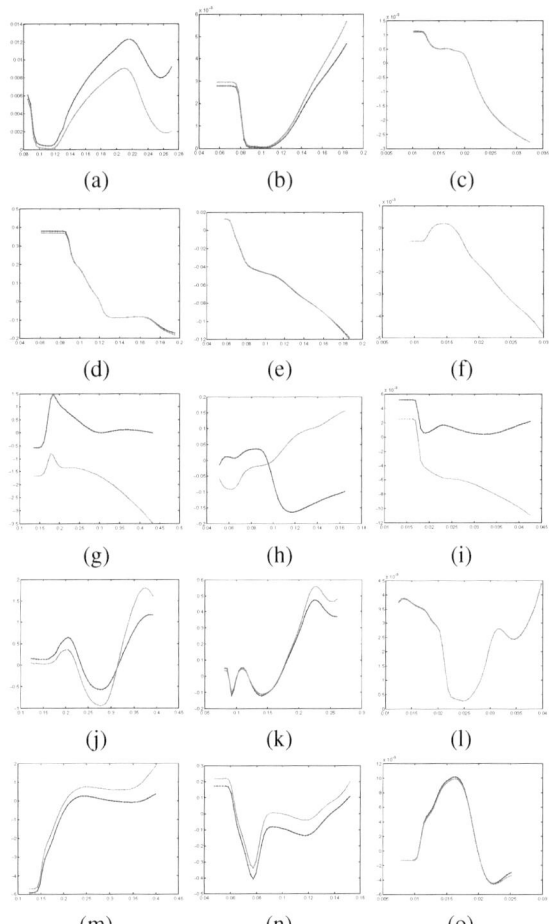

Figure 4: *Comparison between the true error R graphs (blue), and the approximated error $\tilde{R}$ graphs (green) when using quadratics to approximate the test functions using a uniformly distributed random point set. In each graph the x-axis stands for the support size h. In (a), the function F_1 has been used to create the data set, where the density of the points used was 0.25k points per unit square. In (b) the density is 1k points per unit square, and in (c) 25k points per unit square. In (d)-(f) function F_2 has been used. In (g)-(i) function F_3, and since all it's third derivatives vanish at the point of evaluation (origin), the approximation is bad. Using third degree polynomial in (j)-(l) alleviates the problem. Also perturbing the evaluation point by 0.05 in the x coordinate (m)-(o) alleviates the problem.*

merely an approximation of the error function, and it depends on the quality of the approximation of the coefficients f_ν. In the presence of very high derivatives and low sampling density it can perform worse than the data-independent bound. Another drawback appears at points where all the derivatives of order $N+1$ vanish. Then, the approximation of the error function may be damaged, see Figure 4. If we have a prior knowledge about such a point, the problem can be avoided by perturbing the interest point a little, or using higher degree polynomial N' (assuming the $N'+1$ derivatives do not all vanish at that point), in Figure 4, where all the third derivatives of F_3 vanish at the origin (the point of interest in that example) the approximation of the error function is quite inaccurate, however using third degree polynomial or moving the point of interest a little alleviates the problem. The reason of this phenomenon lies in the fact that the sign of the coefficients $D^\nu f(\eta_i(x_i - x) + x)$ are likely to change in the vicinity of x, hence the error function is highly dependent on the values of η_i.

3.5. Noisy Data

In this section we extend our previous error bounds and approximations to optimally handle errors (noise) in the sampled data. We assume that errors ε_i, where $|\varepsilon_i| \leq \varepsilon$ are introduced into the data, that is, $f(x_i) = f^*(x_i) + \varepsilon_i$, where f^* stands for the 'true' sampled function.

It then follows, as in Theorem 3.1, that

$$R^*(x) = R(x) + \alpha! \sum_{i=1}^{I} \varepsilon_i V_{\alpha,i},$$

where $R^*(x) = D^\alpha p(x) - D^\alpha f^*(x)$ and $R(x)$ is given in Eq. (7). Hence,

$$|R^*(x)| \leq |R(x)| + \alpha! \varepsilon \sum_{i=1}^{I} |V_{\alpha,i}|. \tag{13}$$

Note that this bound is again tight since no assumption can be made on the signs of ε_i nor their magnitude, except that $|\varepsilon_i| \leq \varepsilon$.

In Figure 2, using the points inside the black circle (b), which are chosen by the heuristic method, in the MLS approximation leads to $\sum_{i=1}^{I} |V_{\alpha,i}| \approx 15$ for $\alpha = (0,0)$ and ≈ 99 for $\alpha = (1,0)$. Hence, we can suspect that the noise level in the data might be amplified by these factors when approximating the value or the partial derivative ∂_x at the red point, respectively. Indeed, the *true* error in the function evaluation is $\sim 9\varepsilon$ and the error in the derivative approximation is $\sim 106\varepsilon$. Minimizing the bound (13) imply choosing a bigger support size in this case, which results in $\sim 0.17\varepsilon$ and $\sim 3.3\varepsilon$ error in approximation of the value and derivatives respectively. In section 3.6 we discuss another aspect of the error amplification phenomenon.

Next, we integrate the sampling error term into the former error terms. First the data dependent approximation,

$$|R^*(x)| \preceq |\tilde{R}(x)| + \alpha! \varepsilon \sum_{i=1}^{I} |V_{\alpha,i}|,$$

where $\preceq$ stands for $\leq$ up to a term of magnitude $O(h)$. Therefore, we denote our approximated error in the approximation:

$$\tilde{R}_{\alpha,\varepsilon}(x, X_h) = |\tilde{R}_\alpha(x, X_h)| + \alpha! \varepsilon \sum_{i=1}^{I} |V_{\alpha,i}|. \tag{14}$$

For the data-independent bound:

$$|R^*(x)| \leq C|B_\alpha(x, X_h)| + \alpha! \varepsilon \sum_{i=1}^{I} |V_{\alpha,i}|.$$

Since C is unknown, this bound is better presented if we consider relative error, i.e., $f(x_i) = f^*(x_i)(1 + \varepsilon_i)$. In this case by similar consideration as before, the tight error bound in the presence of noise in the data becomes:

$$|R^*(x)| \leq C \left(|B_\alpha(x, X_h)| + \alpha! \varepsilon \sum_{i=1}^{I} |V_{\alpha,i}| \right), \tag{15}$$

where C bounds the relevant derivatives and the function values. In practice we considered the term in the parentheses as the function to be minimized in the presence of noise in the data:

$$B_{\alpha,\varepsilon}(x, X_h) = |B_\alpha(x, X_h)| + \alpha! \varepsilon \sum_{i=1}^{I} |V_{\alpha,i}|. \tag{16}$$

3.6. A Confidence Measure

By Corollary 3.1 and Eq. (15) we have that

$$|D^\alpha p(x) - D^\alpha f(x)| \leq CB_{\alpha,\varepsilon}(x, X_h),$$

where C bounds certain derivatives of the unknown function. Therefore, if we assume that the unknown function is sufficiently smooth with bounded derivatives, $B_{\alpha,\varepsilon}(x, X_h)$ furnishes a tool which justifies an approximation result. It can be understood as a *confidence measure* of the ability of a given set of points X_h to approximate $D^\alpha f(x)$. As an example of this application, assume we want to approximate local curvatures on a mesh. A common way to do it is fitting a local polynomial at each vertex using it's 1 or 2-ring neighborhood, extracting it's derivatives and using some standard classical differential geometry formula. As an easy example, suppose we want to approximate ∂_{xx} at the vertices of a sphere mesh. We use a sphere since we know its derivatives are bounded and are the same everywhere on the sphere, w.r.t the local frame. We define the parameter domain to be the plane perpendicular to the weighted average of the adjacent face's normals. Figure 5 shows a coloring of the sphere mesh, using the two parts of the tight error bound factor (16): In (a), the bound of the error factor caused in the approximation, $|B_\alpha(x, X_h)|$, and in (b) the bound of the error factor related to the noise in the data: $\alpha! \sum_{i=1}^{I} |V_{\alpha,i}|$. Note that the latter means that if one of the 1-ring neighborhoods contains noise in the direction of the normal of the parameter plane, the errors in the approximation might be multiplied by this factor. In this case we see that even 'nice' 1-rings of the sphere might cause ~ 100 times bigger errors than the noise level of the data. In (c) it is shown that irregular triangulation may yield much higher errors.

4. Optimal neighborhoods

In this section we present an algorithm, which finds the optimal support size h_{opt}, which should be used in the ap-

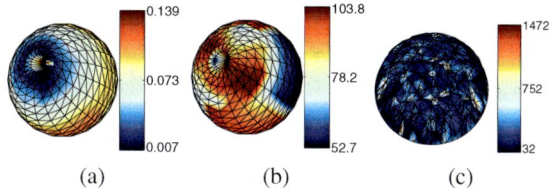

Figure 5: *Color maps of the confidence measures for approximating ∂_{xx} using quadratic polynomial fitted in the least-squares sense to the 1-ring neighborhoods on a sphere mesh. In (a) the (tight) error bound factor B_α caused in the approximation, for non-noisy data. In (b) the factors which multiply the noise ε in the data. Note that small noise level in the data may still cause ~ 100 times bigger errors in the approximation. (c) is the same as (b) for an irregular sphere mesh.*

proximation procedure (1)-(2) to ensure minimal error. We base our algorithm on the error analysis presented in former sections, i.e., equations (14) and (16). More specifically, for a given X, α, N and a point where the approximation is sought, x, we look for the optimal h, which we denote by h_{opt}, which minimizes the bound or the approximation of the error $|D^\alpha f(x) - D^\alpha p(x)|$.

4.1. Finding an Optimal Support Size

We use the same algorithm for both bounds (14) and (16). We look for the support size h which minimizes the bound function $EB_\alpha(x, X, h)$, where for brevity we will use the symbol EB_α for both bounds.

We fix an upper and lower bounds H_{min}, H_{max} for h ,e.g., H_{min} could be set to h_J (which is defined in Section 3.4) and H_{max} to some large support size radius, we used for example $4h_J$. A rough step size Δ is set, e.g., we used $|H_{max} - H_{min}|/50$, and the algorithm traverse $h = H_{min}, H_{min} + \Delta, ..., H_{max}$ where for each h the algorithm calculates the error bound function EB_α, for the data points X_h. Next, after extracting the minimizing support size radius $h_{opt}^{(0)} = argmin_{h=H_{min}+j\Delta}\{EB_\alpha(x, X, h)\}$, we further improve the approximation to h_{opt} by fitting an interpolating quadratic near $h_{opt}^{(0)}$ and minimize it to define $h_{opt}^{(1)}$. We iterate this procedure until $|h_{opt}^{(k+1)} - h_{opt}^{(k)}| \le tolerance$. In our application we actually minimized EB^2, for faster convergence.

For efficient computation the following considerations are employed. First we move the origin to x, i.e., we use the points $x_i := x_i - x$,$i = 1, .., I$. Second, we rearrange $x_i \in X \cap B_{H_{max}}(0)$ with respect to their distance from 0 (x), where now the sub-index i is with respect to this ordering. If E is the Vandermonde matrix based on the data points $x_1, ..., x_I$, then the matrix E' for the data points $x_1, ..., x_{I+1}$ can be written as $E'_{i,j} = E_{i,j}$ for $i \le I$ and $E_{I+1,j} = p_j(x_{I+1})$. This implies that when h changes to include a new point x_{I+1} in X_h, we only need to add a single row to the previous Vandermonde matrix E, to construct the Vandermonde matrix for X_h.

Note that if we have calculated the bound for the data points $x_1, ..., x_I$, then the quantity $\sum_{|\beta|=N+1} |x_i|^\beta$ for $i = 1, .., I$ should be re-used and the only new calculation that should be performed is $\sum_{|\beta|=N+1} |x_{I+1}|^\beta$. Taking into consideration all the above remarks, when h is changed to include a new point, the most time consuming part of the calculation consists of factorization of $J \times J$ matrix (for example for quadratic interpolation we have $J = 6$), and back-substitution for I vectors (the matrix $E^t W$), this leads to an $O(J^3 + J^2 I)$ complexity for each step of the algorithm. This is multiplied by the number of iterations of the algorithm, which in our implementation is ≤ 100. In the case of using the data dependent bound, there is a preprocess step of solving for the coefficients f_ν as explained in Section 3.4. The computational cost of this step is $O(J^4 + J^3 I)$.

An important consequence of the above procedure is that since V is calculated anyway, we actually get for no significant extra calculations the bound of the error for any α, $|\alpha| \le N$, that is, the bound for every possible derivative (and value) approximation.

It should be noted that to get consistent h values, we take the first global minimum (if there is more than one zero). Another delicate point, is that the parameter value h_{min} of the minimum of $EB_\alpha(x, X, h)$, i.e., $h_{min}(x) = argmin_h\{EB_\alpha(x, X, h)\}$, is a piecewise smooth function of x if the neighborhoods used in the approximation of f_ν are smoothly chosen. This implies that the MLS approximation, based on this h field, is only piecewise smooth.

4.1.1. Integrating with the MLS projection operator

All the previous construction dealt with a function over a parameter space. When dealing with a point cloud there is no natural choice of such space. A popular method for choosing this space in the case of surfaces is the MLS projection operator [Lev03, ABCO*01, PKKG03, AK04]. After choosing the parameter space, in this case a plane, we are back to our original functional settings, with a minor difference: the distance to the neighboring points is measured using their actual position in space and not their projection on the parameter space, i.e., p is defined by minimizing

$$\sum_{i=1}^{I} (f(x_i) - p(x_i))^2 \eta_h(\|(x_i, f(x_i)) - (x, z)\|),$$

where (x, z) is chosen by the first step of the MLS projection. This small change can be easily incorporated in our system, one just have to redefine the way distances are measured. We have integrated that into our system and noticed two interesting results: For the test function F_1, F_2 the results were similar to the algorithm which measured the distance on the parameter space (results are demonstrated in Section 6). However, In the case of data taken from F_3, since the function is rapidly oscillating, the new distance measure is likely to prefer points from other periods and not from close parameter values, and the approximation quality deteriorates.

density	L^x	f	N	ε	$\bar{H}$	$\sigma(H)$
.25k	$f(x)$	F_1	2	0	2.24	0.74
.25k	$f(x)$	F_2	2	0	2.15	0.73
1k	$f(x)$	F_2	2	0	2.1	0.69
1k	$f(x)$	F_1	2	0	2.3	0.75
1k	$f(x)$	F_1	3	0	1.88	0.57
1k	$\partial_x f(x)$	F_1	3	0	2.28	0.74
4k	$\partial_{xy} f(x)$	F_2	3	0	2.06	0.74
.5k	$f(x)$	F_3	2	10^{-2}	2.41	0.77
.5k	$f(x)$	F_3	2	10^{-4}	2.35	0.77
.5k	$\partial_x f(x)$	F_2	2	10^{-5}	2.25	0.8

Table 1: *Experiments results used to derive the heuristic support size rule.*

5. Heuristics

In this section we present the derivation of the heuristic method for choosing a support size h, which we later use for comparison with the optimal choice. The method is derived by experiments, in the following way: We consider the *true* error of our MLS approximation procedure as a function of the support size used h. We let h vary from it's minimal value, i.e., the radius h_J of the ball containing the J nearest neighbors up to 4 times this radius. It is observed that the minimum can be predicted as a certain constant times h_J. The heuristic is based on finding the right constant, and we do it by extensive simulation. We define the random variable

$$H = \frac{h_{best}}{h_J},$$

where h_{best} stands for the true optimal support size, i.e., the support size which minimizes the true error function in the interval $[h_J, 4 \times h_J]$. In table 1, we specify several measurements of H, in particular we consider different test functions (4)-(6), point densities, noisy and non-noisy data, quadratic and cubic polynomial and different functionals. The density specify the number of data points per unit square. H was sampled in a grid inside the domain $[-1,1] \times [-1,1]$. The mean $\bar{H}$ and standard deviation $\sigma(H)$ are computed. In general the results of the different scenarios are similar: the $\bar{H}$ is approximately 2.2 and $\sigma(H)$ is generally around 0.75. Therefore, our heuristic choice of support size to be used in the MLS approximation is $h = 2.2 h_J$.

A similar heuristic for choosing the support size h may be obtained by using the error bound B_α, as follows: For a given distribution of data points near the origin we find the optimal h minimizing B_α, and compute h_{opt}/h_J. Averaging these ratios over many randomly chosen distributions of data points, we obtain for example, an average ratio ~ 1.9 for quadratic polynomial approximation of the function value, $N = 2$, $\alpha = (0,0)$, and ratio ~ 2.4 for $N = 2$, $\alpha(1,0)$ with noise level of 10^{-5}. Another option is to compute a different rule for each noise level ε, but we didn't pursue this direction.

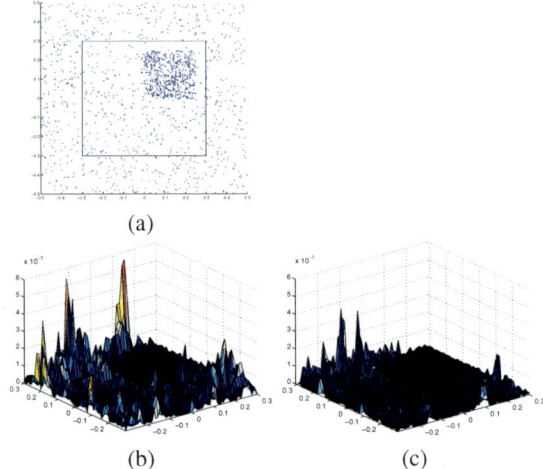

(a)

(b)   (c)

Figure 6: *An example of data set with irregular density. We used quadratic polynomials in the MLS approximation, and resampled in the drawn rectangle (a). In (b) the error resulted using the heuristic support size h. In (c) the optimal h has been used. The error ratio E_1 is 0.37.*

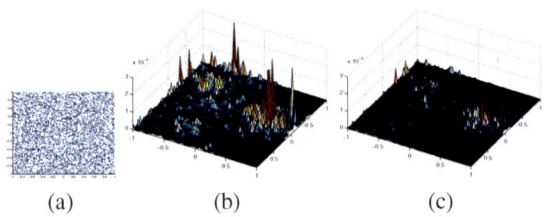

(a)   (b)   (c)

Figure 8: *An example of uniformly distributed data set sampled from F_1. We used cubic MLS approximation, and resampled in the drawn region (a). In (b) we display the error resulting using the heuristic support size. In (c) the error using the optimal support size. Note the errors at the boundaries. The error ratio E_1 is ≈ 0.22.*

6. Numerical experiments

In this section we present numerical experiments performed with the algorithms described in the Section 4. We compare the algorithm for choosing h by minimizing the error bound (16), or the approximation of the error function (14), to the heuristic approach described in Section 5. In general the method based on the data dependent approximation works best, and the method based on the tight bound works slightly better than heuristic method.

We have tested our algorithm in the following three main scenarios: 1. Uniform distributed points. 2. Data with noise. 3. Irregular distributed points (change in density). In our experiments we use quotient of the 1-norms as a measure of error: $E_1 := \|E_{opt}\|_1 / \|E_{heu}\|_1$, and $E_1' := \|E_{bnd}\|_1 / \|E_{heu}\|_1$. We denote by E_{opt} the error of the MLS algorithm using the support size h determined by data dependent error approximation $\tilde{R}$, by E_{bnd} we denote the error of the algorithm when

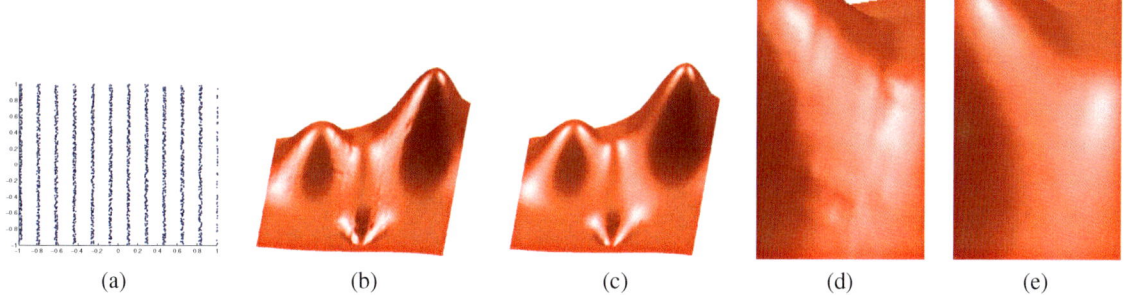

Figure 7: *An example of data set sampled from F_1 with irregular density (a). We used quadratic MLS approximation, and resampled in the drawn region (a). In (b) (zoom-in in (d)) the reconstructed surface using the heuristic support size. In (c) (zoom-in in (e)) optimal h has been used. The resulting error ratio E_1 is ≈ 0.34.*

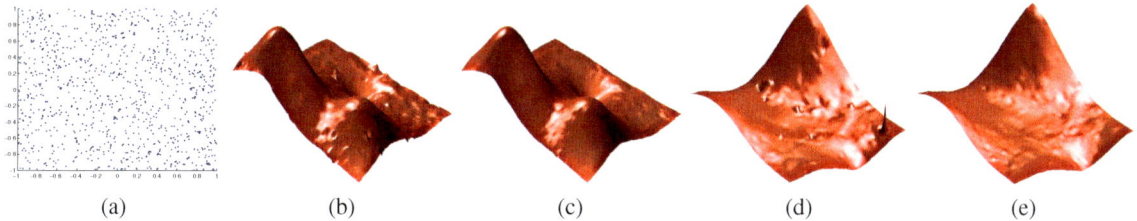

Figure 9: *A very noisy point cloud data (a), for quadratic MLS surface reconstruction. The point cloud was sampled with white noise $\varepsilon = 10^{-1}$ from F_1. In (b) the reconstructed surface (zoom-in in (d)) using the heuristic method, and in (c) (zoom-in in (e)) using the data dependent method.*

using the data-independent bound B_α, and by E_{heu} we denote the error when applying the heuristic approach of choosing the h.

f	density	E_1	E_1'
F_1	0.25k	0.46	0.86
F_1	1k	0.32	0.87
F_1	25k	0.16	0.82
F_2	0.25k	0.53	0.84
F_2	1k	0.38	0.86
F_2	25k	0.19	0.87
F_3	0.25k	0.74	0.82
F_3	1k	0.52	0.89
F_3	25k	0.26	0.89

Table 2: *Experiments with uniform distributed data and no noise.*

Table 2 shows a comparison of the two error analysis based methods to the heuristics, in the case of no noise and uniform distributed points. The approximated functional is point evaluation, i.e., $L^x(f) = f(x)$, and the degree of the polynomial space is $N = 2$. In each experiment a new uniformly distributed data points where taken and the 1000 query points where randomized. The density specify, as before, the number of data points per unit square. Note that method based on the data dependent approximation performs the best, and improves as the density increases. The method based on the data independent bound is working

slightly better than the heuristic. In Figure 8, we show the error graphs when resampling using cubic ($N = 3$) MLS for uniformly distributed data (a), by the heuristic method (b), or by the data dependent method (c). Note that the errors near the boundaries are lower with the latter method.

Table 3 exhibits a similar comparison between the methods for noisy data, approximating various derivative functionals. Figure 9, shows a resampling of F_1 contaminated with high noise level ($\varepsilon = 0.1$) (a), by the heuristic method (b), zoom-in in (d). Similarly (c),(e) exhibits the above cases when using the method based on the data dependent approximation.

Figures 6, 7, exhibit experiments using data set with irregular density. Figure 6 examine the resampling algorithm in region (a) in the vicinity of a high density region. (c) shows the error graph of the heuristic method. (d) is the error graph of the method based on the data dependent approximation. Similarly, Figure 7, shows another configuration of irregular point density (a), which may present itself at scanned point clouds. The reconstructed surfaces are drawn in (b) and (c) and zoom-in at (d),(e).

In Figures 1, 10, we show reconstruction results with scanned data. We compare the heuristic approach to the data-dependent error approximation approach. In Figure 1 we added noise to the data, while in Figure 10 we used the original raw data, and assume error at maximal size $\varepsilon = 10^{-5}$.

f	density	N	L^x	ε	E_1	E_1'
F_1	1k	3	∂_{xy}	10^{-2}	0.32	0.31
F_2	11k	3	∂_x	10^{-4}	0.62	0.71
F_1	1k	3	∂_y	10^{-5}	0.61	2.53
F_2	1k	2	$f(x)$	10^{-1}	0.63	0.6

Table 3: *Experiments with noise.*

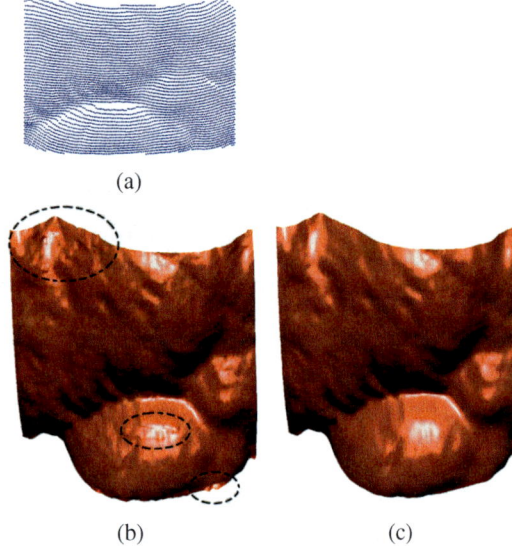

(a)

(b)   (c)

Figure 10: *Resampling part of a raw point cloud of the Bunny taken from the Stanford 3D Scanning Repository (a). The result using a quadratic MLS approximation with the heuristic approach to determine the local support size is shown in (b). The artifacts appearing in (b) are removed by using the approach based on the data dependent error approximation, as shown in (c) (using ε = 10^{-5}).*

7. Conclusions

In this paper we consider the problem of evaluating the approximation quality of the MLS method, and we derive an algorithm which finds the best support size to be used in the approximation. Two methods based on a novel error formula in the MLS approximation were considered: One, based on a conservative tight bound, and second, based on a data dependent approximation to the error function in the MLS approximation. In the process, we have carefully chosen a heuristic, based on the observation that the ratio of the optimal support size h_{best} and the radius of the ball containing the J nearest neighbors h_J, can be fairly well predicted by the constant ~ 2.2.

Comparing our error analysis based methods to the heuristic shows that the method based on the tight bound performs slightly better than the heuristic in presence of very small and very large noise levels. The method based on the data dependent approximation works generally better than the other

methods, and achieves the best approximation and stability properties.

References

[ABCO*01] ALEXA M., BEHR J., COHEN-OR D., FLEISHMAN S., LEVIN D., SILVA C. T.: Point set surfaces. In *VIS '01: Proceedings of the conference on Visualization '01* (Washington, DC, USA, 2001), IEEE Computer Society, pp. 21–28.

[AK04] AMENTA N., KIL Y. J.: The domain of a point set surfaces. *Eurographics Symposium on Point-based Graphics 1*, 1 (June 2004), 139–147.

[Bos91] BOS L.: On certain configurations of points in $\mathbb{R}^n$ which are unisolvent for polynomial interpolation. *Journal of Approximation Theory 64*, 3 (1991), 271–280.

[CP05] CAZALS F., POUGET M.: Estimating differential quantities using polynomial fitting of osculating jets. *Computer Aided Geometric Design 22*, 2 (2005), 121–146.

[FR01] FLOATER M. S., REIMERS M.: Meshless parameterization and surface reconstruction. *Comput. Aided Geom. Des. 18*, 2 (2001), 77–92.

[Fra82] FRANKE R.: Scattered data interpolation: tests of some methods. *Math Comp 38* (1982), 181–200.

[GS00] GASCA M., SAUER T.: Polynomial interpolation in several variables. *Adv. Comput. Math. 12*, 4 (2000), 377–410.

[HDD*92] HOPPE H., DEROSE T., DUCHAMP T., MCDONALD J., STUETZLE W.: Surface reconstruction from unorganized points. *Computer Graphics 26*, 2 (1992), 71–78.

[Lev98] LEVIN D.: The approximation power of moving least-squares. *Mathematics of Computation 67*, 224 (1998), 1517–1531.

[Lev03] LEVIN D.: Mesh-independent surface interpolation. *Geometric Modeling for Scientific Visualization* (2003).

[OBA*03] OHTAKE Y., BELYAEV A., ALEXA M., TURK G., SEIDEL H.-P.: Multi-level partition of unity implicits. *ACM Trans. Graph. 22*, 3 (2003), 463–470.

[PGK02] PAULY M., GROSS M., KOBBELT L. P.: Efficient simplification of point-sampled surfaces. In *VIS '02: Proceedings of the conference on Visualization '02* (Washington, DC, USA, 2002), IEEE Computer Society, pp. 163–170.

[PKKG03] PAULY M., KEISER R., KOBBELT L. P., GROSS M.: Shape modeling with point-sampled geometry. *ACM Trans. Graph. 22*, 3 (2003), 641–650.

[SX95] SAUER T., XU Y.: On multivariate lagrange interpolation. *Math. Comput. 64*, 211 (1995), 1147–1170.

[Wen01] WENDLAND H.: Local polynomial reproduction and moving least squares approximation. *IMA Journal of Numerical Analysis 21* (2001), 285–300.

Eurographics Symposium on Geometry Processing (2006)
Konrad Polthier, Alla Sheffer (Editors)

Spherical Barycentric Coordinates

Torsten Langer, Alexander Belyaev, Hans-Peter Seidel

MPI Informatik

Abstract

We develop spherical barycentric coordinates. Analogous to classical, planar barycentric coordinates that describe the positions of points in a plane with respect to the vertices of a given planar polygon, spherical barycentric coordinates describe the positions of points on a sphere with respect to the vertices of a given spherical polygon. In particular, we introduce spherical mean value coordinates that inherit many good properties of their planar counterparts. Furthermore, we present a construction that gives a simple and intuitive geometric interpretation for classical barycentric coordinates, like Wachspress coordinates, mean value coordinates, and discrete harmonic coordinates.

One of the most interesting consequences is the possibility to construct mean value coordinates for arbitrary polygonal meshes. So far, this was only possible for triangular meshes. Furthermore, spherical barycentric coordinates can be used for all applications where only planar barycentric coordinates were available up to now. They include Bézier surfaces, parameterization, free-form deformations, and interpolation of rotations.

Categories and Subject Descriptors (according to ACM CCS): I.3.5 [Computer Graphics]: Computational Geometry and Object Modeling

1. Introduction

Barycentric coordinates are a special kind of local coordinates that express the location of a point with respect to a given triangle. They were developed by Möbius [Möb27] in the nineteenth century. Wachspress extended the notion of barycentric coordinates to arbitrary convex polygons [Wac75]; another approach is due to Sibson [Sib80]. In recent years, the research on barycentric coordinates has been intensified and led to a general theory of barycentric coordinates and extensions to higher dimensions [FHK06, FKR05, JSWD05, JSW05, JW05]. Barycentric coordinates are natural coordinates for meshes and have many applications including parameterization [DMA02, SAPH04], free-form deformations [SP86, JSW05], finite elements [AO06, SM06], shading [Gou71, Pho75], and elementary geometry.

Our main contribution is the extension of the notion of barycentric coordinates in several directions as follows:

- We introduce spherical barycentric coordinates. They are analogues of the classical barycentric coordinates, but they are defined for polygons on a sphere instead of a plane.

- We construct three-dimensional barycentric coordinates for arbitrary, closed polygonal meshes.
- We show that the vector coordinates in [JSWD05] and the 3D mean value coordinates for triangular meshes in [FKR05, JSW05] are special cases of our constructions.

The extension to polygonal meshes is an important generalization as noted in [JSW05]. In particular, it makes it possible to use barycentric coordinates in conjunction with subdivision surfaces and the recently introduced conical meshes [LPW*06].

Let $P = (\mathbf{v}_j)_{j=1..n}$ be a polygon with vertices $\mathbf{v}_j$. *Barycentric coordinates* $\lambda_i(\mathbf{v}; P) = \lambda_i\left(\mathbf{v}; (\mathbf{v}_j)_{j=1..n}\right)$ of a point $\mathbf{v}$ are continuous functions that satisfy the following properties (2) and (3) for all points $\mathbf{v}$ inside the polygon. If property (1) is additionally fulfilled for all convex polygons, we call them *positive barycentric coordinates.*

$$\forall i \; \lambda_i(\mathbf{v}; P) > 0 \qquad \text{positivity,} \qquad (1)$$

$$\sum_i \lambda_i(\mathbf{v}; P) = 1 \qquad \text{partition of unity,} \qquad (2)$$

$$\sum_i \lambda_i(\mathbf{v}; P)\mathbf{v}_i = \mathbf{v} \qquad \text{linear precision.} \qquad (3)$$

If P is a triangle, its barycentric coordinates are uniquely determined by these conditions. In general, it is only possible to fulfill all three properties for convex polygons. We use also λ_i or $\lambda_i(\mathbf{v})$ if P and the point $\mathbf{v}$ are clear from the context. If a partition of unity (2) is not given, we speak of *homogeneous coordinates*. They can be normalized to satisfy property (2) if and only if $\sum_i \lambda_i(\mathbf{v}) \neq 0$. Property (3) is called "linear precision" since it ensures the correct interpolation of all linear functions f:

$$\sum_i \lambda_i(\mathbf{v}) f(\mathbf{v}_i) = f(\mathbf{v}). \qquad (4)$$

The construction of barycentric coordinates for polygons with more than three vertices had been an open problem for a long time. The coordinates introduced by Wachspress [Wac75] are defined for arbitrary convex polygons $P \subset \mathbb{R}^2$ and satisfy properties (1)–(3), but they are in general only defined inside of convex polygons. This restriction was overcome with the introduction of the mean value coordinates by Floater [Flo03]. They are defined in the whole plane for convex and non-convex polygons and can even be extended to multiple polygons [HF06]. In [FHK06], an overview over all similarity invariant, homogeneous coordinates is given. In particular, it is shown that Wachspress coordinates and mean value coordinates are the only similarity invariant, positive barycentric three-point coordinates. Recently, Wachspress coordinates and mean value coordinates were generalized to polytopes of higher dimensions. However, this was only possible for convex polytopes (3D Wachspress coordinates) [War96, JSWD05] and for polytopes with simplicial boundary (3D mean value coordinates) [FKR05, JSW05].

Spherical barycentric coordinates constitute another variant of barycentric coordinates. They have been studied first by Möbius [Möb46] and were introduced to computer graphics by Alfeld et al. [ANS96]. Spherical barycentric coordinates express the location of a point $\mathbf{v}$ on a sphere with respect to the vertices $\mathbf{v}_i$ of a given spherical triangle P. Since partition of unity (2) and linear precision (3) contradict each other on spheres, Alfeld et al. chose to relax the former condition to

$$\sum_i \lambda_i \geq 1. \qquad (2')$$

Ju et al. [JSWD05] extended spherical barycentric coordinates (they called them "vector coordinates") from spherical triangles to arbitrary convex, spherical polygons. However, non-convex spherical polygons still posed a problem.

In this paper, we show how arbitrary barycentric coordinates from the family of planar barycentric coordinates [FHK06] can be defined on a sphere (Section 2.1). In particular, we introduce spherical mean value coordinates that inherit many good properties of their planar counterparts and are defined for arbitrary spherical polygons (Section 2.2). We show that the vector coordinates by Ju et al. [JSWD05] coincide with spherical Wachspress coordinates from our family (Section 2.3). Furthermore, we show how 3D barycentric coordinates (in particular, mean value coordinates) for arbitrary polygonal meshes can be defined

with the help of our spherical mean value coordinates (Section 3). Finally, we discuss several applications including space deformations and Bézier surfaces (Section 4). In addition, we provide a novel, geometric interpretation of planar barycentric coordinates (Appendix A).

2. Spherical barycentric coordinates

In this section, we deal with the problem of finding (positive) coefficients λ_i for vectors $\mathbf{v}_i \in \mathbb{R}^3$ such that their linear combination is $\mathbf{v} \in \mathbb{R}^3$. If we restrict ourselves to vectors of unit length, the λ_i represent barycentric coordinates for $\mathbf{v}$ with respect to the $\mathbf{v}_i$ on the unit sphere. First, we give a general introduction. Then, we develop spherical mean value coordinates (Equation (8)) and spherical Wachspress coordinates (Section 2.3).

While Equations (1)–(3) are well-chosen to characterize planar barycentric coordinates, it is obviously not possible to fulfill all three conditions if the vertices $\mathbf{v}_i$ and the point $\mathbf{v}$ are located on a sphere instead of a plane. This is especially easy to see if P is a triangle with three vertices. In particular, Equations (2) and (3) contradict each other since the former condition requires all points described by $\sum_i \lambda_i \mathbf{v}_i$ to lie in the triangle plane while the latter condition demands that this sum yields a point $\mathbf{v}$ that lies not in this plane but on the sphere. A similar observation was made in [BW92].

Consequently, we have to relax the above conditions. We follow the suggestion in [ANS96] and replace Equation (2) by (2'). Of course, by dividing by $\sum_j \lambda_j$, we could also obtain coordinates that constitute a partition of unity (2) instead of satisfying the linear precision property (3) if desired. However, for our applications, in particular for constructing barycentric coordinates for meshes, linear precision is more important. Note that this property still implies that linear functions *defined on* $\mathbb{R}^3$ are correctly interpolated (Equation (4)). Constant functions, however, cannot be correctly interpolated if a partition of unity (2) is not given. A different approach that preserves the partition of unity was proposed in [BF01]. We call a set of coordinates λ_i satisfying conditions (2') and (3) *spherical barycentric coordinates*. If P is a triangle, then there exists obviously a unique solution: the unique linear combination of the vectors $\mathbf{v}_1$, $\mathbf{v}_2$, and $\mathbf{v}_3$ such that $\sum_{i=1}^3 \lambda_i \mathbf{v}_i = \mathbf{v}$. A geometric interpretation of these spherical barycentric coordinates was given in [ANS96].

2.1. Definition of spherical barycentric coordinates

In this section, we show how barycentric coordinates can be defined for arbitrary polygons on a sphere.

Definition 2.1 A *spherical polygon* P consists of a set of distinct vertices $\mathbf{v}_i$ located on a sphere and a set of edges $(\mathbf{v}_i, \mathbf{v}_{i+1})$ that connect the vertices $\mathbf{v}_i$ and $\mathbf{v}_{i+1}$ by geodesic lines (these are the arcs of great circles on the sphere).

We consider a spherical polygon on the unit sphere centered at the origin. Let $\mathbf{v}$ be a point on the sphere. Let $\widehat{\mathbf{v}_i}$ be

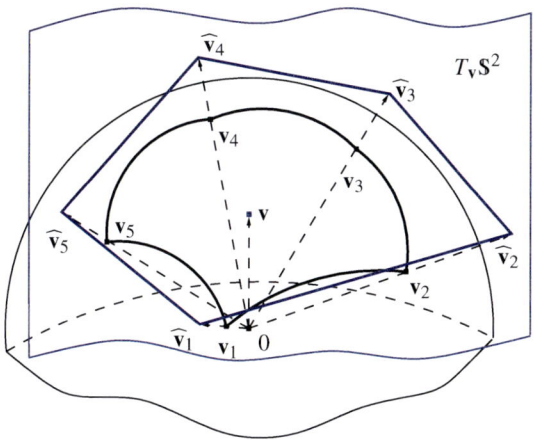

Figure 1: *Construction of spherical barycentric coordinates.*

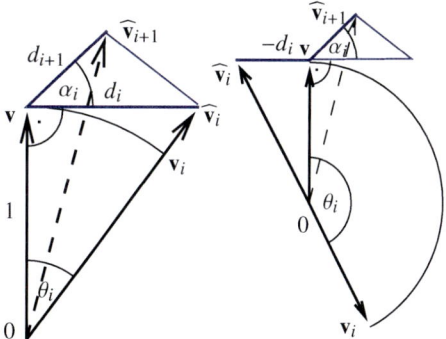

Figure 2: *Notation for spherical barycentric coordinates.*

the intersection points of the line given by $\mathbf{v}_i$ and the tangent plane $T_{\mathbf{v}}\mathbf{S}^2$ at $\mathbf{v}$ to the sphere (see Figures 1 and 2). The points $\widehat{\mathbf{v}}_i$ determine a polygon $\widehat{P}$ (shown in blue) in the plane $T_{\mathbf{v}}\mathbf{S}^2$. (The map $\mathbf{v}_i \mapsto \widehat{\mathbf{v}}_i$ is a gnomonic projection. It is especially useful for our purpose since it projects geodesics to straight lines.) Now, we can compute the planar barycentric coordinates $\widehat{\lambda}_i$ of $\mathbf{v}$ with respect to $\widehat{P}$. The 3D position of $\mathbf{v}$ is an affine linear function on $T_{\mathbf{v}}\mathbf{S}^2$. Consequently, any set $\widehat{\lambda}_i$ of planar barycentric coordinates yields, by Equations (2) and (3),

$$\sum_i \widehat{\lambda}_i \widehat{\mathbf{v}}_i = \mathbf{v}.$$

To obtain spherical barycentric coordinates λ_i of $\mathbf{v}$ that satisfy the linear precision property (3), we define them by

$$\sum_i \lambda_i \mathbf{v}_i = \mathbf{v}, \qquad \lambda_i := \langle \mathbf{v}_i, \widehat{\mathbf{v}}_i \rangle \widehat{\lambda}_i \qquad (5)$$

where $\langle \cdot, \cdot \rangle$ denotes the usual scalar product in $\mathbb{R}^3$. Note that $\langle \mathbf{v}_i, \widehat{\mathbf{v}}_i \rangle$ is just $\pm\|\widehat{\mathbf{v}}_i\|$. Although this value becomes very large and finally undefined if the angle θ_i between $\mathbf{v}$ and $\mathbf{v}_i$ approaches $\frac{\pi}{2}$, this is usually compensated by a shrinkage of $\widehat{\lambda}_i$ such that the definition of λ_i can be extended continuously to the case $\theta_i = \frac{\pi}{2}$. We demonstrate this in Sections 2.2 and 2.3 for spherical mean value and spherical Wachspress coordinates. Note that basically the same construction can be used to obtain planar barycentric coordinates from spherical barycentric coordinates. It follows that there is a bijection between planar barycentric coordinates and spherical barycentric coordinates.

Finally, we remark that these coordinates can also be extended to vectors $\mathbf{v}_i$ and $\mathbf{v}$ of arbitrary length by defining

$$\lambda_i\left(\mathbf{v};(\mathbf{v}_j)_{j=1..n}\right) := \frac{\|\mathbf{v}\|}{\|\mathbf{v}_i\|} \cdot \lambda_i\left(\frac{\mathbf{v}}{\|\mathbf{v}\|};\left(\frac{\mathbf{v}_j}{\|\mathbf{v}_j\|}\right)_{j=1..n}\right). \qquad (6)$$

2.2. Spherical mean value coordinates

We now develop spherical mean value coordinates. They inherit positivity from their planar counterparts: the $\lambda_i(\mathbf{v})$ are positive if $\mathbf{v}$ is contained in the kernel of the polygon given by the $\mathbf{v}_i$. If $\mathbf{v}$ is contained in the convex hull of the $\mathbf{v}_i$, this can always be arranged by reordering the vertices $\mathbf{v}_i$ with respect to their polar angle around $\mathbf{v}$. (The kernel is the region inside the polygon from which the whole polygon is "visible".)

Planar mean value coordinates are given by Floater's formula [Flo03]

$$\widehat{\lambda}_i = \frac{w_i}{\sum_j w_j}, \qquad w_i = \frac{\tan\frac{\alpha_{i-1}}{2} + \tan\frac{\alpha_i}{2}}{d_i} \qquad (7)$$

where α_i is the signed angle between $\widehat{\mathbf{v}}_i - \mathbf{v}$ and $\widehat{\mathbf{v}}_{i+1} - \mathbf{v}$ and d_i is the distance $\|\widehat{\mathbf{v}}_i - \mathbf{v}\|$. As shown in Figure 2, α_i is given by the dihedral, signed angle between the planes determined by $\mathbf{v}$, $\mathbf{v}_i$, and the origin, and $\mathbf{v}$, $\mathbf{v}_{i+1}$, and the origin, respectively. That is, α_i is the signed angle between $\mathbf{v} \times \mathbf{v}_i$ and $\mathbf{v} \times \mathbf{v}_{i+1}$. The distance d_i is given by $d_i = \tan\theta_i$ where θ_i is the angle between $\mathbf{v}$ and $\mathbf{v}_i$. By inserting these terms into Equations (7) and (5) and using $\langle \mathbf{v}_i, \widehat{\mathbf{v}}_i \rangle = \frac{1}{\cos\theta_i}$, we obtain

$$\lambda_i(\mathbf{v}) = \frac{\tan\frac{\alpha_{i-1}}{2} + \tan\frac{\alpha_i}{2}}{\sin\theta_i} \Bigg/ \sum_j \cot\theta_j \left(\tan\frac{\alpha_{j-1}}{2} + \tan\frac{\alpha_j}{2}\right). \qquad (8)$$

This formula gives us a continuous definition of λ_i for arbitrary $\mathbf{v}$. But for a $\mathbf{v}$ with $\theta_i > \frac{\pi}{2}$, it implies that in the projected polygon $\widehat{P}$, the vector $\widehat{\mathbf{v}}_i - \mathbf{v}$ has negative length (and is still oriented like $\mathbf{v} - \widehat{\mathbf{v}}_i$; this is the reason that α_i is measured as the angle between $-(\widehat{\mathbf{v}}_i - \mathbf{v})$ and $\widehat{\mathbf{v}}_{i+1} - \mathbf{v}$ instead of $\widehat{\mathbf{v}}_i - \mathbf{v}$ and $\widehat{\mathbf{v}}_{i+1} - \mathbf{v}$), as indicated on the right of Figure 2. It can be seen that the planar mean value coordinates still satisfy linear precision (3) for this case (for example, this is implied by the construction in Appendix A). Therefore, the derived spherical barycentric coordinates fulfill linear precision as well.

From Equation (8), it is easy to see that the spherical mean value coordinates are well-defined and positive if $\mathbf{v}$ is inside

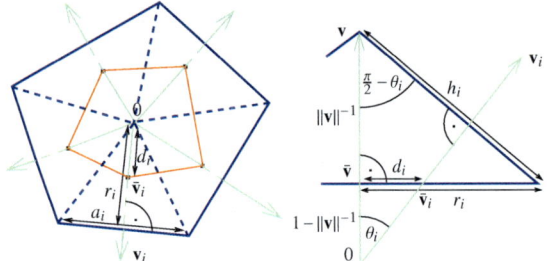

Figure 3: *Bottom and side view of the polar dual.*

a convex spherical polygon and $\theta_i < \frac{\pi}{2}$ for all θ_i. For such polygons, inequality (2′) is implied by the triangle inequality.

2.3. Relation to Ju et al.'s vector coordinates

In this section, we show that the coordinates that were introduced by Ju et al. [JSWD05] to express a vector $\mathbf{v}$ as a linear combination of other vectors $\mathbf{v}_i$ coincide with our spherical Wachspress coordinates. It is sufficient to consider only vectors $\mathbf{v}_i$ and $\mathbf{v}$ of unit length and to show that the coordinates coincide up to a constant factor c. Then, it follows from Equations (6) and (3), which hold for both sets of coordinates, that they coincide for arbitrary vectors and that $c = 1$.

By Equation (5) and Figure 2, spherical Wachspress coordinates are given by

$$\lambda_i = \frac{1}{\cos\theta_i}\widehat{\lambda}_i \qquad (9)$$

where $\widehat{\lambda}_i$ are the planar Wachspress coordinates for the planar polygon with vertices $\widehat{\mathbf{v}}_i$ constructed in Section 2.1. We now show that the same formula holds for the vector coordinates constructed in [JSWD05].

Ju et al.'s vector coordinates are defined for vectors $\mathbf{v}_i$ that are the vertices of a convex spherical polygon. The coordinates are proportional to the area β_i of the triangular faces of the polar dual. The polar dual is the convex polyhedron that is bounded by the planes that have $\mathbf{v}_i$ as normal and pass through the point $\mathbf{v}$ and by the plane perpendicular to $\mathbf{v}$ with distance $\frac{1}{\|\mathbf{v}\|}$ to the point $\mathbf{v}$ (see Figure 3; the polar dual is shown in blue). Let $\bar{\mathbf{v}}$ be the intersection point of $\mathbf{v}$ and the latter plane. Let $\bar{\mathbf{v}}_i$ be the intersection points of the same plane and the rays (green) determined by the $\mathbf{v}_i$. Let $\bar{P}$ be the polygon (shown in red) formed by the $\bar{\mathbf{v}}_i$. Then $\bar{\mathbf{v}}$ and $\bar{P}$ are, by construction, similar to $\mathbf{v}$ and $\widehat{P}$, the polygon defined in Section 2.1. Therefore, the respective Wachspress coordinates coincide: $\widehat{\lambda}_i(\bar{\mathbf{v}};\bar{P}) = \widehat{\lambda}_i(\mathbf{v};\widehat{P})$. Note that the boundary polygon Q (solid blue in Figure 3, left) of the bottom face of the polar dual is dual to $\bar{P}$ (red) with respect to $\bar{\mathbf{v}}$. That is, its edges $\mathbf{a}_i$ are orthogonal to $\bar{\mathbf{v}}_i - \bar{\mathbf{v}}$ (see Appendix A for dual polygons).

The triangle areas β_i can be computed (up to a factor $\frac{1}{2}$) as the product of the length a_i of the edge $\mathbf{a}_i$ and the respective height h_i. The latter can be computed as $h_i = \frac{\|v\|^{-1}}{\cos(\frac{\pi}{2}-\theta_i)} = \frac{\|v\|^{-1}}{\sin\theta_i}$. In the bottom face, the distance r_i of the edge $\mathbf{a}_i$ to the center $\bar{\mathbf{v}}$ is given by $\|v\|^{-1}\tan(\frac{\pi}{2}-\theta_i) = \|v\|^{-1}\cot\theta_i$ and the distance of the intersection points $\bar{\mathbf{v}}_i$ to $\bar{\mathbf{v}}$ is given by $d_i = (1 - \|v\|^{-1})\tan\theta_i$. The product $r_id_i = \frac{\|v\|-1}{\|v\|^2}$ is independent of i. Therefore, Q is (up to scaling) that dual polygon of $\bar{P}$ such that the distance r_i of the edges $\mathbf{a}_i$ to the center $\bar{\mathbf{v}}$ is inverse to the distance d_i between $\bar{\mathbf{v}}_i$ and $\bar{\mathbf{v}}$ (up to a constant factor). The edges of such a polygon have lengths a_i proportional to $d_i\widehat{\lambda}_i$ (see Appendix A). If we put everything together, we obtain the following formula for the vector coordinates

$$\beta_i = \frac{1}{2}a_ih_i = c_1 d_i h_i \widehat{\lambda}_i = c\frac{\widehat{\lambda}_i}{\cos\theta_i}$$

with some constants c and c_1. A comparison with Equation (9) concludes our proof.

3. Barycentric coordinates for closed meshes with polygonal faces

First, we present the approach to compute barycentric coordinates for triangular meshes that was introduced in [JW05]. Then, we show how barycentric coordinates for arbitrary closed polygonal meshes can be obtained by using the spherical barycentric coordinates proposed in Section 2.1. 3D mean value coordinates are computed by Equation (10) if $\mathbf{v}_F$ is calculated by Equation (11) and λ_i by Equation (12).

Let $\mathbf{x} \in \mathbb{R}^3$ be a point. Its barycentric coordinates with respect to a mesh with vertices $\mathbf{v}_i$ consist of coordinate functions $\lambda_i^{3D}(\mathbf{x})$ that fulfill Equations (2) and (3). To define them, we need the following notation: $F(\mathbf{v}_i)$ denotes the set of faces incident to $\mathbf{v}_i$. For a face F, let $V(F)$ denote the set of indices i such that $\mathbf{v}_i$ is incident to F and let $P_F := (\mathbf{v}_i - \mathbf{x})_{i\in V(F)}$ be the boundary polygon of F, relative to $\mathbf{x}$. Using Equation (2), Equation (3) is equivalent to

$$\sum_i \lambda_i^{3D} \cdot (\mathbf{v}_i - \mathbf{x}) = \mathbf{0}. \qquad (3')$$

Thus, it is sufficient to find a solution λ_i^{3D} for Equation (3′). This is done in two steps [JW05]:

- A face vector $\mathbf{v}_F$ is assigned to each face F of the mesh such that $\sum_F \mathbf{v}_F = \mathbf{0}$. (Think of $\mathbf{v}_F$ as some kind of face normal.)
- The face vectors are distributed to their respective face vertices by finding λ_i such that $\sum_{i\in V(F)} \lambda_i \cdot (\mathbf{v}_i - \mathbf{x}) = \mathbf{v}_F$.

While this procedure was proposed only for triangular meshes in [JW05] (in this case, the latter step has a unique solution), we can extend it to arbitrary meshes by employing spherical barycentric coordinates λ_i for the distribution of the face vectors $\mathbf{v}_F$. Using them, we can assign barycentric coordinates to each vertex $\mathbf{v}_i$:

$$\lambda_i^{3D}(\mathbf{x}) := \frac{w_i(\mathbf{x})}{\sum_j w_j(\mathbf{x})}, \quad w_i(\mathbf{x}) := \sum_{F \in F(\mathbf{v}_i)} \lambda_i(\mathbf{v}_F; P_F). \quad (10)$$

Note that P_F is in general not a spherical polygon. Therefore, the λ_i refer to the generalized spherical barycentric coordinates (6). It is clear from the construction that the w_i satisfy Equation (3′). It follows that the λ_i^{3D} satisfy (2) and (3).

We can now vary our 3D barycentric coordinates by the choice of $\mathbf{v}_F$ and the choice of the spherical barycentric coordinates λ_i. In [JW05] is described (for triangular meshes) how $\mathbf{v}_F$ has to be chosen to obtain Wachspress coordinates, mean value coordinates, discrete harmonic coordinates, or any other barycentric coordinates. In the next section, we describe the necessary choices to obtain 3D mean value coordinates for polygonal meshes.

3.1. Mean value coordinates for closed meshes

Mean value coordinates are the most flexible since they can be computed for non-convex meshes with non-convex (planar) faces. The following construction of the associated face vector is due to [FKR05, JSW05]. Let $\mathbf{x}$ be a point inside the body bounded by the mesh. For a face F, let P_F be the boundary polygon with respect to $\mathbf{x}$ as above, and let Q_F be the spherical polygon that is obtained by projecting the vertices of P_F to the unit sphere centered at $\mathbf{x}$. We know from Stokes' theorem that the integral over the unit normals $\mathbf{v} - \mathbf{x}$ of this sphere is zero:

$$\int_{\mathbf{v}-\mathbf{x} \in \mathbb{S}^2} (\mathbf{v}-\mathbf{x})dS = \mathbf{0}.$$

The Q_F induce a polygonal tessellation of this sphere and we define

$$\mathbf{v}_F := \int_{\mathbf{v}-\mathbf{x} \in Q_F} (\mathbf{v}-\mathbf{x})dS, \quad \sum_F \mathbf{v}_F = \mathbf{0}.$$

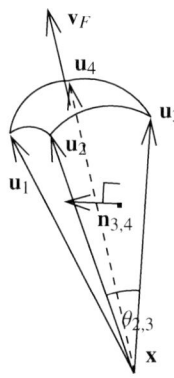

Let the vertices of Q_F be $\mathbf{u}_1, \ldots \mathbf{u}_n$. Another application of Stokes' theorem yields the formula

$$\mathbf{v}_F = \sum_{i=1}^n \frac{1}{2} \theta_{i,i+1} \mathbf{n}_{i,i+1} \quad (11)$$

where $\theta_{i,i+1}$ is the angle between $\mathbf{u}_i$ and $\mathbf{u}_{i+1}$ and $\mathbf{n}_{i,i+1} := \frac{\mathbf{u}_i \times \mathbf{u}_{i+1}}{\|\mathbf{u}_i \times \mathbf{u}_{i+1}\|}$ is the oriented unit normal to the plane determined by these vectors (Figure 4). To distribute the face vector to the incident vertices, we choose spherical mean value coordinates. This choice of the λ_i leads to 3D mean value coordinates whose restriction to the faces of the mesh yields the planar mean value coordinates of the respective faces (Section 3.2). For each face F, the $\lambda_i(\mathbf{v}_F; P_F)$ can

Figure 4: *The face normal $\mathbf{v}_F$.*

easily be computed with Equations (8) and (6). We obtain

$$\lambda_i = \frac{\|\mathbf{v}_F\|}{\|\mathbf{v}_i - \mathbf{x}\|} \cdot \frac{\tan\frac{\alpha_{i-1}}{2} + \tan\frac{\alpha_i}{2}}{\sin\theta_i} \bigg/ \sum_j \cot\theta_j \left(\tan\frac{\alpha_{j-1}}{2} + \tan\frac{\alpha_j}{2}\right). \quad (12)$$

Since the vertices $\mathbf{v}_i$, $i \in V(F)$ of P_F are the boundary vertices of the planar face F, it follows from the construction that the spherical mean value coordinates λ_i are well-defined and positive for convex faces. Consequently, our 3D mean value coordinates are well-defined and positive for points $\mathbf{x}$ in the interior of convex polyhedra. By construction, our mean value coordinates coincide with the mean value coordinates from [FKR05, JSW05] if they are computed for meshes with triangular faces.

3.2. Behavior of the mean value coordinates on the faces

The denominator of Equation (12) becomes zero if $\mathbf{x}$ is contained in a face F of the mesh. In this case, the face vector $\mathbf{v}_F$ is orthogonal to F and Q_F lies on a great circle since its vertices $\frac{\mathbf{v}_i - \mathbf{x}}{\|\mathbf{v}_i - \mathbf{x}\|}$ lie in the plane determined by F. We show now that the 3D mean value coordinates have nevertheless a continuous extension to the faces and that this extension coincides with the 2D mean value coordinates. Assume that $\mathbf{x}$ approaches a point located on the face F. Then $\mathbf{v}_F$ approaches the face normal. It follows that the denominator of the $\lambda_i(\mathbf{v}_F; P_F)$ defined in Equation (12) approaches infinity. Therefore, in the limit (due to the normalization)

$$\lambda_i^{3D}(\mathbf{x}) = \begin{cases} \dfrac{w_i}{\sum_{j \in V(F)} w_j}, \ w_i = \lambda_i(\mathbf{v}_F), & i \in V(F) \\ 0, & \text{otherwise.} \end{cases}$$

This approaches the usual 2D mean value coordinates.

4. Applications

4.1. Interpolation and extrapolation

The most direct application of mean value coordinates is their use for interpolation and extrapolation using Equation (4). In Figure 5, color values are specified on the eight vertices of the cube. In the top row, the values are interpolated on the faces. In the bottom row, the color values are interpolated and extrapolated on a plane that passes through the cube. In the left column, the cube was triangulated before the interpolation. The piecewise linear structure of the interpolation on the triangles is clearly visible. With our 3D mean value coordinates, a triangulation is no longer necessary, and the resulting interpolation is much smoother.

4.2. Space deformations with 3D mean value coordinates

Figure 6 shows an example how mean value coordinates can be used for space deformations. We determine the mean value coordinates of the vertices of the tube with respect to the black control mesh with vertices $\mathbf{v}_i$. Then we deform the

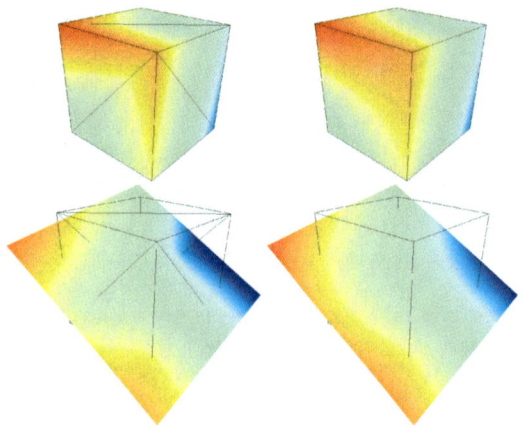

Figure 5: *An example of interpolation of color values using 3D mean value coordinates. The color values are specified at the vertices of the cube. They are interpolated on the faces (top) and on a plane passing through the cube (bottom). If the cube is triangulated beforehand (left), the interpolation is less smooth than with our method (right).*

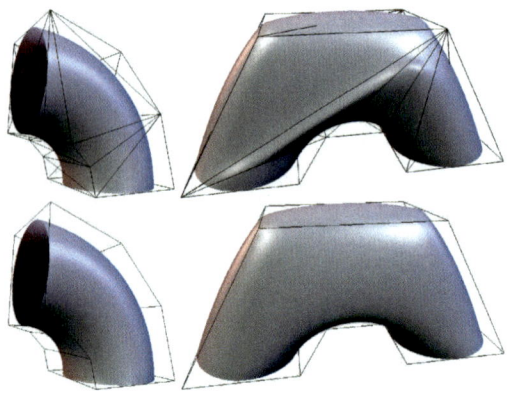

Figure 6: *An example of a space deformation using 3D mean value coordinates with respect to the polygonal control mesh. If the control mesh is triangulated beforehand, strong artifacts may be introduced (top). No triangulation is necessary with our method (bottom).*

control mesh by moving the vertices to points $\mathbf{w}_i$ and calculate the new location of the tube by $\mathbf{x} = \sum_i \lambda_i^{3D}\left(\mathbf{x};(\mathbf{v}_j)_j\right)\mathbf{w}_i$. Note that we can compute these coordinates for non-convex (control) meshes with non-convex faces (bottom row). If all faces are triangulated, the number of faces is nearly tripled, the result depends on the chosen triangulation, and large artifacts may be introduced (top row).

Although this approach is very simple, it is possible to obtain pleasing results. A more sophisticated framework could use Bernstein polynomials on polyhedra (Section 4.3) to generate free-form deformations [SP86].

Figure 7: *The quadratic mean value Bernstein polynomials B_{2000}^2, B_{1100}^2, and $B_{1010}^2 + B_{0101}^2$.*

4.3. Bernstein polynomials on polygons and polyhedra

Bézier surfaces are defined by a linear combination of Bernstein polynomials which are polynomials in barycentric coordinates. Using classical barycentric coordinates, this was only possible for triangles. Using tensor product polynomials, Bernstein polynomials can be defined on quadrangular domains as well, but this leads to a higher degree of the polynomial. The only approaches for general polygons that we are aware of are restricted to convex polygons [LD89, Gol02].

We can define mean value Bernstein polynomials for arbitrary polygons and polyhedra. For a polygon or polyhedron with k vertices, the general form for the Bernstein polynomials in the coordinates $\lambda = (\lambda_1, \ldots \lambda_k)$ is

$$B_\alpha^n(\mathbf{x}) = \frac{n!}{\alpha!}\lambda^\alpha(\mathbf{x})$$

where we use multi-indices $\alpha = (\alpha_1, \ldots \alpha_k) \in \mathbb{N}^k$ with the notation $\alpha! := \alpha_1! \cdots \alpha_k!$ and $\lambda^\alpha := \lambda_1^{\alpha_1} \cdots \lambda_k^{\alpha_k}$. In Figure 7, we show some quadratic Bernstein polynomials on a square using mean value coordinates [Flo03].

Important properties of classical Bézier surfaces like the convex hull property and the de Casteljau algorithm still hold in this extended setup.

4.4. Bézier surfaces on spherical polygonal domains

In the above applications, spherical barycentric coordinates were only indirectly used to construct 3D barycentric coordinates. But they can also be used directly to construct Bézier surfaces on spherical domains. For a triangulation of the sphere, this has been done in [ANS96]. Our extension to arbitrary spherical polygons makes it possible to handle arbitrary tessellations of a sphere. Again, the de Casteljau algorithm can be used for computations of the spherical Bézier patches.

5. Conclusions

We have introduced spherical barycentric coordinates that generalize Ju et al.'s vector coordinates. Our spherical mean value coordinates are defined not only for convex spherical polygons but for arbitrary polygons that are contained in a single hemisphere. We have shown that spherical mean value coordinates can be used to construct 3D mean value coordinates for meshes with arbitrary polygonal faces while so far only 3D mean value coordinates for triangular meshes were known. This concludes the generalization of mean value coordinates from two to three dimensions.

The examples in Section 4 demonstrate that these 3D mean value coordinates are well-defined for arbitrary polyhedra. This is proven for convex polyhedra in this paper. We intend to give a proof for the general case in the near future.

With barycentric coordinates for arbitrary polyhedra in $\mathbb{R}^3$, we can construct spherical barycentric coordinates for the three-dimensional sphere. These can then be used to obtain barycentric coordinates for arbitrary polytopes in $\mathbb{R}^4$ and successively in higher dimensions. It would be interesting to find a general theory for barycentric coordinates for arbitrary polytopes similar to the one given in [FHK06,JW05]. It should shed light on the relationship between "Euclidean" and spherical coordinates. To construct the general 3D mean value coordinates, we used the construction for 3D mean value coordinates for triangular meshes together with the spherical mean value coordinates. However, should we use Wachspress coordinates in both cases to obtain the general Wachspress coordinates? Or should rather the spherical mean value coordinates be used again, due to their better properties? These topics need to be addressed in future work.

Acknowledgements

We are grateful to Tao Ju and Joe Warren for providing us with their manuscripts prior to publication. We would like to thank the anonymous reviewers for their comments and suggestions. The research of the authors has been supported in part by the EU-Project "AIM@SHAPE" FP6 IST Network of Excellence 506766.

References

[ANS96] ALFELD P., NEAMTU M., SCHUMAKER L. L.: Bernstein-Bézier polynomials on spheres and sphere-like surfaces. *Comput. Aided Geom. Des. 13*, 4 (1996), 333–349.

[AO06] ARROYO M., ORTIZ M.: Local *maximum-entropy* approximation schemes: a seamless bridge between finite elements and meshfree methods. *Int. J. Numer. Meth. Engng 65*, 13 (2006), 2167–2202.

[BF01] BUSS S. R., FILLMORE J. P.: Spherical averages and applications to spherical splines and interpolation. *ACM Trans. Graph. 20*, 2 (2001), 95–126.

[BW92] BROWN J. L., WORSEY A. J.: Problems with defining barycentric coordinates for the sphere. *Mathematical Modelling and Numerical Analysis 26* (1992), 37–49.

[DMA02] DESBRUN M., MEYER M., ALLIEZ P.: Intrinsic parameterizations of surface meshes. *Computer Graphics Forum 21* (2002), 209–218.

[FHK06] FLOATER M. S., HORMANN K., KÓS G.: A general construction of barycentric coordinates over convex polygons. *Adv. Comp. Math. 24*, 1–4 (Jan. 2006), 311–331.

[FKR05] FLOATER M. S., KÓS G., REIMERS M.: Mean value coordinates in 3D. *Comp. Aided Geom. Design 22* (2005), 623–631.

[Flo03] FLOATER M. S.: Mean value coordinates. *Computer Aided Geometric Design 20*, 1 (2003), 19–27.

[Gol02] GOLDMAN R.: *Pyramid Algorithms: A Dynamic Programming Approach to Curves and Surfaces for Geometric Modeling.* Morgan Kaufmann, 2002.

[Gou71] GOURAUD H.: Continuous shading of curved surfaces. *IEEE Trans. Computers C-20*, 6 (1971), 623–629.

[HF06] HORMANN K., FLOATER M. S.: Mean value coordinates for arbitrary planar polygons. *ACM Transactions on Graphics* (Mar. 2006). Accepted.

[JSW05] JU T., SCHAEFER S., WARREN J.: Mean value coordinates for closed triangular meshes. *ACM Trans. Graph. 24*, 3 (2005), 561–566.

[JSWD05] JU T., SCHAEFER S., WARREN J., DESBRUN M.: A geometric construction of coordinates for convex polyhedra using polar duals. In *Proceedings of the Symposium on Geometry Processing* (2005), pp. 181–186.

[JW05] JU T., WARREN J.: *General Constructions of Barycentric Coordinates in a Convex Triangular Polyhedron.* Tech. rep., Washington University in St. Louis, Nov. 2005.

[LD89] LOOP C. T., DEROSE T. D.: A multisided generalization of Bézier surfaces. *ACM Trans. Graph. 8*, 3 (1989), 204–234.

[LPW*06] LIU Y., POTTMANN H., WALLNER J., YANG Y.-L., WANG W.: Geometric modeling with conical meshes and developable surfaces. *ACM Transactions on Graphics, Proceedings of SIGGRAPH 2006* (2006). To appear.

[MBLD02] MEYER M., BARR A., LEE H., DESBRUN M.: Generalized barycentric coordinates on irregular polygons. *J. Graph. Tools 7*, 1 (2002), 13–22.

[Möb27] MÖBIUS A. F.: *Der barycentrische Calcul.* Johann Ambrosius Barth, Leipzig, 1827.

[Möb46] MÖBIUS A. F.: Ueber eine neue Behandlungsweise der analytischen Sphärik. In *Abhandlungen bei Begründung der Königl. Sächs. Gesellschaft der Wissenschaften.* Jablonowski Gesellschaft, 1846, pp. 45–86.

[Pho75] PHONG B. T.: Illumination for computer generated pictures. *Communications of ACM 18*, 6 (1975), 311–317.

[PP93] PINKALL U., POLTHIER K.: Computing discrete minimal surfaces and their conjugates. *Experimental Mathematics 2*, 1 (1993), 15–36.

[SAPH04] SCHREINER J., ASIRVATHAM A., PRAUN E., HOPPE H.: Inter-surface mapping. *ACM Trans. Graph. 23*, 3 (2004), 870–877.

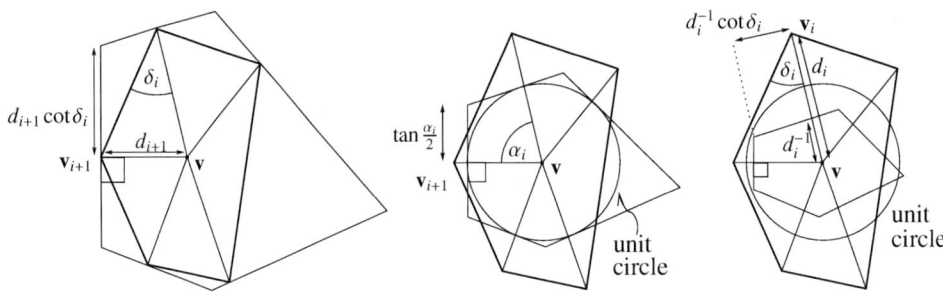

Figure 8: *A geometric construction of the discrete harmonic, mean value, and Wachspress coordinates (from left to right).*

[Sib80]　Sibson R.: A vector identity for the Dirichlet tessellation. *Math. Proc. Cambridge Phil. Soc. 87* (1980).

[SJW06]　Schaefer S., Ju T., Warren J.: A unified, integral construction for coordinates over closed curves. *Computer Aided Geometric Design* (2006). Submitted.

[SM06]　Sukumar N., Malsch E. A.: Recent advances in the construction of polygonal interpolants. *Arch. Comput. Meth. Engng. 13*, 1 (2006), 129–163.

[SP86]　Sederberg T. W., Parry S. R.: Free-form deformation of solid geometric models. In *Computer Graphics, Proceedings of ACM SIGGRAPH* (1986), pp. 151–160.

[Wac75]　Wachspress E. L.: *A Rational Finite Element Basis*, vol. 114. Academic Press, New York, 1975.

[War96]　Warren J.: Barycentric coordinates for convex polytopes. *Adv. Comp. Math. 6*, 2 (1996), 97–108.

Appendix A: A geometric interpretation for planar barycentric coordinates

Here, we present a unified, geometric, and intuitive construction that explains the "linear precision" property of an especially interesting one-parameter family of barycentric coordinates that was introduced in [FHK06]. A different, but equivalent, approach was recently presented in [SJW06]. With this construction, we can derive analogues of the discrete harmonic, mean value, and Wachspress coordinates for arbitrary dimensions. In this paper, we used it in Section 2.3 to show that our spherical Wachspress coordinates and Ju et al.'s vector coordinates coincide. Nevertheless, this construction constitutes an independent contribution on its own.

Our construction is indicated in Figure 8. It is based on a theorem of Minkowski which states that the sum over the edge normals of a polygon, weighted with the respective edge lengths, is zero. Consider a polygon with vertices $\mathbf{v}_i$. It is always possible to construct a dual polygon (that may have self-intersections) with respect to a vertex $\mathbf{v}$ whose edges are orthogonal to the edges $\mathbf{v}\mathbf{v}_i$ and whose vertices are given by the intersection point of two consecutive edges. In fact, there are even infinitely many dual polygons since we can choose the intersection point of the dual edges with the line given

by $\mathbf{v}\mathbf{v}_i$ freely. Since the normals of the dual edges are given by the edges $\mathbf{v}\mathbf{v}_i$, the lengths a_i of the dual edges yield homogeneous coordinates w_i for $\mathbf{v}$ that satisfy property (3'). Since the edges $\mathbf{v}\mathbf{v}_i$ don't have unit length in general, the exact relationship between a_i and $\widehat{\lambda}_i = \frac{w_i}{\sum_j w_j}$ is $a_i = d_i w_i$ where $d_i = \|\mathbf{v}\mathbf{v}_i\|$. Negative weights correspond to inversely oriented dual edges.

In Figure 8, three particular choices for the intersection point of the dual edges are depicted. On the left, the dual edges pass through the points $\mathbf{v}_i$, in the middle, the dual edges have constant distance to $\mathbf{v}$, and on the right, the distance of the dual edges to $\mathbf{v}$ is d_i^{-1}. Using a little trigonometry, it is easy to show that these choices correspond to the standard formulae for discrete harmonic, mean value, and Wachspress coordinates [PP93, Flo03, MBLD02]. In fact, this construction had been used to derive the discrete harmonic coordinates.

Now, it is natural to ask what kind of coordinates are obtained if the distance of the dual edges to $\mathbf{v}$ is chosen as d_i^p, $p \in \mathbb{R}$. The answer, given as a formula, is

$$w_{i,p} = \frac{1}{d_i}\left(\frac{d_{i+1}^p - d_i^p \cos\alpha_i}{\sin\alpha_i} + \frac{d_{i-1}^p - d_i^p \cos\alpha_{i-1}}{\sin\alpha_{i-1}}\right).$$

If we compare this to the one-parameter family of barycentric coordinates $\widehat{w}_{i,p}$ from [FHK06], we see that $w_{i,p} = \frac{1}{2}\widehat{w}_{i,p+1}$. Therefore, both families generate the same barycentric coordinates (after normalization), and the analysis of Floater et al. applies to our family as well:

Corollary A.1 ([FHK06]) The only members of the one-parameter family $w_{i,p}$ which are positive for all convex polygons are the Wachspress and the mean value coordinates.

Another appealing property of this geometric construction is that it easily generalizes to higher dimensions, and barycentric coordinates for polytopes with simplicial boundary can be derived. In the three-dimensional case, the analogous weights $w_{i,p}$ lead to three-dimensional Wachspress coordinates for $p = -1$, but for $p = 0$ and $p = 1$, they do not correspond to the mean value and discrete harmonic coordinates constructed in [JW05] unlike in the two-dimensional case.

Eurographics Symposium on Geometry Processing (2006)
Konrad Polthier, Alla Sheffer (Editors)

On Transfinite Barycentric Coordinates

Alexander Belyaev

Max Planck Institut für Informatik, Saarbrücken, Germany

Abstract

A general construction of transfinite barycentric coordinates is obtained as a simple and natural generalization of Floater's mean value coordinates [Flo03, JSW05b]. The Gordon-Wixom interpolation scheme [GW74] and transfinite counterparts of discrete harmonic and Wachspress-Warren coordinates are studied as particular cases of that general construction. Motivated by finite element/volume applications, we study capabilities of transfinite barycentric interpolation schemes to approximate harmonic and quasi-harmonic functions. Finally we establish and analyze links between transfinite barycentric coordinates and certain inverse problems of differential and convex geometry.

1. Introduction

Design and analysis of transfinite interpolation schemes has been a core research topic in geometric modeling since seminal works of CAGD pioneers [Far02]. More recent interest in transfinite interpolation stems from inventing transfinite versions of classical and modern barycentric coordinates [JSW05b, WSHD06, HF06, SJW06].

The contribution of this paper is threefold. First, we invent a general construction of transfinite barycentric coordinates as a weighted version of transfinite mean value coordinates [Flo03, JSW05b] (Section 2). Next, we analyze relationships between transfinite barycentric interpolation schemes and PDE-based interpolation (Sections 3,4,5). A special attention is paid to approximating harmonic functions (Sections 3,4). In particular, we study and generalize simple and elegant transfinite barycentric coordinates proposed twelve years ago by Gordon and Wixom [GW74] and now almost forgotten (Sections 4,5). Finally, we reveal interesting links between the barycentric interpolation schemes and the famous Christoffel-Minkowski problems studied in differential and convex geometry (Section 6).

After this paper was submitted, we became aware of a recent study [SJW06] where a similar approach to a general construction of 2D transfinite barycentric coordinates was proposed.

While our work is pure theoretical, we believe that the presented results may find applications

in computational mechanics and fluid dynamics where numerical methods based on barycentric interpolation schemes gain more and more popularity [AO06, SM06, KFCO06]. Free-form shape deformations constitute another potential area of applications [JSW05b, DM06].

The starting point for our study is a general construction of transfinite barycentric coordinates introduced in [WSHD06]. Given a convex domain Ω, let us consider a smooth function $b(\boldsymbol{x}, \boldsymbol{y})$, $\boldsymbol{x} \in \Omega$ and $\boldsymbol{y} \in \partial\Omega$, satisfying the following three properties

$$\text{Non-negativity} \quad b(\boldsymbol{x}, \boldsymbol{y}) \geq 0, \tag{1}$$

$$\text{Partition of unity} \quad \int_{\partial\Omega} b(\boldsymbol{x}, \boldsymbol{y})\, ds_y = 1, \tag{2}$$

$$\text{Linear precision} \quad \int_{\partial\Omega} \boldsymbol{y}\, b(\boldsymbol{x}, \boldsymbol{y})\, ds_y = \boldsymbol{x}. \tag{3}$$

Now interpolation of function $f(\boldsymbol{y})$ defined on $\partial\Omega$ into Ω is given by

$$f(\boldsymbol{x}) = \int_{\partial\Omega} f(\boldsymbol{y})\, b(\boldsymbol{x}, \boldsymbol{y})\, ds_y. \tag{4}$$

2. Weighted mean value coordinates

The main idea behind the Floater mean value coordinates [Flo03] consists of applying the mean value property for harmonic functions to piecewise linear functions. Consider a convex bounded domain Ω in $\mathbb{R}^n$ and a function $f(\boldsymbol{x})$ harmonic inside Ω

$$\Delta f \equiv \frac{\partial^2 f}{\partial x_1^2} + \ldots + \frac{\partial^2 f}{\partial x_n^2} = 0.$$

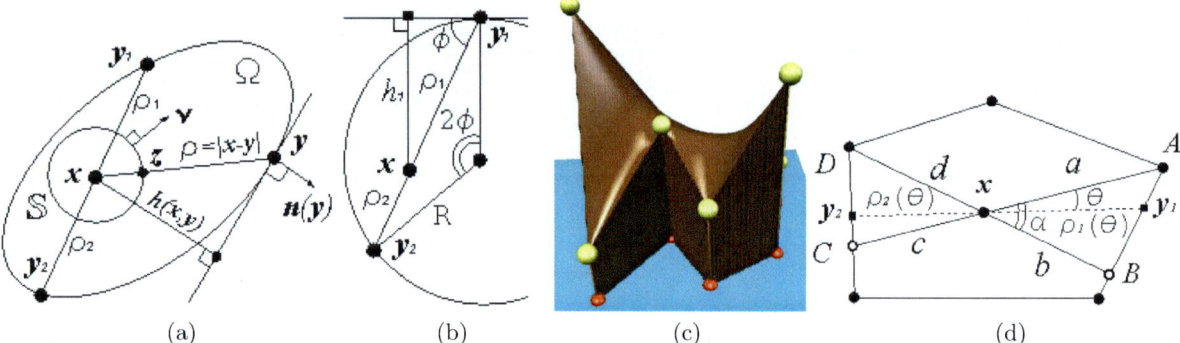

(a)   (b)   (c)   (d)

Figure 1: *(a) Main notations employed throughout the paper. (b) Notations used in the proof that certain weighted Gordon-Wixom interpolation schemes are pseudo-harmonic (c) Interpolating inside of a non-convex polygon with the Gordon-Wixom coordinates adapted for non-convex domains. (d) Notations used to derive the Gordon-Wixom coordinates for polygons.*

Let $\boldsymbol{x}$ be a point inside Ω and $S(\boldsymbol{x},r)$ be the sphere of radius r centered at $\boldsymbol{x}$ and contained inside Ω. Then according to the mean value theorem for harmonic functions

$$f(\boldsymbol{x}) = \frac{1}{|S(\boldsymbol{x},r)|} \int_{S(\boldsymbol{x},r)} f(\boldsymbol{z})\, d\boldsymbol{\nu}.$$

The authors of [JSW05b] also introduced an elegant continuous version of the the mean value coordinates. Below we give a simple derivation of the continuous mean value coordinates.

Consider the unit sphere $\mathbb{S} = S(\boldsymbol{x},1)$ centered at $\boldsymbol{x}$ and parameterized by the outer unit normal $\boldsymbol{\nu}$ (see Fig. 1(a) for a visual feedback on some notations used below). Let $\boldsymbol{y}$ be a point on $\partial\Omega$ and $\rho = |\boldsymbol{x}-\boldsymbol{y}|$. Denote by $\boldsymbol{z}$ the intersection point between the ray $[\boldsymbol{x}\boldsymbol{y})$ and $\mathbb{S}$. Assume that we know the values of function $f(\cdot)$ on $\partial\Omega$ and at $\boldsymbol{x}$. Then we can estimate $f(\boldsymbol{z})$ using the linear interpolation:

$$f(\boldsymbol{z}) \approx \frac{(\rho-1)f(\boldsymbol{x}) + f(\boldsymbol{y})}{\rho}$$

Now let us apply $\mathbb{S}$-averaging to the left and right sides of the above equation:

$$\int_{\mathbb{S}} f(\boldsymbol{z})\, d\boldsymbol{\nu} = |\mathbb{S}|\, f(\boldsymbol{x}) - f(\boldsymbol{x})\int_{\mathbb{S}} \frac{d\boldsymbol{\nu}}{|\boldsymbol{x}-\boldsymbol{y}|} + \int_{\mathbb{S}} \frac{f(\boldsymbol{y})\, d\boldsymbol{\nu}}{|\boldsymbol{x}-\boldsymbol{y}|}$$

and, assuming that $f(\cdot)$ is harmonic, we arrive at the transfinite mean value interpolation scheme of [JSW05b]

$$f(\boldsymbol{x}) = \int_{\mathbb{S}} \frac{f(\boldsymbol{y})}{|\boldsymbol{x}-\boldsymbol{y}|}\, d\boldsymbol{\nu} \bigg/ \int_{\mathbb{S}} \frac{d\boldsymbol{\nu}}{|\boldsymbol{x}-\boldsymbol{y}|}. \qquad (5)$$

This interpolation construction has a natural generalization

$$f(\boldsymbol{x}) = \int_{\mathbb{S}} \frac{f(\boldsymbol{y})w(\boldsymbol{x},\boldsymbol{\nu})}{|\boldsymbol{x}-\boldsymbol{y}|}\, d\boldsymbol{\nu} \bigg/ \int_{\mathbb{S}} \frac{w(\boldsymbol{x},\boldsymbol{\nu})}{|\boldsymbol{x}-\boldsymbol{y}|}\, d\boldsymbol{\nu}, \qquad (6)$$

where $w(\boldsymbol{x},\boldsymbol{\nu})$ is a weighting function associated with interpolated location $\boldsymbol{x}$.

Obviously (6) is just another form of (4) whose kernel satisfies the partition of unity property (2). One can also consider (6) as a transfinite version of the basic Shepard interpolation [She68].

Let us check whether (6) satisfies the linear precision property. For $f(\boldsymbol{x}) \equiv \boldsymbol{x}$ we have $f(\boldsymbol{y}) \equiv \boldsymbol{y} = \boldsymbol{x}+\rho\boldsymbol{\nu}$. Substituting the latter into (6) gives

$$0 = \int_{\mathbb{S}} \boldsymbol{\nu}\, w(\boldsymbol{x},\boldsymbol{\nu})\, d\boldsymbol{\nu} \quad \text{for each} \quad \boldsymbol{x} \in \Omega \qquad (7)$$

which is necessary and sufficient for linear precision.

One can see that (7) is satisfied if the weighting function is centrally-symmetric, i.e., coincides at each pair of antipodal points:

$$w(\boldsymbol{x},\boldsymbol{\nu}) = w(\boldsymbol{x},-\boldsymbol{\nu}). \qquad (8)$$

Consider an important case of the planar interpolation: $n = 2$, $\boldsymbol{\nu} = (\cos\theta, \sin\theta)$,

$$f(\boldsymbol{x}) = \int_0^{2\pi} \frac{f(\boldsymbol{y})w(\boldsymbol{x},\theta)}{|\boldsymbol{x}-\boldsymbol{y}|}\, d\theta \bigg/ \int_0^{2\pi} \frac{w(\boldsymbol{x},\theta)}{|\boldsymbol{x}-\boldsymbol{y}|}\, d\theta. \qquad (9)$$

The orthogonality conditions

$$\int_0^{2\pi} w(\boldsymbol{x},\theta)\cos\theta\, d\theta = 0 = \int_0^{2\pi} w(\boldsymbol{x},\theta)\sin\theta\, d\theta \qquad (10)$$

for each $\boldsymbol{x} \in \Omega$ are necessary and sufficient for linear precision.

It turns out that any barycentric interpolation scheme can be represented in the form of (6) with the weight function (which may be a measure of a generalized function) satisfying orthogonality conditions (7). Indeed, according to the Riesz representation theorem, given $\boldsymbol{x} \in \Omega$, a linear transfinite interpolation scheme can be considered as a linear functional $T_{\boldsymbol{x}}$ defined on an appropriate space of functions on $\partial\Omega$

$$T_{\boldsymbol{x}} : f|_{\partial\Omega} \to f(\boldsymbol{x})$$

and, therefore, can be represented by integration against a certain density μ_x

$$f(\boldsymbol{x}) \equiv T_x(f) = \int_{\partial\Omega} f(\boldsymbol{y})\mu_x(\boldsymbol{y})\,ds_y. \qquad (11)$$

Denote by $h(\boldsymbol{x}, \boldsymbol{y})$ the distance from $\boldsymbol{x}$ to the supporting plane at $\boldsymbol{y} \in \partial\Omega$, as seen in Fig. 1. The area (length) element ds_y of $\partial\Omega$ at $\boldsymbol{y} \in \partial\Omega$ can be expressed in polar coordinates $(\rho, \boldsymbol{\nu})$ centered at $\boldsymbol{x} \in \Omega$ as

$$ds_y = \rho^n d\boldsymbol{\nu}/h. \qquad (12)$$

Therefore we can rewrite (11) as follows

$$T_x(f) = \int_{\mathbb{S}} f(\boldsymbol{y})\mu_x(\boldsymbol{y})\frac{|\boldsymbol{x}-\boldsymbol{y}|^n}{h(\boldsymbol{x},\boldsymbol{y})}\,d\boldsymbol{\nu}.$$

It remains to set $w(\boldsymbol{x}, \boldsymbol{\nu}) = \mu_x(\boldsymbol{y})|\boldsymbol{x}-\boldsymbol{y}|^{n+1}/h(\boldsymbol{x},\boldsymbol{y})$.

Below we consider two important examples.

Transfinite Laplace coordinates. The transfinite Laplace coordinates are a continuous version of the discrete harmonic coordinates which are used widely in computational mechanics [SM06]. The discrete Laplace interpolation scheme was probably first time proposed in [CFL82] and then reinvented many times in connection with research on finite element methods [BVIK*97] (the so-called non-Sibsonian interpolation), computational geometry [Sug99], and discrete minimal surfaces [PP99] (the so-called cotangent weights). A particular case of 2D transfinite Laplace coordinates was studied in [HS00] where interpolation of data defined on straight line segments was considered.

Following [SJW06] one can define the transfinite Laplace coordinates in a way similar to the derivation of the transfinite mean value coordinates if, instead of satisfying the mean value property, we search for a solution minimizing the Dirichlet energy. Let us assume that $f(\boldsymbol{x})$ is known and consider a ruled surface patch generated by straight segments connecting inner point $(\boldsymbol{x}, f(\boldsymbol{x}))$, $\boldsymbol{x} \in \Omega$, with all boundary points $(\boldsymbol{y}, f(\boldsymbol{y}))$, $\boldsymbol{y} \in \partial\Omega$. Then $f(\boldsymbol{x})$ is defined such that the Dirichlet energy attains its minimal value.

Let $r = \rho(\theta) \equiv |\boldsymbol{x}-\boldsymbol{y}|$ be the graph of $\partial\Omega$ in the polar coordinates (r, θ) centered at $\boldsymbol{x}$. Denote by $F_x(\boldsymbol{z})$, $\boldsymbol{z} \in \Omega$, the ruled surface described in the previous paragraph

$$F_x(\boldsymbol{z}) = \frac{(\rho-r)f(\boldsymbol{x})+rf(\boldsymbol{y})}{\rho}.$$

Here $r = |\boldsymbol{x}-\boldsymbol{z}|$ and $\boldsymbol{y} \in \partial\Omega$ is the intersection point between $\partial\Omega$ and the ray from $\boldsymbol{x}$ through $\boldsymbol{z}$. We arrive at the following minimization problem

$$\min \leftarrow \iint_\Omega |\nabla F_x|^2 \, dz = \int_0^{2\pi} d\theta \int_0^\rho r\,dr \left\{ \left[\frac{f(\boldsymbol{x})-f(\boldsymbol{y})}{\rho}\right]^2 + \right.$$

$$\left. + \left[(f(\boldsymbol{y})-f(\boldsymbol{x}))\left(\frac{1}{\rho}\right)'_\theta + \frac{1}{\rho}f'_\theta(\boldsymbol{y})\right]^2\right\} =$$

$$= \frac{1}{2}\int_0^{2\pi} d\theta \left\{ [f(\boldsymbol{x})-f(\boldsymbol{y})]^2 + \left[f'_\theta(\boldsymbol{y})+(f(\boldsymbol{y})-f(\boldsymbol{x}))\frac{\rho'_\theta}{\rho}\right]^2\right\},$$

where the last integral is a quadratic function w.r.t. $f(\boldsymbol{x})$. Thus the optimal value of $f(\boldsymbol{x})$ is given by

$$f(\boldsymbol{x}) = \frac{\int_0^{2\pi}\left\{f(\boldsymbol{y})-f'_\theta(\boldsymbol{y})\left[\rho'_\theta/\rho\right]+f(\boldsymbol{y})\left[\rho'_\theta/\rho\right]^2\right\}d\theta}{\int_0^{2\pi}\left\{1+\left[\rho'_\theta/\rho\right]^2\right\}d\theta}$$

$$= \int_0^{2\pi} f(\boldsymbol{y})\frac{\rho''_{\theta\theta}+\rho}{\rho}\,d\theta \Big/ \int_0^{2\pi}\frac{\rho''_{\theta\theta}+\rho}{\rho}\,d\theta, \quad (13)$$

where integrations by parts are used to derive (13).

Notice that (13) corresponds to (9) with

$$w(\boldsymbol{x},\theta) = \rho''_{\theta\theta}+\rho. \qquad (14)$$

A simple integration by parts shows that (14) satisfies the orthogonality conditions (10).

Probably the simplest way to define the transfinite Laplace coordinates in $\mathbb{R}^n$, $n \geq 3$, consists of expressing the coordinates via the gradient of area (volume), see [MBLD02, DMA02], for the relation between the discrete harmonic coordinates and the gradient of an area functional. Let $\boldsymbol{n}(\boldsymbol{y})$ be the unit outer normal of $\partial\Omega$ at $\boldsymbol{y} \in \partial\Omega$. Denote by $V(\boldsymbol{x})$ the volume of Ω as a function of $\boldsymbol{x}$. Then, similar to the discrete case, we have

$$0 = \nabla_x V(\boldsymbol{x}) = \int_{\partial\Omega} \boldsymbol{n}(\boldsymbol{y})\,ds_y \equiv \int_{\mathbb{S}} \frac{\boldsymbol{n}\,d\boldsymbol{n}}{K(\boldsymbol{n})},$$

where $K(\boldsymbol{n})$ is the Gaussian curvature of $\partial\Omega$ parameterized by its unit normal. Thus, assuming that $\partial\Omega$ is parameterized by its Gauss map, the transfinite Laplace coordinates are given by (6) with

$$w(x,\boldsymbol{\nu}) = 1/K(\boldsymbol{\nu}), \qquad (15)$$

where $K(\boldsymbol{\nu})$ is the Gaussian curvature of $\partial\Omega$ at the point with normal $\boldsymbol{\nu}$.

In 2D, (15) is obviously reduced to (14) (see also Section 6).

Transfinite Wachspress-Warren coordinates. Motivated by FEM applications, Wachspress [Wac75] proposed a construction of affine-invariant barycentric coordinates for convex polygons in 2D. Recently Warren and co-workers [War96, WSHD06] extended the Wachspress coordinates to convex polyhedra in $\mathbb{R}^n$ and presented a transfinite version of the coordinates

Let us denote by $\boldsymbol{n}(\boldsymbol{y})$ the outer unit normal to Ω at $\boldsymbol{y} \in \partial\Omega$. The transfinite Wachspress-Warren coordinates are given by [WSHD06]

$$f(\boldsymbol{x}) = \int_\Omega f(\boldsymbol{y})\frac{K(\boldsymbol{y})\rho(\boldsymbol{x},\boldsymbol{y})}{h(\boldsymbol{x},\boldsymbol{y})^n}\,ds_y \Big/ \!\!\int_\Omega \frac{K(\boldsymbol{y})\rho(\boldsymbol{x},\boldsymbol{y})}{h(\boldsymbol{x},\boldsymbol{y})^n}\,ds_y,$$

where $K(\boldsymbol{y})$ is the Gaussian curvature of $\partial\Omega$ at $\boldsymbol{y}$ and, as before, $h(\boldsymbol{x}, \boldsymbol{y})$ is the distance from $\boldsymbol{x}$ to the supporting plane at $\boldsymbol{y} \in \partial\Omega$.

Thus, in view of (12), the transfinite Wachspress-Warren coordinates are given by (6) with

$$ w(\boldsymbol{x}, \boldsymbol{\nu}) = K(\boldsymbol{y}) \left[\rho(\boldsymbol{x}, \boldsymbol{y}) / h(\boldsymbol{x}, \boldsymbol{y}) \right]^{n+1} . \quad (16) $$

In 2D, straightforward computations show that (16) can be written as

$$ w(\boldsymbol{x}, \theta) = (1/\rho)''_{\theta\theta} + 1/\rho. \quad (17) $$

A geometric proof of (17) is given in Section 6. Similar to (14), orthogonality conditions (10) for (17) are easily derived via integration by parts.

Notice a similarity between (17) and (14). It will be explained in Section 6.

3. Pseudo-harmonic interpolation

Let us call an transfinite interpolation scheme *pseudo-harmonic* if it reproduces the harmonic functions in a ball.

Surprisingly, transfinite mean value interpolation (5) does not approximate harmonic functions. Indeed, let us consider the following pseudo-harmonic interpolation scheme

$$ f(\boldsymbol{x}) = \int_{\partial\Omega} \frac{f(\boldsymbol{y}) \, ds_y}{|\boldsymbol{x} - \boldsymbol{y}|^n} \bigg/ \int_{\partial\Omega} \frac{ds_y}{|\boldsymbol{x} - \boldsymbol{y}|^n} \quad (18) $$

which was introduced for $d = 2$ in [CMS98] as a transfinite version of a variant of the well known Shepard interpolation method [She68]

$$ f(\boldsymbol{x}) = \frac{\sum_{i=1}^{N} f(\boldsymbol{y}_i) |\boldsymbol{x} - \boldsymbol{y}_i|^{-n}}{\sum_{i=1}^{N} |\boldsymbol{x} - \boldsymbol{y}_i|^{-n}}. $$

It is easy to see that (18) coincides with the Poisson Integral Formula for the Laplace equation when Ω is a ball.

According to (12) we can rewrite (18) as follows

$$ f(\boldsymbol{x}) = \int_{\mathbb{S}} \frac{f(\boldsymbol{y}) \, d\boldsymbol{\nu}}{h(\boldsymbol{x}, \boldsymbol{y})} \bigg/ \int_{\mathbb{S}} \frac{d\boldsymbol{\nu}}{h(\boldsymbol{x}, \boldsymbol{y})} \quad (19) $$

which is another, quite elegant, form of the Poisson Integral Formula. [†]

In general, (19) (or, equivalently, (18)) does not give linear precision.

It is interesting to observe a similarity and difference between (19) and transfinite mean value coordinates

---

[†] Certainly representations (18) and (19) of the solution to the Laplace equation in a ball are deserved to be mentioned in PDE textbooks.

(5). It also explains why (5) does not enjoy the property of being pseudo-harmonic.

Simple computations show that already in 2D the transfinite Laplace coordinates are not pseudo-harmonic. This is not surprising since in general the discrete harmonic coordinates lead to only zeroth-order consistency of the corresponding finite difference approximation of the Laplacian [Suk03]. However the discrete harmonic coordinates deliver a good approximation of the Laplace operator in Sobolev spaces of negative order [War05].

As shown in [FHK06], the discrete harmonic and Wachspress coordinates are the same for a circumscribable polygon. This implies that their transfinite versions coincide for a 2D circle. Thus the transfinite Wachspress-Warren coordinates are not pseudo-harmonic as well.

4. Weighted Gordon-Wixom coordinates

For 2D convex shapes, a simple and elegant construction for transfinite pseudo-harmonic barycentric coordinates was proposed by Gordon and Wixom in [GW74]. Below we describe their multidimensional analog.

Given a point $\boldsymbol{x}$ inside Ω, consider the unit sphere $\mathbb{S}$ centered at $\boldsymbol{x}$ and a unit normal $\boldsymbol{\nu} \in \mathbb{S}$. Let the straight line through $\boldsymbol{x}$ determined by $\boldsymbol{\nu}$ intersect the boundary $\partial\Omega$ in two points $\boldsymbol{y}_1$ and $\boldsymbol{y}_2$. Denote by ρ_1 and ρ_2 the distances from $\boldsymbol{x}$ to $\boldsymbol{y}_1$ and $\boldsymbol{y}_2$, respectively. Now we estimate $f(\boldsymbol{x})$ using the linear interpolation between $f(\boldsymbol{y}_1)$ and $f(\boldsymbol{y}_2)$

$$ f(\boldsymbol{x}, \theta) = \frac{\rho_2}{\rho_1 + \rho_2} f(\boldsymbol{y}_1) + \frac{\rho_1}{\rho_1 + \rho_2} f(\boldsymbol{y}_2) = \quad (20) $$

$$ = \left[\frac{f(\boldsymbol{y}_1)}{\rho_1} + \frac{f(\boldsymbol{y}_2)}{\rho_2} \right] \bigg/ \left[\frac{1}{\rho_1} + \frac{1}{\rho_2} \right] $$

$\mathbb{S}$-averaging w.r.t. $\boldsymbol{\nu}$ defines the Gordon-Wixom interpolation

$$ f(\boldsymbol{x}) = \frac{1}{|\mathbb{S}|} \int_{\mathbb{S}} \left[\frac{\rho_2}{\rho_1 + \rho_2} f(\boldsymbol{y}_1) + \frac{\rho_1}{\rho_1 + \rho_2} f(\boldsymbol{y}_2) \right] d\boldsymbol{\nu}. \quad (21) $$

As shown below for a more general situation, the Gordon-Wixom interpolation scheme is pseudo-harmonic and has linear precision.

Similar to weighted transfinite mean value coordinates (6), let us introduce a weighted version of the Gordon-Wixom interpolation scheme

$$ f(\boldsymbol{x}) = \frac{\int_{\mathbb{S}} \left[\left(\frac{f(\boldsymbol{y}_1)}{\rho_1} + \frac{f(\boldsymbol{y}_2)}{\rho_2} \right) \bigg/ \left(\frac{1}{\rho_1} + \frac{1}{\rho_2} \right) \right] W(\boldsymbol{x}, \boldsymbol{\nu}) \, d\boldsymbol{\nu}}{\int_{\mathbb{S}} W(\boldsymbol{x}, \boldsymbol{\nu}) \, d\boldsymbol{\nu}}, \quad (22) $$

where $W(\boldsymbol{x}, \boldsymbol{\nu})$ is centrally-symmetric (8).

To show that (22) enjoys linear precision let us set $f(\boldsymbol{x}) \equiv \boldsymbol{x}$. Then $f(\boldsymbol{y}_1) \equiv \boldsymbol{y}_1 = \boldsymbol{x} + \rho_1 \boldsymbol{\nu}$ and $f(\boldsymbol{y}_2) \equiv \boldsymbol{y}_2 = \boldsymbol{x} - \rho_2 \boldsymbol{\nu}$ and (22) is obviously satisfied.

Observe that if weighting function of (6) satisfies (8) (centrally-symmetric), then the weighting functions of (22) and (6) are related to each other by

$$W(\boldsymbol{x}, \boldsymbol{\nu}) = w(\boldsymbol{x}, \boldsymbol{\nu}) \left(\frac{1}{\rho_1} + \frac{1}{\rho_2} \right).$$

In particular, (22) with $W(\boldsymbol{x}, \boldsymbol{\nu}) = 1/\rho_1 + 1/\rho_2$ is reduced to mean value coordinates (5).

Now let us consider a particular case of (22) obtained when $W(\boldsymbol{x}, \boldsymbol{\nu})$ depends on distances ρ_1 and ρ_2 only. Since $W(\boldsymbol{x}, \boldsymbol{\nu})$ is centrally-symmetric, it can be written as a function of $\rho_1 \cdot \rho_2$ and $\rho_1 + \rho_2$

$$W(\boldsymbol{x}, \boldsymbol{\nu}) = W(\rho_1 + \rho_2, \rho_1 \rho_2). \tag{23}$$

Let us show that if (23) depends only on the product of ρ_1 and ρ_1, then barycentric coordinates (22) are pseudo-harmonic. Indeed if Ω is a ball then according to the Intersecting Chords Theorem [‡]

$$\rho_1 \rho_2 \equiv c(\boldsymbol{x}),$$

where $c(\boldsymbol{x})$ depends only on $\boldsymbol{x}$. Thus it is sufficient to demonstrate that (21) is pseudo-harmonic. Notice that we can exploit symmetry properties of (21) and rewrite it as

$$f(\boldsymbol{x}) = \frac{2}{|\mathbb{S}|} \int_{\mathbb{S}/2} \frac{\rho_2}{\rho_1 + \rho_2} f(\boldsymbol{y}_1) \, d\boldsymbol{\nu}, \tag{24}$$

where by $\mathbb{S}/2$ we denote a unit semisphere. Let Ω be a ball of radius R. Consider the 2D disc obtained as the intersection between Ω and the plane formed by $\boldsymbol{y}_1$, $\boldsymbol{y}_2$, and the center of the ball. In notations of Fig. 1(b), we obviously have

$$h_1 = \rho_1 \sin \varphi, \quad h_2 = \rho_2 \sin \varphi, \quad \rho_1 + \rho_2 = 2R \sin \varphi,$$

$$\frac{h_1 h_2}{\sin^2 \varphi} = c(\boldsymbol{x}), \quad \frac{\rho_2}{\rho_1 + \rho_2} = \frac{h_2}{2R \sin^2 \varphi} = \frac{c(\boldsymbol{x})}{2R} \cdot \frac{1}{h_1}.$$

Thus (24) is reduced to (19). A slightly more complex proof of this result for the 2D version of (21) is given in [GW74] and attributed to W. W. Meyer.

Now we can see why the original Gordon-Wixom interpolation scheme is good in approximating harmonic functions. The second-order directional derivative of $f(\boldsymbol{x})$ in the $\boldsymbol{\nu}$-direction can be approximated by

$$f''_{\nu\nu}(\boldsymbol{x}) \approx D_{\nu\nu}[f(\boldsymbol{x})] \equiv \tag{25}$$
$$\equiv \frac{2}{\rho_1 \rho_2} \left(\left[\frac{f(\boldsymbol{y}_1)}{\rho_1} + \frac{f(\boldsymbol{y}_2)}{\rho_2} \right] \Big/ \left[\frac{1}{\rho_1} + \frac{1}{\rho_2} \right] - 2f(\boldsymbol{x}) \right).$$

[‡] The Intersecting Chords Theorem is usually formulated for a 2D circle. However its extension onto the multidimensional case is straightforward.

Since

$$2\Delta f(\boldsymbol{x}) = \frac{1}{|\mathbb{S}|} \int_{\mathbb{S}} f''_{\nu\nu}(\boldsymbol{x}) \, d\boldsymbol{\nu},$$

we obtain weighted Gordon-Wixom interpolation (22) with weight

$$W(\boldsymbol{x}, \boldsymbol{\nu}) = \frac{1}{\rho_1 \rho_2} \tag{26}$$

and, according to the Intersecting Chords Theorem, arrive at (21) when Ω is a ball.

As mentioned before, weighting functions $w(\boldsymbol{x}, \boldsymbol{\nu})$ in (6) and $W(\boldsymbol{x}, \boldsymbol{\nu})$ in (22) may be generalized functions. In particular, Gordon and Wixom [GW74] considered a transfinite interpolation scheme which can be obtained from (22) if we set $W(\boldsymbol{x}, \boldsymbol{\nu}) = \delta(\boldsymbol{\nu} - \boldsymbol{\nu}_0)$, where $\delta(\cdot)$ is the Dirac delta function and $\boldsymbol{\nu}_0$ is a given direction.

Non-convex domains. So far we have assumed that Ω is convex. It turns out that it is very easy to extend the weighted Gordon-Wixom interpolation scheme (22) to generic non-convex domains. Consider the straight line $l(\boldsymbol{x}, \boldsymbol{\nu})$ through $\boldsymbol{x}$ at direction $\boldsymbol{\nu}$ intersecting $\partial\Omega$ in m points $\boldsymbol{y}_1, \ldots, \boldsymbol{y}_m$. Let us set $\varepsilon_i = 1$ if the ray $[\boldsymbol{x}, \boldsymbol{y}_i)$ arrives at $\boldsymbol{y}_i$ from inside of Ω, $\varepsilon_i = -1$ if the ray approaches $\boldsymbol{y}_i$ from outside of Ω, and $\varepsilon_i = 0$ if the ray is tangent to $\partial\Omega$ at $\boldsymbol{y}_i$. Define

$$f(\boldsymbol{x}, \boldsymbol{\nu}) = \sum_{i=1}^{m} f(\boldsymbol{y}_i) w(\boldsymbol{x}, \boldsymbol{y}_i) \frac{\varepsilon_i}{\rho_i} \Big/ \sum_{i=1}^{m} w(\boldsymbol{x}, \boldsymbol{y}_i) \frac{\varepsilon_i}{\rho_i}, \tag{27}$$

where $\rho_i = |\boldsymbol{x} - \boldsymbol{y}_i|$. This simple one-dimensional construction belongs to the family of barycentric rational interpolation schemes studied widely in constructive approximation [BBM05]. Now $\mathbb{S}$-averaging w.r.t. $\boldsymbol{\nu}$

$$f(\boldsymbol{x}) = \frac{1}{|\mathbb{S}|} \int_{\mathbb{S}} f(\boldsymbol{x}, \boldsymbol{\nu}) \, d\boldsymbol{\nu} \tag{28}$$

gives an extension of (22) to generic non-convex domains.

If Ω is not generic, it may happen that $l(\boldsymbol{x}, \boldsymbol{\nu}) \cap \partial\Omega$ contains linear segments. It is natural to treat such a situation as the tangency case and set $\varepsilon = 0$ in (27) for all the points of those linear segments on $l(\boldsymbol{x}, \boldsymbol{\nu})$.

Instead of (27) one can use high-order 1D barycentric interpolation schemes introduced very recently by Floater and Hormann [FH06].

An example of interpolating inside of a non-convex polygon with (27) is shown in Fig.1(c). Similar to [GW74] we evaluate (28) numerically.

Gordon-Wixom coordinates for polygons. Let Ω be a 2D convex polygon and $f(\boldsymbol{y})$, $\boldsymbol{y} \in \partial\Omega$, be given by its values at the polygon vertices. Assume that $f(\boldsymbol{y})$ is linear on the edges of the polygon. Then (21) allows for a closed-form solution. For a planar domain, one can rewrite (21) as

$$\frac{1}{\pi} \int_0^{2\pi} \frac{\rho_2}{\rho_1 + \rho_2} f(\boldsymbol{y}_1)\, d\theta. \qquad (29)$$

Consider point $\boldsymbol{x}$ inside polygon Ω and place new vertices on the boundary of the polygon such that the straight line connecting x with an original vertex intersects $\partial\Omega$ in another, possibly new, vertex. See Fig. 1(d) where A and D are original polygon vertices and B and C are new vertices. Then we repeat trigonometric calculations of [Flo03] and, in the notations of Fig. 1(d), arrive at

$$\rho_1(\theta) = \frac{ab\sin\alpha}{a\sin\theta + b\sin(\alpha - \theta)},$$

$$\lambda_1 = \frac{b\sin(\alpha - \theta)}{a\sin\theta + b\sin(\alpha - \theta)}, \quad \lambda_2 = \frac{a\sin(\theta)}{a\sin\theta + b\sin(\alpha - \theta)},$$

where $\lambda_1(\theta)$ and $\lambda_2(\theta)$, $\lambda_1 + \lambda_2 = 1$, are the weights for the linear interpolation inside the segment AB. We compute $\rho_2(\theta)$ in a similar way and get the following formula for the part of (29) corresponding to AB.

$$\frac{1}{\pi} \int_0^{\alpha} \frac{cd\,[f(A)b\sin(\alpha - \theta) + f(B)a\sin\theta]}{ac(b + d)\sin\theta + bd(a + c)\sin(\alpha - \theta)}\, d\theta. \quad (30)$$

Although (30) looks quite complex, it can be evaluated in closed form (we have used Maple to express (30) in terms of elementary functions). Unfortunately the result is too lengthy to present here.

5. Barycentric coordinates and PDEs

Quasi-Laplacian. Let us consider the Dirichlet boundary value problem for a quasi-Laplacian operator:

$$\nabla\cdot(a(\boldsymbol{x})\nabla f(\boldsymbol{x})) = 0 \text{ in } \Omega, \quad f(\boldsymbol{x}) \text{ is known on } \partial\Omega, \quad (31)$$

where $a(\boldsymbol{x}) > 0$ is a known conductivity coefficient. Integration by parts gives

$$\int_{S(\boldsymbol{x},r)} a\frac{\partial f}{\partial\boldsymbol{\nu}}\, d\boldsymbol{\nu} = 0, \qquad (32)$$

where, as before, $S(\boldsymbol{x}, r)$ is the sphere of radius r centered at $\boldsymbol{x}$ and contained inside Ω and $\boldsymbol{\nu}$ is outward unit normal to $S(\boldsymbol{x}, r)$. Let us group opposite points of $\mathbb{S}_r$ together in (32):

$$a\frac{\partial f}{\partial\boldsymbol{\nu}}\bigg|_{\boldsymbol{x}+r(\boldsymbol{\nu})} - a\frac{\partial f}{\partial\boldsymbol{\nu}}\bigg|_{\boldsymbol{x}-r(\boldsymbol{\nu})} \approx 2r\frac{\partial}{\partial\boldsymbol{\nu}}\left(a(\boldsymbol{x})\frac{\partial f(\boldsymbol{x})}{\partial\boldsymbol{\nu}}\right) =$$
$$= 2r\left[a_\nu'(\boldsymbol{x})f_\nu' + a(\boldsymbol{x})f_{\nu\nu}''\right] \qquad (33)$$

We estimate the first- and second-order directional derivatives of $f(\cdot)$ by

$$f_\nu'(\boldsymbol{x}) \approx \frac{f(\boldsymbol{y}_1) - f(\boldsymbol{x})}{\rho_1},$$

and (25), respectively. Substituting these approximations into the right hand-side of (33), integrating over a half-sphere $S(\boldsymbol{x}, r)/2$, and taking into account (32) yields

$$0 \approx \int_{\mathbb{S}/2} \frac{f(\boldsymbol{y}_1) - f(\boldsymbol{x})}{\rho_1} a_\nu'(\boldsymbol{x})\, d\boldsymbol{\nu} + a(\boldsymbol{x})\int_{\mathbb{S}/2} D_{\nu\nu}[f(\boldsymbol{x})]\, d\boldsymbol{\nu} \qquad (34)$$

with $D_{\nu\nu}[f(\boldsymbol{x})]$ defined in (25). We rewrite the integrals in the right-hand side of (34) as integrals over the whole unit sphere $\mathbb{S}$ and arrive at

$$f(\boldsymbol{x})\left[\int_{\mathbb{S}} \frac{a_\nu'(\boldsymbol{x})}{\rho}\, d\boldsymbol{\nu} + 2a(\boldsymbol{x})\,|\mathbb{S}|\right] \approx$$
$$\approx \int_{\mathbb{S}} \frac{f(\boldsymbol{y})}{\rho} a_\nu'(\boldsymbol{x})\, d\boldsymbol{\nu} + a(\boldsymbol{x})\int_{\mathbb{S}} D_{\nu\nu}[f(\boldsymbol{x})]\, d\boldsymbol{\nu}$$

which can be considered as a combination of the (26)-weighted Gordon-Wixom interpolation scheme and $a_\nu'(x)$-weighted transfinite mean value coordinates.

Note that it does not satisfy the linear precision property because of weighting function $a_\nu'(x)$ in the mean value component. Notice however that linear functions do not satisfy (31) unless the conductivity coefficient $a(\boldsymbol{x})$ is constant.

Of course, linear PDE (31) is rather a toy example. However a similar approximation approach can be applied to quasi-linear PDE operators in the form

$$\frac{2}{|\mathbb{S}|} \int_{\mathbb{S}/2} \partial_\nu\left(g\left(|\partial_\nu f|\right)\partial_\nu f\right)\, d\boldsymbol{\nu}, \qquad (35)$$

where $g(\cdot)$ is a positive function. A 2D version of (35) was used in [Wei94] for anisotropic nonlinear image diffusion purposes.

AMLE. According to an axiomatic approach to image interpolation developed in [CMS98] and numerical experiments conducted in [ACGR02], absolutely minimizing functions [ACJ04] satisfy many properties desirable for a feature-preserving interpolation of height data.

Weighted Gordon-Wixom coordinates (21) approximate the solution to the AMLE equation if we set $W(\boldsymbol{x}, \boldsymbol{\nu}) = \delta(\boldsymbol{\nu} - \boldsymbol{\nu}_0(\boldsymbol{x}))$, where $\boldsymbol{\nu}_0(\boldsymbol{x})$ is a direction for which the right-hand side of

$$f_\nu'(\boldsymbol{x}) \approx \frac{f(\boldsymbol{y}_1) - f(\boldsymbol{y}_2)}{\rho_1 + \rho_2}, \qquad (36)$$

attains its maximal absolute value. Similar to the AMLE interpolation, (36) is capable of interpolating isolated values. For example, let Ω be the unit disk without its center (a punctured disk). Define $f(\boldsymbol{y}) = 1$ at the disk center and $f(\boldsymbol{y}) \equiv 0$ on the outer boundary of the punctured disk. Then the solution to (36) is given by cone $f(\boldsymbol{x}) = 1 - |\boldsymbol{x}|$. Exactly the same result is delivered by AMLE.

6. Christoffel-Minkowski type problems and barycentric coordinates

So far we have studied abilities of various barycentric interpolation schemes to approximate solutions to second-order elliptic PDEs. In this section, we reveal and discuss surprising links between the barycentric coordinates and classical inverse problems of differential and convex geometry.

Minkowski problem. The classical Minkowski problem is an inverse problem in differential geometry and concerns reconstruction of a closed convex hypersurface from its Gaussian curvature given as a function of the outer surface normal [Min03]. Given a positive function $K(\boldsymbol{\nu})$ defined over the unit sphere $\mathbb{S}$, a necessary and sufficient condition of the Minkowski problem is

$$\int_{\Sigma} \boldsymbol{\nu} \, dl \equiv \int_{\mathbb{S}} \frac{\boldsymbol{\nu} \, d\boldsymbol{\nu}}{K(\boldsymbol{\nu})} = 0, \qquad (37)$$

where Σ is the reconstructed hypersurface, dl its area element and $dl = K \, d\boldsymbol{\nu}$ by definition of the Gaussian curvature. The necessity of (37) follows immediately from the divergence theorem of vector calculus. The sufficiency is non-trivial and was proven, under various assumptions of smoothness of Σ, by Minkowski himself (1903), Alexandrov (1938), Lewy (1938), Miranda (1939), Pogorelov (1952), Nirenberg (1953), Cheng and Yau (1976), and others.

Exploring the similarity between (37) and (7), we get for free the following result delivering a geometric description of barycentric interpolation schemes. For each $\boldsymbol{x} \in \Omega$, consider a family of hyperplanes in $\mathbb{R}^n_z$

$$(\boldsymbol{z} - \boldsymbol{x}) \cdot \boldsymbol{\nu} = p(\boldsymbol{x}, \boldsymbol{\nu}) \qquad (38)$$

parameterized by $\boldsymbol{\nu} \in \mathbb{S}$. For a given $\boldsymbol{\nu}$, $p(\boldsymbol{x}, \boldsymbol{\nu})$ stands for the signed distance from the plane defined by $\boldsymbol{\nu}$ to $\boldsymbol{x}$. We assume that $p(\boldsymbol{x}, \boldsymbol{\nu})$ is sufficiently smooth. Denote by Σ_x the envelope of family (38) and let $K_x(\boldsymbol{\nu})$ be the Gaussian curvature of Σ_x. Assuming that Σ_x is convex (and, therefore, its Gaussian curvature is positive), we obtain transfinite barycentric coordinates (6) with $w(\boldsymbol{x}, \boldsymbol{\nu}) = 1/K_x(\boldsymbol{\nu})$. Vice versa, given barycentric coordinates (6) whose weighting function $w(\boldsymbol{x}, \boldsymbol{\nu})$ is positive at $\boldsymbol{x}$, a convex hypersurface Σ_x is reconstructed from its Gaussian curvature

$$K_x(\boldsymbol{\nu}) = 1/w(\boldsymbol{x}, \boldsymbol{\nu}).$$

In order to use the Minkowski problem for a geometric description of barycentric coordinates (6) whose weighting function $w(\boldsymbol{x}, \boldsymbol{\nu})$ is not always positive, one can consider the so-called hedgehogs, closed surfaces parameterized by their Gauss maps and described as the envelops of their tangent planes. Unfortunately, if $n \geq 3$, (37) is not sufficient in the case of envelopes forming non-convex surfaces [MM01](Proposition 7).

Christoffel-Minkowski problem. A generalization of the Minkowski problem, the so-called Christoffel-Minkowski problem, consists of finding a convex hypersurface Σ with a prescribed elementary symmetric polynomial of the surface principal radii [GG02]. For $\boldsymbol{\lambda} = (\lambda_1, \ldots, \lambda_{n-1}) \in \mathbb{R}^{n-1}$, let $S_k[\boldsymbol{\lambda}]$ be an elementary symmetric polynomial of degree k

$$S_k[\boldsymbol{\lambda}] = \sum_{i_1 < \ldots < i_{n-1}} \lambda_{i_1} \ldots \lambda_{i_{n-1}},$$

where the sum is taken over all permutations of the indices $\{1, \ldots, n-1\}$. Denote by $R_i(\boldsymbol{\nu}) = 1/k_i(\boldsymbol{\nu})$ the surface principal radii parameterized by the Gauss map of the surface. Direct computations (see, for example, [Bla30, BF34, Bus58, Sch93]) show that $R_i(\boldsymbol{\nu})$, $i = 1, \ldots, n-1$, are the eigenvalues of the matrix

$$\nabla^2_{\nu} p(\boldsymbol{\nu}) + p(\boldsymbol{\nu}) I,$$

where ∇^2_{ν} is the Hessian operator w.r.t. a local orthonormal frame on $\mathbb{S}$ and I is the identity matrix. It can be shown [BF34, Bus58, Sch93] that

$$\int_{\mathbb{S}} \boldsymbol{\nu} S_k[R_1, \ldots, R_{n-1}](\boldsymbol{\nu}) \, d\boldsymbol{\nu} = 0. \qquad (39)$$

The Christoffel-Minkowski problem consists of determining a convex surface whose curvature radii $R_1(\boldsymbol{\nu}), \ldots, R_{n-1}(\boldsymbol{\nu})$ satisfy

$$S_k[R_1, \ldots, R_{n-1}](\boldsymbol{\nu}) = \varphi(\boldsymbol{\nu}), \qquad (40)$$

where $\varphi(\boldsymbol{\nu})$ is a given function such that

$$\int_{\mathbb{S}} \boldsymbol{\nu} \varphi(\boldsymbol{\nu}) \, d\boldsymbol{\nu} = 0. \qquad (41)$$

Orthogonality condition (41) generalizes (37) and can be used for a geometric characterization of transfinite barycentric coordinates.

A substantial progress in solving the general Christoffel-Minkowski problem for convex bodies was recently achieved in [GG02, STW04].

The Monge-Ampère equation

$$K(\boldsymbol{\nu}) \equiv \det \left[\nabla^2_{\nu} p(\boldsymbol{\nu}) + p(\boldsymbol{\nu}) I \right] = \varphi(\boldsymbol{\nu})$$

used to solve the Minkowski problem is a particular case of (40) which arises when the product of the curvature radii is considered:

$$S_{n-1}[R_1, \ldots, R_{n-1}] = R_1 \cdot \ldots \cdot R_{n-1} \equiv 1/K.$$

Christoffel problem. Another particular case of the Christoffel-Minkowski problem is obtained from (40) when $k = 1$. It gives the so-called Weingarten formula [Wei84] (as cited in [Bla30])

$$\text{trace} \left[\nabla^2_{\nu} p(\boldsymbol{\nu}) + p(\boldsymbol{\nu}) I \right] \equiv$$
$$\equiv \Delta_{\nu} p(\boldsymbol{\nu}) + (n-1) p(\boldsymbol{\nu}) = R(\boldsymbol{\nu}), \qquad (42)$$

where Δ_ν is the spherical Laplacian and

$$R(\nu) = \sum_{i=1}^{n-1} R_i(\nu)$$

is the sum of the principal curvature radii (the reciprocal of the sum is the so-called *harmonic curvature*). Since $\lambda = (n-1)$ is the first eigenvalue of Δ_ν and ν are the corresponding $n-1$ eigenfunctions, the necessary and sufficient condition of solvability (42) is given by

$$\int_\Sigma \nu R(\nu)\, ds = 0. \qquad (43)$$

In 3D, this particular case of the Christoffel-Minkowski problem was proposed and studied by Christoffel in 1865 [Chr65] and seems to be the earliest inverse problem in differential and convex geometry. For a relatively recent account, see, for example, [Sch93](Section 4.3),

Given barycentric coordinates (6) with weighting function $w(x, \nu)$ satisfying (7), for each $x \in \Omega$ we set

$$R(x, \nu) = w(x, \nu)$$

and solve linear second-order elliptic PDE (42) Then surface Σ_x is reconstructed from support function $p(x, \nu)$. (The simplest way to solve (42) consists of expanding $w(x, \nu)$ into spherical harmonics and constructing support function $p(x, \nu)$ of Σ_ν as a spherical harmonic series.)

Christoffel-Minkowski problems in 2D and 3D. In 2D, the Christoffel and Minkowski problems coincide and lead to the following second-order differential equation

$$p''(\theta) + p(\theta) = R(\theta) \equiv 1/k(\theta) \qquad (44)$$

for the support function $p(\theta)$. Curve Σ is then obtained from its support function $p(\theta)$ in the following parametric form

$$z(\theta) \equiv \begin{bmatrix} z_1 \\ z_2 \end{bmatrix}(\theta) = \begin{bmatrix} x_1 \\ x_2 \end{bmatrix} + \begin{bmatrix} p(\theta)\cos\theta + p'(\theta)\sin\theta \\ p(\theta)\sin\theta + p'(\theta)\cos\theta \end{bmatrix} \qquad (45)$$

For our purposes, for each $x \in \Omega$ we determine Σ_ν by solving

$$p''(\theta) + p(\theta) = w(x, \theta) \qquad (46)$$

and then using (45). If $w(x, \theta)$ is a positive function of θ, then Σ_x forms a convex curve. If $w(x, \theta)$ changes its sign, Σ_x has cusps at the points corresponding to the zeros of the curvature radius $R(\theta) = w(x, \theta)$.

Fig. 2 shows an example of a planar curve generated from its support function.

In 3D, we have two types of the Christoffel-Minkowski problems corresponding to the Gaussian curvature (the Minkowski problem) and harmonic curvature (the Christoffel problem). For the Christoffel problem, Σ_x is reconstructed by solving

$$\Delta_\nu p(\nu) + 2p(\nu) = w(x, \nu). \qquad (47)$$

Then one can reconstruct Σ_x from its support function $p(\nu)$ using formulas derived in [VF92]. Similar to the 2D case, $w(x, \nu)$ may change its sign and Σ_x form cuspidal edges at the points where its harmonic curvature $1/w(x, \nu)$ becomes infinite.

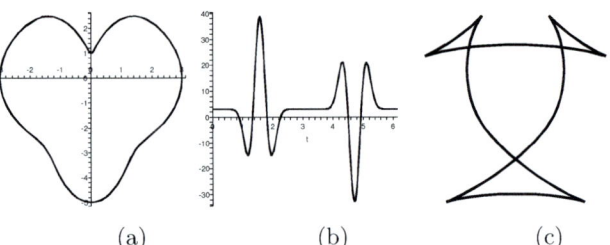

(a)   (b)   (c)

Figure 2: *(a) Polar plot of a given support function $p(\theta)$. (b) Graph of the curvature radius $R(\theta) = p''(\theta) + p(\theta)$. (c) The curve reconstructed from its support function $p(\theta)$. The curve has six cusps corresponding to the zeros of its curvature radius $R(\theta)$.*

Pedal and negative pedal surfaces. Pedal curves and surfaces are differential geometry objects which have a special significance in classical geometric optics [Her00] and are often mentioned in textbooks on classical differential geometry. Given a surface Σ and a point x, the *pedal* of Σ w.r.t. x is defined as the locus of the foots of the perpendiculars from x to the tangent planes to the surface. The negative pedal curves and surfaces are much less mentioned in the differential geometry literature in spite of the fact that they can be used for designing mirrors with prescribed properties [FMR01]. The *negative pedal* of Σ w.r.t. x is constructed as the envelope of planes passing through the points of Σ and perpendicular to the segments connecting x with the points of Σ. The pedal and negative pedal curves and surfaces are also studied in the projective geometry framework [CG67](Chapter 6).

Pedals and negative pedals find applications in materials science [POMZ99]. Following [POMZ99] let us consider two Legendre transformations of positive functions defined on the unit sphere $\mathbb{S}$. Let $r : \mathbb{S} \to \mathbb{R}_+$ be a continuous function. The *polar plot* of $r(\nu)$, $\nu \in \mathbb{S}$, is the surface (curve in 2D) defined in the (r, ν) polar coordinates by $r = r(\nu)$. The first Legendre transform $r_*(\omega)$ is the function corresponding to the negative pedal of the polar plot of $r(\nu)$. The second Legendre transform $r^*(\nu)$ is the function corresponding to the pedal of the polar plot of $r(\omega)$.

As demonstrated in [POMZ99], the following duality relations are hold

$$1/r^* = [1/r]_*, \quad 1/r_* = [1/r]^*. \quad (48)$$

In addition, if $r(\boldsymbol{\nu})$ satisfies certain convexity conditions, then

$$[r_*]^*(\boldsymbol{\nu}) \equiv r(\boldsymbol{\nu}) \equiv [r^*]_*(\boldsymbol{\nu}). \quad (49)$$

Geometry of Wachspress-Warren coordinates.
Now we are ready to establish a link between the transfinite Wachspress-Warren coordinates and the Minkowski problem. As before, given a convex bounded domain Ω in $\mathbb{R}^n$ and point $\boldsymbol{x} \in \Omega$, let $r = \rho(\boldsymbol{\nu})$ be the polar plot of $\partial\Omega$ w.r.t. $\boldsymbol{x}$, $K(\boldsymbol{y})$ denote the Gaussian curvature of $\partial\Omega$ at $\boldsymbol{y} \in \partial\Omega$, and $h(\boldsymbol{x}, \boldsymbol{y}) = (\boldsymbol{y} - \boldsymbol{x}) \cdot \boldsymbol{n}(\boldsymbol{y})$, where $\boldsymbol{n}(\boldsymbol{y})$ is outer unit normal to $\partial\Omega$ at $\boldsymbol{y}$, be the support function of $\partial\Omega$ w.r.t. $\boldsymbol{x}$. Consider surface $\Sigma_{\boldsymbol{x}}$ whose support function w.r.t. $\boldsymbol{x}$ is given by $1/\rho(\boldsymbol{\nu})$. Using duality relations (48) and (49) it is easy to show that the reciprocal of the Gaussian curvature of $\Sigma_{\boldsymbol{x}}$ is given by $K\rho^{n+1}/h^{n+1}$. Thus, according to (37), we have proven the linear precision property of transfinite Wachspress-Warren coordinates (6), (16). In the 2D case, (17) now follows from (44).

An equivalent interpretation of the transfinite Wachspress-Warren coordinates in terms of pedal surfaces (polar duals) was considered in [WSHD06, SJW06] (see also [JSW05a] where a similar description of the discrete Wachspress-Warren coordinates is presented).

Geometry behind Laplace coordinates. According to (15), the geometry of the transfinite Laplace coordinates is simpler than that of Wachspress-Warren coordinates. It is reduced to the Minkowski problem for $\partial\Omega$ (therefore, it can be called by the Minkowski extension of the 2D Laplace coordinates). For each $\boldsymbol{x} \in \Omega$, the support function of $\Sigma_{\boldsymbol{x}}$ is given by $r = \rho(\boldsymbol{\nu})$, the polar plot of $\partial\Omega$ w.r.t. $\boldsymbol{x}$. The linear precision is guaranteed by (37).

It seems, however, that the Christoffel extension of the 2D Laplace coordinates is more appropriate for approximating harmonic functions than the Minkowski extension considered in the previous paragraph. Indeed, let us assume that functions $p(\boldsymbol{\nu})$ and $R(\boldsymbol{\nu})$, $\boldsymbol{\nu} \in \mathbb{S}$, in (43) are defined in $\mathbb{R}^n$ as homogeneous of degree one: $p(t\boldsymbol{z}) \equiv tp(\boldsymbol{z})$ and $R(t\boldsymbol{z}) \equiv tR(\boldsymbol{z})$, $\boldsymbol{z} \in \mathbb{R}^n$. Then (43) can be rewritten as

$$\Delta_{\boldsymbol{z}} p(\boldsymbol{z}) = R(\boldsymbol{z}) \quad (50)$$

and the left-hand side of (50) with $p(\boldsymbol{z}) = \rho(\boldsymbol{z})$, where $\rho(\boldsymbol{\nu})$ is the polar plot of $\partial\Omega$, defines the weighting function

$$w(\boldsymbol{x}, \boldsymbol{\nu}) = R(\boldsymbol{z}/|\boldsymbol{z}|) \quad (51)$$

corresponding to the Christoffel extension of the 2D

Laplace coordinates. The Laplacian from the left-hand side of (50) indicates that (6) with (51) is a proper choice for approximating harmonic functions.

Weighted Gordon-Wixom coordinates. As mentioned before, weighted Gordon-Wixom coordinates (22) are equivalent to (6) with centrally-symmetric weighting: $w(\boldsymbol{x}, \boldsymbol{\nu}) = w(\boldsymbol{x}, -\boldsymbol{\nu})$. Therefore, surfaces $\Sigma_{\boldsymbol{x}}$ constructed as solutions to a Christoffel-Minkowski problem are also centrally-symmetric. Centrally-symmetric convex bodies possess interesting properties and are widely studied in convex geometry [BF34, Sch93].

Geometry behind mean value coordinates. Since for the mean value coordinates $\Sigma_{\boldsymbol{x}}$ is always the unit sphere $\mathbb{S}$, these coordinates play a distinguish role in the above inverse problem constructions and correspond to the case when curvature function $S_k[R_1, \ldots, R_{n-1}](\boldsymbol{\nu})$ from (40) is constant.

Algebra of barycentric coordinates. Given two sets defined by their support functions, say $p(\boldsymbol{\nu})$ and $q(\boldsymbol{\nu})$, $\boldsymbol{\nu} \in \mathbb{S}$, one can consider algebraic operations over the sets by adding (Minkowski addition), multiplying, and convolving the support functions.

As observed before, the support functions of surfaces $\Sigma_{\boldsymbol{x}}$ corresponding to the Laplacian and Wachspress-Warren coordinates are given by $\rho(\boldsymbol{\nu})$ and $1/\rho(\boldsymbol{\nu})$, respectively, where $\rho(\boldsymbol{\nu})$ is the polar plot of $\partial\Omega$. Thus their product gives us the support function corresponding to the mean value coordinates and we can say that the Laplacian and Wachspress-Warren coordinates are dual w.r.t. to the multiplication of their corresponding support functions.

As noted in [MM06](Section 6, see also referneces to works of H. Görtler therein), the convolution of two support functions inherits properties of the factors. For the sake of simplicity, let us consider the 2D case. The convolution of $p(\theta)$ and $q(\theta)$ is given by

$$(p \otimes q)(\theta) = \frac{1}{2\pi} \int_0^{2\pi} p(\theta - \alpha) q(\alpha) \, d\alpha. \quad (52)$$

If $p(\theta)$ is centrally-symmetric, i.e., $p(\theta + \pi) \equiv p(\theta)$, so is convolution $p \otimes q$. If $q(\theta)$ is of constant-width (Fig. 2 gives an example of a constant-width curve generated from its support function),

$$q(\theta + \pi) + q(\theta) \equiv \text{const},$$

then $p \otimes q$ is also of constant width. Notice that a circle is the only centrally-symmetric set of constant width.

According to (46), if support function $p(\theta)$ of $\Sigma_{\boldsymbol{x}}$ is centrally-symmetric / constant-width, so is weighting function $w(x, \theta)$, and vice versa. Since the support functions corresponding to the weighted Gordon-Wixom coordinates are centrally-symmetric, then "constant-width coordinates" are dual to the weighted Gordon-Wixom ones w.r.t. convolution (52).

7. Directions for further work

As shown in [SJW06], transfinite barycentric interpolation schemes can be viewed as limiting cases of their corresponding discrete versions, barycentric coordinates on polyhedral domains. Certainly the discrete case is more complex than the continuous one considered in this paper. For example, several different discrete counterparts of the transfinite barycentric coordinates are studied in [FHK06, JW05]. Establishing links between discrete barycentric coordinates and discrete Christoffel-Minkowski problems constitutes an interesting topic for further work.

In this paper, we haven't even touched on the transfinite Sibson coordinates [GF99] and Möobius-invariant interpolation [BE03]. Following [GW74], it would be also interesting to consider the Hermite data interpolation problem and study its relationships with fourth-order PDEs.

Another fascinating theme for future research consists of studying links between transfinite barycentric coordinates and continuous valuations on convex sets. [Had57, Sch93]. Some links between the Hadwiger characterization theorem of rigid motion invariant valuations [Had57] and barycentric coordinates are pointed out in [DMA02]. An interesting problem here is to study relations between barycentric coordinates and a more general class of translation invariant valuations [Sch96].

Acknowledgements

I am grateful to Kai Hormann, Tao Ju, and Joe Warren for fruitful discussions and for providing me with their manuscripts prior to publication. I would like to thank the anonymous reviewers for their comments and suggestions.

References

[ACGR02] Almansa A., Cao F., Gousseau Y., Rougé: Interpolation of digital elevation models using AMLE and related methods. *IEEE Transactions on Geoscience and Remote Sensing 40*, 2 (2002), 314–325. 6

[ACJ04] Aronsson G., Crandall M. G., Juutinen P.: A tour of the theory of absolutely minimizing functions. *Bull. Amer. Math. Soc. 41*, 4 (2004), 439–505. 6

[AO06] Arroyo M., Ortiz M.: Local maximum-entropy approximation schemes: a seamless bridge between finite elements and meshfree methods. *International Journal for Numerical Methods in Engineering 65*, 13 (2006), 2167–2202. 1

[BBM05] Berrut J.-P., Baltensperger R., Mittelmann H. D.: Recent developments in barycentric rational interpolation. In *Trends and Applications in Constructive Approximation*, D. H. Mache and J. Szabados and M. G. de Bruin, (Ed.), vol. 151 of *International Series of Numerical Mathematics*. Birkhäuser, Basel, 2005, pp. 27–52. 5

[BE03] Bern M. W., Eppstein D.: Möobius-invariant natural neighbor interpolation. In *Proc. 14th Symp. Discrete. Algorithms* (2003), pp. 128–129. 10

[BF34] Bonnesen T., Fenchel W.: *Theorie der Konvexen Körper*. Springer-Verlag, Berlin, 1934. 7, 9

[Bla30] Blaschke W.: *Vorlesung über Differentialgeometrie I*, dritte erweiterte auflage ed. Springer-Verlag, Berlin, 1930. 7

[Bus58] Busemann H.: *Convex Surfaces*. Interscience, New York, 1958. 7

[BVIK*97] Belikov V. V., V.D. Ivanov V. D., Kontorovich V. K., Korytnik S. A., Semenov A. Y.: The non-Sibsonian interpolation: a new method of interpolation of the values of a function on an arbitrary set of points. *Comput. Math. Math. Phys. 37*, 1 (1997), 9–15. 3

[CFL82] Christ N. H., Friedberg R., Lee T. D.: Weights of links and plaquettes in a random lattice. *Nucl. Phys. B 210*, 3 (1982), 337–346. 3

[CG67] Coxeter H. S. M., Greitzer S. L.: *Geometry Revisited*. Math. Assoc. Amer., Toronto - New York, 1967. 8

[Chr65] Christoffel E. B.: Uber die Bestimmung der Gestalt einer krummen. Oberfläche durch lokale messungen auf derselben. *J. Reine Angew. Math. 64* (1865), 193–209. 8

[CMS98] Caselles V., Morel J.-M., Sbert C.: An axiomatic approach to image interpolation. *IEEE Transactions on Image Processing 7*, 3 (1998), 376–386. 4, 6

[DM06] DeRose T., Meyer M.: *Harmonic coordinates*. Tech. Rep. Pixar Technical Memo #06-02, PIXAR, January 2006. 1

[DMA02] Desbrun M., Meyer M., Alliez P.: Intrinsic parameterizations of surface meshes. *Computer Graphics Forum 21*, 2 (2002), 209–218. Proc. Eurographics 2002. 3, 10

[Far02] Farin G.: A history of curves and surfaces in CAGD. In *Handbook of Computer Aided Geometric Design*, Farin G., Hoschek J., Kim M.-S., (Eds.). Elsevier, 2002, ch. 1. 1

[FH06] Floater M. S., Hormann K.: *Barycentric rational interpolation with no poles and high rates of approximation*. Tech. Rep. IfI-06-06, Department of Informatics, Clausthal University of Technology, May 2006. 5

[FHK06] Floater M. S., Hormann K., Kós G.: A general construction of barycentric coordinates over convex polygons. *Advances in Computational Mathematics* (2006). 4, 10

[Flo03] Floater M. S.: Mean value coordinates. *Computer Aided Geometric Design 20*, 1 (2003), 19–27. 1, 6

[FMR01] FAROUKI R. T., MOON H. P., RAVANI B.: Minkowski geometric algebra of complex sets. *Geometriae Dedicata 85* (2001), 283–315. 8

[GF99] GROSS L., FARIN G. E.: A transfinite form of Sibson's interpolant. *Discrete Applied Mathematics 93*, 1 (1999), 33–50. 10

[GG02] GUAN B., GUAN P.: Convex hypersurfaces of prescribed curvature. *Annals of Mathematics 156* (2002), 655–674. 7

[GW74] GORDON W., WIXOM J.: Pseudo-harmonic interpolation on convex domains. *SIAM J. Numer. Anal. 11*, 5 (1974), 909–933. 1, 4, 5, 10

[Had57] HADWIGER H.: *Vorlesungen über Inhalt, Oberfläche und Isoperimetrie*. Springer, Berlin, 1957. 10

[Her00] HERMAN R. A.: *A Treatise on Geometrical Optics*. Cambridge University Press, 1900. 8

[HF06] HORMANN K., FLOATER M. S.: Mean value coordinates for arbitrary planar polygons. *ACM Transactions on Graphics* (2006). 1

[HS00] HIYOSHI H., SUGIHARA K.: An interpolant based on line segment Voronoi diagrams. In *JCDCG '98: Revised Papers from the Japanese Conference on Discrete and Computational Geometry* (2000), Springer-Verlag, pp. 119–128. 3

[JSW05a] JU T., SCHAEFER S., WARREN J.: A geometric construction of coordinates for convex polyhedra using polar duals. In *Third Eurographics Symposium on Geometry Processing* (Vienna, Austria, July 2005), pp. 181–186. 9

[JSW05b] JU T., SCHAEFER S., WARREN J.: Mean value coordinates for closed triangular meshes. *ACM Transactions on Graphics 24*, 3 (2005), 561–566. Proceedings of ACM SIGGRAPH 2005. 1, 2

[JW05] JU T., WARREN J.: *General Constructions of Barycentric Coordinates in a Convex Triangular Polyhedron*. Tech. rep., Washington University in St. Louis, November 2005. 10

[KFCO06] KLINGNER B. M., FELDMAN B. E., CHENTANEZ N., O'BRIEN J. F.: Fluid animation with dynamic meshes. *ACM Transactions on Graphics 25*, 3 (2006). Proceedings of ACM SIGGRAPH 2006. 1

[MBLD02] MEYER M., BARR A., LEE H., DESBRUN M.: Generalized barycentric coordinates on irregular polygons. *Journal of Graphics Tools 7*, 1 (2002), 13–22. 3

[Min03] MINKOWSKI H.: Volumen und Oberfläche. *Math. Ann. 57* (1903), 447–495. 7

[MM01] MARTINEZ-MAURE Y.: Hedgehogs and zonoids. *Advances in Mathematics 158*, 1 (2001), 1–17. 7

[MM06] MARTINEZ-MAURE Y.: Geometric study of Minkowski differences of plane convex bodies. *Canadian Journal of Mathematics* (2006). 9

[POMZ99] PENG D., OSHER S., MERRIMAN B., ZHAO H.-K.: The geometry of wulff crystal shapes and its relations with Riemann problems. *Contemporary Mathematics 238* (1999), 251–303. 8

[PP99] PINKALL U., POLTHIER K.: Computing discrete minimal surfaces and their conjugates. *Experimental Mathematics 2*, 1 (1999), 15–36. 3

[Sch93] SCHNEIDER R.: *Convex Bodies: The Brunn-Minkowski Theory*. Cambridge Univ. Press, 1993. 7, 8, 9, 10

[Sch96] SCHNEIDER R.: Simple valuations on convex sets. *Mathematika 43* (1996), 32–39. 10

[She68] SHEPARD D.: A two-dimensional interpolation function for irregularly-spaced data. In *Proceedings of the 1968 23rd ACM national conference* (New York, NY, USA, 1968), ACM Press, pp. 517–524. 2, 4

[SJW06] SCHAEFER S., JU T., WARREN J.: A unified, integral construction for coordinates over closed curves. *Computer-Aided Geometric Design* (2006). A Special Issue on Discrete Geometry. Submitted. 1, 3, 9, 10

[SM06] SUKUMAR N., MALSCH E. A.: Recent advances in the construction of polygonal interpolants. *Archives of Computational Methods in Engineering 13*, 1 (2006), 129–163. 1, 3

[STW04] SHENG W., TRUDINGER N., WANG X.-J.: Convex hypersurface of prescribed Weingarten curvatures. *Comm. in Analysis and Geometry 12*, 1 (2004), 213–232. 7

[Sug99] SUGIHARA K.: Surface interpolation based on new local coordinates. *Computer-Aided Design 31* (1999), 51–58. 3

[Suk03] SUKUMAR N.: Voronoi cell finite difference method for the diffusion operator on arbitrary unstructured grids. *International Journal for Numerical Methods in Engineering 57*, 1 (2003), 1–34. 4

[VF92] VAILLANT R., FAUGERAS O. D.: Using extremal boundaries for 3-d object modeling. *IEEE Transactions on Pattern Analysis and Machine Intelligence 14*, 2 (1992), 157–173. 8

[Wac75] WACHSPRESS E. L.: *A Rational Finite Element Basis*. Academic Press, New York, 1975. 3

[War96] WARREN J.: Barycentric coordinates for convex polytopes. *Advances in Computational Mathematics 6*, 2 (1996), 97–108. 3

[War05] WARDETZKY M.: Convergence of the cotan formula - an overview. Preprint, ZIB, 2005. To appear in 'Discrete Differential Geometry (A. I. Bobenko, J. M. Sullivan, P. Schröder, G. Ziegler, eds), Oberwolfach Seminars, Birkhäuser Basel. 4

[Wei84] WEINGARTEN J.: *Über die Theorie der aufeinander abwickelbaren Oberflächen, Festschrift der technischen Hochschule*. Berlin, 1884. 7

[Wei94] WEICKERT J.: Anisotropic diffusion filters for image processing based quality control. In *Proc. Seventh European Conf. on Mathematics in Industry* (Stuttgart, 1994), Teubner, pp. 355–362. 6

[WSHD06] WARREN J., SCHAEFER S., HIRANI A. N., DESBRUN M.: Barycentric coordinates for convex sets. *Advances in Computational and Applied Mathematics* (2006). To appear. 1, 3, 9

Eurographics Symposium on Geometry Processing (2006)
Konrad Polthier, Alla Sheffer (Editors)

A Decomposition-based Representation for 3D Simplicial Complexes

Annie Hui[1], Lucas Vaczlavik[1], Leila De Floriani[1,2]

[1] Department of Computer Science, University of Maryland, College Park, MD (USA)
[2] Department of Computer Science, University of Genova, Genova (Italy)

Abstract

We define a new representation for non-manifold 3D shapes described by three-dimensional simplicial complexes, that we call the Double-Level Decomposition (DLD) *data structure. The DLD data structure is based on a unique decomposition of the simplicial complex into nearly manifold parts, and encodes the decomposition in an efficient and powerful two-level representation. It is compact, and it supports efficient topological navigation through adjacencies. It also provides a suitable basis for geometric reasoning on non-manifold shapes. We describe an algorithm to decompose a 3D simplicial complex into nearly manifold parts. We discuss how to build the DLD data structure from a description of a 3D complex as a collection of tetrahedra, dangling triangles and wire edges, and we present algorithms for topological navigation. We present a thorough comparison with existing representations for 3D simplicial complexes.*

1. Introduction

Simplicial complexes are widely used representation for 3D shapes in computer graphics, Computer Aided Design (CAD) and finite element simulation, because of their inner simplicity and of the availability of algorithms for generating such representations effectively. In this work, we consider the problem of modeling non-manifold 3D shapes described by three-dimensional simplicial complexes. A lot of work has been developed on modeling 3D shapes by decomposing their boundary into triangle meshes, or into more general simplicial 2-complexes [DH05]. These latter are used to model non-manifold shapes, which are subsets of the Euclidean space that can be regarded as combinations of wire frame, surface, solid and cellular decompositions. Informally, a *manifold* object is a subset of the Euclidean space for which the neighborhood of each internal point is homeomorphic to an open ball. Objects that do not fulfill this property at one or more points are called *non-manifold* objects.

Three-dimensional shapes are often discretized as tetrahedral meshes mainly in finite element simulations [CDM04]. Most of the work in the literature, however, has been focused on representations for tetrahedral meshes partitioning manifold shapes. On the other hand, when generating a finite element mesh from a CAD model to meet simulation requirements, several simplification operations need to be per-

formed, such as removal of details, topology modification, e.g. hole removal, or reduction in the dimensionality of some parts, which produce non-manifold geometries (see, for instance, [FRL00]). Thus, the need arises for representing non-manifold 3D shapes discretized as general 3D simplicial complexes, i.e., consisting of tetrahedra, but also of dangling triangles and wire edges describing lower-dimensional geometric entities. A suitable approach would consist of partitioning a 3D non-manifold shape, which has parts of different dimensionalities, into uniformly-dimensional components which are manifold, or nearly manifold. The objective is to be able to understand the structure of a shape, to identify parts of the shape that define characteristic features which are relevant in a specific application environment, like protrusions, or depressions, or parts defining through-holes or handles.

In this work, we address the problem of modeling non-manifold 3D shapes discretized as 3D simplicial complexes through a decomposition-based approach. In [DMMP03], a theory has been proposed addressing the criteria for a sound decomposition of arbitrarily dimensional abstract simplicial complexes, not necessarily embeddable in the Euclidean space. Intuitively, a sound decomposition should remove as many non-manifold singularities as possible, without breaking the complex at manifold parts. Naturally, such decomposition cuts the complex at all non-manifold simplexes. The resulting components have been shown to be almost, but not exactly, manifold. This approach has been shown to produce a unique decomposition of an abstract simplicial complex. In this paper, we consider this decomposition for 3D simplicial complexes embedded in 3D Euclidean space, and we propose dimension-specific algorithms for generating and navigating on such decomposition.

We define an efficient and effective representation for a 3D simplicial complex, that we call the *Double-Level Decomposition (DLD)* data structure. The DLD data structure provides a two-level description of the complex, where the upper level describes the decomposition of the complex into simpler uniformly dimensional components as a graph, while the lower-level representation describes the tetrahedra, dangling triangle and wire edges in each component, with connectivity information and mutual adjacency relations. The DLD data structure is compact, it scales very well to the manifold case (that is, it requires almost the same amount of storage compared with a data structure of the same class specific for manifold 3D complexes) and it supports efficient navigation within the complex. We compare the DLD data structure with a highly optimized representation for 3D simplicial complexes presented in [DH03] as well as with 3D instances of dimension-independent data structures for simplicial complexes.

The remainder of this paper is organized as follows. Section 2 provides the background notions on simplicial complexes and on entities and topological relations in a simplicial complex. Section 3 reviews related work in the areas of topological data structures, and on shape decomposition. In particular, it reviews the theory behind the decomposition of abstract simplicial complexes mentioned above. Section 4 presents an algorithm for the decomposition of a 3D simplicial complex into nearly manifold components. Section 5 describes the DLD data structure, and analyses its storage costs and scalability. Section 6 presents algorithms for retrieving topological relations from the DLD data structure. Section 7 presents a comparison of the DLD data structure with existing representations. Finally, Section 8 draws some concluding remarks.

2. Background Notions

In this Section, we review the notion of simplicial complexes and related definitions as well as the formal definition of topological relations.

2.1. Simplicial Complexes

A Euclidean *simplex* σ of dimension k is the convex hull of $k+1$ linearly independent points in the n-dimensional Euclidean space E^n, $0 \leq k \leq n$. We simply call a *Euclidean simplex* of dimension k a k-simplex. k is called the *dimension* of σ and is denoted $dim(\sigma)$. Any Euclidean p-simplex σ', with $0 \leq p < k$, generated by a set $V_{\sigma'} \subseteq V_\sigma$ of cardinality $p+1 \leq d$, is called a *p-face* of σ. Whenever no ambiguity

arises, the dimensionality of σ' can be omitted, and σ' is simply called a *face* of σ. Any face σ' of σ such that $\sigma' \neq \sigma$ is called a *proper face* of σ.

A finite collection Σ of Euclidean simplexes forms a *Euclidean simplicial complex* if and only if (i), for each simplex $\sigma \in \Sigma$, all faces of σ belong to Σ, and (ii), for each pair of simplexes σ and σ', either $\sigma \cap \sigma' = \emptyset$ or $\sigma \cap \sigma'$ is a face of both σ and σ'. If d is the maximum of the dimensions of the simplexes in Σ, we call Σ a *d-dimensional simplicial complex*, or a *simplicial d-complex*. The *domain*, or *carrier*, of a Euclidean simplicial d-complex Σ embedded in E^n, with $0 \leq d \leq n$, is the subset of E^n defined by the union, as point sets, of all the simplexes in Σ.

The *boundary* of a simplex σ is the set of all faces of σ in Σ, different from σ itself. The *star* of a simplex σ is the set of simplexes in Σ that have σ as a face. Any simplex σ such that the star of σ contains only σ is called a *top* simplex. The *link* of a simplex σ is the set of all the faces of the simplexes in the star of σ which are not incident in σ.

We call h-simplex σ in a d-complex Σ, $0 \leq h \leq d-1$, a *manifold h-simplex* if and only if there are at most two $(h+1)$-simplexes incident at σ. We call an *h-path* any path between two $(h+1)$-simplexes formed by an alternating sequence of h-simplexes and $(h+1)$-simplexes. An h-path, $0 \leq h \leq d-1$, such that every h-simplex in the path is a manifold simplex is called a *manifold path*. Two simplexes are *h-connected* if and only if there exists an h-path joining them. Two $(h+1)$-simplexes are *h-manifold connected* if and only if there exists a *manifold h-path* connecting them. We call a d-complex *densely connected* if it is $(d-1)$-manifold connected. A regular $(d-1)$-connected d-complex in which all $(d-1)$-simplexes are manifold is called a *(combinatorial) pseudo-manifold* complex (possibly with boundary).

A d-dimensional pseudo-manifold in which the link of each vertex is homeomorphic to the unit d-sphere (or to the unit $(d-1)$-dimensional open disk) is called a *manifold complex*. We call a k-simplex σ, with $k < d - 1$, a *non-manifold simplex* if and only if $lk(\sigma)$ consists of more than one connected component. A $(d-1)$-simplex σ is called a *non-manifold simplex* if three or more d-simplexes are incident at σ. Figure 1 shows examples of non-manifold simplexes in a simplicial 3-complex. Figure 1(a) shows a non-manifold edge e, the link of e is highlighted in Figure 1(b). Figure 1(c) shows an example of a non-manifold vertex v.

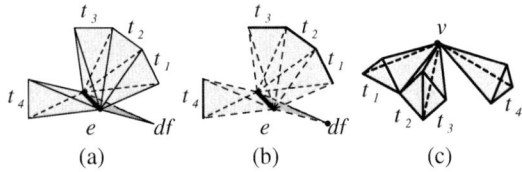

Figure 1: *Singularities in 3D simplicial complexes*

2.2. Topological Relations

We introduce here a formalization of the topological relations among the entities in a simplicial complex. Topological relations are an effective framework for defining, analyzing and comparing the data structures for cell and simplicial complexes [De 05]. They describe the connectivity information among the cells. The choice of cells and relations to be encoded determines the effectiveness of the data structure for a specific application. Topological relations have been defined among the entities in a cell complex. Relations in a simplicial complex can be defined in the same way. Let Σ be a simplicial d-complex and let $\sigma \in \Sigma$ be a p-simplex, with $0 \leq p \leq d$. We define the following *topological relations*:

- For $0 \leq q \leq p-1$, *boundary relation* $R_{p,q}(\sigma)$ consists of the set of q-simplexes in the set of faces of σ.
- For $p+1 \leq q \leq d$, *coboundary relation* $R_{p,q}(\sigma)$ consists of the set of q-simplexes in the star of σ.
- For $p > 0$, *adjacency relation* $R_{p,p}(\sigma)$ is the set of p-simplexes in Σ that are $(p-1)$-adjacent to σ. *Adjacency relation* $R_{0,0}(\sigma)$, where σ is a vertex, consists of the set of vertices σ' such that $\{\sigma, \sigma'\}$ is a 1-simplex of Σ.

Boundary and coboundary relations are called *incidence relations*. In the remainder, we also define various partial relations. Partial relation, generally denoted by $R^*_{p,q}$, is a subset of the complete $R_{p,q}$ relation. We call *constant* any relation which involves a constant number of entities. Relations, which involve a variable number of entities, are called *variable*. Co-boundary and adjacency relations are variable relations in general, while boundary relations are constant. We consider an algorithm for retrieving a topological relation R to be *optimal* if it retrieves a given relation R in time linear with respect to the number of entities involved in R.

3. Related Work

In this Section, we review related work on representations of 3D shapes, and on shape decomposition techniques. In particular, we review the topological decomposition for abstract simplicial complexes proposed in [DMMP03].

3.1. Representations of 3D shapes

Various approaches have been proposed in the literature for representing non-manifold 3D shapes. Most of the works have been on representing a 3D shape through a decomposition of its boundary by a 2D cell or simplicial complex embedded in the 3D Euclidean space. The first proposal for a boundary representation of non-manifold 3D shapes is provided by the Radial-Edge data structure [Wei88]. Several variants of the Radial-edge data structure exist, such as, for instance, the Partial Entities [LL01], and the Loop Edge-use [MH01] data structures. The Directed Edge [CKS98] extends the Half-Edge data structure, proposed for manifold 2D cell complexes, to arbitrary 2D simplicial complexes. All such data structures are verbose and do not scale well to the manifold case [DH05].

An alternative to the previous data structures, with the same expressive power, is the Incidence Graph [Ede87] and its variants [DGH04, VL97]. Such data structures describe a complex by capturing the incidence relations among the cells in the complex. They are simpler to implement and definitely more compact. However, to achieve high compactness, it is necessary to minimize the encoded information by encoding only a subset of the topological entities and relations. In this view, the Indexed data structure with Adjacencies (IA) [Nie97] is a pioneer work, which however is limited to pseudo-manifold complexes. The Triangle-Segment data structure [DMPS04] is the first data structure that extends the indexed data structure to handle non-manifold 2D simplicial complex. It is very compact and allows retrieving topological relations in optimal time (see [DH05] for an analysis and comparison with other representations for 2D simplicial complexes).

There are few representations for describing 3D shapes discretized as 3D simplicial complexes. Most such representations are limited to the manifold domain. Examples are the Facet-Edge [DL89, Muc93] and the Handle-Face [LT97]. The Non-manifold Indexed Data Structure with Adjacencies (NMIA) proposed in [DH03] is suitable for general 3D simplcial complexes.This data structure explicitly represents all vertices and top simplexes, specifically, tetrahedra, dangling faces and wire edges. It is highly compact and supports the efficient retrieval of topological relations.

3.2. Decompositions of 3D shapes

Another approach to represent non-manifold shapes consists of decomposing the shape into manifold components. Some approaches have been proposed in the literature for uniformly-dimensional non-manifold shapes.

Desaulniers and Stewart [DS92] propose a representation scheme based on a decomposition of solid object into regular parts (r-sets). Such a decomposition provides interesting topological information about an object. In [FR92], Falcidieno and Giannini discuss the problem of identifying form features from the r-set decomposition of a non-manifold object. In [GTLH98], Gueziec et al. propose a decomposition-based technique to convert a non-manifold object into a manifold one. Pesco et al. [PTL04] propose a decomposition of a 2D cell complex based on a combinatorial stratification of the complex, inspired by Whitney stratification. They propose a data structure and a set of operators based on such representation, but they do not provide an algorithm for building it from a given (non-decomposed) complex. Selective Geometric Complexes (SGCs) [RO90] can describe objects through cell complexes whose cells can be either open, or not simply connected. In SGCs, cells and their mutual adjacencies are encoded in an incidence graph [Ede87]. Lopes et al. [LNPT00] define a stratification of 3D cell complexes, but limited to manifold ones, and propose a data structure and editing operators for manipulating it.

3.3. Initial Quasi-Manifold Decomposition

A decomposition of a non-manifold shape into simpler parts can be obtained by splitting the shape at those elements (vertices, edges, faces, etc.) where singularities occur. In order to be effective, the decomposition process should remove as many singularities as possible, without introducing artificial, or arbitrary, "cuts" through manifold parts. Under these assumptions, a decomposition into manifold components is possible, in general, only for two-dimensional complexes. In three or higher dimensions, a decomposition into manifold components may need to introduce artificial cuts through the object. In six or higher dimensions, a decomposition into manifold components is not feasible in general, since the class of d-manifolds has been proven to be not decidable for $d \geq 6$ [Nab96]. A decomposition has been defined in [DMMP03] for abstract simplicial complexes, and it is described here in the context of complexes embeddable in the Euclidean space. This decomposition is *unique*, since it does not make any arbitrary choice in deciding where the object has to be decomposed, and *natural*, since it removes singularities by splitting the complex at non-manifold simplexes only.

A complex Σ' is a decomposition of another complex Σ whenever Σ' can be obtained by *cutting* Σ along some faces. If Σ' is a decomposition of Σ, then any other decomposition of Σ' will also be a decomposition for Σ. This fact induces a partial order over the set of all possible decompositions of a complex, in which Σ is the minimum and the complex obtained by decomposing Σ into the collection of its top simplexes is the maximum. This latter complex is called the *totally exploded* decomposition of Σ, and it is denoted with $\Sigma^{\top}$. Any decomposition of Σ can be seen as obtained by *pasting* together simplexes in $\Sigma^{\top}$. Pasting occurs through atomic operations that identify two vertices of the form v_n and v_m at a time. The set of all possible decompositions of a complex Σ forms a lattice. Two complexes adjacent in the lattice can be transformed into each other by an atomic split or join involving just a pair of vertices. Figure 2 gives an example of all possible decompositions of a complex and the resultant lattice. The complex in the root (top) of the lattice is the totally exploded complex. The complex in the sink (bottom) is the original complex. Each edge in the lattice connects two adjacent complexes, indicating that one complex can be obtained from another by cutting (if moving up the lattice) or pasting (if moving downwards).

The standard decomposition of a complex is a specific element of the lattice, which is obtained by discarding the whole set of "non-interesting" decompositions, and taking the "most general" of the remaining decompositions. Usually, one perceives non-manifold simplexes as "joints" between manifold parts, and it might seem reasonable to build a decomposition by splitting the complex just at them. On the other hand, it does not seem desirable to introduce cuts along manifold simplexes. Such a decomposition Σ' is considered in some sense "essential". A decomposition Σ' is an

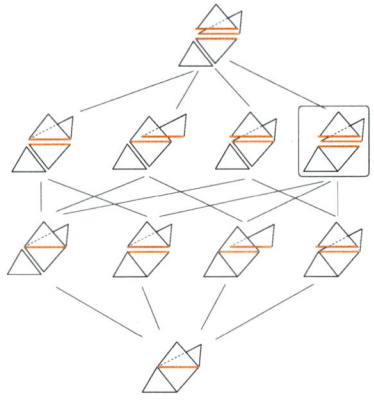

Figure 2: *An example of the lattice of all possible decompositions of a 2D simplicial complex. The standard decomposition is marked. All the essential decompositions are those belonging to the sub-lattice rooted at the standard decomposition*

essential decomposition of Σ if and only if Σ' is obtained by splitting Σ only at non-manifold simplexes.

Among all the essential decompositions, the *standard* decomposition $\Delta\Sigma$ is the most decomposed. Thus, it has been decomposed at all singularities that can be eliminated by cutting only through non-manifold faces. It is easy to see that $\Delta\Sigma$ must be a complex with regular connected components. Moreover, all connected components belong to the class of *Initial Quasi-Manifolds (IQMs)*. A regular d-complex Σ is called an *initial quasi-manifold* if and only if every pair of d-simplexes in the star of every vertex of Σ is $(d-1)$-manifold-connected. Up to dimension two, the class of initial quasi-manifolds coincides with that of manifolds. In general, in three or higher dimensions, an IQM is not always a manifold (and not even a pseudo-manifold). However, in dimension $d \geq 3$, if an IQM is embeddable in E^d, it must be a pseudo-manifold complex. Figure 3 shows an example of a 3D initial quasi-manifold, which is not a manifold complex.

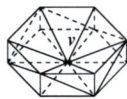

Figure 3: *A pinched-pie, which is an 3D initial qusai-manifold but not a manifold*

In [DMP03], a representation for abstract simplicial complexes, not necessarily embeddable in the Euclidean space, has been designed. Any h-dimensional IQM component is described by an adjacency-based data structure, in which relation $R_{h,h}(\sigma)$ for h-simplex σ is encoded, that, however, involves an arbitrary number of h-simplexes adjacent to σ, since it is not guaranteed that a component is a pseudo-manifold. This representation has not been implemented, and it can be heavily simplified when dealing with d-complexes embedded in E^d for a specific value of d.

4. Decomposition of 3D simplicial complex

In this Section, we present an algorithm for computing the IQM decomposition of a 3D simplicial complex. In a 3D simplicial complex, non-manifold singularities may occur at edges and vertices. The IQM decomposition can be obtained by cutting the complex along all non-manifold vertices and non-manifold edges. For an efficient computation of such decomposition, we need: adjacency relations $R_{3,3}$ for all tetrahedra, adjacency relations $R_{2,2}$ for all dangling faces, and the stars of all the vertices. The decomposition algorithm performs the following five steps, that are detailed in the rest of this Section:

1. Compute adjacency relation $R_{3,3}$ for all tetrahedra.
2. Compute adjacency relation $R_{2,2}$ for all dangling faces.
3. Compute the stars of all vertices, where each star is described as the set of all top simplices incident at that vertex.
4. Identify non-manifold edges and non-manifold vertices through a traversal of the star of each vertex.
5. Decompose the complex at non-manifold simplices and identify IQM components.

Step 1: Compute adjacency relation $R_{3,3}$ for all tetrahedra.

An efficient way to compute it is to sort the tetrahedra by their four faces. It can be done as follows:

1. The faces of the tetrahedra are not explicit in the input. Each such face can be described through a 4-tuple (u_1, u_2, u_3, t), where u_1, u_2, u_3 are three vertices that describe one face of t, and are sorted in the increasing order of their indices. Each 4-tuple not only identifies a unique face, but also associates the face with a tetrahedron bounded by it. For each tetrahedron t four 4-tuples are created.
2. After sorting all the 4-tuples in lexicographical order, adjacent 4-tuples of the form (u_1, u_2, u_3, t_1) and (u_1, u_2, u_3, t_2) indicate that tetrahedra t_1 and t_2 are face-adjacent.

The time complexity for this step is $O(m_3 log(m_3))$, where m_3 denotes the number of tetrahedra in the complex.

Step 2: Compute adjacency relation $R_{2,2}$ for all dangling faces.

The technique is the same as the computation of relation $R_{3,3}$ for tetrahedra described above. The complexity of this step is, thus, $O(d_2 log(d_2))$, where d_2 denotes the number of dangling faces in the complex.

Step 3: Compute the stars of all vertices.

This is performed as follows:

1. For each vertex v and each h, create empty sets, $b(v, h)$, which we call *buckets*, for collecting all the top simplices of dimension h incident at v.
2. For each top h-simplex σ described by vertices $\{v_1, \cdots, v_{h+1}\}$, add σ to buckets $b(v_i, h)$, for $i = 1, \cdots, h+1$.

This step is performed in time linear with respect to the number of vertices and the number of top simplices, i.e., $O(v_0 + w_1 + d_2 + m_3)$, where v_0 is the number of vertices, w_1 the number of wire edges, d_2 the number of dangling faces and m_3 the number of tetrahedra.

Step 4: Identify non-manifold edges and non-manifold vertices.

Non-manifold vertices and edges are identified through a traversal of the star of each vertex. This traversal is done by using the information stored in the buckets $b(v, h)$, plus the relations $R_{3,3}$ for tetrahedra, and $R_{2,2}$ for dangling faces. During the traversal, the top simplices in the star of v are grouped into densely $(h-1)$-connected components. Each component found is assigned a unique label, which we call the *component index*. All vertices (except for v) in a component C are labeled with the index of C. These labels are used for identifying non-manifold edges in the star of v. If a vertex u in the link of v has more than one label, then edge (u, v) is a non-manifold edge. If the star of v consists of more than one component, then v is a non-manifold vertex; it is a manifold vertex otherwise. Algorithm 1 provides a pseudo-code description of the traversal strategy.

Algorithm 1 FindComponentsInStar(v, b)

```
 1:  j ← 1
 2:  for h from 3 downto 1 do
 3:      while b(v, h) is not empty do
 4:          Remove the unvisited top simplex σ from b(v, h)
 5:          Create new component C_j for v
 6:          Enqueue(Q, σ)
 7:          while not empty(Q) do
 8:              σ ← Dequeue(Q)
 9:              C_j ← C_j ∪ σ
10:              for each γ in R_{h,h}(σ) do
11:                  if the (h−1)-face between σ and γ is manifold
                          and visited(γ)=0 then
12:                      visited(γ) ← 1
13:                      Enqueue(Q, γ)
14:                  end if
15:              end for
16:          end while
17:          for each σ in C_j do
18:              Add label j to all vertices of σ (except v)
19:          end for
20:      end while
21:  end for
```

We illustrate the labeling of the star of v through the example in Figure 4. In this example, the four tetrahedra form two densely 2-connected components and the three dangling faces three densely 1-connected components in the star of vertex v. The vertices in the link of v are labeled according to the component(s) to which they belong, thus exposing the non-manifold edges in the star of v.

The traversal of the star of each vertex is a linear process with respect to the number of top simplices in that star. The

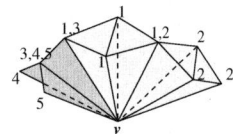

Figure 4: *(a) Labeling densely connected components in the star of vertex v*

time complexity for Step 4 is thus $O(\sum_{v \in \Sigma} |st(v)|)$, where $|st(v)|$ denotes the size of the star of vertex v in Σ. Since each h-simplex belongs to the stars of exactly $h+1$ vertices, the time complexity results to be linear in the number of top simplexes in the complex.

Step 5: Decompose non-manifold simplexes and identify IQM components. To complete the decomposition, the complex is cut at the non-manifold simplexes. For each non-manifold vertex v, one vertex copy v_i is created for each IQM component in the star of v. After the cutting, the whole complex is traversed once, but the traversal does not pass through non-manifold edges and non-manifold vertices. All tetrahedra that are 2-connected belong to the same IQM component. All the dangling faces that are 1-connected form a separate manifold component. Likewise for all wire edges that are 0-connected.

The star of each non-manifold vertex is partitioned when copies are created for the non-manifold vertex. The subsequent traversal of the whole complex takes linear time with respect to the size of the complex. Thus this step takes $O(\sum_{v_s \in \Sigma} |st(v_s)|) + O(v_0 + w_1 + d_2 + m_3 + k_0)$, where $|st(v_s)|$ denotes the size of the star of non-manifold vertex v_s in Σ, k_0 is the number of IQM components at all non-manifold vertices, and v_0, w_1, d_2, m_3 denote the number of vertices, wire edges, dangling faces and tetrahedra respectively.

Both Steps 1 and 2 involve sorting, while all the other steps perform operations that are linear in terms of the total number of top simplexes in the complex. For a typical 3-complex that is mostly 3-manifold with few dangling faces and wire edges, the time consumption of the decomposition is dominated by Step 1.

5. A Decomposition-based Data Structure for a Simplicial 3-complex

In this Section, we present the *Double-Level Decomposition (DLD) data structure*, which is based on the IQM decomposition and is generated through the algorithm described in Section 4. The DLD data structure is a two-layer representation in which the upper level describes the connectivity of the IQM components through their non-manifold simplexes, while the lower level describes the entities, their connectivity and adjacency relation inside the IQM components $C_1, \cdots, C_k$. This is similar in concept to the representation proposed in [DMMP03] for decomposed abstract simplicial complexes which is still a two-level data structure, but the

description of the single IQM component is more complex since it may not necessarily be pseudo-manifold.

The connectivity of the components in the decomposition is represented as a hypergraph $G = <N, A>$, where N is a set of nodes representing the IQM components $C_1, \cdots, C_k$, and A is a set of hyperarcs. There are two kinds of hyperarcs: *hyperarcs of type vertex*, which represent non-manifold vertices, and *hyperarcs of type edge* which represent non-manifold edges. A hyperarc representing a non-manifold vertex v connects all components which contain copies of vertex v. Similarly, a hyperarc representing a non-manifold edge e connects all components which contain copies of edge e.

Figures 5(a)-(c) give an example of the IQM decomposition of a simple 3-complex and the hypergraph that represents the decomposition. The 3-complex shown in Figure 5(a) consists of two tetrahedra that share the non-manifold edge e which is incident at non-manifold vertices u and v, and two wire edges that are incident at vertex v. The decomposition of this complex consists of four components: C_1 and C_2 are the two tetrahedra, C_3 and C_4 are the wire edges. Figure 5(b) shows all the components of the decomposition. The non-manifold edge e and non-manifold vertices u and v are copied for each component. Figure 5(c) is a full description of the decomposition graph G. The nodes are $C_1, \cdots, C_4$ and the hyperarcs are e, u and v. In the hypergraph, the solid lines connecting C_i and the hyperarcs are the copies of the non-manifold joints. The dashed lines between u, v and e indicate their incidence.

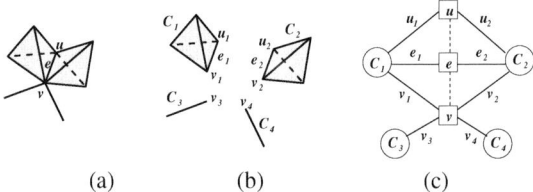

(a)	(b)	(c)

Figure 5: *(a) a 3D complex; (b) its IQM decomposition; and (c) the hypergraph describing the decomposition*

All non-manifold singularities are thus explicitly represented only in the upper level, which encodes hypergraph G. The following information are encoded:

- For each node representing IQM component C_i:

  - dimension of the component;
  - and a pointer to one top simplex in this component.

- For each hyperarc representing non-manifold edge e: (We consider hyperarc e in Figure 5(c) to illustrate the following)

  - a pointer each to its extreme vertices, which are hyperarcs in G (in our illustration, they are hyperarcs u and v for e);
  - two lists of pointers: each pointer references a representative top simplex for each 2D or 3D IQM compo-

nent in the star of e. One list collects the 2D representatives and the other the 3D representatives (in our illustration, the 2D list of e is empty while its 3D list consists of the two tetrahedra in Figure 5(b)).

- For each hyperarc representing non-manifold vertex v: (See hyperarc v in Figure 5(c) as an example)

  - A list of pointers, each to the vertex copy of v in each IQM component in the star of v, (for the example, the list of v consists of copies $v_1, \cdots, v_4$);
  - a list of pointers, one for each non-manifold edge in graph G, that is incident at v (for the same example, the list of v contains hyperarc e).

The lower level describes the IQM components. For any h-dimensional IQM component C_i, we use the Indexed data structure with Adjacencies which encodes all the h-simplexes, the vertices and the following relations:

- For each vertex v in the component, relation $R_{0,h}^*(v)$ which consists of one h-simplex in the star of v;
- For each h-simplex σ of C_i, relation $R_{h,0}(\sigma)$ and relation $R_{h,h}(\sigma)$

The low level data structure is implemented through the following constructs:

- For each vertex v in component C_i:

  - A 1-bit flag to indicate whether v is manifold;
  - One pointer for relation $R_{0,h}^*(v)$;

- For each wire edge σ of C_i: a pointer array of size 2 for relation $R_{1,0}(\sigma)$
- For each top h-simplex σ of C_i, $h > 1$:

  - A pointer array of size $(h+1)$ for relation $R_{h,0}(\sigma)$
  - A pointer array of size $(h+1)$ for relation $R_{h,h}(\sigma)$
  - A bit flag of size $\binom{h+1}{1}$ to indicate whether each edge of σ is manifold;

- A hash table H_v that stores the pointers from the vertex copies to the node that corresponds to v in graph G.

The storage cost required by the DLD data structure can be evaluated in terms of the following quantities:

m_0 : number of manifold vertices;

n_0 : number of non-manifold vertices;

k_0 : total number of IQM components at all non-manifold vertices;

n_1 : number of non-manifold edges;

w_1 : number of wire edges;

k_1 : total number of IQM components at all non-manifold edges;

d_2 : number of dangling faces;

m_3 : number of tetrahedra;

C : total number of IQM components in the whole complex.

In the lower level data structure, the total number of vertices (including all manifold vertices and copies of non-manifold vertices) is $m_0 + k_0$. Assuming that the hash tables are 10% full in order to support constant access time, the size

of the hash table H_v is $20n_0$ pointers. The storage cost of the DLD data structure for various domains is shown below:

- For general non-manifold complexes:

  - lower level: $m_0 + k_0 + 2w_1 + 4d_2 + 8m_3 + 20n_0 + 30n_1$ pointers and $m_0 + k_0 + 3d_2 + 6m_3$ bits,
  - upper level: $2C + 6n_1 + 2n_0 + k_1 + k_0$ pointers,
  - hash table: $20n_0$ pointers.

- For manifold complexes, $d_2 = w_1 = n_0 = n_1 = k_0 = k_1 = 0$ and $C = 1$

  - lower level: $m_0 + 8m_3$ pointers and $m_0 + 6m_3$ bits,
  - upper level: 2 pointers,
  - hash table: 0 pointers.

A comparison with the extended Indexed data structure with Adjacencies (IA)[Nie97] for manifolds gives us a measure of the scalability of the DLD data structure. When encoding manifolds, the DLD data structure is reduced to the IA with just some additional bit flags. Thus, the overhead of the DLD data structure in encoding manifold is $m_0 + 6m_3$ bits and 2 pointers.

6. Navigation in the DLD data structure

In this Section, we discuss how to retrieve topological relations from the DLD data structure. These algorithms are the basic building blocks for any algorithm which navigates or updates the complex.

Boundary relation can be retrieved both for top simplexes and for faces of top simplexes. For any top p-simplex, σ ($p = 2, 3$), the set of q-faces of σ are described as $\binom{p+1}{q+1}$ combinations of $(q+1)$ vertices of σ. Thus, to retrieve boundary relation $R_{p,q}(\sigma)$, relation $R_{p,0}(\sigma)$ is retrieved, and the combinations describing the q-faces are generated.

We retrieve coboundary relation $R_{p,q}(\sigma)$ of a p-simplex σ through a traversal of the star of σ, ($p = 0, 1, 2$). In the case in which σ is a manifold simplex, all q-simplexes incident at σ belong to the same IQM component. Thus, the traversal of the star of σ is performed within the lower level data structure. When σ is non-manifold, the star of σ is distributed among several components. Therefore, it is necessary to access the upper level data structure to retrieve all the components incident at σ.

Relation $R_{0,h}(v)$, $h = 2, 3$, in an h-dimensional IQM component C is retrieved by traversing the star of v in C through relations $R_{0,h}^*$, $R_{h,h}$ and $R_{h,0}$. The traversal is performed by starting with h-simplex $\sigma = R_{0,h}^*(v)$. The h-simplexes $(h-1)$-adjacent to σ are found through $R_{h,h}(\sigma)$, and those which are incident at v are found by considering $R_{h,0}$ for such h-simplexes. This process is linear in the number of h-simplexes incident at v. If we want to retrieve $R_{0,q}(v)$ in an h-dimensional IQM component with $q < h$, we perform the same traversal described above, but we collect as result the q-faces of the h-simplexes found in the retrieval. The time complexity is still linear in the number of q-simplexes incident at v, since the number of h-simplexes in the star of v is linear in the number of q-simplexes incident at v because of Euler' formula.

Next, we consider how to retrieve relation $R_{0,q}(v)$, $0 < q \leq h$, for a non-manifold vertex u. The star of u is the union of the stars of all its vertex copies. The vertex copies of u are retrieved from the upper level data structure. Relation $R_{0,q}$ for each vertex copy is retrieved from the lower level data structure as though the vertex copy u was a manifold vertex. Given an arbitrary vertex v in a given component C of dimension h, the bit-flag indicates whether v is a vertex copy. If it is, the reference to the hyperarc representing the non-manifold vertex is retrieved from the hash table H_v. From the hyperarc, we retrieve all the other copies of the same vertex, and then the q-simplexes incident at each such copy. Thus, all $R_{0,q}(v)$ relations, where $0 < q \leq h$, can be retrieved in time linear in the number of q-simplexes in the star of v.

We consider how to retrieve co-boundary relation of type $R_{1,q}(e)$, $0 \leq q \leq h$, for an edge e. If e is a manifold edge, we consider a tetrahedron or triangle σ containing it. (Note that since the edges are not encoded in the DLD data structure for the IQM component, we consider all edges to be specified through a top simplex containing it.) For a 3-component, we retrieve all tetrahedra, or triangles, depending on whether $q = 2$ or 3, incident at e, by traversing the star of e starting from σ, and retrieving all the other tetrahedra or triangles, by using $R_{h,h}$ and $R_{h,0}$ relations. For a 2-component, $R_{1,2}(e)$ is retrieved by simply considering $R_{2,2}$ of triangle σ.

If e is a non-manifold edge, we get access from the hyperarc describing it to a top simplex in each component containing it. For each component, we repeat the process discussed above for the manifold edge. The time complexity of this algorithm is linear in the number of q-simplexes incident at e.

Coboundary relation $R_{2,3}(f)$ for a triangle f is retrieved through the $R_{3,3}$ relation of a tetrahedron that shares f. Adjacency relations $R_{p,p}$ $(p = 0, 1, 2)$ are retrieved as a combination of boundary and coboundary relations, and are not elaborated here. Their time complexity is linear in the number of p-simplexes produced as result of retrieval. Thus, all topological relations can be extracted in optimal time from the DLD data structure.

7. Comparisons

In this Section, we analyze and highlight the distinctive features of the DLD data structure by comparing it with existing data structures for representing non-manifold simplicial complexes.

7.1. Non-manifold Indexed data structure with Adjacencies (NMIA)

The *Non-Manifold Indexed data structure with Adjacencies (NMIA)* [DH03] is a highly compact data structure for simplicial 3-complexes. The NMIA data structure encodes vertices and top simplexes, and the following complete relations: $R_{h,0}(\sigma)$, for each h-dimensional top simplex σ,

$R_{3,3}(t)$, each tetrahedron t. In addition, partial coboundary relations are encoded for vertices and non-manifold edges. Vertex-based partial relations encoded by the NMIA associates with each vertex v one q-dimensional connected component in $st(v)$. In the example of Figure 6(a) each of the top simplexes is a connected component by itself, so the partial coboundary relations encoded at v consist of we, df and t. Edge-based partial relations encoded by the NMIA associate each non-manifold edge e with one top q-simplex in each q-dimensional component incident at e. In Figure 6(b), there are three connected components at edge e, so the partial relations by the NMIA for edge e consist of df, t_1 and t_4.

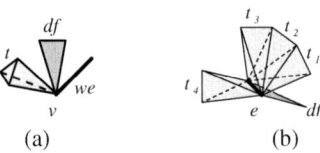

Figure 6: *Partial coboundary relations encoded by the NMIA at (a) non-manifold vertex v and (b) non-manifold edge e*

Both the DLD and the NMIA data structures encode only vertices and top simplexes. Their primary difference is that the DLD data structure encodes the complex as an IQM decomposition, thus allowing the non-manifold singularities to be explicitly addressable, while the NMIA encodes the complex as a single piece with non-manifold singularities distributed inside the complex. Both the NMIA and the DLD data structure are comparable to the extended IA when the domain is manifold. Also, both data structures support an efficient retrieval of topological relations. The retrieval algorithms for relations $R_{p,3}$, for $p = 0, 1$, are sub-optimal for the NMIA data structure, but still linear in the number of top simplexes in the star of a vertex for $p = 0$, or edge for $p = 1$.

7.2. Incidence Graph and Simplified Incidence Graph

The *Incidence Graph (IG)* [Ede87] is a data structure for representing cell complexes of any dimensions. We consider it for simplicial complexes here. The IG encodes the following topological relations:

- for each p-simplex σ, where $0 < p \leq d$, boundary relations $R_{p,p-1}(\sigma)$,
- for each p-simplex σ, where $0 \leq p < d$, coboundary relations $R_{p,p+1}(\sigma)$

Thus, for each p-simplex σ, the IG encodes its immediate boundary, and its immediate coboundary.

The *Simplified Incidence Graph (SIG)* [DGH04] simplifies the IG by encoding the coboundary partially as follows: for each simplex σ, the SIG encodes one top h-simplex for each $(h-1)$-connected component of top h-simplexes in the star of σ. Figure 7 shows two examples of the coboundary relations encoded at a vertex. For the example of Figure 7(a), the SIG encodes just one triangle as partial coboundary relation $R_{0,2}$ of v. In Figure 7(b), the partial coboundary relation encoded by SIG at v consists of two triangles.

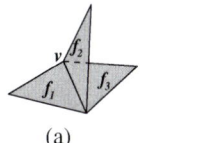 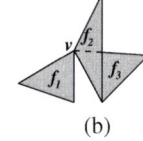

(a)   (b)

Figure 7: *Two examples of partial coboundary relations encoded by the SIG at vertex v*

The SIG encodes the same boundary relations as the IG. It simplifies the IG by encoding a subset of the coboundary instead of the complete immediate coboundary. We have shown in [DH05] that the SIG is almost always more compact than the IG. The main difference with respect to the DLD data structure is that both the IG and the SIG are dimension-independent, and they encode all simplexes in a complex. Both the IG and the SIG support topological navigation in optimal time, as does the DLD data structure.

7.3. Comparison on storage costs

In this Subsection, we provide a comparison of the storage costs of the four data structures on synthetic data sets with high degree of non-manifoldness. Table 1 summarizes the characteristics of each data set. The storage costs of the data structures for each of them are shown in Table 2. The storage cost is measured in terms of the number of entities and relations encoded, which is independent of each individual implementation.

From Table 2, we can make three remarks. First, the number of entities encoded by the NMIA and the DLD data structures are $\frac{1}{4}$ to $\frac{1}{3}$ of that encoded by the IG and by the SIG. Second, the DLD data structure is more compact than the SIG and IG, since it occupies only less than $\frac{1}{2}$ the size of the SIG and even less than that of the IG. Finally, the storage cost of the DLD data structure is almost the same as that of the NMIA data structure, especially when there is a restricted number of non-manifold singularities with respect to number of simplexes in the complex (as it is almost always the case in practical applications).

Model	T	F	E	V	DF	WE	V_n	E_n
3D13	28	82	84	31	0	0	1	0
3D15	40	116	112	37	0	0	5	4
3DDe	30	97	101	37	8	2	5	4
3D25	48	208	234	79	56	0	7	6

Table 1: *Four 3D data sets with non-manifold properties: T=#tetrahedra, F=#faces, E=#edges, DF=#dangling faces, WE=#wire edges, V=#vertices, V_n=#non-manifold vertices, E_n=#non-manifold edges,*

Model	IG		IS		NMIA		DLD	
	Ent	Rel	Ent	Rel	Ent	Rel	Ent	Rel
3D13	225	1052	225	756	59	258	59	263
3D15	305	1464	305	1045	77	373	77	405
3DDe	265	1226	265	879	77	327	77	389
3D25	569	2568	569	1815	183	695	183	943

Table 2: *Storage costs of four data structures on 3D data sets with non-manifold properties: Ent=#entities, Rel=#relations*

8. Concluding Remarks

We have addressed the issue of modeling non-manifold 3D shapes discretized as 3D simplicial complexes through a decomposition-based approach. To this aim, we have described a data structure for 3D simplicial complexes embedded into 3D Euclidean space based on unique and sound decomposition of such complexes into nearly manifold components, that we termed the DLD data structure. The structure of the decomposition is encoded as the upper level in the DLD data structure, while each nearly manifold component is described as an extended indexed data structure with adjacencies by encoding connectivity and adjacency relations. The DLD data structure is compact, since it explicitly encodes only vertices and top simplexes, it is highly scalable to the manifold case, and it supports efficient retrieval of topological relations. The DLD data structure is more expressive than the NMIA data structure [DH03], since it describes the non-manifold entities explicitly, and has almost the same storage cost. Also, navigation algorithms are simpler to implement on the DLD data structure since non-manifold singularities are kept distinct from the lower-level representation of the IQM components. The DLD data structure, its construction and navigation algorithms have been implemented and tested on synthetic data sets. Figures 8(a)-(f) show a 3D complex described by the DLD data structure.

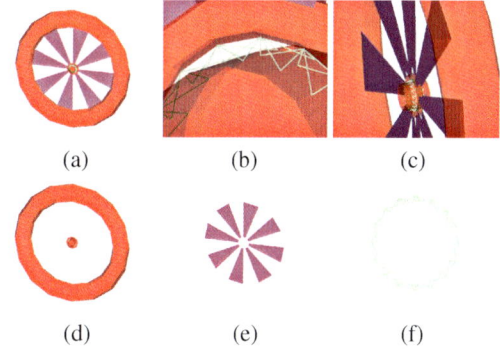

(a)   (b)   (c)

(d)   (e)   (f)

Figure 8: *(a) A wheel model with 1000 tetrahedra, 32 dangling faces and 96 wire edges; (b) a zoomed-in view of its center; (c) a side-way view; (d) the tetrahedral parts form three 3D manifold components; (e) the parts described by dangling faces form eight 2D components; and (f) the wire parts (enlarged)*

A further important advantage of the DLD data structure is that it provides a unique decomposition of a 3D shape and thus is a very suitable basis for performing geometric reasoning of a 3D shape by identifying interesting topological features, and for shape understanding and recognition. The decomposition also makes it easier to extract topological invariants, such as Betti numbers, which can be used as a shape signature for reasoning and recognition purposes.

9. Acknowledgement

This work has been partially supported by the European Network of Excellence AIM@SHAPE under contract number 506766, and by Project FIRB-MIUR SHALOM (SHape modeLing and reasOning: new Methods and tools) under contract number RBIN04HWR8 .

References

[CDM04] CUTLER B., DORSEY J., MCMILLAN L.: Simplification and improvement of tetrahedral models for simulation. In *Proc. 2nd Eurographics SGP* (2004).

[CKS98] CAMPAGNA S., KOBBELT L., SEIDEL H.-P.: Directed edges - a scalable representation for triangle meshes. *J. of Graphics Tools 3*, 4 (1998), 1–12.

[De 05] DE FLORIANI L.: Combinatorial characteristics of shapes. *Lect. notes on Geometric and Solid Modeling* (Fall 2005). Dept of Computer Science, Univ. of Maryland.

[DGH04] DE FLORIANI L., GREENFIELDBOYCE D., HUI A.: A data structure for non-manifold simplicial *d*-complexes. In *Proc. 2nd Eurographics SGP* (2004), ACM Press.

[DH03] DE FLORIANI L., HUI A.: A scalable data structure for three-dimensional non-manifold objects. In *Proc. 1st Eurographics SGP* (2003), pp. 73–83.

[DH05] DE FLORIANI L., HUI A.: Data structures for simplicial complexes: an analysis and a comparison. In *Proc. 3rd Eurographics SGP* (2005), pp. 119–128.

[DL89] DOBKIN D., LASZLO M.: Primitives for the manipulation of three-dimensional subdivisions. *Algorithmica 5*, 4 (1989), 3–32.

[DMMP03] DE FLORIANI L., MESMOUDI M. M., MORANDO F., PUPPO E.: Non-manifold decompositions in arbitrary dimensions. *CVGIP: Graphical Models 65*, 1/3 (2003), 2–22.

[DMP03] DE FLORIANI L., MORANDO F., PUPPO E.: Representation of non-manifold objects in arbitrary dimension through decomposition into nearly manifold parts. In *8th ACM Symposium on Solid Modeling and Applications* (June 2003), ACM Press, pp. 103–112.

[DMPS04] DE FLORIANI L., MAGILLO P., PUPPO E., SOBRERO D.: A multi-resolution topological representation for non-manifold meshes. *Computer-Aided Design J. 36*, 2 (February 2004), 141–159.

[DS92] DESAULNIER H., STEWART N.: An extension of manifold boundary representation to R-sets. *ACM Trans. on Graphics 11*, 1 (1992), 40–60.

[Ede87] EDELSBRUNNER H.: *Algorithms in Combinatorial Geometry*. Springer Verlag, Berlin, 1987.

[FR92] FALCIDIENO B., RATTO O.: Two-manifold cell-decomposition of R-sets. In *Proc. Computer Graphics Forum* (1992), vol. 11, pp. 391–404.

[FRL00] FINE L., REMONDINI L., LÉON J.-C.: Automated generation of FEA models through idealization operators. *Int'l J. for Numerical Methods in Engineering 49* (2000), 83–108.

[GTLH98] GUEZIEC A., TAUBIN G., LAZARUS F., HORN W.: Converting sets of polygons to manifold surfaces by cutting and stitching. In *Conference abstracts and applications: SIGGRAPH 98* (1998), ACM Press, pp. 245–245.

[LL01] LEE S. H., LEE K.: Partial-entity structure: a fast and compact non-manifold boundary representation based on partial topological entities. In *Proc. 6th ACM Solid Modeling and App.* (2001), ACM Press, pp. 159–170.

[LNPT00] LOPES H., NONATO L. G., PESCO S., TAVARES G.: Dealing with topological singularities in volumetric reconstruction. In *Curve and Surface Design* (2000), Vanderbilt University Press, pp. 229–238.

[LT97] LOPES H., TAVARES G.: Structural operators for modeling 3-manifolds. In *Proc. 4th ACM Solid Modeling and App.* (1997), ACM Press, pp. 10–18.

[MH01] MCMAINS S., HELLERSTEIN C. S. J.: Out-of-core building of a topological data structure from a polygon soup. In *Proc. 6th ACM Solid Modeling and App.* (2001), pp. 171–182.

[Muc93] MUCKE E.: *Shapes and Implementations in Three-Dimensional Geometry*. PhD thesis, Dept of Computer Sci., Univ. of Illinois at Urbana-Champaign, Urbana, 1993.

[Nab96] NABUTOVSKY A.: Geometry of the space of triangulations of a compact manifold. *Commun. Math. Phys. 181* (1996), 303–330.

[Nie97] NIELSON G. M.: Tools for triangulations and tetrahedralizations and constructing functions defined over them. In *Scientific Visualization: overviews, Methodologies and Techniques*. IEEE Computer Society, 1997, ch. 20, pp. 429–525.

[PTL04] PESCO S., TAVARES G., LOPES H.: A stratification approach for modeling two-dimensional cell complexes. *Computers and Graphics 28* (2004), 235–247.

[RO90] ROSSIGNAC J. R., O'CONNOR M. A.: SGC: a dimension-independent model for point-sets with internal structures and incomplete boundaries. In *Geometric Modeling for Product Engineering*. Elsevier Sci. Publishers B. V., 1990, pp. 145–180.

[VL97] VÉRON P., LÉON J. C.: Static polyhedron simplification using error measurements. *Computer-Aided Design 29*, 4 (1997), 287–298.

[Wei88] WEILER K.: The radial-edge data structure: a topological representation for non-manifold geometric boundary modeling. In *Geometric Modeling for CAD Applications*. Elsevier Sci. Publishers B. V., 1988, pp. 3–36.

Eurographics Symposium on Geometry Processing (2006)
Konrad Polthier, Alla Sheffer (Editors)

Folding meshes: Hierarchical mesh segmentation based on planar symmetry

Patricio Simari, Evangelos Kalogerakis and Karan Singh[†]

DGP Lab, Department of Computer Science, University of Toronto

Abstract

Meshes representing real world objects, both artist-created and scanned, contain a high level of redundancy due to (possibly approximate) planar reflection symmetries, either global or localized to different subregions. An algorithm is presented for detecting such symmetries and segmenting the mesh into the symmetric and remaining regions. The method, inspired by techniques in Computer Vision, has foundations in robust statistics and is resilient to structured outliers which are present in the form of the non symmetric regions of the data. Also introduced is an application of the method: the folding tree *data structure. The structure encodes the non redundant regions of the original mesh as well as the reflection planes and is created by the recursive application of the detection method. This structure can then be* unfolded *to recover the original shape. Applications include mesh compression, repair, skeletal extraction of objects of known symmetry as well as mesh processing acceleration by limiting computation to non redundant regions and propagation of results.*

Categories and Subject Descriptors (according to ACM CCS): I.3.5 [Computational Geometry and Object Modeling]: Geometric algorithms, languages, and systems. Curve, surface, solid, and object representations. Hierarchy and geometric transformations.

1. Introduction

Symmetry plays a fundamental role in nature, manifested both in the form and function of living organisms. Visually, symmetry is important to humans, as it influences our perceptual understanding of objects in the world. Symmetric patterns, not surprisingly, are an important design principle in guiding the aesthetic and construction of synthetic objects [Arn54, Gom69]. Neuroscience research goes so far as to indicate that aspects of symmetry in humans may be hardwired into our visual processing system [NCP*02]. Symmetries are ubiquitous in humans, our environment and our creations of art and architecture.

The classification, understanding and intelligent representation of shape is an active area of research in geometry processing. Recognizing the common presence of symmetries in many real world objects can greatly assist in solutions to various shape representation problems such as simplification, repair, noise removal and skeletal extraction. Of the various types of symmetries found, planar symmetry is perhaps the most commonplace and is thus the focus of this paper. While planar symmetries have been recognized to be an important feature in shape understanding, there has been little work in shape representations that are defined as a structured assembly of symmetric parts. In this paper we present the concept of a *folding tree* (see figure 4), where an object is defined as a hierarchical union of planar symmetric and asymmetric parts. Each nested detection of a symmetric part reduces the complexity of representation of said part by half, greatly simplifying the overall representation complexity of many objects.

For folding trees to be useful beyond an academic concept, we must be able to automatically construct them from geometric data such as meshes as well as regenerate the object from its folding tree. Central to folding tree construction is the problem of automatically finding a maximally symmetric part of the object. We observe that most organic ob-

[†] {psimari, kalo, karan}@dgp.toronto.edu

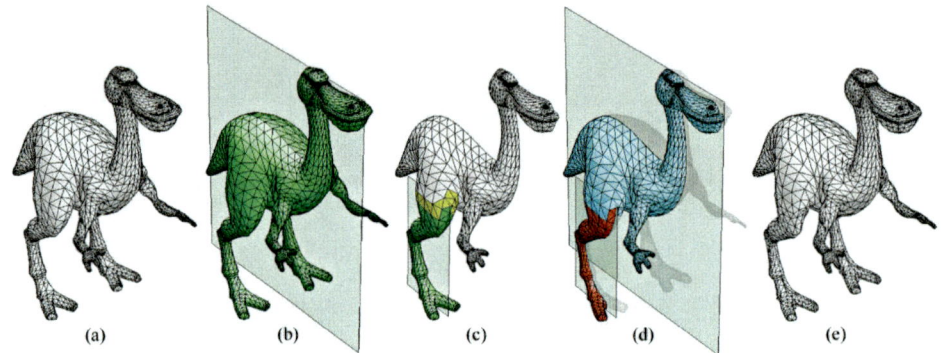

Figure 1: *(a) Original dino mesh. (b) Detected global symmetry plane. (c) Remaining half dino mesh with a detected local symmetry. Green indicates the region of symmetry and yellow indicates faces still in the support region but not included in the symmetric region. (d) Remaining leaf geometry (the two leaves shaded a different color). (e) Dino reconstructed from the leaf geometry and symmetry planes.*

jects with planar symmetry are articulated figures with coherent symmetric parts that are connected at the joints of an underlying skeletal structure. Most synthetic objects show a construction history involving symmetric primitives, symmetry creating operations such as reflections, planar symmetry preserving operations like revolves and coherent combinations of various symmetric parts often with some asymmetric refinement. Motivated by these observations we additionally constrain our problem to finding a maximal symmetric part that is a single connected surface component of the object. The constraint has several advantages, including simplifying reconstruction. Multiple surface components with the same planar symmetry are just as easily represented as multiple symmetry nodes that have the same symmetry plane.

Overview: Our approach to finding maximally symmetric parts is based on robust M-estimation using an iteratively reweighted least squares (IRLS) algorithm. We simultaneously solve for the reflection plane as well as the region of surface that is symmetric with respect to it. Upon convergence, the symmetric region found is separated from the rest. The algorithm is then applied to the remaining regions to find other local symmetries, as well as recursively to half of the symmetric region to find nested symmetries. This process leads to a hierarchical folding tree representation of the object where geometry is only stored in tree leaves. The original surface can then be reconstructed from its folding tree in a bottom up fashion by reflecting symmetric geometry and reconnecting segmentation boundaries.

Contribution: This paper presents two principal contributions. Firstly, we introduce a method capable of detecting *global* as well as *local* approximate planar symmetries in 3D meshes. Our algorithm has solid foundations on statistical methods for robust M-estimation [HRRS86] which has also been used in many recent computer vision applica-

tions [SAG95, BA96, Ste99]. Secondly, we exploit our symmetry detection approach for *mesh folding*: the elimination of planar symmetry redundancy from the mesh data. Our algorithm is orthogonal to other existing methods for mesh compression [Hop96, GBTS99, KG00, IA02] which do not explicitly take advantage of repeating symmetric areas in 3D meshes. (For a recent survey, see [AG05].)

The rest of the paper is organized as follows: section 2 describes the related work on symmetry detection and applications; section 3 describes our method for detecting global and local symmetries on mesh data; section 4 introduces folding trees and describes their construction; section 5 shows our results; and finally, section 6 presents our conclusions and future work.

2. Related Work

Although the computation of symmetries in shapes has been an intriguing area of research in computer vision and computational geometry literature for the last 30 years at least, to our knowledge, there has been comparatively little research on symmetry detection in 3D meshes mainly due to the increased complexity of the existing algorithms when extended from the 2D to the 3D case. Our work differs from most approaches for 3D symmetry detection in meshes in that we aim at the robust detection of not only global but also local reflection symmetries, i.e. those present only in parts of a 3D mesh, and we exploit these symmetries in order to achieve mesh compression by eliminating faces implied by the discovered symmetries.

The detection of symmetries in 2D and 3D models has mainly been applied in object classification, recognition and reconstruction. Early approaches dealt with symmetries of planar point sets by applying pattern recognition algorithms that search for matches in circular strings representing the graphs of polyhedral objects [Ata85,WWV85,Hig86, JYB96].

Despite the optimality of these algorithms that could also detect all the possible symmetries in a shape, they were only able to recover perfect symmetries in 2D and 3D shapes making them useless in the presence of small perturbations, imprecision and noise which is very common in meshes. The problem of approximate symmetries was addressed in [AMWW88] that considered the problem of computing generic geometric transformations between two point sets. The paper gives a detailed theoretical analysis of the developed algorithm for symmetry, however, it deals with global symmetry and it is unclear if the given algorithm could be extended to three dimensions. Such is also the case of other methods for finding symmetries of symmetric or almost symmetric 2D planar images [Mar89,GK96,SICT99].

An interesting extension of that early work which introduced the notion of symmetry distance, meaning how much of a given symmetry an object possesses, was developed in [ZPA95]. The approach can evaluate symmetries in the presence of noise and also find locally symmetric regions in 2D images and 3D range data. The reflection plane of the image is determined by minimizing the symmetry value over all possible reflection planes using a gradient descent algorithm to locate the plane of maximal symmetry. More recently, an approach also taking 3D range images as input was presented in [TW05]. A probabilistic measurement model is used to detect symmetries in order to reconstruct partially occluded 3D shape models. Although both methods can find local symmetries, they follow a greedy technique searching in growing localized regions. In contrast, our method is top down, leading to the gradual removal of asymmetric outliers from the region for the robust detection of the maximal symmetric area using an M-estimator approach. In this regard, our approach tends to find the largest areas of symmetry first, avoiding over-segmentation.

Another original approach that detects the dominant hyperplane of bilateral symmetry in range images of 3D objects with a linear time algorithm is presented in [MO96]. The hyperplane is uniquely defined by the centroid and eigenvectors of the covariance matrix of the object. This method is limited to the detection of the plane of global symmetry and is not robust to outliers or imprecision in the 3D object. Sun *et. al* [SS97,SS99] address the symmetry detection problem by, in the first case, searching for correlations in the 3D object's gaussian image, and in the second, by using the image's orientation histogram in a similar fashion. More recently, Martinet *et al.* [MSHS05] recover symmetries by examining the extrema and spherical harmonic coefficients of the object's generalized moments. These approaches, however, also focus strictly on global symmetry detection.

Symmetry shape descriptors are introduced in [KCD*04, KFR04] where a collection of spherical functions are used to describe the measure of rotational and reflective sym-

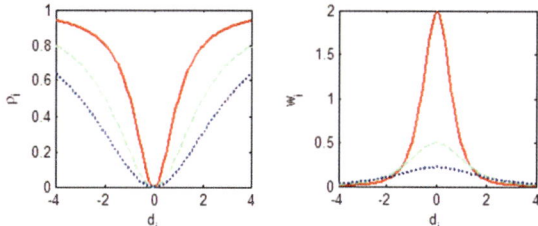

Figure 2: *GM estimator cost* ρ *(left) and associated weight* w *(right) as a function of distance for scale parameter* $\sigma = 1$ *(red solid),* $\sigma = 2$ *(green dashed) and* $\sigma = 3$ *(blue dotted).*

metry present in a mesh with respect to every axis passing through its center of mass. The descriptors had several desirable properties such as robustness and stability. However, the approach aimed at using global symmetry information as a shape descriptor and not at extracting local symmetries.

In parallel with our work, two very interesting symmetry detection algorithms have been developed. In the first, Podolak *et al.* propose the planar reflective symmetry transform (PRST) [PSG*06] as a shape descriptor. For any given plane, the PRST indicates the degree of symmetry which the object exhibits with respect to it. For efficiency of computation, the authors propose a Monte Carlo framework, in which pairs of randomly selected points vote for the plane between them. These votes are accumulated in discrete bins over polar parameters in a manner reminiscent of the Hough transform, and local maxima are later refined. These local maxima correspond to planes for which the object exhibits a degree of local or global symmetry. In the second, Mitra *et al.* [MGP06] also propose to consider pairs of points and their determined symmetry. Rotational, and translational symmetries are also considered, as well as scaling. Instead of a bin counting scheme, representative symmetries are extracted from the transform space through mean shift clustering. Both proposed methods seem robust and efficient. On the other hand, our approach represents an easy-to-implement alternative symmetry detection method which is also based on robust statistics along with a hierarchical simplification scheme that can be directly applied to mesh compression and reconstruction, in both cases through our folding tree structure.

3. Symmetric region detection

Given a mesh, we wish to find a connected region S of faces that exhibit planar symmetry within a tolerance parameter ε. In the case of global symmetry, this region should be the entire mesh. We approach the problem as a model fitting scenario, in which the model consists of the sought plane, and the connected region of symmetry.

Distance metric: given a plane p, we denote the distance from a point r_i^p, which is the reflection of v_i with respect to p, to a mesh M as $d_i^p = dist(r_i^p, M)$. For notational simplicity we will avoid including the p superscript in the following.

We compute the distance function *dist* from a reflected vertex to the mesh by taking the minimum point-to-triangle distance from the point to the closest compatible face on mesh M [RL01]. We consider a face to be compatible with a given query vertex if the angle between the interpolated normal at the closest point on the face and the vertex's reflected normal is less than 45 degrees.

Given the presence of structured outliers in the form of the non symmetric regions of the mesh, we interleave solving for the symmetric region S and the plane p based on an iteratively reweighted least squares (IRLS) approach, using the Geman-McClure (GM) robust M-estimator [HRRS86, FP02]. In essence, the GM estimator maps error values to an associated cost. This cost approaches constant (and the associated weight approaches zero) as error values approach infinity, thus mitigating the influence of outliers on the minimization process. The GM estimator exhibits excellent behavior in rejecting structured outliers with the appropriate choice of the scale factor σ [SAG95]. This parameter essentially controls the rate with which weight decreases as error increases. (See figure 2.)

Solving for the plane: Given the current distances d_i, the GM cost estimator ρ_i and associated weight w_i for each vertex are given by

$$\rho_i = \frac{d_i^2}{\sigma^2 + d_i^2} \qquad w_i = \frac{1}{d_i}\frac{\partial \rho_i}{\partial d_i} = \frac{2\sigma^2}{(\sigma^2 + d_i^2)^2}$$

In addition, in order to be robust in the presence of tessellations with varying face sizes, we multiply the obtained weights by their associated vertex areas, i.e. $w_i \leftarrow w_i \frac{1}{3}\sum_{j=1}^{k} area(f_j)$, where $f_1, ..., f_k$ are the faces incident on vertex v_i.

For a body which exhibits planar symmetry it is known that its plane of symmetry is perpendicular to a principal axis and contains the object's center of mass [MIK92]. This lets us solve for the current plane of maximum symmetry in a closed form manner by considering the center of mass m and weighted covariance matrix C relative to the weights w_i.

$$m = \frac{1}{s}\sum_{i=1}^{n} w_i v_i \qquad C = \frac{1}{s}\sum_{i=1}^{n} w_i(v_i - m)(v_i - m)^T$$

where $s = \sum_i w_i$.

We compute the eigenvectors of C and consider the three planes determined by these vectors and m. For each of these planes we compute the distances d_i and associated costs ρ_i retaining the one of minimum sum cost. This now lets us solve for the support region.

Support region: Given the current ρ values and a candidate face $f = (v_1, v_2, v_3)$ we consider it to be a *support* face if it holds that $\forall_{i \in \{1,2,3\}} d_i \leq 2\sigma$ [HRRS86]. We then find the

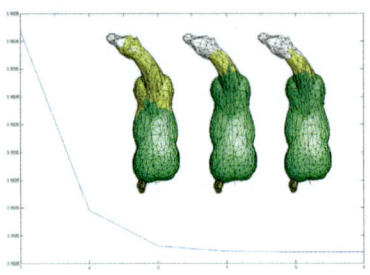

Figure 3: *Illustration of algorithm convergence. The plot shows the objective function $\sum_i \rho_i$ for support vertices v_i. The placement of the estimate of the symmetry plane along with support region (yellow) and symmetric region (green) are shown for the horse model. Left to right, iterations 1, 5 and 10 respectively, using the precise distance metric.*

largest connected region of support faces, taking this as our new estimate of the support region and set the weights for all vertices outside this region to be 0, thus controlling leverage.

The estimation and region finding steps are iterated until convergence.

Initialization and details: Initially we simply define w_i to be the mesh area associated with vertex v_i as defined above, and the initial support region contains all faces.

In order to accelerate and improve convergence we initially use a discrete approximation of the above distance function. For a given vertex v_i and face f_j this distance function $dist_{discrete}(v_i, f_j)$ is defined as the Euclidean distance from the reflected vertex r_i to the face plane of f_j if the angle between the reflected normal and f_j is less than 90 degrees, and infinite otherwise. The distance $dist_{discrete}(v_i, M)$ is defined as the distance to the f_j whose centroid is closest to v_i. During these initial iterations we set $\sigma = 1.4826 \cdot median(d_i)$, which is a popular estimate of scale [FP02], not letting it fall below 2ε to avoid instability.

This discrete distance function is first used until convergence, after which the more precise distance function *dist* is used. At this point we set $\sigma = 2\varepsilon$. This setting allows for near-symmetric vertices to be included in the support region albeit with lower weight.

Upon convergence, the symmetric region S is extracted as the largest connected region of faces whose vertex distance values are all below ε. We detect convergence by comparing the current plane estimate with that of the previous iteration checking for a sufficiently small difference.

Convergence: In our experiments both distance functions exhibit very good convergence behavior (to their respective minima). Figure 3 illustrates an example of the convergence properties of our approach.

4. Folding trees

4.1. Definitions

We consider a region R of a mesh M to be a connected subset of the faces of M.

A segmentation $\{R_1, R_2, ..., R_n\}$ of a mesh M is a set of mutually exclusive regions whose union results in M.

A folding tree T representing a mesh M is inductively defined as one of the following:

- a *leaf* node, which contains mesh data for M.
- a *folding* node, with one subtree S, and a plane of symmetry p.
- a *segmentation* node, with n subtrees $T_1, T_2, ..., T_n$, where T_i is a folding tree for region R_i such that regions $R_1, R_2, ..., R_n$ are a segmentation of the mesh represented by T.

The *unfold* operation can now be defined on a folding tree T as follows.

- The mesh data M if T is a leaf.
- $unfold(S) \cup reflect(unfold(S), p)$, where S is the unique subtree of T, p is the reflection plane,
- $\cup_{i=1}^{n} unfold(T_i)$ where $T_1, T_2, ..., T_n$ are the subtrees of T.

Here, *reflect* indicates the mesh resulting from planar reflection of the argument mesh's vertices with respect to the argument plane.

4.2. Folding tree construction and unfolding

Given a mesh M, a folding tree T that represents it can be constructed in preorder through repeated application of the segmentation method of section 3. First, we apply the segmentation algorithm to M to find a subregion of planar symmetry, S, also obtaining the plane of symmetry p. We remove S from M and consider the set of remaining connected components $\{R_1, R_2, ..., R_n\}$. Note that in the case of a global symmetry, this set will be empty. We now construct a folding tree T with $n + 1$ children $T_0, T_1, ..., T_n$, each representing $S, R_1, R_2, ..., R_n$ respectively. We know S to be symmetric with respect to plane p, so we can now *fold* S, retaining half of its surface S'. In particular, T_0 will be a folding node, labeled with p, and its child T_0' will represent S'. The resulting structure is illustrated in figure 4. The subtrees $T_0', T_1, ..., T_n$ can now be created recursively with regions $S', R_1, ..., R_n$ respectively as inputs.

When discarding half of the faces of a particular region, it must be decided which half to keep, which to discard, and which to modify, if any. Because of varying tessellation and the provided tolerance, both sides need not be identical. In our implementation we keep the side with the most faces in order to preserve detail. Alternatively, we could keep the half with the least faces in order to minimize storage. Faces with all vertices on the discarded side of the plane are removed. In addition, faces that straddle the symmetry plane are clipped.

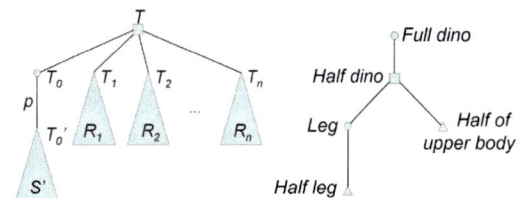

Figure 4: *Left: Folding tree construction structure when a local symmetry is detected. Right: Example of a complete folding tree for the dino mesh decomposition of figure 1. Circles represent folding nodes, squares segmentation nodes, small triangles are leaf geometry and large triangles represent (recursive) folding trees.*

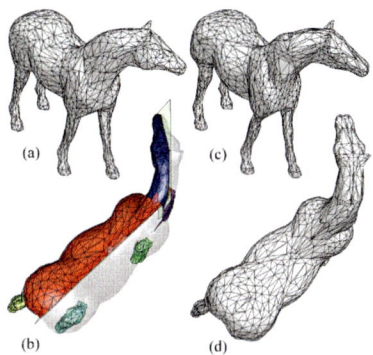

Figure 5: *(a) Original horse model. (b) Resulting tree leaves after folding and symmetry planes. Note that the local symmetries of the body and the articulated head were detected. (c) and (d) The reconstructed horse model.*

The recursive construction of the tree may be stopped, triggering the creation of a leaf node, by using one or more criteria: when the number of faces in the mesh is below a given threshold, when the area of the mesh is below a certain percentage of that of the original mesh, or when the number of recursive folds exceeds a given maximum. Our implementation allows for any or all three.

The procedure for unfolding a tree consists of a postorder traversal according to the definition of subsection 4.1. Upon reconstruction, because of the tolerance parameter of the region-finding algorithm, as well as differences in tessellation, the resulting mesh may have vertex misalignments. These can be corrected automatically using known techniques [GTLH98, TL94, Ju04] or software. The unfolding drives this repair, the union of the definition of subsection 4.1 between unfolded parts indicating that these should be processed in this manner. Figures 1, 4 and 5 illustrate the concepts.

Figure 6: *1st column: original mesh data.* **2nd column:** *All folding planes present in the constructed tree.* **3rd column:** *model resulting from unfolding of folding-tree representation.* **4th column:** *Folding tree. Circles represent folding nodes, squares segmentation nodes, and triangles are leaf geometry. Leaf nodes corresponding to asymmetric regions are not shown for clarity.*

Mesh	# Orig. f's	#Nonred. f's	Mesh bytes	Tree bytes	Comp.	Tree height	Rec. error	Time
Dino	6638	3142	265864	130916	50.76%	3	0.0358%	152 sec
Horse	3306	2672	120312	93932	21.93%	3	0.0004%	24 sec
Chair	5736	2460	229880	104352	54.61%	4	0.1600%	58 sec
Hammer	4360	677	174712	53120	69.60%	7	0.1764%	174 sec
Triceratops	5660	2447	226712	123696	45.44%	7	0.0147%	202 sec
Eagle	33072	15808	160440	96516	39.84%	5	0.0067%	936 sec
Queen	3360	600	134712	31952	76.28%	4	0.0001%	120 sec

Table 1: *Results for seven characteristic meshes. Columns from left to right: mesh name, number of faces in the original mesh, number of non redundant faces stored in folding tree leaves, original mesh storage in bytes, tree storage in bytes, compression achieved according to one minus the ratio of the previous two columns, height of the folding tree, reconstruction error as a percentage of the bounding box diagonal and running time, as reported by our Matlab 7 implementation.*

5. Results

The implementation of the symmetry detection algorithm and the folding tree representation of meshes, as described in sections 3 and 4, has been developed in Matlab 7. The user defines the tolerance parameter of the algorithm and the criteria for stopping the hierarchical segmentation of the mesh. The default tolerance value is 2% of the bounding box diagonal of the mesh. The default criterium for terminating the recursion is that the area of the current region is less than 5% of the total mesh area.

We present characteristic results, concerning mesh compression, the depth of the hierarchical segmentation, average reconstruction error as a percentage of the bounding box diagonal, as well as running times in table 1. Figure 6 shows initial and reconstructed meshes, complementing the results of figures 1 and 5, as well as illustrating all folding plane positions and the folding trees. In the chair model, we find the vertical plane of global symmetry then each cushion, which was a separate connected component, was folded through three perpendicular planes. The hammer is firstly folded in half through a vertical global plane of symmetry. Then, the handle is divided twice more. The cylindrical portion of the head is also subdivided twice more recursively. In the case of the triceratops and eagle models we find a global plane of symmetry and then local symmetries in the legs, tail and body for the triceratops model, and in the wings and upper legs of the eagle. Finally, in the case of the octagonal queen chess piece, all planar symmetries are recursively discovered resulting in one eighth of the original surface being stored.

Figure 7 further illustrates the symmetry detection approach. In the woman, we firstly find the dominant partial symmetry that includes her body and legs. Searching for nested symmetry, the algorithm detects the symmetry of the leg. Proceeding to the remaining regions, we find the local symmetries of the head and arms. Analogously, in the dragon, we find the symmetry of its body and then legs, head and arms. In the bull, we detect the local symmetry of its body including the back left leg, then symmetries in the other three legs and the head and finally another weak nested

symmetry found in the middle of its body. We detect symmetries on the body, and separately in head and the ears of the bunny and also find two other weak nested symmetries in its body. Lastly, in the octopus, its head is found to be symmetric, also containing a nested symmetry, and multiple local symmetries are found in different parts of its tentacles. We would like to note that all meshes with the exception of the chair are originally a single connected component.

Our tests were run on an Intel Pentium M 2.13GHz processor under Matlab 7.

6. Conclusions and future work

We have proposed a novel approach for finding global planar symmetries in 3D meshes based on robust M-estimation. In addition, we have presented a new compact representation of meshes, called folding trees, which represent the original mesh by only encoding the non redundant regions as well as the planes of symmetry and can be used to recover the original object through unfolding.

Given the fact that real objects, both organic and synthetic, often exhibit this type of data redundancy and human perception is strongly related to the notion of symmetry, a significant number of applications based on our methodology can further be developed. The elimination of faces which are repeated in redundant areas of global and local symmetries leads to new mesh compression schemes that can be used for mesh storage, processing, and transmission. Automatic segmentation, reconstruction and repairing of the meshes, driven by the extracted symmetries, is also another interesting field of application of our method. The folding tree representation could also facilitate skeleton extraction and advanced editing operations which preserve symmetries.

Our future research will be focused on both the development of such applications as well as the exploitation of other types of symmetries in 3D meshes that can open up new implementations and extensions of our proposed methodology.

In cases where there is no strong symmetry in the neighborhood of the principal axes, our approach may fail to find

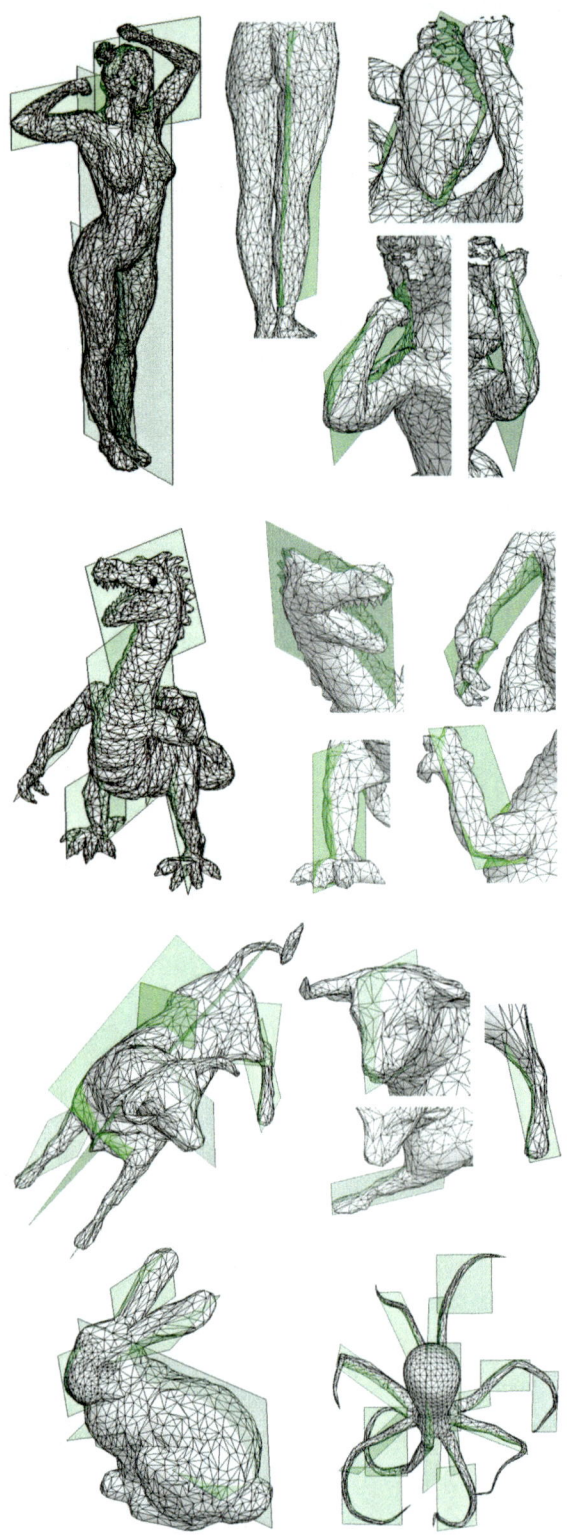

Figure 7: *Local symmetry planes identified by the method.*

existing local symmetries in small regions (for example, in the hands of the dino mesh of figure 1). In cases such as these in which our initial guess does not provide a large enough support, an alternative initialization can be sought through random sampling or perhaps through a voting scheme like the ones described by Mitra *et al.* [MGP06] or Podolak *et al.* [PSG*06].

Acknowledgements: Research funded in part by MITACS. http://www.mitacs.ca.

References

[AG05]   ALLIEZ P., GOTSMAN C.:   Recent advances in compression of 3D meshes. In *Advances in Multiresolution for Geometric Modelling* (2005), Dodgson N., Floater M., Sabin M., (Eds.), Springer-Verlag, pp. 3–26.

[AMWW88]   ALT H., MEHLHORN K., WAGENER H., WELZL E.: Congruence, similarity, and symmetries of geometric objects. *Discrete Compututational Geometry 3*, 3 (1988), 237–256.

[Arn54]   ARNHEIM R.: *Art and Visual Perception: A Psychology of the Creative Eye.* University of California Press, Berkeley, 1954.

[Ata85]   ATALLAH M.:   On symmetry detection.   *IEEE Trans. Computers 34*, 7 (1985), 663–666.

[BA96]   BLACK M. J., ANANDAN P.:   The robust estimation of multiple motions: parametric and piecewise-smooth flow fields. *Computer Vision and Image Understanding 63*, 1 (1996), 75–104.

[FKS*04]   FUNKHOUSER T., KAZHDAN M., SHILANE P., MIN P., KIEFER W., TAL A., RUSINKIEWICZ S., DOBKIN D.: Modeling by example. *ACM Trans. Graph. 23*, 3 (2004), 652–663.

[FP02]   FORSYTH D. A., PONCE J.: *Computer Vision: A Modern Approach*, first ed. Prentice Hall, 2002.

[GBTS99]   GUEZIEC A., BOSSEN F., TAUBIN G., SILVA C.:   Efficient compression of non-manifold polygonal meshes. In *Proceedings of the 10th IEEE Visualization 1999 Conference (VIS '99)* (Washington, USA, 1999), IEEE Computer Society.

[GK96]   GOFMAN Y., KIRYATI N.:   Detecting symmetry in grey level images: The global optimization approach. In *Proceedings of the 13th International Conference on Pattern Recognition* (1996), pp. 951–956.

[Gom69]   GOMBRICH E. H.: *Art and Illusion: A Study in the Psychology of Pictorial Representation.* Princeton University Press, Princeton, 1969.

[GTLH98]   GUEZIEC A., TAUBIN G., LAZARUS F., HORN W.: Converting sets of polygons to manifold surfaces by cutting and stitching. In *VIS '98: Proceedings of the conference on Visualization '98* (Los Alamitos, CA, USA, 1998), IEEE Computer Society Press, pp. 383–390.

[Hig86] HIGHNAM P. T.: Optimal algorithms for finding the symmetries of a planar point set. *Inf. Process. Lett. 22*, 5 (1986), 219–222.

[Hop96] HOPPE H.: Progressive meshes. In *SIGGRAPH '96: Proceedings of the 23rd annual conference on Computer graphics and interactive techniques* (New York, USA, 1996), ACM Press, pp. 99–108.

[HRRS86] HAMPEL F. R., RONCHETTI E. M., ROUSSEEUW P. J., STAHEL W. A.: *Robust Statistics: The Approach Based on Influence Functions.* Wiley-Interscience, New York, USA, 1986.

[IA02] ISENBURG M., ALLIEZ P.: Compressing polygon mesh geometry with parallelogram prediction. In *Proceedings of the conference on Visualization '02* (Washington, USA, 2002), IEEE Computer Society, pp. 141–146.

[Ju04] JU T.: Robust repair of polygonal models. *ACM Trans. Graph. 23*, 3 (2004), 888–895.

[JYB96] JIANG X., YU K., BUNKE H.: Detection of rotational and involutional symmetries and congruity of polyhedra. *Visual Computing 12*, 4 (1996), 193–201.

[KCD*04] KAZHDAN M., CHAZELLE B., DOBKIN D., FUNKHOUSER T., RUSINKIEWICZ S.: A reflective symmetry descriptor for 3d models. *Algorithmica 38*, 1 (2004), 201–225.

[KFR04] KAZHDAN M., FUNKHOUSER T., RUSINKIEWICZ S.: Symmetry descriptors and 3d shape matching. In *SGP '04: Proceedings of the 2004 Eurographics/ACM SIGGRAPH symposium on Geometry processing* (New York, NY, USA, 2004), ACM Press, pp. 115–123.

[KG00] KARNI Z., GOTSMAN C.: Spectral compression of mesh geometry. In *SIGGRAPH '00: Proceedings of the 27th annual conference on computer graphics and interactive techniques* (New York, USA, 2000), ACM Press/Addison-Wesley Publishing Co., pp. 279–286.

[Mar89] MAROLA G.: On the detection of the axes of symmetry of symmetric and almost symmetric planar images. *IEEE Trans. Pattern Anal. Mach. Intell. 11*, 1 (1989), 104–108.

[MGP06] MITRA N. J., GUIBAS L. J., PAULY M.: Partial and approximate symmetry detection for 3d geometry. *To appear in SIGGRAPH '06* (2006).

[MIK92] MINOVIC P., ISHIKAWA S., KATO K.: *Three Dimensional Symmetry Identification, Part I: Theory.* Tech. Rep. 21, Kyushu Institute of Technology, Japan, 1992.

[MO96] MARA D. O., OWENS R.: Measuring bilateral symmetry in digital images. In *In Proceedings of IEEE TENCON - Digital Signal Processing Applications* (1996), vol. 1, pp. 151–156.

[MSHS05] MARTINET A., SOLER C., HOLZSCHUCH N., SILLION F.: *Accurately Detecting Symmetries of 3D Shapes.* Tech. Rep. RR-5692, INRIA, September 2005.

[NCP*02] NORCIA A., CANDY R., PETTET M., VILDAVSKI V., TYLER C.: Temporal dynamics of the human response to symmetry. *Journal of Vision 2*, 2 (3 2002), 132–139.

[PSG*06] PODOLAK J., SHILANE P., GOLOVINSKIY A., RUSINKIEWICZ S., FUNKHOUSER T.: A planar-reflective symmetry transform for 3d shapes. *To appear in SIGGRAPH '06* (2006).

[RL01] RUSINKIEWICZ S., LEVOY M.: Efficient variants of the icp algorithm. *In proceedings of the Third International Conference on 3-D Digital Imaging and Modeling* (2001), 145.

[SAG95] SAWHNEY H. S., AYER S., GORKANI M.: Model-based 2d&3d dominant motion estimation for mosaicing and video representation. In *Proceedings of the Fifth International Conference on Computer Vision* (Washington, USA, 1995), IEEE Computer Society, p. 583.

[SICT99] SHEN D., IP H. S., CHEUNG K. T., TEOH E. K.: Symmetry detection by generalized complex moments: A close-form solution. *IEEE Trans. Pattern Anal. Mach. Intell. 21*, 5 (1999), 466–476.

[SS97] SUN C., SHERRAH J.: 3d symmetry detection using the extended gaussian image. *IEEE Trans. Pattern Anal. Mach. Intell. 19*, 2 (1997), 164–168.

[SS99] SUN C., SI D.: Fast reflectional symmetry detection using orientation histograms. *Real-Time Imaging 5*, 1 (1999), 63–74.

[Ste99] STEWART C. V.: Robust parameter estimation in computer vision. *SIAM Rev. 41*, 3 (1999), 513–537.

[TL94] TURK. G., LEVOY M.: Zippered polygon meshes from range images. In *SIGGRAPH '94: Proceedings of the 21st annual conference on Computer graphics and interactive techniques* (New York, NY, USA, 1994), ACM Press, pp. 311–318.

[TW05] THRUN S., WEGBREIT B.: Shape from symmetry. In *Tenth International Conference on Computer Vision* (2005), vol. 2, IEEE Computer Society, pp. 1824–1831.

[WWV85] WOLTER J. D., WOO T. C., VOLZ R. A.: Optimal algorithms for symmetry detection in two and three dimensions. *Visual Compututer 1* (1985), 37–48.

[ZPA95] ZABRODSKY H., PELEG S., AVNIR D.: Symmetry as a continuous feature. *IEEE Trans. Pattern Anal. Mach. Intell. 17* (1995), 1154–1166.

Probabilistic Fingerprints for Shapes

Niloy J. Mitra [†]   Leonidas Guibas[†]   Joachim Giesen[‡]   Mark Pauly[§]

[†] Stanford University   [‡] Max-Planck Institut für Informatik   [§] ETH Zurich

Abstract

We propose a new probabilistic framework for the efficient estimation of similarity between 3D shapes. Our framework is based on local shape signatures and is designed to allow for quick pruning of dissimilar shapes, while guaranteeing not to miss any shape with significant similarities to the query model in shape database retrieval applications. Since directly evaluating 3D similarity for large collections of signatures on shapes is expensive and impractical, we propose a suitable but compact approximation based on probabilistic fingerprints *which are computed from the shape signatures using Rabin's hashing scheme and a small set of random permutations. We provide a probabilistic analysis that shows that while the preprocessing time depends on the complexity of the model, the fingerprint size and hence the query time depends only on the desired confidence in our estimated similarity. Our method is robust to noise, invariant to rigid transforms, handles articulated deformations, and effectively detects partial matches. In addition, it provides important hints about correspondences across shapes which can then significantly benefit other algorithms that explicitly align the models. We demonstrate the utility of our method on a wide variety of geometry processing applications.*

Categories and Subject Descriptors (according to ACM CCS): I.3.5 [Computer Graphics]: Computational Geometry and Object Modeling.

1. Introduction

There has been great progress in recent years in the areas of shape acquisition and modeling, resulting in large collections of digital geometric models. Classification, navigation, and usability of such shape databases largely hinge on the following operation: Given two shapes in arbitrary poses, how can we meaningfully define their similarity and evaluate it efficiently? For database applications, off-line preprocessing of each shape is typically acceptable, if it results in fast query handling. In the same context, it is important to determine, quickly and reliably, when two shapes are *dissimilar*.

At a fine level, global shape similarity is traditionally estimated by comparing optimally aligned models: Translation is factored out by aligning the respective centroids, while rotation is handled using principal component analysis (PCA) for the final alignment. Alternatively, at a coarser level and so as to avoid the alignment step, global shape descriptors can be computed that are invariant under rigid transformations.

Shape similarity is then estimated indirectly by comparing these descriptors.

Global approaches fail for partial similarity, which is much more challenging, as factoring out arbitrary rigid transforms without any knowledge of regions of overlap between shapes is difficult. In such cases, we can use more specialized techniques like geometric hashing [GCO06], or explicitly determine correspondence between models and use it for alignment [GMGP05]. However, these methods require high storage or processing time, even if the two shapes are very different.

In this paper we propose an efficient method to define probabilistic fingerprints for 3D shapes and use it to estimate partial similarity. Our approach is complementary to existing work on shape descriptors and signatures, as we can make use of available shape descriptors to define partial similarity across multiple shapes in arbitrary poses. We compress these descriptors using a probabilistic hashing scheme motivated by ideas from the database community. Our fingerprints are such that if they largely disagree, then we

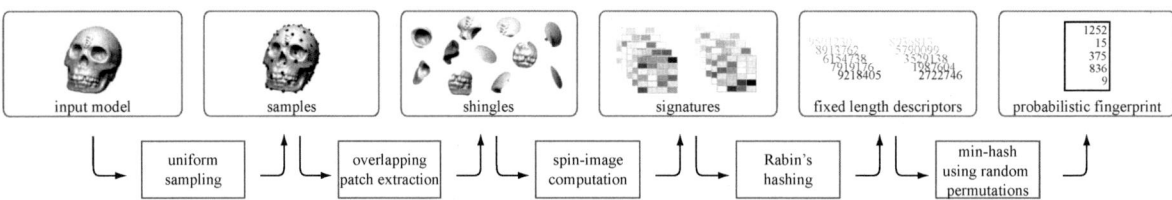

Figure 1: Fingerprint generation. *We first cover an object with ρ-radius balls spaced δ apart with $\rho \gg \delta$. The intersection of each ball with the surface defines a* shingle. *For each shingle, we compute a descriptor, spin-image in this case, which is hashed using Rabin's scheme. Then, in the min-hashing phase, according to m random permutations, we select a small subset of descriptors and store them as the* probabilistic fingerprint.

can claim with high certainty that the corresponding shapes are dissimilar. This yields an efficient way to quickly filter large shape collections when searching for objects matching a particular model.

In our approach to partial shape similarity, we first cover a given 3D shape with a large collection of overlapping patches. Each patch is mapped to a point in a high dimensional space using a compact, local descriptor that is invariant to rigid transformations. We do not preserve any information about the relative spatial ordering of the patches. The shape is thus mapped to an unordered point set in a high dimensional signature space. We select descriptors that are robust to perturbations, so that patches which are very similar are likely to be mapped to the same locale of this signature space. This is important, as similar regions may not be covered by patches in exactly the same way across two shapes. Clearly, if two shapes are similar, then the corresponding point sets will have proximal regions in proportion to the partial similarity between the original objects. However, since we lose relative patch ordering, it is possible that two largely different shapes have a significant overlap in signature space. Statistically this is a rare event and leads to only a few false positives. It is made even more unlikely by ensuring large overlap between neighboring patches. Motivated by this intuition, we define similarity between two shapes in terms of the similarity between the signature sets. Our definition is invariant to rigid transforms, handles partial matching, and is robust to local deformations and articulated motion.

However, these large high-dimensional point sets have high storage requirements and are difficult to compare efficiently. We therefore compress signature information using a technique called min-hashing [Bro97] to generate a short *probabilistic fingerprint* for each signature set. Subsequently, fingerprints of multiple shapes are compared to estimate similarity between the signature sets, and hence between the original objects. We first map the signatures to a finite universe of numbers using Rabin's hashing scheme [Rab81]. Then, during min-hashing, we use a

random permutation to assign a complete ordering to all elements of this finite universe of numbers. We can think of this ordering as the ranking of an 'expert', asked to evaluate the patches according to her criteria. According to the expert ranking, we then select the winner among the set of hashed signatures corresponding to an object. For each object, we collect the winners of m randomly chosen permutations and save them as the *probabilistic fingerprint* of the shape. The same random permutations are used for all shapes. This ensures that patches from different shapes are consistently ordered, according to each of the m chosen 'experts'.

We can efficiently detect if two shapes are similar using our shape fingerprints. However, as mentioned before, we can get a few false positive matches. In practice, the number of such false hits is very small and can be handled by match verification using more expensive partial similarity methods. Further, we can show that if two fingerprints are different, then with high probability the shapes are also different. Thus both false positives and false negatives are bounded.

We provide a probabilistic analysis of our scheme to show the attractive property that the cost of our algorithm depends on the confidence we want from our estimates, and not on the complexity of the shapes themselves. The storage required for preprocessing and query stages depends on the length of the fingerprint m. Finally, though our algorithm only assigns similarity scores between multiple objects without explicitly determining an alignment, it gives important hints about the regions of overlap and correspondences. We apply our fingerprints to address a variety of geometry processing applications, including shape retrieval, automatic scan alignment, adaptive feature point selection, and mesh authentication.

Contributions

We propose a new statistical approach to efficiently estimate partial or total shape similarity. We introduce the concept of probabilistic fingerprints for 3D

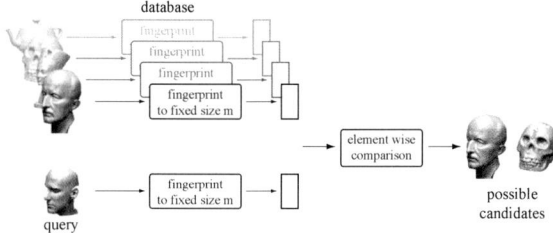

Figure 2: Query Processing. *A query object is first processed to generate its fingerprint using the same parameters used to pre-process the database shapes. Objects with fingerprints similar to fingerprint of the query are returned as possible candidates.*

shapes, provide a statistical analysis on its effectiveness for partial matching, and show its practical use on a number of different applications. The key insight is that the similarity of two shapes can be estimated by comparing signatures derived from very sparse sets of local patches generated on each model. Based on probabilistic arguments, we show how to *pre-select* such patches, *without first needing to establish explicit correspondence relations between the models.* This approach produces a fixed-length fingerprint and avoids costly explicit alignment of the models. Our similarity measure is invariant to rigid transforms, robust to perturbations, handles articulations, and most importantly, detects partial matches.

2. Related Work

The problem of shape similarity and retrieval has been extensively studied in computer vision and graphics. It has been addressed in great detail by the extensive work done by the Princeton Shape Retrieval and Analysis Group [FKS*04, FKMS05]. Meaningful similarity between two 3D shapes, partial or whole, has to be invariant to rigid transforms. Global shape descriptors, invariant to rigid transformations, include spherical harmonics [KFR03], shape distributions [OFCD02], reflective symmetry descriptors [KCD*04], and Laplace-Beltrami Spectra [RWP05]. Similarity estimation between two models is then reduced to a comparison of the corresponding global shape descriptors. Alternatively, an object can be canonically oriented using principal component analysis (PCA) and descriptors computed on the rotation normalized shape – examples include extended Gaussian images [Hor84] and shape histograms [AKKS99].

However, all of these global methods are less suitable for detecting partial matches. This problem can be addressed by establishing an explicit correspondence across feature points of the models to compute a good alignment [GMGP05]. Such a solution involves exhaustively considering the various correspondence assignments and is thus computationally expensive. Gal and Cohen-Or [GCO06] proposed a different method for determining partial similarity using geometric hashing techniques. Briefly, in a pre-processing stage, geometric hashing encodes all the possible candidate transforms in a large hash table. While this approach is more efficient, it trades computation time for memory, leading to space requirements of multiple gigabytes even for moderately complex models.

In a different setting, the problem of identifying text or web documents with partial similarity has been extensively studied. Effective solutions to this problem involve clever combinations of hashing and random sampling techniques [Blo70, SGM98, Bro00]. In these schemes, a text document is first converted to a set of overlapping text segments. Similarity between two documents is assigned based on the size of the intersection of the segment-sets which is efficiently estimated using random sampling techniques. Some of these concepts motivated our approach. Our problem, however, is significantly more challenging, as digital 3D shapes have neither the linear ordering nor the canonical decomposition into discrete tokens that is exploited in the text document case.

3. Shape Fingerprinting

Our goal is to reliably and efficiently estimate (partial) similarity between two shapes. The similarity measure should be invariant to rigid transforms and robust to small perturbations. Here we define such a similarity measure for a restricted class of shapes, namely surfaces in $\mathbb{R}^3$ whose normal is defined almost everywhere, e.g., a smooth surface (implicit or explicitly parameterized) or a triangle mesh. The measure is based on surface signatures that allow for effective compression using hashing. We call a hashed signature a fingerprint and start by listing the properties we expect in general from a shape fingerprint.

Fingerprint properties. A *probabilistic fingerprint* is a function f that assigns to each admissible shape a fixed size bit string, i.e. a string in $\{0, 1\}^m$. The main purpose of fingerprints is to allow for efficient comparison of shapes. Given a definition of shape similarity (or dissimilarity), any meaningful fingerprint function should have the following properties.

1. Given two shapes S_1 and S_2, we want the following relations to hold with high probability:

   a. If $f(S_1) \neq f(S_2)$, then S_1 and S_2 are dissimilar.
   b. If $f(S_1) = f(S_2)$, then S_1 and S_2 are similar.

2. The number of bits m is small compared to the number of bits needed to encode the actual shapes.

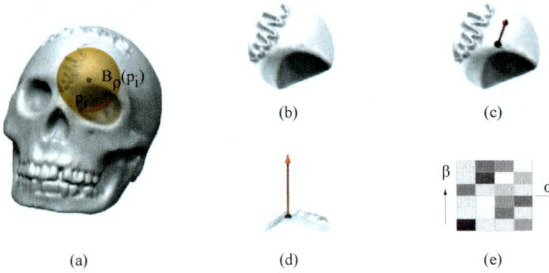

Figure 3: Shingle generation. *(a) $B_\rho(p_i)$ is the neighborhood ball for a point p_i. (b) The selected surface patch (shingle) T_i around p_i. (c) The patch along with the surface normal at p_i. (d) The normal oriented along the z-axis. (e) Computed spin image for patch T_i. The signature is invariant to rigid transforms, and robust to sampling and small surface perturbations.*

3. The function f is efficiently computable. Also $f(S_1) = f(S_2)$ can be quickly checked.

In the following we describe in detail a fingerprint and the corresponding notion of similarity/dissimilarity it is based on. The pipeline for computing the fingerprint is depicted in Figure 1.

Our fingerprint relies on the concepts of *sample, shingle, signature, resemblance,* and *hashing*. Next we describe these concepts and how we use them for defining and computing a probabilistic fingerprint.

Sample. In the first step, we generate a set of (approximately) uniformly sampled points on the input shape S. Let $P = \{p_1, \ldots, p_n\}$ be the set of sample points with sampling spacing δ. For our fingerprint to work well we assume that for any $p \in S$, $\exists p_i \in P$ such that $\|p_i - p\| \le \delta$. Further, the number of such neighboring points is bounded by a small constant, preventing the sampling from being arbitrarily dense.

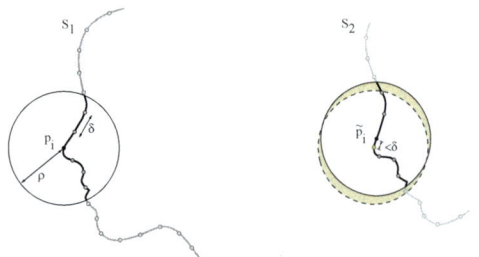

Figure 4: Overlapping Shingles. *Shingles for two shapes S_1 and S_2 are computed using ρ-radius balls spaced roughly δ apart. If $\tilde{p}_i$ lies in a matching region between S_1 and S_2, then with high probability the shingle at $\tilde{p}_i$ will have a corresponding shingle from S_1 with a significant overlap as $\rho \gg \delta$.*

There is a simple and efficient process for generating such a sample set: Let A be the surface area of the shape. On the surface of the object, we randomly place $n = \lceil A/\pi\delta^2 \rceil$ samples and uniformly spread them out using particle repulsion [Tur92].

Shingles. For each sample point $p_i \in P$, we define a neighborhood of radius ρ where $\rho \gg \delta$. A surface patch $T_i \subset S$ corresponding to point p_i is obtained as $T_i = S \cap B_\rho(p_i)$ where $B_\rho(p_i)$ denotes the ball of radius ρ around p_i (Figure 3). If multiple components are present in T_i, we retain only the component containing p_i (the surface is assumed to be a manifold). We refer to these patches as *shingles* and denote the multi-set of all shingles by $\mathcal{P}$. Keep in mind that $\mathcal{P}$ depends on the sample P. Given two shapes S_1 and S_2, with high probability, any shingle from the matching region has a corresponding shingle on the other shape with significant overlap (Figure 4).

Signatures. We compute a signature σ_i for each shingle $T_i \in \mathcal{P}$ and denote the multi-set of all signatures by $\mathcal{S}$. A signature σ_i is essentially a string that represents a shingle T_i. Any signature that is invariant to rigid transforms and robust to sampling and local perturbations can be used to this end. Here we use spin images [Joh97] which are defined as follows: Let the surface normal at any sample point $p_i \in P$ be n_i. For any point x in the corresponding shingle T_i, its spin-map is defined as:

$$(\alpha, \beta) = \left(\sqrt{\|\bar{y}\|^2 - \langle n_i, \bar{y}\rangle^2}, \langle n_i, \bar{y}/\|\bar{y}\|\rangle \right)$$

where $\bar{y} = x - p_i$. The spin-image s_i of T_i is simply the quantized version of the (α, β) space recording the spin-map of the points of T_i falling into a set of discrete bins (Figure 3). Since spin images are robust to perturbations, if two shingles have significant overlap then they are likely to have the same signatures.

Resemblance. Now we introduce our similarity/dissimilarity measure. Given two surfaces S_1 and S_2 we define their resemblance r with respect to their corresponding signatures $\mathcal{S}_1$ and $\mathcal{S}_2$. Remember that $\mathcal{S}_1$ and $\mathcal{S}_2$ are multi-sets. For each $\sigma \in \mathcal{S}_i$ let $m_i(\sigma)$ denote its multiplicity in $\mathcal{S}_i$. The resemblance of S_1 and S_2 is defined as:

$$r(S_1, S_2) = \frac{|\mathcal{S}_1 \cap \mathcal{S}_2|}{|\mathcal{S}_1 \cup \mathcal{S}_2|},$$

where $|\mathcal{S}_1 \cap \mathcal{S}_2|$ denotes

$$\sum_{\sigma \in \mathcal{S}_1 \cap \mathcal{S}_2} \min(m_1(\sigma), m_2(\sigma))$$

and $|\mathcal{S}_1 \cup \mathcal{S}_2|$ denotes

$$\sum_{\sigma \in \mathcal{S}_1 \cup \mathcal{S}_2} \max(m_1(\sigma), m_2(\sigma)).$$

Since resemblance $0 \leq r(S_1, S_2) \leq 1$ is higher when two shapes are similar, we define distance between shapes as $1 - r(S_1, S_2)$. Observe that this definition is based on the signature sets, and hence depends on the scale parameter ρ used to define the shingles.

Finally, we want to estimate the resemblance using a much sparser representation for the shapes than their signatures, namely their fingerprints. To this end each signature is first hashed into a finite set $\mathcal{U}$ using Rabin's hashing scheme.

Hashing. Rabin's hashing scheme [Rab81] gives a low collision probability for a fixed bit budget by working with irreducible degree k polynomials over $\mathbb{Z}_2$. Let the number of bits required to represent a signature be t. For instance, if a spin image is computed using b bins and each bin is of length l bits, then t is upper bounded by the maximum length of the spin images which is $\log(b2^l)$. If Rabin's hashing scheme h maps each of the n signatures σ_i corresponding to a shingle T_i down to k bits, then the probability of collision is bounded by

$$\Pr[h(\sigma_i) = h(\sigma_j) | \sigma_i \neq \sigma_j] \leq n^2 t/2^k. \quad (1)$$

For example, if $n = 10^8$ and $t = 128$ then for $k = 80$ the probability of collision is less than 10^{-6}. Thus even with 10 bytes for each signature, we get low collision probability. Further, Rabin's hashing scheme can be very efficiently computed using simple bit arithmetic [Bro93, CL01]. For each signature we store only k bits corresponding to the coefficients of the degree k polynomial in $\mathbb{Z}_2$. We denote the universe of all k-bit numbers by $\mathcal{U}$ and we denote the multi-set of all hashed signatures as $\mathcal{I}$.

We can define the analog of our resemblance function r for multi-sets of hash values as

$$r'(S_1, S_2) = \frac{|\mathcal{I}_1 \cap \mathcal{I}_2|}{|\mathcal{I}_1 \cup \mathcal{I}_2|},$$

where $\mathcal{I}_i$ is the multi-set of hash values corresponding to surface S_i. Evaluating this function instead of $r(S_1, S_2)$ remains impractical, as the involved multi-sets are still too large even though we need less bits to store their elements than we need to store the original signatures. Moreover, set operations between these large unordered multi-sets require $O(n_i \log n_i)$ time where n_i is the number of set elements. As a solution, we further compress each of the multi-sets $\mathcal{I}$ of hash values to generate a small fingerprint. This is done by *min-hashing* using random permutations on the universe $\mathcal{U}$.

Probabilistic Fingerprint. Let $\pi_1, \ldots, \pi_m$ be m random permutations on $\mathcal{U}$, the universe of k-bit numbers. Intuitively, each permutation is like an 'expert' assigning an ordering to $\mathcal{U}$ according to her criteria. Given a multi-set $\mathcal{I}$ of hash values we use the random permutations π_i to compress the set as follows: We replace the multi set $\mathcal{I}$ by the length m sequence of k-bit strings obtained as

$$f(S) = \left(\pi_1^{-1}\big(\min\{\pi_1(\mathcal{I})\}\big), \ldots, \pi_m^{-1}\big(\min\{\pi_m(\mathcal{I})\}\big)\right),$$

where the corresponding multiplicities are propagated in the obvious way. This sequence is our definition of a *fingerprint* for a surface S. To generate the permutations π_i, we apply 2-universal hashing [MR00] as an approximation for random permutations, using a random pair of numbers as parameters.

Based on the fingerprints we estimate the resemblance $r(S_1, S_2)$ by

$$\hat{r}(S_1, S_2)$$
$$= \sum_{j=1}^{m} \frac{\min\{m_{1j}, m_{2j}\}}{\max\{m_{1j}, m_{2j}\}} \chi(f(S_1)_j = f(S_2)_j),$$

where $\chi(\cdot)$ is the characteristic function taking value 1 if the condition of its argument evaluates to *true*, and 0 otherwise. For the surface S_i, the j-th component of its fingerprint is $f(S_i)_j$ with multiplicity m_{ij}. Notice that to compare two fingerprints, we simply need to compare them element-wise without any need to solve for correspondences.

In the next section we show that choosing a large enough m gives, with high probability, a good estimate of resemblance. In practice $m \approx 1000$ is sufficient and hence the probabilistic fingerprints, in the order of 10KBytes, are very compact.

4. Analysis

In this section we analyze the performance of our fingerprints — as mentioned before our goal is to approximate the resemblance of two surfaces effectively and efficiently.

Rabin's hashing scheme maps any signature σ to a number with bit length k. This mapping obviously results in some collisions that can be quantified as:

$$\sigma_i = \sigma_j \Rightarrow h(\sigma_i) = h(\sigma_i)$$
$$\sigma_i \neq \sigma_j \Rightarrow \Pr[h(\sigma_i) = h(\sigma_j)] \leq p, \quad (2)$$

where p is $n^2 t/2^k$, see Equation 1. Let $\mathcal{S}_1$ and $\mathcal{S}_2$ be multi-sets of signatures for the surfaces S_1 and S_2, respectively. Let

$$\mathcal{A} = \mathcal{S}_1 \cap \mathcal{S}_2, \mathcal{B} = \mathcal{S}_1 \setminus \mathcal{S}_2, \text{ and } \mathcal{C} = \mathcal{S}_2 \setminus \mathcal{S}_1,$$

where the set operations again are defined in the multi-set setting, i.e. the \ operation respects the multiplicities. We can relate $a = |\mathcal{A}|$ (size of a multi-set is the sum of the multiplicities of its elements), $b = |\mathcal{B}|$, and $c = |\mathcal{C}|$ to the resemblance of S_1 and S_2 as follows:

$$r(S_1, S_2) = a/(a + b + c). \qquad (3)$$

Let $\mathcal{I}_1$ and $\mathcal{I}_2$ denote the multi-sets of hashed signatures for S_1 and S_2, respectively. Having set the terminology we now quantify the errors incurred by Rabin's hashing scheme.

Using Equation 2 and the definitions of $a, b,$ and c in expectation we get:

$$\begin{array}{ccccc} a & \leq & |\mathcal{I}_1 \cap \mathcal{I}_2| & \leq & a + d \\ a + b + c - d & \leq & |\mathcal{I}_1 \cup \mathcal{I}_2| & \leq & a + b + c, \end{array}$$

where $d = (2bc + ac + ab)p$ is obtained by a simple union bound argument: For any element in $\mathcal{B}$ the probability to participate in a collision that affects the set operations is upper bounded by $(a + c)p$. Thus by linearity of expectation the expected number of such collisions contributed by $\mathcal{B}$ is upper bounded by $(a + c)bp$. Similarly, the expected number of collisions contributed by $\mathcal{C}$ is upper bounded by $(a + b)cp$. Adding these two bounds gives the bound d.

For the approximation quality of the resemblance by $r'(S_1, S_2) = |\mathcal{I}_1 \cap \mathcal{I}_2|/|\mathcal{I}_1 \cup \mathcal{I}_2|$ we get in expectation the following bounds:

$$r(S_1, S_2) = \frac{a}{a + b + c} \leq r'(S_1, S_2) \leq \frac{a + d}{a + b + c - d}.$$

Crucial is only the upper bound which we can also write as

$$\frac{a + d}{a + b + c - d} = \frac{r(S_1, S_2) + \epsilon}{1 - \epsilon},$$

if we set $\epsilon = d/(a + b + c)$. By increasing k, i.e. the number of bits each signature gets mapped to, we can make ϵ arbitrarily small. For small enough ϵ and assuming $r(S_1, S_2) > 0$ we get in expectation

$$r'(S_1, S_2) \leq \frac{r(S_1, S_2) + \epsilon}{1 - \epsilon} \leq (r(S_1, S_2) + \epsilon)(1 + \epsilon)$$
$$\leq r(S_1, S_2)(1 + \sqrt{\epsilon}).$$

Using Markov's inequality this yields

$$\begin{aligned} \Pr[r'(S_1, S_2) & \geq \lambda(1 + \sqrt{\epsilon})r(S_1, S_2)] \qquad (4) \\ & \leq \Pr[r'(S_1, S_2) \geq \lambda \mathrm{E}[r'(S_1, S_2)]] \\ & \leq 1/\lambda. \end{aligned}$$

Finally, we have to check how our estimate behaves under the random permutations that we used to compute the fingerprints. We make use of the following fact, see [Bro97],

$$\Pr[f(S_1)_j = f(S_2)_j] = r'(S_1, S_2),$$

model	# vts.	uniform samp.	spin image	Rabin hash	min hash
skull	54k	0.8	7.5	0.05	4.5
Caesar	65k	1.4	7.3	0.08	10.3
bunny	121k	1.8	13.8	0.04	2.9
horse	8k	0.7	5.7	0.05	7.3

Table 1: Performance. *Timing in seconds for the different stages of the fingerprint computation ($m = 1000$) on a 3 GHz Pentium 4 with 2 GB RAM. Caesar and bunny refer to the complete models. Average query time is roughly 15 msec.*

for all $j = 1, \ldots, m$. Hence estimating $r'(S_1, S_2)$ by using m random permutations is equivalent to performing m coin tosses to evaluate the bias of the coin. Using strong Chernoff bounds we can bound the estimated resemblance $\hat{r}(S_1, S_2)$ as

$$\begin{aligned} \Pr[(1 - \delta)r'(S_1, S_2) \leq \hat{r}(S_1, S_2) & \leq (1 + \delta)r'(S_1, S_2)] \\ & \geq 1 - \eta, \end{aligned} \qquad (5)$$

if $m \geq 4\ln(2/\eta)/(\delta^2 r'(S_1, S_2))$.

Combining our probabilistic bounds by taking a union bound for the event we dealt with in Equation 5 and the complement of the event we dealt with in Equation 4 we conclude that

$$(1 - \delta)r(S_1, S_2) \leq \hat{r}(S_1, S_2) \leq \lambda(1 + \delta)(1 + \sqrt{\epsilon})r(S_1, S_2)$$

with probability at least $1 - (\eta + 1/\lambda)$. That is, with high probability $\hat{r}(S_1, S_2)$ is very close to $r(S_1, S_2)$. Observe that the size m of the fingerprint depends on the desired confidence in our estimated resemblance.

5. Results and Applications

We have implemented the framework shown in Figure 1. Along with each fingerprint, we store some additional header information: a seed for the random number generator, sample spacing δ, shingle radius ρ, parameters for computing spin-images, and k, the degree of the polynomial used in Rabin's hashing scheme. The choice of these parameters is not critical for the success of our scheme provided the conditions given in Section 4 are satisfied. However, we can only compare fingerprints computed using consistent sets of parameters. Typical time requirements for the various stages are shown in Table 1.

Partial Matching. Our scheme is tailored to detect partial matches efficiently. In an experiment we take a bust of Caesar along with its three partial scans (Figure 5). The triangulations of the models are very different and thus test the robustness of our scheme. For

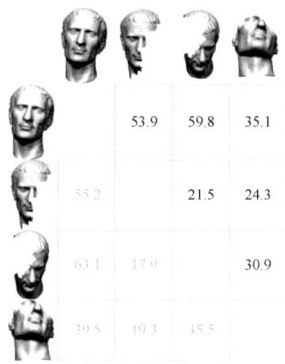

	53.9	59.8	35.1
55.2		21.5	24.3
63.1	17.9		30.9
39.5	19.3	35.5	

Figure 5: Resemblance between Partial Scans. *In black, resemblance (in %) computed using fingerprints. In yellow, approximate ground truth computed from spin-images. Our resemblance definition being symmetric, difference between diagonally opposite elements quantifies the corresponding estimation error.*

each model, we independently compute their probabilistic fingerprint. Then, for each model pair, we compute its resemblance using the corresponding fingerprints (shown in black). For comparison, we compute the ground truth resemblance (shown in yellow) via spin image signatures. Since our resemblance measure is symmetric, the difference between diagonally opposite elements in the resemblance matrix quantifies our estimation error.

Articulated Motion. Resemblance, as measured by our scheme, is robust to articulated deformations. If large chunks of a model are rigidly deformed across two poses, then the corresponding shingles and their hashed descriptors are also preserved. Results on two articulated poses of a horse model are shown in Figure 6. The size of the shingle, determined by ρ, affects the resemblance score: smaller ρ gives higher resemblance and vice versa. The ground truth (yellow curve), determined using the spin-image signatures, is within $\pm 5\%$ of our estimated values.

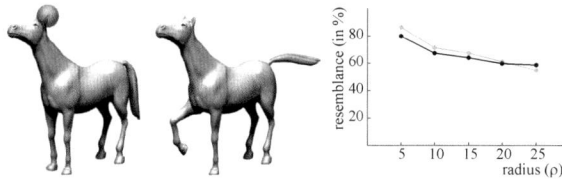

Figure 6: Resemblance between Articulated Shapes. *The yellow ball shows a neighborhood with $\rho = 10$ used for defining shingles. At low values of ρ, we get a high resemblance, since the effect of articulation is felt only by few of the shingles. As ρ increases, resemblance goes down. In yellow, we show the ground truth.*

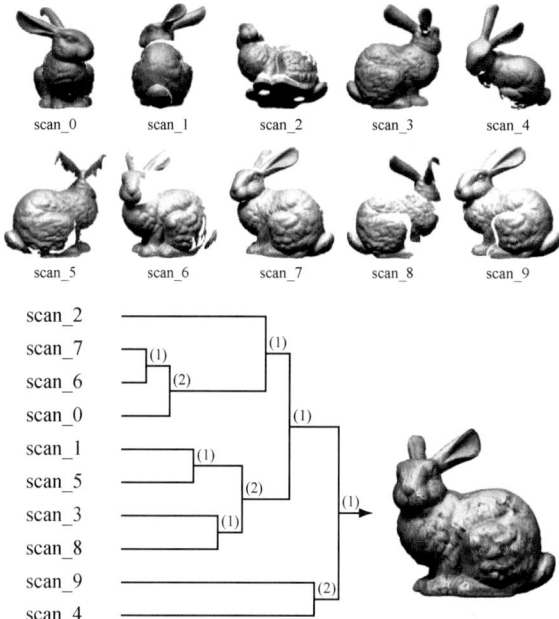

Figure 7: Automatic Scan Alignment. *Given ten initial scans of the Stanford bunny in arbitrary poses, for each scan we compute its probabilistic fingerprint with $m = 1000$. Scan pairs with highest resemblance are picked, and then using [GMGP05] their alignment verified. If the scan-pair align, the fingerprint for the merged scan is estimated. The process continues until no more scans can be combined. The number of required global alignment steps is shown in parenthesis.*

Automatic Scan Alignment. The problem of automatic scan alignment has been previously addressed by Huber and Hebert in [HH03]. Their system can be made significantly faster using our scheme. We explain our method with reference to scans (in arbitrary initial orientations) of the Stanford bunny (Figure 7). In the pre-processing stage, for each of the ten scans, we independently compute its fingerprint. Now for each pair of models, we estimate their resemblance using the respective fingerprints and store the edge joining that pair in a heap with the largest element on top. We then extract the top edge, try to explicitly align the corresponding patches using a global aligner [GMGP05], and if the alignment is not valid we just pick the next largest edge. If an alignment is valid, we merge the respective patches, and need to compute the fingerprint for the merged patch. However, given two patch fingerprints $(a_1, \ldots, a_m)$ and $(b_1, \ldots, b_m)$, we can very efficiently estimate the fingerprint for the merged scan simply as $(\pi_1^{-1}(\min\{\pi_1(a_1), \pi_1(b_1)\}), \ldots, \pi_m^{-1}(\min\{\pi_m(a_m), \pi_m(b_m)\}))$ without explicitly computing the fingerprint for the merged scans. Using this estimated fingerprint, the heap can be efficiently

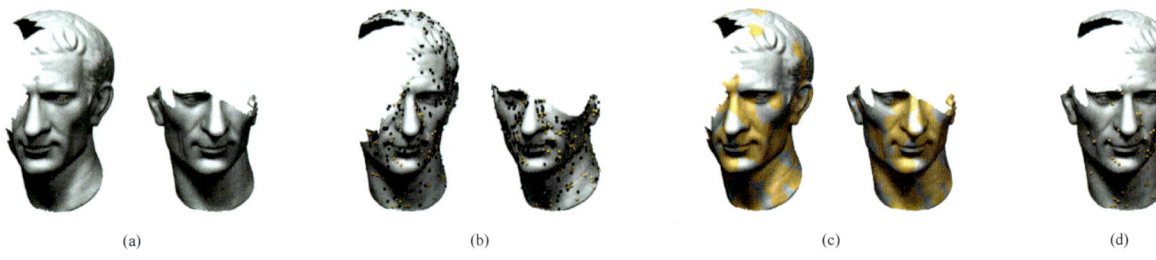

| (a) | (b) | (c) | (d) |

Figure 8: Adaptive Feature Point Selection. *(a) Two shapes in arbitrary poses. (b) For each scan, black and yellow balls denote shingle centers chosen by min-hash. In yellow, shingle centers whose min-hashes agree across the two models. (c) Hints about possible overlap regions obtained by mapping matching min-hashes back onto the objects. These are used for adaptive feature point selection. (d) Final alignment using chosen feature points. The set of features with correct correspondence is shown in yellow.*

updated by re-evaluating only the affected edges. In the figure, the number of required global registration or verification steps are shown in parenthesis. Since global registration is much more costly compared to fingerprint matching, our method, by quickly pruning away non-matching scans, greatly speeds up the whole process.

Adaptive Feature Selection. As shown previously, we can even identify shapes that match only partially. However, with a bit more effort we also get very good hints about regions of overlap. While comparing two fingerprints, we identify the min-hashes that agree and map them back to the original surface shingles. The union of these shingles give us a very good estimate of the region of overlap (Figure 8). Subsequent global registration algorithms benefit significantly from this stage, since the adaptive feature points, given by matching shingle patches, very likely lie in areas of overlap and have correct correspondences. In cases when fingerprints are computed independently, we can similarly identify potential overlap regions across multiple shapes, if we additionally store shingle locations for the min-hashed patches along

with the fingerprints. Timing complexity and storage requirements still remain O(m).

In order to identify partial complementary matches between two shapes S_i and S_j, we can use a similar method. The fingerprint for S_i is computed as usual. For S_j, when computing spin-images, we flip the point normals to take care of complementary shapes. More generally, for each shape we can compute its fingerprints and its complement fingerprint. A possible application is automatic alignment of broken fragments as shown in Figure 9. In this special scenario, using prior information, the flat surface of the scans can be automatically removed as proposed by Huang et al. [HFG*06] as these regions are known not to be in overlap areas.

Database Classification and Retrieval. We use resemblance between pairs of shapes to efficiently classify a shape database and retrieve models from it (Figure 2). Our database comprises of models, in arbitrary initial positions, from the Princeton shape benchmark [FKMS05]. For each shape, we first compute its fingerprint. A shape distance matrix is then build using $1 - r(S_i, S_j)$ as the distance between any pair of shapes S_i and S_j. We extract a 2D embedding of the fingerprint shape space using multi-dimensional scaling [CC94] on the computed distance matrix. Figure 10 shows a selection of models in the embedded shape space. We get meaningful clustering of shapes even in the presence of articulations and partial matches. A typical query result from the processed database is shown in Figure 11. The resemblance scores for this query with any of the tables, planted pots, furniture, or car models is less than 2%. Most of the models in our database being degenerate meshes, we expect a volumetric representation coupled with a suitable signature like spherical harmonics [KFR03] will further improve our performance. Our method is in complement with existing algorithms for shape matching and hence we can use many of the

| Scan A | Scan B | Final Alignment |

Figure 9: Complementary Shapes. *Given two complementary scans in arbitrary poses, we find their alignment using our adaptive feature selection. To detect complementary shapes, flipped normals are used for computing fingerprint of scan B.*

Figure 10: Database Classification. *Shape classification result according to our probabilistic fingerprints. Given any two shapes S_i and S_j, distance between them is defined as $1 - r(S_i, S_j)$. Using this notion of distance, we compute the full distance matrix for a database of shapes. The 2D projection of the fingerprint shape space is computed using classical multi-dimensional scaling (MDS).*

popular shape descriptors in our framework. However, a careful study has to be done to fully evaluate these benefits of our algorithm.

Mesh Authentication. Our scheme can be modified for authenticating geometric models. In the signature computation phase we increase the number of bins making the spin-images sensitive to minor perturbations. Then given a mesh and a partially modified copy, we can use fingerprints to probabilistically identify regions that remain unchanged. For example, if we compute such fragile fingerprints for the original skull model and one corrupted with 1% (of the bounding box) noise, their resemblance is 1.8%. However, if we reorder the vertices, or apply any rigid transform to the original mesh, the resemblance is almost 100%. Local deformations or partial matches can be detected as before. Moreover, for authentication, only a small fingerprint needs to be transmitted and compared with the fingerprint computed from a copy of the mesh. Further investigation needs to be done for quantifying immunity against other types of attacks [WC05].

5.1. Improvements and Limitations

Rabin's method can hash similar signatures to very different values. Such error can be reduced by using locality sensitive hashing (LSH) [IMRV97] where probability of collision between any two signatures is inversely related to their distance. In practice, this may improve the resemblance estimates.

As seen in Figure 6, in some cases the scale ρ has a significant effect on resemblance. To deal with this, we can compute a multi-scale fingerprint over ν different choices of ρ. Storage requirement increases ν fold.

In the current form, we cannot handle scaling. Though pre-scaling of objects may be done using anisotropic scaling [KFR04], such a method fails for partial similarities. We are currently researching other possibilities to handle this scenario.

6. Conclusions

In this paper we have shown how to build compact probabilistic fingerprints for digital geometry models that allow efficient model comparison for partial or total similarity. It is interesting that our scheme re-

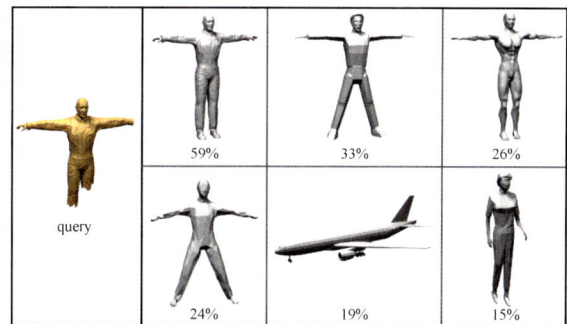

Figure 11: Database Retrieval. *Given a query shape, we show the models retrieved by our algorithm from a database of shapes in arbitrary poses. Our scheme handles partial matches, and is robust to articulations. Corresponding resemblance scores are shown.*

lies on randomness for selecting the 'shape features' through the presence or absence of which similarity is estimated. We give provable bounds on the quality of our shape comparisons demonstrating the power of randomization in a geometric context.

The compactness of our fingerprints may also enable distributed geometry processing tasks in sensor network settings, where geometry acquisition, storage, and retrieval may be required over geographically dispersed deployments.

Acknowledgements. This research was supported in parts by DARPA grant 32905, NSF grants FRG-0354543 and ITR 0205671, NIH grant GM-072970, Max-Planck Center for Visual Computing and Communication and a Stanford Graduate fellowship. The authors thank Natasha Gelfand, Marc Levoy, Daniel Russel, and Afra Zomorodian for their helpful comments and suggestions at various stages of the work. We thank the Aim@SHAPE Shape Repository for the Caesar model, the Stanford 3D repository for the bunny scans. Special thanks to the anonymous reviewers for their valuable comments.

References

[AKKS99] ANKERST M., KASTENMÜLLER G., KRIEGEL H. P., SEIDL T.: 3D shape histograms for similarity search and classification in spatial databases. In *SSD* (1999), pp. 207–228.

[Blo70] BLOOM B.: Space/time trade-offs in hash coding with allowable errors. In *Communications of the ACM* (1970), vol. 13, pp. 422–426.

[Bro93] BRODER A. Z.: Some applications of rabins fingerprinting method. In *Sequences II: Methods in Communications* (1993).

[Bro97] BRODER A. Z.: On the resemblance and containment of documents. In *Sequences* (1997).

[Bro00] BRODER A. Z.: Identifying and filtering near-duplicate documents. In *CPM* (2000), pp. 1–10.

[CC94] COX T., COX M.: *Multidimensional Scaling.* Chapman and Hall, London, 1994.

[CL01] CHAN C., LU H.: *Fingerprinting Using Polynomial (Rabin's Method).* Tech. rep., University of Alberta, 2001.

[FKMS05] FUNKHOUSER T., KAZHDAN M., MIN P., SHILANE P.: Shape-based retrieval and analysis of 3d models. *Comm. of the ACM* (2005), 58–64.

[FKS*04] FUNKHOUSER T., KAZHDAN M., SHILANE P., MIN P., KIEFER W., TAL A., RUSINKIEWICZ S., DOBKIN D.: Modeling by example. *ACM Transactions on Graphics 23*, 3 (2004), 652–663.

[GCO06] GAL R., COHEN-OR D.: Salient geometric features for partial shape matching and similarity. vol. 25(1), pp. 130–150.

[GMGP05] GELFAND N., MITRA N. J., GUIBAS L., POTTMANN H.: Robust global registration. In *Symp. on Geometry Processing* (2005), pp. 197–206.

[HFG*06] HUANG Q., FLORY S., GELFAND N., HOFER M., POTTMANN H.: Reassembling fractured objects by geometric matching. In *Siggraph* (2006), p. to appear.

[HH03] HUBER D. F., HEBERT M.: Fully automatic registration of multiple 3D data sets. In *Image and Vision Computing* (2003), vol. 21(7), pp. 637–650.

[Hor84] HORN B.: Extended gaussian images. In *Proc. of IEEE* (1984), vol. 72(12), pp. 1671–1686.

[IMRV97] INDYK P., MOTWANI R., RAGHAVAN P., VEMPALA S.: Locality-preserving hashing in multidimensional spaces. pp. 618–625.

[Joh97] JOHNSON A.: *Spin-Images: A Representation for 3-D Surface Matching.* PhD thesis, Robotics Inst., Carnegie Mellon Univ., August 1997.

[KCD*04] KAZHDAN M., CHAZELLE B., DOBKIN D., FUNKHOUSER T., RUSINKIEWICZ S.: A reflective symmetry descriptor for 3d models. In *Algorithmica* (2004), pp. 201–225.

[KFR03] KAZHDAN M., FUNKHOUSER T., RUSINKIEWICZ S.: Rotation invariant spherical harmonic representation of 3d shape descriptors. In *Symp. on Geometry Proc.* (2003), pp. 167–175.

[KFR04] KAZHDAN M., FUNKHOUSER T., RUSINKIEWICZ S.: Shape matching and anisotropy. *ACM TOG 23*, 3 (2004), 623–629.

[MR00] MOTWANI R., RAGHAVAN P.: *Randomized Algorithms.* Cambridge University, 2000.

[OFCD02] OSADA R., FUNKHOUSER T., CHAZELLE B., DOBKIN D.: Shape distributions. In *ACM Transactions on Graphics* (2002), vol. 21(4).

[Rab81] RABIN M. O.: Fingerprinting by random polynomials. In *Center for Research in Computing Tech., Harvard Univ., Report TR-15-81* (1981).

[RWP05] REUTER M., WOLTER F.-E., PEINECKE N.: Laplace-spectra as fingerprints for shape matching. In *SPM* (2005), pp. 101–106.

[SGM98] SHIVAKUMAR N., GARCIA-MOLINA H.: Finding near-replicas of documents on the web. In *Proceedings of Workshop on Web Databases* (1998).

[Tur92] TURK G.: Re-tiling polygonal surfaces. In *SIGGRAPH* (1992), pp. 55–64.

[WC05] WU H.-T., CHEUNG Y.-M.: A fragile watermarking scheme for 3d meshes. In *Multimedia and Security* (2005), pp. 117–124.

Eurographics Symposium on Geometry Processing (2006)
Konrad Polthier, Alla Sheffer (Editors)

Partial Matching of 3D Shapes with Priority-Driven Search

T. Funkhouser and P. Shilane

Princeton University, Princeton, NJ

Abstract

Priority-driven search is an algorithm for retrieving similar shapes from a large database of 3D objects. Given a query object and a database of target objects, all represented by sets of local 3D shape features, the algorithm produces a ranked list of the c best target objects sorted by how well any subset of k features on the query match features on the target object. To achieve this goal, the system maintains a priority queue of potential sets of feature correspondences (partial matches) sorted by a cost function accounting for both feature dissimilarity and the geometric deformation. Only partial matches that can possibly lead to the best full match are popped off the queue, and thus the system is able to find a provably optimal match while investigating only a small subset of potential matches. New methods based on feature distinction, feature correspondences at multiple scales, and feature difference ranking further improve search time and retrieval performance. In experiments with the Princeton Shape Benchmark, the algorithm provides significantly better classification rates than previously tested shape matching methods while returning the best matches in a few seconds per query.

1. Introduction

Large databases of 3D models are becoming available in a number of disciplines, including computer graphics, mechanical CAD, molecular biology, and medicine. As these databases grow, shape-based similarity search is emerging as a valuable tool for analysis and discovery.

The goal of our work is to develop effective methods to retrieve from a database a ranked list of 3D models most similar in shape to a 3D model provided as a query. This problem is difficult because objects of the same type may not have exactly the same sets of parts (e.g., some chairs have arms and others don't), and some parts that distinguish object types may be relatively small (e.g., the ears of a bunny). Shape representations that account only for global shape properties do not perform well at recognizing shapes in these situations.

A common method for addressing this problem is to represent every object by a set of local shape features centered on points sampled from the object's surface and then to compute a similarity metric for every pair of objects based on a cost function measuring the quality of matches in the optimally aligned set of feature correspondences (e.g., [BMP01, CJ96, JH99]). This approach is attractive because it is robust to shape variations within a class – as long as a few key shape features match, then the objects will

match. The challenge is that the number of possible feature correspondence sets grows exponentially with the set size – naively checking all possible sets of k feature correspondences among n features on two objects takes $O(n^k)$ operations. In practice, searching the space of potential feature correspondences for a single pair of surfaces can take several seconds or minutes, and using these methods to find the best matches in a large database is impractical.

In this paper, we introduce a priority-driven algorithm for searching all objects in a database at once. The algorithm is given a query object and a database of target objects, all represented by sets of local shape features, and its goal is to produce a ranked list of the best target objects sorted by how well any subset of k features on the query match features on the target object. To achieve this goal, the system maintains a priority queue of potential sets of feature correspondences (partial matches) sorted by a cost function accounting for both feature dissimilarity and geometric deformation. Initially, all pairwise correspondences between the features of the query and features of target objects are loaded onto the priority queue. Then, at every step, the best partial match m is popped off the priority queue, new partial matches are created by extending m to include compatible feature correspondences, and those new partial matches are added to the priority queue. This process is iterated until the desired number of full matches with k feature correspondences have been popped off the priority queue.

The advantage of this approach is that the algorithm provably finds the optimal set of matches over the entire database while investigating only a small subset of the potential matches. Like any priority-driven backtracking search (e.g., Dijkstra's shortest path algorithm), the algorithm considers only the partial matches that can possibly lead to the lowest cost match (Figure 1). Although some poor partial matches are generated, they never rise to the top of the priority queue, and thus they incur little computational overhead. By using a single priority queue to store partial matches for all objects in the database at once, we achieve great speedups when retrieving only the top matches – if a small set of target objects match the query well, their feature correspondences will be discovered quickly and the details of other potential matches will be left unexplored. This approach largely avoids the combinatorial explosion of searching for multi-feature matches in dissimilar objects.

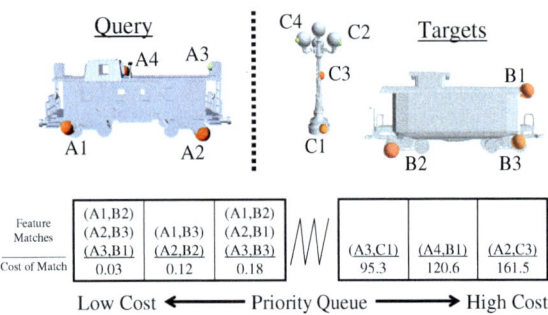

Figure 1: *Priority driven search: a priority queue (bottom) stores potential matches of features (labeled dots) on a query to features of all target objects at once. Matches are extended only when they reach the top of the priority queue (the leftmost entry), and thus high cost feature correspondences sit deep in the priority queue and incur little computational expense.*

This paper makes several research contributions. In addition to the idea of priority-driven search, we explore ways of improving computational efficiency and retrieval performance of multi-feature matching algorithms: 1) we use ranks rather than L^2 differences to measure feature similarity; 2) we use a measure of class distinction to select features; and, 3) we match features at multiple scales. Finally, we provide a working shape-based retrieval system and analyze its performance over a wide range of options and parameter settings. We find that our system provides significantly better retrieval performance than previous shape matching approaches on the Princeton Shape Benchmark [SMKF04] while using increased, but reasonable, processing and storage costs.

The organization of the paper is as follows. The next section contains a summary of related work on matching of 3D surfaces. Section 3 contains an overview of the priority-driven search algorithm followed by a detailed description for every algorithmic step. Section 4 compares the performance of the priority-driven search approach to other state-of-the-art shape matching methods and investigates how modifying several aspects of the algorithm impacts its performance. Finally, Section 5 provides a brief discussion of limitations and topics for future work.

2. Background and Related Work

There has been a large amount of research on algorithms for shape-based retrieval of 3D surface models. In this section, we focus on the previous work most closely related to ours and refer the reader to survey articles for broad overviews of prior work in related areas [BKS*05, IJL*05, TV04].

The most common approach to shape-based retrieval of 3D objects is to represent every object by a single global *shape descriptor* representing its overall shape. Shape Histograms [AKKS99], the Light Field Descriptor [COTS03], and the Depth Buffer Descriptor [HKSV02] are a few examples. These descriptors can be searched efficiently, and thus they are suitable for queries into large databases of 3D shapes. However, retrieval precision is generally poor when objects within the same class have different overall shapes – e.g., due to articulated motions, missing parts, or extra parts.

Recently, several researchers have investigated approaches to partial shape matching based on feature correspondences (e.g., [BMP01, CJ96, GCO06, JH99, NDK05]). The general strategy is to compute multiple local shape descriptors (shape features) for every object, each representing the shape for a region centered at a point on the surface of the object. Then, the similarity of any pair of objects is determined by a cost function determined by the optimal set of feature correspondences at the optimal relative transformation, where the optimal match minimizes the differences between corresponding shape features and the geometric distortion implied by the feature correspondences. This approach has been used for recognizing objects in 2D images [BMP01, BBM05], recognizing range scans [JH99], registering medical images [AFP00], aligning point sets [CR00], aligning 3D range scans [GMGP05, LG05], and matching 3D surfaces [NDK05, SMS*04].

The challenge is to find an optimal set of feature correspondences efficiently. One approach is to consider an association graph containing a node for every possible feature correspondence and an edge for every compatible pair of correspondences [BB76]. If each node is weighted by the dissimilarity of its associated features and each edge is weighted by the cost of the geometric deformation implied by its associated pair of correspondences, then finding the optimal set of k feature correspondences reduces to finding a minimum weight k-clique in the association graph. Researchers have approached this problem with algorithms based on branch-and-bound [GMGP05], integer quadratic programming [BBM05], etc. However, previous work in this area has been aimed at pairwise alignment of objects, and current solution methods are generally too slow for search of large databases.

Another approach is based on the RANSAC algorithm [FB81, SMS*04]. Sets of k feature correspondences are generated, where k is large enough to determine an aligning transformation, and the remaining features are used to score how well the objects match after the implied alignment. For example, [JH99] finds small sets of compatible feature correspondences, computes the alignment providing a least-squared best fit of corresponding features, and then "verifies" the alignment with an iterative closest point algorithm [BM92]. [SMS*04] proposed a "Batch RANSAC" version of this algorithm that considers matches to all target objects in a database all at once, generating candidate matches preferentially for the target objects with features compatible with ones in the query. However, their evaluation focused on recognition of vehicles from a small set of range scans, and studies have not been done to show how well it works for large databases of surface models.

Several researchers have considered methods for accelerating database searches using discrete approximations. For example, geometric hashing [LW88] uses a grid-based hash table to store every feature of every target object in n choose k hash cells. For every query, k features are used to determine a mapping into the hash, and then other features vote for object transformations wherever there are hash collisions. This approach is quite popular in computer vision, molecular biology, and partial surface matching (e.g., [GCO06]). However, it requires a lot of memory to store hash tables and produces approximate matches, since it discretizes both the set of possible transformations (it only considers transformations induced by combinations of features) and Euclidean space (features match only if they fall in the same grid cell).

Alternatively, "bag of words" approaches can be used to discretize feature space. For example, [MBM01] clusters features into "shapemes," builds a histogram of shapemes for every object, and then approximates the similarity of two objects by the similarity of their histograms, and [GD05] extends this approach to consider pyramids of clusters. However, these methods make little or no use of the geometric arrangements of features, and thus they do not provide as distinguishing matches as possible.

Our approach is to use priority-driven search to find the objects in a database that have sets of local feature correspondences minimizing a continuous cost function. The key idea is to use a priority queue to focus a backtracking search on sets of feature correspondences with lowest matching cost among all objects in the database all at once. This approach provides a significant efficiency improvement over more expensive algorithms that compute pairwise matches between the query and all objects in the database indepen-

dently (e.g., [BBM05]) – i.e., it avoids computing the optimal set of correspondences for the target objects that do not appear at the top of the retrieval list. It can also provide an accuracy improvement over discrete or greedy approximate algorithms, since it guarantees optimal matches with respect to a continuous cost function.

3. System Execution

Execution of our system proceeds in two phases: a preprocessing phase and a query phase.

During the preprocessing phase, we build a multi-feature representation of every object in the database. First, we generate for each object a set of spherical regions covering its surface at different scales. Second, for every region, we compute a descriptor of the shape within that region. Third, we compute differences between all pairs of descriptors at the same scale and associate with every descriptor a mapping from rank to difference. Finally, we select a subset of features to represent each object based on how distinctive they are of their object class. The result of this preprocessing is a set of "shape features" (or "features," for short) for every object, each with an associated position (p), normal ($\vec{n}$), radius (r), and shape descriptor (a feature vector of numbers representing a local region of shape), and a description of how discriminating its shape descriptor is with respect to others in the database.

For every query, our matching procedure proceeds as shown in Figure 2. The inputs are: 1) a query object, *query*, 2) a database of target objects, *db*, each represented by a set of shape features, 3) a cost function, *cost*, measuring the quality of a proposed set of feature correspondences, 4) a constant, k, indicating the number of feature correspondences that should be found for a complete match, and 5) a constant, c, indicating the number of objects for which to retrieve optimal matches. The output is a list of the best matching target objects, M, along with a description of the feature correspondences and cost for each one.

Initially, a priority queue, Q, is created to store partial matches, and an array, M, is created to store the best match to every target object. Then, all pairwise correspondences between the features of the query and features of the target objects are created, stored in lists associated with the target objects, and loaded onto the priority queue. The priority queue then holds all possible matches of size 1. Then, until c complete matches have been found, the best partial match, m, is popped off the priority queue. If it is a complete match (i.e., the number of feature correspondences satisfies k), then the search of that target object is complete, and the priority queue is cleared of partial matches to that object. Otherwise, for every feature correspondence between the query and the target of m, the match is extended by one feature correspondence to form a new match, m'. The best match for every target object is retained in an array, M, when it is added to

```
PriorityDrivenSearch(Object query, Database db,
      Function cost, int k, int c)
   # Create correspondences
   foreach Object target in db
      foreach Feature q in query
         foreach Feature t in target
            p = CreatePairwiseCorrespondence(q, t, cost)
            if (IsPlausible(p))
               AddToPriorityQueue(Q, p)
               AddToList(C[target], p)
               if (cost(p) < cost(M[target]))
                  M[target] = p


   # Expand matches until find complete ones
   complete_match_count = 0
   while (complete_match_count < c)
      # Pop match off priority queue
      m = PopBestMatch(Q)
      target = GetTargetObject(m)

      # Check for complete match
      if (IsMatchComplete(m, k))
         RemoveMatchesFromPriorityQueue(Q, target)
         complete_match_count++
         continue;

      # Extend match
      foreach PairwiseCorrespondence p in C[target]
         m' = ExtendMatch(m, p, cost)
         if (IsPlausible(m'))
            AddToPriorityQueue(Q, m')
            if (cost(m') < cost(M[target]))
               M[target] = m'

   # Return result
   return M
```

Figure 2: *Pseudo-code for priority-driven search.*

the priority queue. This process is iterated until at least *c* full matches with *k* feature correspondences have been popped off the priority queue for *c* distinct target objects, and the array of the best matches to every target object, *M*, is returned as the result.

The computational savings of this procedure come from two sources. First, matches are considered from best to worst, and thus, poor pairwise correspondences are never considered for extension and add little to the execution time of the algorithm. Second, after complete matches for at least *c* target objects have been added to the priority queue, it is possible to determine an upper-bound on the cost of matches that can plausibly lead to one of the best matches. If the score computed for an extended match, m', is higher than that upper bound, then there is no reason to add it to the queue, and it can be ignored. Similarly, if a match, *m*, is popped off the queue, then it is provably the best remaining match – i.e., no future match can be considered with a lower cost. Thus, the algorithm can terminate early (immediately after *c* best matches have been popped off the priority queue) while still guaranteeing an optimal solution.

Of course, there are many design decisions that impact the efficacy of this search procedure, including how regions are constructed, how shape descriptors are computed, what cost function is used, how implausible matches are culled, and so on. The following subsections describe our design decisions in detail, and Section 4 provides the results of experiments aimed at evaluating the impact of each one on search speed and retrieval performance.

3.1. Constructing Regions

The first step of the process is to define a set of local regions covering the surface of every object. In theory, the regions could be volumetric or surface patches; they could be disjoint or overlap; and, they could be defined at any scale.

In our system, we construct overlapping regions defined by spherical volumes centered on points sampled from the surface of an object [KPNK03, NDK05]. We have experimented with two different point sampling methods, one that selects points randomly with uniform distribution with respect to surface area, and another that selects points at vertices of the mesh with probability equal to the surface area of the vertices' adjacent faces. However, they do not give significantly different performance, and so we consider only random sampling with respect to surface area for the remainder of this paper. Of course, other sampling methods that sample according to curvature, saliency, or other surface properties would be possible as well.

We have experimented with regions at four different scales. The smallest scale has radius 0.25 times the radius of the entire object and the other scales are 0.5, 1.0, and 2.0 times, respectively. These scales are chosen because the smallest scale is approximately the size of most "distinguishing features of an object" and the largest scale is just big enough to cover the entire object for spheres centered at the most extreme positions on the surface (Figure 3).

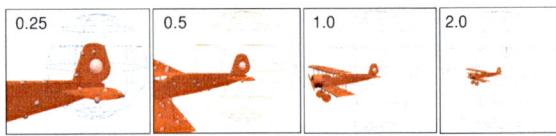

Figure 3: *Shape regions at four different scales.*

3.2. Computing Shape Descriptors

The second step of the process is to generate and store a representation of the shape for each spherical region (a shape descriptor). There will be many such regions for every surface, so the shape descriptors must be quick to compute, concise to store, and fast to compare, in addition to being as discriminating as possible. There are many shape descriptors that meet some or all of these criteria (see surveys in [BKS*05, IJL*05, TV04]). Examples include shape contexts [BMP01, MBM01, FHK*04], spin images [JH99, SMS*04], harmonic shape descriptors [FHK*04, NDK05], curvature profiles [CJ96, GCO06], and volume integrals [GMGP05].

In our system, we have experimented with three different shape descriptors based on spherical harmonics. All three decompose a sphere into concentric shells of different radii and then describe the distributions of shape within those shells using properties of spherical harmonics. The first ("SD") simply stores the amplitude of all shape within each shell (the zero-th order component of spherical harmonics) – it is a one-dimensional descriptor equivalent to the "Shells" shape histogram of [AKKS99]. The second ("HSD") stores the amplitude of spherical harmonic coefficients within each frequency – it is equivalent to the Harmonic Shape Descriptor of [FMK*03, KFR03]. The last ("FSD") descriptor stores the amplitude of every spherical harmonic coefficient separately – it is similar to the Harmonic Shape Contexts of [FHK*04]. In all of our experiments, we utilize 32 spherical shells and 16 harmonic frequencies for each descriptor.

We chose these shape representations for several reasons. First, they are well-known descriptors that have been shown to provide good performance in previous studies [FHK*04, SMKF04]. Second, they are reasonably robust, concise, and fast to search. Finally, they provide a nested continuum with which to investigate the trade-offs between verbosity and discrimination – SD is very concise (32 values), but not that discriminating; HSD is more verbose (512 values) and more discriminating; and FSD is the most verbose (4352 values) and the most discriminating. The three descriptors are related in that each of the more concise descriptors is simply a subset or aggregation of terms in the more verbose ones (e.g., the SD descriptor stores the amplitude of only the zero-th order spherical harmonic frequencies). Thus, the L^2 difference of each descriptor provides a lower bound on the L^2 difference between the more verbose ones, which enables progressive refinement of descriptor differences, as proposed in Section 5.

Our method for computing the descriptors for all regions of a single surface starts by computing a 3D grid containing the Euclidian Distance Transform of the surface. The grid resolution is chosen to match the finest sampling rate required by the HSD for regions at the smallest scale; the triangles of the surface are rasterized into the grid; and the squared distance transform is computed and stored. Then, for every spherical region centered on a point sampled from the surface, a spherical grid is constructed by aligning a sphere with the normal to the surface; a Gaussian function of the distance transform (GEDT) [KFR03] is sampled at regular intervals of radius and polar angles; the Spharmonickit software is used to compute the spherical harmonic decomposition for each radius; the amplitudes of the harmonic coefficients (or frequencies, depending on the type of shape descriptor) are computed; the shape descriptors are compressed using principal component analysis (PCA); and, the dimensions associated with the top C eigenvalues ($C \sim 10\%$) are stored as a shape descriptor.

For each 3D object, computing the three types of shape descriptors centered at 128 points for 4 scales (0.25, 0.5, 1.0, and 2.0) takes approximately four minutes overall and generates around 1MB of data per object. One minute is spent rasterizing the triangles and computing the squared distance transform at resolution sufficient for the smallest scale descriptors, almost two minutes are spent computing the spherical grids, and a few seconds are spent decomposing the grids into spherical harmonics for each object. Compression amortizes to approximately one minute per object for SFDs and approximately 1 second per object for SHDs.

3.3. Selecting Distinctive Features

The third step of our process is to characterize the differences between shape descriptors and to select a subset of the shape features to be used for matching for each target object. Our goal is to augment the features with information about how discriminating they are and to select a subset of the most distinctive features in order to improve processing speed and retrieval precision.

Selecting a subset of local shape descriptors is a well known technique for speeding up retrieval, and several researchers have proposed different methods for this task. The simplest technique is to select features randomly [JH99, FHK*04, MBM01]. Other methods have considered selecting features based on surface curvature [YF02], saliency [GCO06], likelihood within the same shape [GMGP05, JH99], persistence across scales [GMGP05], number of matches to another shape [SMS*04], likelihood within the database [SF06], and distinction of its object's class [SF06].

In our system, we follow the ideas of [SF06]. For every feature, we compute the L^2 difference of its shape descriptor to the best match of every other object in the database, sort the differences from best to worst, and save them in a *rank-to-difference mapping* (RTD). To save space, we store an approximation to the RTD containing log(N) values by sampling distances at exponentially larger ranks. We then use the RTD to estimate the distinction of every shape feature. Distinctive features are ones that are both similar to features in few other objects (ones in the same class) and different from the rest (objects in other classes). When given a classification for the target objects, we quantify this notion by using a measure of retrieval performance as our model for feature distinction [SF06].

Once the distinction of every feature has been computed, we employ a greedy algorithm to select a small set of features to represent every target object during the query phase (Figure 4). The selection algorithm iteratively chooses the feature with highest DCG whose position is not closer than a Euclidean distance threshold, *minlength*, to the position of any previously selected feature. This process avoids selecting features nearby each other on the mesh and provides an easy way to vary the subset size by adjusting the distance threshold.

(a) All Features　　(b) Feature Distinction　　(c) Selected Features

Figure 4: *Feature selection: (a) positions sampled randomly on surface, (b) computed DCG values used to represent feature distinction (red is highest, blue is lowest), and (c) features selected to represent object during matching.*

The net result of this process is a small set of features for every target object, each with an associated position (p), normal ($\vec{n}$), radius (r), a set of shape descriptors (SD, HSD, and FSD), a rank-to-difference-mapping (RTD), and a retrieval performance score (DCG). In our implementation, computing the RTD and the distinction for each feature takes a little less than 1 second, and selecting the most distinctive features takes less than a second for each object. The storage required for the resulting data required at query time is approximately 100KB per object.

3.4. Creating Pairwise Feature Correspondences

When given a query object to match to a database of target objects, the first step is to compute the cost of pairwise correspondences between features of the query to features of the target. The key to this step is to develop a cost function that provides low values only when two features are compatible and gradually penalizes pairs that are less similar. The simplest and most common approach is to use the L^2 difference between their associated shape descriptors. This approach forms the basis for our implementation, but we augment it in three ways.

First, given features F_1 and F_2, we compute the L^2 difference, D, between their shape descriptors. Then, we use the rank-to-difference mappings (RTD) of each feature to convert D into a rank (i.e., where that distance falls in the ranked list associated with each feature). The new difference measure (C_{rank}) is the sum of the ranks computed for F_1 with respect to the RTD of F_2, and vice versa:

$$C_{rank} = Rank(RTD_1, D) + Rank(RTD_2, D)$$

This feature rank cost (which we believe is novel) avoids the problem that very common features (e.g., flat planar regions) can provide indistinguishing matches (false positives) when L^2 differences are small. Our approach considers not the absolute difference between two features, but rather their

difference relative to the best matching features of other objects in the database. Thus, a pair of features will only be considered similar if both rank highly in the retrieval list of the other.

Second, we augment the cost function with geometric terms. For part-in-whole object matching, we can take advantage of the fact that features are more likely to be in correspondence if they appear at the same relative position and orientation with respect to the rest of their objects. Thus, for each feature, we compute the distance between its position and the center of mass of its object (R), scaled by the average of R for all features in the object ($RAVG$), and we add a distance term C_{radius} to the cost function accounting for the difference between these distances:

$$C_{radius} = |\frac{R_1}{RAVG_1} - \frac{R_2}{RAVG_2}|$$

We also compute a normalized vector $\vec{r}$ from the object's center of mass to the position of each feature and store the dot product of that vector with the surface normal ($\vec{n}$) associated with the feature. The absolute value of the dot product is taken to account for the possibility of backfacing surface normals. Then, the difference between dot products for any pair of features is used to form a normal consistency term to the cost function:

$$C_{normal} = ||\vec{r_1} \cdot \vec{n_1}| - |\vec{r_2} \cdot \vec{n_2}||$$

Overall, the cost of a feature correspondence is a simple function of these three terms:

$$C_{correspondence} = \alpha_{rank}C_{rank}^{\gamma_{rank}} + \alpha_{radius}C_{radius}^{\gamma_{radius}} + \alpha_{normal}C_{normal}^{\gamma_{normal}}$$

where the α coefficients and γ exponents are used to normalize and weight the terms with respect to each other.

Of course, computing all potential pairwise feature correspondences between a query object and a database of targets is very costly. If the query has M_Q features and each of N targets has M_T selected features, then the total number of potential feature correspondences is $N \times M_Q \times M_T$. To accelerate this process, we utilize conservative thresholds on each of the three terms (*maxrank*, *maxradius*, *maxnormal*) to throw away obviously poor feature correspondences. The terms are computed and the thresholds are checked progressively in order of how expensive they are to compute (e.g., C_{rank} is last), and thus there is great opportunity for trivial rejection of poor matches with little computation. Indexing and progressive refinement could further reduce the compute time as described in Section 5.

3.5. Searching for the Optimal Multi-Feature Match

The second step of the query process is to search for the best multi-feature matches between the query object and the target objects. This is the main step of priority-driven search.

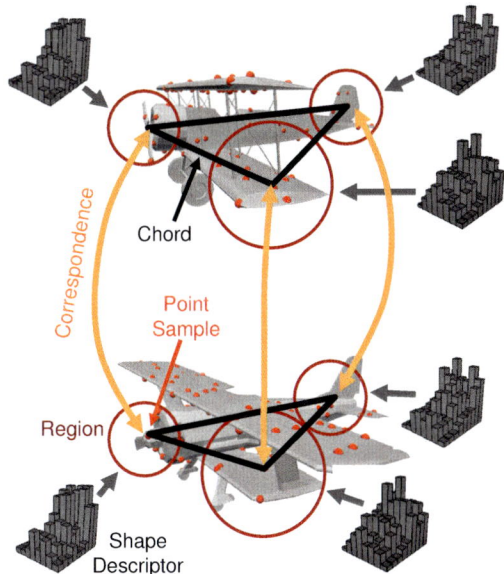

Figure 5: *A 3-feature match for two airplanes. Red points represent feature positions on the surface. For three features, red circles represent regions, gray histograms represent shape descriptors, orange lines represent feature correspondences, and black lines represent lengths between features of same object. Consistency of all shape descriptors, lengths, and angles is required for a good match.*

A priority queue is used to store incomplete sets of feature matches during a backtracking search. Initially, all pairwise correspondences (computed as described in the previous subsection) are loaded onto the priority queue. Then, the best partial match, m, is repeatedly popped off the priority queue, and then extended matches are created for every compatible feature correspondence and loaded onto the priority queue. This process is iterated until at least c full matches with k feature correspondences have been popped off the priority queue for distinct target objects.

As a partial match is extended to include one more feature correspondence, two extra terms are added to the cost function to account for geometric deformations implied by multiple pairwise feature correspondences (Figure 5). First, a chord length term C_{length} is added to penalize matches with inconsistent inter-feature lengths. Specifically, for every pair of feature correspondences in m, we compute the length of the chord between feature positions in the same object (L), scaled by the average of L over all features in the object ($LAVG$). Then, we compute the difference between these distances and normalize by the greater of the two to produce the length term of the cost function:

$$C_{length} = \frac{|\frac{L_1}{LAVG_1} - \frac{L_2}{LAVG_2}|}{max(\frac{L_1}{LAVG_1}, \frac{L_2}{LAVG_2})}$$

Second, a surface orientation term is added to penalize matches with pairs of feature correspondences whose surface normals are inconsistent. This term penalizes both mismatches in the relative orientations of the two pairs of normals with respect to one other and mismatches in the orientations of the normals with respect to the chord between the features. If $\vec{v_1}$ is the normalized vector between features 1a and 1b with normals $\vec{n_{1a}}$ and $\vec{n_{1b}}$ in object 1, and similar variables describe the relative orientations of features in object 2, then the orientation term of the cost function can be computed as follows:

$$C_{orient} = \quad ||\vec{n_{1a}} \cdot \vec{n_{1b}}| - |\vec{n_{2a}} \cdot \vec{n_{2b}}|| + \\ ||\vec{v_1} \cdot \vec{n_{1a}}| - |\vec{v_2} \cdot \vec{n_{2a}}|| + \\ ||\vec{v_1} \cdot \vec{n_{1b}}| - |\vec{v_2} \cdot \vec{n_{2b}}||$$

These terms are also weighted and raised to exponents to provide normalization when added to the overall scoring function computed for a match with k feature correspondences:

$$C_{chord} = \alpha_{length}C_{length}^{\gamma_{length}} + \alpha_{orient}C_{orient}^{\gamma_{orient}}$$

As in the previous section, we utilize conservative thresholds on C_{length} and C_{orient} (*maxlength* and *maxorientation*) to throw away obviously poor feature correspondences. We also utilize a threshold on the minimum distance between features within the same object (*minlength*) in order to avoid matches comprised of features in close proximity to one another.

The overall cost of a match is the sum of the terms representing differences in the k feature correspondences and the geometric differences between the k(k-1)/2 chords spanning pairs of features:

$$C_{match} = \sum_{i<k}C_{correspondence}(i) + \sum_{i,j<k,i<j}C_{chord}(i,j)$$

4. Results

In this section, we present results of experiments with priority-driven search. We investigate the performance of the method in relation to the state of the art in shape-based retrieval and investigate the impact of several design choices on the speed and quality of retrieval results.

All experiments were based on the 3D data provided in the Princeton Shape Benchmark [SMKF04]. It contains 907 polygonal models partitioned into 92 classes (sedans, race cars, commercial jets, fighter jets, dining room chairs, etc.). This database was chosen because it has been used in several previous 3D shape retrieval studies (e.g., [BKS*05, SMKF04]) and thus forms a basis for comparison with competing methods.

In a representative preprocessing phase, we generated features at 128 surface points with 4 different scales for every object. For every feature, we computed its shape descriptors,

RTDs, and DCGs, and then we selected the most distinctive set of descriptors using the methods described in Sections 3.1-3.3. The total preprocessing time for all 907 objects was 70 hours and the total size of all data generated was 1GB, of which 64MB represents the selected features that had to be stored in memory for target objects during the query phase.

During the query phase, we performed a series of "leave-one-out" classification tests. In each test, every object of the database was used as a *query object* to search databases containing the remaining $N - 1$ target objects. Standard information retrieval metrics, such as precision, recall, nearest neighbor classification rate (1-NN), first-tier percentage (1-tier), second-tier percentage (2-tier), and discounted cumulative gain (DCG), were computed to measure how many objects in the query's class appear near the top of its ranked retrieval list, and those metrics were averaged for all queries.

Unless otherwise stated, experiments were run on a x86_64 processor with 12GB of memory running Linux. Parameters for the "base configuration" of the system were set as follows: $c = 1$, $k = 3$, number of features per object = 128, number of feature scales = 4 (0.25, 0.5, 1.0, and 2.0), shape descriptor type = HSD, compression ratio = 10X, $maxradius = maxnormal = maxlength = maxorientation = 0.25$, $minlength = 0.3 \cdot RAVG$, $\alpha_{rank} = 0.01$, $\alpha_{radius} = \alpha_{normal} = \alpha_{length} = \alpha_{orient} = 1$, and $\gamma_{rank} = 4$, $\gamma_{radius} = \gamma_{normal} = \gamma_{length} = \gamma_{orient} = 2$. These parameters were determined empirically and used for all experiments without adjustment, except in Section 4.1 where the FSD shape descriptor was used, and in Section 4.3 where the impact of specific parameter settings was studied.

4.1. Comparison to Previous Methods

The goal of the first experiment was to evaluate the retrieval performance of the proposed priority-driven search (PDS) approach with respect to previous state-of-the-art shape-based retrieval methods:

- **Depth Buffer Descriptor (DSR740B)**: this shape descriptor achieved the highest retrieval performance in the study of [BKS*06]. It describes an object by six depth buffer images captured from orthogonal parallel projections [HKSV02]. Images are stored as Fourier coefficients of the lowest frequencies, and differences between Fourier coefficients provide a measure of object dissimilarity. We use Dejan Vranic's implementation of this method [Vra06] without modification and ran it on a 2GHz Pentium4 running WindowsXP.
- **Light Field Descriptor (LFD)**: this shape descriptor achieved the highest retrieval performance in the study of [SMKF04]. It represents an object as a collection of images rendered from uniformly sampled positions on a view sphere [COTS03]. The dissimilarity of two objects is defined as the minimum L_1-difference between aligned

images of the light field, taken over all rotations and all pairings of vertices on two dodecahedra. We use the original implementation provided by Chen et al. without modification and ran it on a 2GHz Pentium4 running WindowsXP.

- **Global Harmonic Shape Descriptor (GHSD)**: this is the shape descriptor currently used in the Princeton 3D Search Engine [FMK*03]. It describes an object by a single HSD feature positioned at the center of mass with radius *RAVG*. We include it in this study to provide an apples-to-apples comparison to a method that matches a single global shape descriptor of the same type used in our study.
- **Random**: This method provides a baseline for retrieval performance. It produces a random retrieval list for every query.

Figure 6 shows a precision-recall plot comparing the average retrieval performance for all queries for each of these shape matching methods. Briefly, precision and recall are metrics used to evaluate ranked retrieval lists. If one considers the top M matches for any query, *recall* measures the fraction of the query's class found, and *precision* measures the fraction of objects found from the query's class – higher curves represent better retrieval performance.

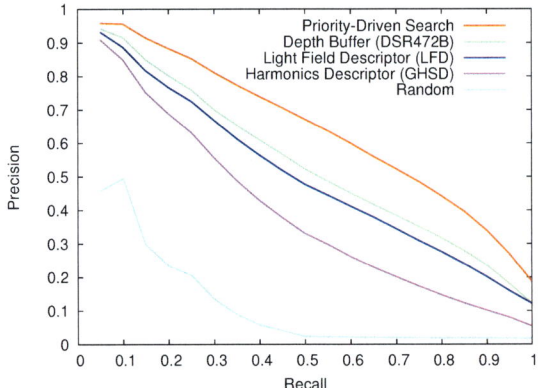

Figure 6: *Precision-recall plot comparing priority-driven search (PDS) to other state-of-the-art shape matching methods using the Princeton Shape Benchmark.*

Timing statistics and standard retrieval performance measures are also shown in Table 1. The leftmost column indicates the shape matching method (PDS is the one described in this paper). The remaining columns list the average time required for one query into the database (in seconds), the average classification rate achieved with a nearest neighbor classifier (1-NN), the average percentages of the query's class that appear in the first-tier (1-Tier) and second-tier (2-Tier), and the average discounted cumulative gain (DCG) computed from the ranked retrieval lists.

From these statistics, we see that the priority-driven search algorithm provides the best retrieval performance of

Method	Time	1-NN	1-Tier	2-Tier	DCG
PDS	2.4	83.4	51.7	63.4	75.9
DSR740B	0.005	66.5	40.3	51.2	66.3
LFD	-	65.0	37.2	47.4	63.6
GHSD	0.003	55.6	30.9	41.1	58.4
Random	0	1.7	1.6	3.4	26.1

Table 1: *Comparison of retrieval statistics between priority-driven search (PDS) and other methods on the Princeton Shape Benchmark (times are in seconds).*

the tested methods on this data set. The improvement in nearest neighbor classification rate over the Depth Buffer Descriptor is 25.4% (83.4% vs. 66.5%) and the improvement over the Light Field Descriptor is 28.3% (83.4% vs. 65.0%). These are remarkable improvements for this data set – typical differences between algorithms found in other studies are usually a couple of percentage points [SMKF04].

However, the PDS algorithm takes considerably more compute time to preprocess the database (4-5 minutes per object), more memory per object (100KB per target object), and more time to find matches (2.4 seconds per query) than the other tested shape descriptors. Almost all of the query processing time is spent establishing the cost of feature correspondences, and less than a tenth of a second is spent finding the optimal multi-feature match with priority driven search. Thus, we believe that simple improvements to the basic algorithm (e.g., compression, indexing, etc.) will significantly improve the processing speed and that query processing times less than a second are possible in this framework (Section 5).

In any case, it seems that priority-driven search is well-suited for batch applications where retrieval accuracy is premium. Often, query results can be computed off-line and cached for later interactive analysis – e.g., for discovery of relationships in mechanical CAD, molecular biology, etc. Even interactive search engines can benefit from off-line preprocessing with high-accuracy matching methods, for example, to preprocess queries that find a shape similar to another in the database (over 90% of the 3D queries to the Princeton 3D Search Engine are of this type [MHKF03]).

4.2. Evaluation of Algorithmic Contributions

The goal of the second experiment is to understand which algorithmic features of the priority-driven search algorithm contribute most to its timing and retrieval performance. To study this question, we started with the "base configuration" and ran the system multiple times on the Princeton Shape Benchmark with different aspects of the system enabled and disabled.

- **Rank (R):** If enabled, the cost of two corresponding shape descriptors (C_{rank}) was the sum of the two ranks in their respective retrieval lists, as described in Section 3.4. Otherwise, it was the direct L^2 distance between shape de-

scriptors (the most common measure of descriptor difference in other systems).

- **Multi-Scale (S):** If enabled, the costs of the best matches found at all four scales were summed. Otherwise, the cost of the best match found among features at scale 0.5 was used (the scale that gave the best retrieval performance on its own).

- **Distinction (D):** If enabled, a small subset of features ($\sim$ 7) was selected for matching within every target object, as described in Section 3.3. Otherwise, all features were included within the target objects.

Results of this experiment are shown in Figure 7 and Table 2. The first three columns of Table 2 indicate whether each of the three algorithmic features (R, M, and D) are enabled (Y) or disabled (N), and the remaining columns provide retrieval performance statistics (note that the top row repeats the performance statistics for PDS with all its algorithmic features enabled: Y Y Y).

R	S	D	1-NN	1-Tier	2-Tier	DCG
Y	Y	Y	74.3	45.5	57.0	70.6
N	Y	Y	67.9	37.0	47.7	64.1
Y	N	Y	67.0	36.6	47.5	62.8
Y	Y	N	66.6	37.2	48.7	64.4
N	Y	N	63.4	32.7	42.6	60.6
Y	N	N	63.0	30.2	41.0	58.3
N	N	Y	54.0	28.2	38.3	56.0
N	N	N	57.0	26.7	36.2	54.7

Table 2: *Results of experiments to investigate the individual and combined value of three algorithmic features of priority-driven search (PDS). The top row represents the base PDS algorithm (Y Y Y). Other rows represent variants of the algorithms with three algorithmic features (R = rank, S = multi-scale, and D = distinctive feature selection) enabled (Y) or disabled (N). Differences in the results achieved with these variants provide insights into which aspects of the PDS algorithm contribute most to its results.*

From these results, we see that the retrieval performance of our system comes from several sources. That is, all three algorithmic features tested contribute a modest but significant improvement to the overall result. Specifically, if we consider the incremental improvements in nearest neighbor classification rates (1-NN) of the combinations shown in Figure 7, we find that multi-scale features provide 11% improvement over using the best single scale (63.4% vs. 57.0%); selecting distinctive features of target objects further boosts performance by another 7% (67.9% vs. 63.4%); and, using descriptor ranks rather than L^2 differences provides a further 9% improvement (74.3% vs. 67.9%). These three algorithmic features combine to contribute a cumulative 30% improvement in retrieval performance over the basic version of our multi-feature matching algorithm (74.3% vs. 57.0%).

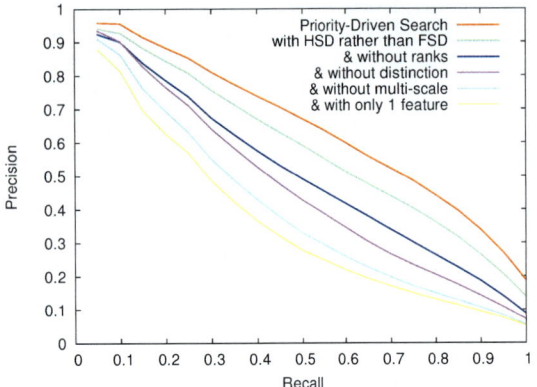

Figure 7: *Precision-recall plot showing the relative contributions of different algorithmic features of priority-driven search. The top curve (red) shows the retrieval performance of the best performing set of options for the PDS algorithm (it is the same as the red curve in Figure 6). The second curve (green) shows the result of using HSDs rather than FSDs as shape descriptors (it represents the "base configuration" for the study in Section 4.3). The third curve (blue) shows the results of using L^2 differences instead of ranks to measure feature correspondence costs; the fourth curve (magenta) shows the same, but without selecting a subset of distinctive features on target objects; the next-to-bottom curve (cyan) also disables multi-scale feature matching (all features are matched only at scale 0.5); and, the bottom curve (yellow) shows the results when finding only one point per match rather than 3. Note how the retrieval performance degrades significantly when each of these algorithmic features is disabled.*

With respect to timing, the main expense of the priority driven search implementation is establishing the initial set of pairwise feature correspondences (~0.3 seconds per query per scale). By comparison, the time required to search for the best multi-feature match is negligible (<0.1 seconds). So, the timing results are currently dominated by the number of features considered for each target object and the number of scales considered for each feature.

Overall, we find that choosing distinctive features (D) improves both precision and speed significantly; using ranks rather than L^2 differences (R) improves precision with negligible extra compute time; and, using features at four scales (S) improves precision, but incurs four times the computational expense.

4.3. Investigation of Parameter Settings

The goal of the third experiment is to investigate in detail how various options of the priority-driven search system affect the timing and retrieval performance. Of course, there is a large space of possible options, and thus we are forced to focus our discussion on small "slices" through this space.

Our approach is to center our investigation on the "base configuration" set of options described in the beginning of this section and to study how timing and retrieval statistics are affected independently as one option is varied at a time.

The results of this study are shown in Table 3(a-d) – each table studies the impact of a different option, and different rows represent a different setting for that option. Please note that rows marked with an '*' represent the same data – they provide results for the base configuration through which slices of option space are being studied.

Descriptor	Time	1-NN	1-Tier	2-Tier	DCG
SD	1.1	75.5	44.2	56.1	71.1
HSD *	1.2	74.3	45.5	57.0	70.6
FSD	2.4	83.4	51.7	63.4	75.9

(a) Shape descriptor type

Radius	Time	1-NN	1-Tier	2-Tier	DCG
0.25	0.3	62.6	31.0	41.2	58.8
0.5	0.3	67.0	36.6	47.5	62.8
1.0	0.3	63.9	37.2	48.5	63.0
2.0	0.3	60.0	33.2	43.4	59.5
Multi-scale *	1.2	74.3	45.5	57.0	70.6
All	0.6	71.3	40.6	54.2	68.0

(b) Scales used for matching shape features

# Points	Time	1-NN	1-Tier	2-Tier	DCG
64	0.6	71.9	42.4	54.2	68.6
128 *	1.2	74.3	45.5	57.0	70.6
256	4.0	75.5	47.3	59.2	71.9
512	17.6	76.6	48.6	60.2	72.6

(c) Number of sample points per object

k	Time	1-NN	1-Tier	2-Tier	DCG
1	1.2	72.9	43.4	54.9	68.9
2	1.2	72.1	44.3	55.8	69.5
3 *	1.2	74.3	45.5	57.0	70.6
4	1.2	72.8	45.4	56.8	70.3
5	1.2	71.2	44.9	56.3	69.7

(d) Number of feature correspondences per match (k)

Table 3: *Results of experiments to investigate the impact of several options on the query time (in seconds) and retrieval performance of priority-driven search.*

Impact of shape descriptor type (Table 3(a)): more verbose descriptors generally provide better retrieval performance, albeit at higher storage and compute costs. For example, the Fourier shape descriptor (FSD) provides better nearest neighbor classification rates (83.4%) than the Harmonic shape descriptor (HSD) (74.3%). However, it is also eight times bigger, and thus eight times more expensive to compare. Interestingly, the Shells shape descriptor (SD) provides retrieval performance similar that of the HSD in this test. Further study is required to determine which descriptors provide the best "bang for the buck" for specific applications and how multiple descriptors can be combined to provide the accuracy of the most verbose ones while incurring query times of the smaller ones (Section 5).

Impact of feature scale (Table 3(b)): medium scale features (radius = 0.5-1.0) provide better retrieval performance than small and large scales in this test, and multi-scale features perform the best of all (nearest neighbor classification rates are 74.3% with multi-scale versus 67.0% with the best single scale (0.5)). Interestingly, summing the cost functions computed for matches at all four scales separately ("Multi-scale") provides better retrieval performance than matching features at all scales simultaneously ("All"). The difference is that the same set of features must match at all 4 scales in "All," while different features can be selected independently for each scale in "Multi-scale." This result seems to suggest that features persistent across multiple scales are not necessarily as useful for classification as ones that are very distinctive at a particular scale.

Impact of the number of sample points per object (Table 3(c)): including more sample points for each object improves retrieval performance in this test, at least up to 512 points. The nearest neighbor classification rate is 76.6% for 512 points per object, while it is 75.5% for 256 points, 74.3% for 128 points, and 71.9% for 64 points. Although a small set of distinctive features are ultimately selected for every target object during a preprocess, features centered at all sample points of the query object are candidates for a match, and thus the compute time for each query should be proportional to the number of points (the quadratic growth observed in this experiment is an artifact of our implementation).

Impact of number of feature correspondences (Table 3(d)): matching large numbers of features does not improve retrieval performance in this study. In fact, matching more than 3 features seems to degrade performance. This result may be because features are quite large scale and spread apart, and thus 3 features may describe the shape as well as is possible with the HSD feature representation. Interestingly, matching larger numbers of features also does not increase query times – this is because the priority-driven search algorithm is able to find good matches in time that is largely independent of the number of possible matches – it investigates only the good matches and ignores the rest.

5. Conclusion and Future Work

This paper describes an algorithm for multi-feature matching of 3D shapes with priority-driven search. The main contribution is an algorithm for searching a database for the best multi-feature matches without computing complete matches for every object. Perhaps just as valuable is the investigation of factors that contribute to speed and retrieval performance improvements in a multi-feature matching system. We find that: 1) using ranks to measure the cost of a feature correspondence is more effective that using L^2 differences directly; 2) selecting target features based on how distinctive

they are of their object's class can improve both search speed and retrieval performance significantly; and, 3) matching features at different scales independently and then adding the resulting costs is an effective way to combine shape information from multiple scales.

This work suggests several areas for improvement and future work. In particular, there are three main computational bottlenecks in the system: 1) constructing shape descriptors, 2) determining the distinction of shape descriptors, and 3) generating pairwise feature correspondences. There are many simple ways to speed up these steps, including random sampling, compression, and indexing. For example, the time required to establish the best pairwise correspondences between features could be improved with standard multi-dimensional indexing schemes. We have focused our efforts in this paper on the priority-driven search algorithm, and thus we have not yet investigated these options in detail.

Another interesting option is to compute the cost of feature correspondences progressively – i.e., initially compute a conservative lower bound on the difference between shape descriptors (e.g., using SD), and only refine it for the best matches. When the correspondence rises to the top of the priority queue, the lower bound on the descriptor difference can be refined a little further (e.g., using HSD) and loaded back onto the priority queue. After the feature correspondence has reached the top of the priority queue and been refined a number of times, the full correspondence cost will be computed (e.g., using FSD) and the PDS algorithm could proceed as usual with that correspondence. This approach would utilize the priority driven strategy not only for extending matches, but also for computing correspondences in the first place.

Perhaps the most interesting question for further study is to investigate how best to recognize 3D objects from their parts. Of course, this is an active topic in computer vision, but the issues for 3D shapes are different than they are for 2D images. Our study seems to suggest that just a few shape features are sufficient to recognize most 3D objects. It will be interesting to see whether other object types follow this pattern, and whether effective algorithms can be developed using even fewer features.

References

[AFP00] AUDETTE M. A., FERRIE F. P., PETERS T. M.: An algorithmic overview of surface registration techniques for medical imaging. *Medical Image Analysis 4*, 3 (2000), 201–217.

[AKKS99] ANKERST M., KASTENMÜLLER G., KRIEGEL H.-P., SEIDL T.: 3D shape histograms for similarity search and classification in spatial databases. In *Proc. SSD* (1999).

[BB76] BARROW H., BURSTALL R.: Subgraph isomorphism, matching relational structures and maximal cliques. *Inf. Process. Lett. 4* (1976), 83–84.

[BBM05] BERG A. C., BERG T. L., MALIK J.: Shape matching and object recognition using low distortion correspondence. In *IEEE Computer Vision and Pattern Recognition (CVPR)* (2005).

[BKS*05] BUSTOS B., KEIM D., SAUPE D., SCHRECK T., VRANIĆ D.: Feature-based similarity search in 3D object databases. *ACM Computing Surveys 37*, 4 (2005), 345–387.

[BKS*06] BUSTOS B., KEIM D., SAUPE D., T.SCHRECK, VRANIĆ D.: An experimental effectiveness comparison of methods for 3D similarity search. *International Journal on Digital Libraries, Special issue on Multimedia Contents and Management in Digital Libraries 6*, 1 (2006), 39–54.

[BM92] BESL P., MCKAY N.: A method for registration of 3D shapes. *IEEE Transactions on Pattern Analysis and Machine Intelligence 14*, 2 (1992), 239–256.

[BMP01] BELONGIE S., MALIK J., PUZICHA J.: Matching shapes. *ICCV* (2001).

[CJ96] CHUA C., JARVIS R.: Point signatures: A new representation for 3D object recognition. *International Journal of Computer Vision 25*, 1 (1996), 63–85.

[COTS03] CHEN D.-Y., OUHYOUNG M., TIAN X.-P., SHEN Y.-T.: On visual similarity based 3D model retrieval. *Computer Graphics Forum* (2003), 223–232.

[CR00] CHUI H., RANGARAJAN A.: A new algorithm for non-rigid point matching. *IEEE Conference on Computer Vision and Pattern Recognition (CVPR)* (2000), 44–51.

[FB81] FISCHLER M., BOLLES R.: Random sample consensus: a paradigm for model fitting with application to image analysis and automated cartography. *Commun. Assoc. Comp. Mach. 24* (1981), 381–395.

[FHK*04] FROME A., HUBER D., KOLLURI R., BULOW T., MALIK J.: Recognizing objects in range data using regional point descriptors. In *European Conference on Computer Vision (ECCV)* (May 2004), pp. 224–237.

[FMK*03] FUNKHOUSER T., MIN P., KAZHDAN M., CHEN J., HALDERMAN A., DOBKIN D., JACOBS D.: A search engine for 3D models. *Transactions on Graphics 22*, 1 (2003), 83–105.

[GCO06] GAL R., COHEN-OR D.: Salient geometric features for partial shape matching and similarity. *ACM Transaction on Graphics* (January 2006).

[GD05] GRAUMAN K., DARRELL T.: The pyramid match kernel: Discriminative classification with sets of image features. In *IEEE International Conference on Computer Vision (ICCV)* (2005).

[GMGP05] GELFAND N., MITRA N. J., GUIBAS L. J., POTTMANN H.: Robust global registration. In *Symposium on Geometry Processing* (2005).

[HKSV02] HECZKO M., KEIM D., SAUPE D., VRANIĆ D.: Methods for similarity search on 3D databases. In *Datenbank-Spektrum* (2002), vol. 2, pp. 54–63. (In German).

[IJL*05] IYER N., JAYANTI S., LOU K., KALYANARAMAN Y., RAMANI K.: Three dimensional shape searching: State-of-the-art review and future trends. *Computer-Aided Design 37*, 5 (2005), 509–530.

[JH99] JOHNSON A., HEBERT M.: Using spin-images for efficient multiple model recognition in cluttered 3-D scenes. *IEEE PAMI 21*, 5 (1999), 433–449.

[KFR03] KAZHDAN M., FUNKHOUSER T., RUSINKIEWICZ S.: Rotation invariant spherical harmonic representation of 3D shape descriptors. In *Symposium on Geometry Processing* (June 2003), pp. 167–175.

[KPNK03] KORTGEN M., PARK G.-J., NOVOTNI M., KLEIN R.: 3D shape matching with 3D shape contexts. In *7th Central European Seminar on Computer Graphics* (April 2003).

[LG05] LI X., GUSKOV I.: Multi-scale features for approximate alignment of point-based surfaces. In *Symposium on Geometry Processing* (2005).

[LW88] LAMDAM Y., WOLFSON H.: Geometric hashing: a general and efficient model-based recognition scheme. In *Proc. ICCV* (December 1988).

[MBM01] MORI G., BELONGIE S., MALIK J.: Shape contexts enable efficient retrieval of similar shapes. In *IEEE Computer Vision and Pattern Recognition (CVPR)* (December 2001).

[MHKF03] MIN P., HALDERMAN J., KAZHDAN M., FUNKHOUSER T.: Early experiences with a 3D model search engine. In *Proceeding of the eighth international conference on 3D web technology* (2003), pp. 7–18.

[NDK05] NOVOTNI M., DEGENER P., KLEIN R.: *Correspondence Generation and Matching of 3D Shape Subparts*. Tech. Rep. CG-2005-2,ISSN 1610-8892, Friedrich-Wilhelms-Universität Bonn, June 2005.

[SF06] SHILANE P., FUNKHOUSER T.: Selecting distinctive 3D shape descriptors for similarity retrieval. In *Shape Modeling International* (June (to appear) 2006).

[SMKF04] SHILANE P., MIN P., KAZHDAN M., FUNKHOUSER T.: The Princeton Shape Benchmark. In *Shape Modeling International* (June 2004), pp. 167–178.

[SMS*04] SHAN Y., MATEI B., SAWHNEY H. S., KUMAR R., HUBER D., HEBERT M.: Linear model hashing and batch ransac for rapid and accurate object recognition. In *IEEE International Conference on Computer Vision and Pattern Recognition* (2004).

[TV04] TANGELDER J., VELTKAMP R.: A survey of content based 3D shape retrieval methods. In *Shape Modeling International* (June 2004), pp. 145–156.

[Vra06] VRANIĆ D.: Tools for 3d model retrieval, 2006. http://merkur01.inf.uni-konstanz.de/3Dtools/.

[YF02] YAMANY S., FARAG A.: Surfacing signatures: An orientation independent free-form surface representation scheme for the purpose of objects registration and matching. *IEEE Trans. Pattern Anal. Mach. Intell. 24* (2002), 1105–1120.

Eurographics Symposium on Geometry Processing (2006)
Konrad Polthier, Alla Sheffer (Editors)

Defining and Computing Curve-skeletons with Medial Geodesic Function

Tamal K. Dey and Jian Sun [†]

Abstract

Many applications in geometric modeling, computer graphics, visualization and computer vision benefit from a re-duced representation called curve-skeletons *of a shape. These are curves possibly with branches which compactly represent the shape geometry and topology. The lack of a proper mathematical definition has been a bottleneck in developing and applying the the curve-skeletons. A set of desirable properties of these skeletons has been identi-fied and the existing algorithms try to satisfy these properties mainly through a procedural definition. We define a function called medial geodesic on the medial axis which leads to a methematical definition and an approximation algorithm for curve-skeletons. Empirical study shows that the algorithm is robust against noise, operates well with a single user parameter, and produces curve-skeletons with the desirable properties. Moreover, the curve-skeletons can be associated with additional attributes that follow naturally from the definition. These attributes capture shape eccentricity, a local measure of how far a shape is away from a tubular one.*

Categories and Subject Descriptors (according to ACM CCS): I.3.3 [Computer Graphics]: Line and Curve Genera-tion

1. Introduction

The problem of representing a three dimensional shape with a one dimensional geometry (curves) appears in various ap-plications of geometric modeling, computer graphics, visu-alization and computer vision. For example, in animation and tracking a 'stick-figure' is immensely useful which rep-resents the main geometric entities of a shape with their con-nectivities. It allows registrations, deformations, matching and other operations in a more controlled manner because of the reduced dimension. The concept of curve-skeleton was born from these various needs which, roughly speak-ing, should be a curve possibly with branches in the 'center' of the shape. A related and much more well defined concept is the medial axis which is also referred as the skeleton. The medial axis consists of the centers of the maximal balls in-scribed inside the shape. For a three dimensional shape, the medial axis, in general, has two dimensional components of-ten referred as medial surface. Therefore, medial axis cannot be a substitute for one dimensional skeletons.

A main problem with computing curve-skeletons is that

they are not well defined. Although desirable properties of these skeletons have been identified based on different ap-plications, no mathematical definition has been formulated. To fill this void different procedural definitions leading to different methods have been proposed for computing curve-skeletons. They include, to name a few, topological thinning [BNB99], distance field based methods [ZT99] [BKS01] [BST03] [HF05], potential field based methods [CSYB05], and others [OK95] [Cos99] [VL00]. Cornea et al. [CSM05] give a comprehensive survey of these techniques. Although many of these methods produce curve-skeletons with a set of desirable properties, they are not completely satisfactory as pointed out in Cornea et al. [CSM05]. We believe that this limitation stems from the lack of a proper mathematical definition of curve-skeletons.

In this paper, we give a mathematical definition of the curve-skeletons. Since the curve-skeleton should be in the 'middle' of the shape it is natural to define it as some sub-set of the medial axis. What we aim for is to determine the 'middle' of the medial axis. Algorithms to thin the medial axis based on the distances from the boundary have been de-signed on this principle. The main problem in this approach is that a large part or the entire medial axis may have the same distance from the boundary; e.g., the medial surface of

---

[†] Dept. of CSE, The Ohio State University, Columbus, OH 43210.
{tamaldey,sunjia}@cse.ohio-state.edu

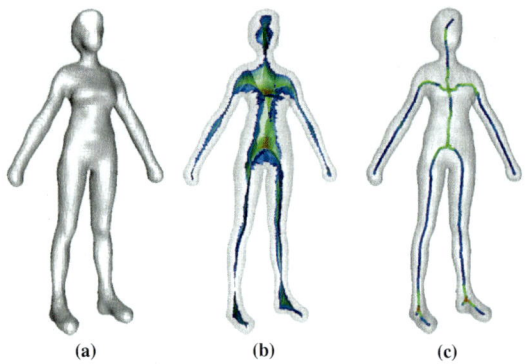

Figure 1: *(a) Female model, (b) approximated medial axis rendered with MGF values, (c) extracted curve-skeleton rendered with the eccentricity values, Coloring scheme: red tone : big values, blue tone : small values, green tone : medium values.*

a thin plate. It is not clear how the thinning process should proceed in such cases. One of our main contributions is to define a function on the medial axis whose singularity brings out its 'middle'. The function is based on the geodesic distances between points where the maximal balls defining the medial axis touch the shape boundary. We call it the *medial geodesic function*(MGF). In a sense, the medial geodesic function combines the intrinsic property of the bounding surface (by geodesic distances) with its embedding in three space (by the medial axis) thereby capturing the shape information comprehensively. Just as the singular points of the standard distance function gives the medial axis, the singular points of this function gives the curve-skeleton. Figure 1(b) and (c) show the medial geodesic function and the curve-skeleton of the Female model respectively.

Our definition allows additional shape information to be associated with the curve-skeleton. First, the medial geodesic function values given by the shortest geodesic distances between the points where the maximal balls touch the surface give the size information of the shape. Second , the ratios between the geodesic and the Euclidean circles passing through these touching points tell how far the shape is locally away from a tubular one. We refer to this ratio as *eccentricity*. The coloring in Figure 1(c) shows the eccentricity values associated with the curve-skeleton. Furthermore, Our definition allows to map the curve-skeleton back to the surface easily. These extra features are useful for various applications.

2. Definition

Let $O \subset \mathbb{R}^3$ be a space called *shape* bounded by a *connected manifold* surface S. The *medial axis* $M \subset O$ is the set of centers of the maximal balls inscribed in O.

Giblin and Kimia [GK04] show that, generically, M con-

sists of five types of points giving it a stratified structure. One type form two dimensional sheets, two of the types form curves and the rest two types remain as isolated points on the medial axis. Figure 2(a) shows four of these types.

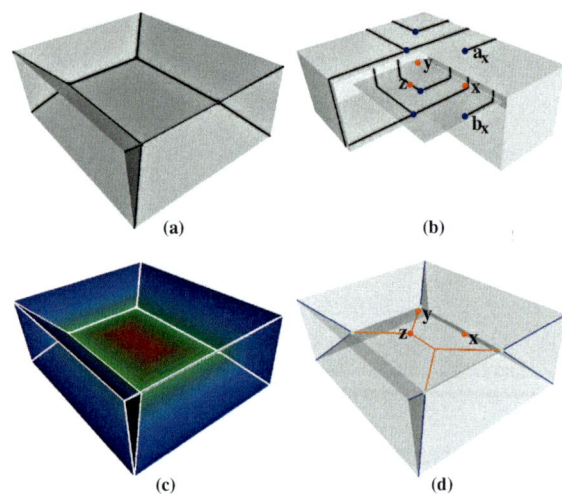

Figure 2: *(a) The stratified structure of the medial axis of a rectangular block. (b) One sheet of the medial axis. Red points x, y, z are on this sheet. Green points are their corresponding contact points on the surface and black paths are the shortest geodesic paths between these contact points. (c) Medial axis rendered with MGF values following the coloring scheme of the title figure. (d) Red lines, blue lines and green points are Sk_2, Sk_3 and the limit points of their union, respectively.*

First we focus on the points that form sheets; shown with grey in Figure 2(a). The maximal inscribed balls of such points touch the surface S at exactly two distinct points. Let $M_2 \subseteq M$ be the set of such points. Each point in M_2 has a neighborhood which is an open disk and hence M_2 is a 2-manifold. Since $M \setminus M_2$ has measure zero in M, in general, M_2 covers most of M.

2.1. Medial Geodesic Function (MGF)

For a point $x \in M_2$, let B_x be the maximal inscribed ball centered at x and a_x and b_x be the two points of S where B_x meets S. Define $f(x)$ to be the length of the shortest geodesic path on S between a_x and b_x. We call f the *medial geodesic function*, MGF, of O. Figure 2(b) shows the corresponding shortest geodesic paths for several points on M_2. If a point $x \in M_2$ have more than one shortest geodesic paths between a_x and b_x, their lengths are equal. Hence MGF is well defined on M_2. Figure 2(c) shows the rendering of M_2 based on the MGF value. We will define the curve-skeleton in M_2 as the singular set of f. Several properties of this singular set (some observed and some proved) motivates this definition.

By our assumption the surface S is connected, compact, and without boundary. Further, we assume that S is *smooth* (C^∞). The function f is defined on M_2. Define $\phi: M_2 \to S \times S$, $x \mapsto (a_x, b_x)$. Using a technique similar to Attali et al. [AJDL03], it can be shown that M_2 is a smooth manifold and ϕ is differentiable provided S is smooth. We denote the lengths of the shortest geodesic between two points x and y in S, M_2, and $\mathbb{R}^2$ by $d_S(x,y)$, $d_M(x,y)$, and $d(x,y)$ respectively. Considering d_S as a function from $S \times S$ to $\mathbb{R}$ we have $f = d_S \circ \phi$. Let $\alpha: M_2 \to \mathbb{R}^2$ be the local coordinate function for $x \in M_2$. The map α is a diffeomorphism since M_2 is a smooth manifold. We use α to define $\tilde{f}: \mathbb{R}^2 \to \mathbb{R}$ so that $\tilde{f}(\alpha(x)) = f(x)$. From standard definition in differential geometry f is called differentiable at x if and only if $\tilde{f}$ is differentiable at $\alpha(x)$.

We argue that the singularities of f, i.e., the points in M_2 where f is not differentiable has a Lebesgue *measure* zero. This would mean that the singularities of f constitute curves or isolated points on M_2, which we define as the curve-skeleton in M_2.

Property 1 The singularity of f has measure zero in M_2.

To prove that the singularity of f has measure zero, we show that $\tilde{f}$ is locally k-Lipschitz (defined below) for some $k > 0$. For then the singular set of $\tilde{f}$ has measure zero by Rademacher's theorem [Fed96]. It follows that the singular set of f has measure zero by local coordinate maps.

A function $g: \mathbb{R}^n \to \mathbb{R}$ is called locally k-Lipschitz near a point $x \in \mathbb{R}^n$ if for some $\epsilon > 0$

$$g(x) - g(y) \le k\|x - y\|$$

for any $y \in \mathbb{R}^n$ where $\|x - y\| \le \epsilon$.

Observation 1 The function $\tilde{f}: \mathbb{R}^2 \to \mathbb{R}$ is locally k-Lipschitz for some $k > 0$.

Proof For some $\epsilon > 0$ and a point $x \in M_2$, let y be any point where $d(\alpha(x), \alpha(y)) \le \epsilon$. Consider the shortest geodesics γ_x between a_x and b_x, and γ_y between a_y and b_y. The lengths $|\gamma_x|$ and $|\gamma_y|$ satisfy

$$|\gamma_x| \le |\gamma_y| + d_S(a_x, a_y) + d_S(b_x, b_y).$$

Since ϕ is differentiable, there is some $k_1 > 0$ so that we have $\max\{d_S(a_x, a_y), d_S(b_x, b_y)\} \le k_1 d_M(x,y)$ when x and y are sufficiently close. Also since the local coordinate function α is differentiable, we have $d_M(x,y) \le k_2 d(\alpha(x), \alpha(y))$ for some $k_2 > 0$. Therefore,

$$\tilde{f}(\alpha(x)) = |\gamma_x| \le |\gamma_y| + 2k_1 k_2 d(\alpha(x), \alpha(y))$$
$$= \tilde{f}(\alpha(y)) + 2k_1 k_2 d(\alpha(x), \alpha(y))$$

proving that $\tilde{f}$ is locally $2k_1 k_2$-Lipschitz. $\square$

We do not have rigorous mathematical proofs for the next two properties though we conjecture them to be true. We have observed the properties from experiments as well.

Property 2 There is no local minimum of f in M_2.

This property can be argued roughly as follows. Since M_2 is a manifold, any point x in M_2 has a neighborhood $N \subset M_2$, which is a disk. Let γ be the geodesic path between a_x and b_x. In any small enough neighborhood of x, there is a point y such that a_y is on γ. If b_y is also on γ then $f(x) > f(y)$ and we are done. However, in general, b_y may not be on γ. But we observe that b_y is close to γ and hence it is likely that $f(x) > f(y)$ still holds.

Property 3 At each singular point x of f there are more than one shortest geodesic paths between a_x and b_x.

A rough argument why the above property is true may proceed as follows. We have $f = d_S \circ \phi$ and ϕ is differentiable on M_2. Therefore, f is differentiable at a point $x \in M_2$ if d_S is differentiable at $\phi(x)$. Suppose that there is only a single shortest geodesic path γ between a_x and b_x. Then in a sufficiently small neighborhood N of (a_x, b_x) in $S \times S$, all geodesic paths between a and b for $(a,b) \in N$ smoothly converge to γ as (a,b) approaches to (a_x, b_x). This means d_S is smooth at (a_x, b_x) and so is f at x contradicting that f is singular at x.

2.2. Skeleton definition

We observe that the behavior of the medial geodesic function is like a distance function. First, MGF is continuous and differentiable everywhere on M_2 except at a measure zero set in M_2. Second, we have observed that property 2 and 3 hold in practice. This means MGF has no local minimum on M_2 which is open in M and the singularity of f occurs roughly in the 'middle' of M_2.

We define the curve-skeleton in M_2, denoted by Sk_2, as the set of singular points of MGF on M_2. To extend the definition beyond M_2, we use a different characterization of the singular points by means of divergence. It is reminiscent of the use of divergence for defining medial axis by Siddiqi et al. [SBTZ99]. The gradient of MGF, ∇f, defines a vector field on M_2 except at the singular points. The divergence of the vector field at point x, $div(x)$, is the net outward flux per unit area on M_2 taken over a neighborhood D shrinking to zero, i.e.,

$$div(x) := lim_{A \to 0} \frac{\int_C < \nabla f, n > dC}{A}$$

where A is the area of D, C is the boundary of D and n is the outward normal at a point on C as shown in Figure 3(a). The divergence is negative at the singular points but 0 everywhere else. In other words, Sk_2 consists of the points where the gradient flow of MGF sinks into.

Next we consider the set of points $M_3 \subseteq M$ where M_3 constitutes curves lying at the intersection of the closure of three sheets in M_2. The thick black lines in Figure 2(a) are such curves. The maximal inscribed ball of such a point touches S at three points. Although MGF is not well defined for these points, we consider MGF defined on their neighborhoods.

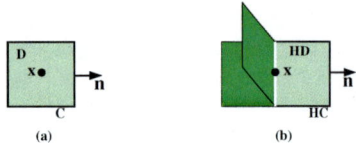

Figure 3: *(a) Neighborhood of a point in M_2. (b) Neighborhood of a point in M_3.*

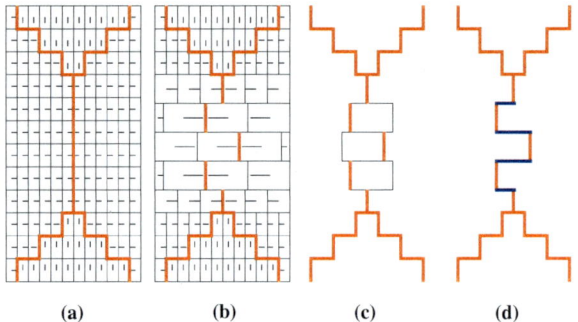

Figure 4: *(a) and (b) two different polygonal approximation of the medial axis. The small black lines starting from the centers of the polygons represent the gradient vector of MGF at these centers. Red line segments are the marked skeleton edges where the divergence of the gradient of MGF is negative, (c) eroded medial axis, (d) blue segments are the skeleton edges collected during erosion.*

Let x be such a point. The neighborhood of x in M_2 consists of three topological half disks. Consider one of them, say HD. We define the divergence with respect to HD for x, $div(x)|_{HD}$, as follows.

$$div(x)|_{HD} := lim_{HA \to 0} \frac{\int_{HC} < \nabla f, n > \mathrm{d}HC}{HA}$$

where HA is the area of the half disk HD, HC is the boundary of HD on M_2 as in Figure 3(b). The point x is on the curve-skeleton if $div(x)|_{HD}$ is negative for all three half disks in the neighborhood of x. Basically, x is on the curve-skeleton if the gradient flow of MGF from all three local neighboring sheets sink into it. Let Sk_3 denote such set of points.

Now consider the rest of the types of points in M. One type of points form the boundary curves of M where two contact points of the maximal inscribed ball with the surface coincide. In case of the rectangular block shown in Figure 2(a), these curves are the twelve edges of the block. The rest two types of points are the isolated points on the medial axis where at least two curves meet. We do not explicitly define the curve-skeleton for these three types of points since they are either the boundary or the isolated points on the medial axis. A point of one of these three types is on the curve-skeleton if it is the limit point of Sk_2 or Sk_3.

Finally, we define the *curve-skeleton* of O, Sk_O as the closure of $Sk_2 \cup Sk_3$. Figure 2(d) shows the curve-skeleton for a rectangular block.

3. Algorithm Overview

In general it is extremely hard to compute the curve-skeleton exactly as we defined. It is well known that exact medial axis computation is hard due to numerical instability associated with the computations. Performing an exact computation based on the analysis of a function defined on the medial axis would be even harder. We bypass this difficulty by computing an *approximation* of the curve-skeleton. Extensive experiments show that the algorithm is effective in practice.

Assume that the input is a shape represented by a polygonal surface. Ideally, the approximation algorithm can be described in the following three steps. First, compute a polygonal approximation of the medial axis. Second, compute the gradient of MGF for the center of each medial axis facet

(polygon), which approximates the gradient for all points in that facet. Third, mark the edges of the medial axis facets with negative divergence as skeleton edge. The output curve-skeleton consists of all marked skeleton edges. Figure 4(a) illustrates a curve-skeleton computed with this strategy. This result, however, is an accident. In practice, the polygonal approximation of the medial axis will most likely be worse and hence the computation of the gradient of MGF will be less accurate. As a result, the curve-skeleton computed by the above three steps will most likely be disconnected as Figure 4(b) illustrates. To overcome this problem caused by discretization and approximation error, we introduce an additional step of medial axis erosion to recover the missing part. The medial geodesic function guides the erosion, which is the reason why the curve-skeleton is roughly in the middle of the medial axis. Specifically, the erosion only removes pieces of the medial axis from the boundary of the subset that has yet not been eroded. Also, among all erodable elements, the one with the smallest MGF value is eroded first. At the same time the edges marked in the third step are never allowed to be eroded. Figure 4(c) shows a stage where three facets still need to be eroded. Figure 4(d) shows the extracted curve-skeleton.

Extraction Algorithm:

step 1: (MA approximation) Compute a polygonal approximation of the medial axis using the input polygonal surface.

step 2: (MGF approximation) Approximate the medial geodesic function and its gradient for the points inside each medial axis facet.

step 3: (Marking) Mark the edges of the medial axis facets with negative divergence as skeleton edges.

step 4: (Erosion) Erode the medial axis in the increasing or-

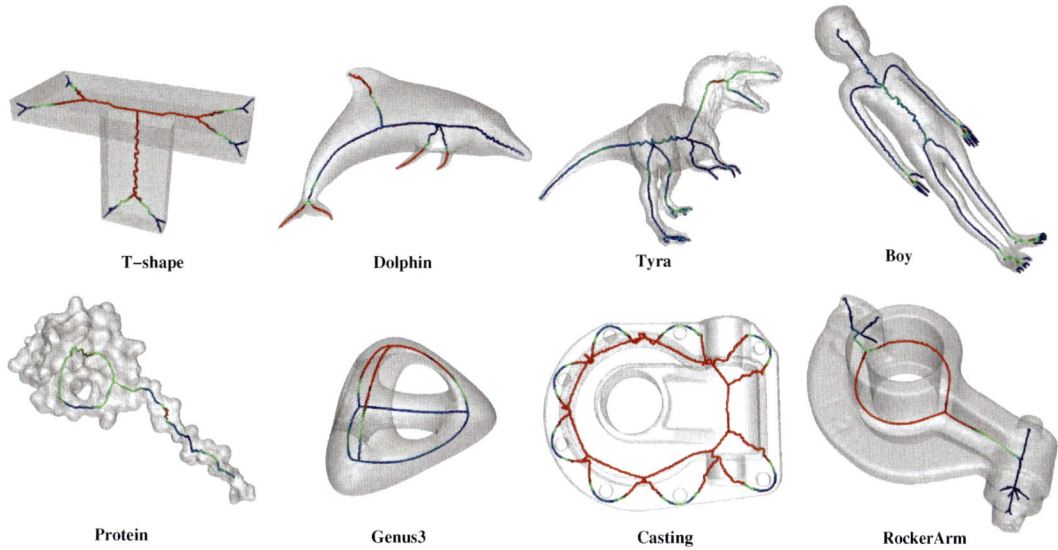

Figure 5: *The first row shows the extracted curve-skeletons for surfaces with genus 0. The second row shows the extracted curve-skeletons for surfaces with genus more than 1. The skeleton edges are colored based on their eccentricity values.*

der of its MGF value from the boundary while keeping the edges marked in the step (3) intact. Output the edges of the medial axis that survive the erosion as the curve-skeleton.

Figure 5 shows the curve-skeleton extracted by the above algorithm for a number of shapes.

Before we detail each step of the extraction algorithm in section 5, we illustrate several properties of the curve-skeleton extracted by our algorithm.

4. Properties

In a nice survey, N.D. Cornea et al. [CSM05] compiled a list of desirable properties for the curve-skeletons based on numerous applications. In general, it is desirable for a curve-skeleton to be homotopy equivalent to the shape, invariant under isometric transformations, thin, centered, junction detective, stable (robust) and connected. Our curve-skeleton enjoys all of these properties.

It is obvious that our algorithm guarantees that the extracted curve-skeleton is invariant under isometric transformation, connected and thin.

The homotopy equivalence follows from the following observation. First of all the medial axis is a deformation retract of the shape. Second, the erosion is implemented with a *collapse* operation that gives a deformation retract of the medial axis (see the detailed description in section 5.4). Hence the curve-skeleton is actually a deformation retract of the shape. Figure 5 shows that the curve-skeletons have the same number of loops as the number of tunnels in their corresponding shapes.

The extracted curve-skeleton is centered because of the following two reasons. First of all, the curve-skeleton is a subset of the approximated medial axis and hence is centered with respect to the distance field defined by the surface. Second, by property 3, a point x in M_2 is on the curve-skeleton only if there are multiple shortest geodesic paths between two touching points, a_x and b_x. This means that the point x is in the middle of M_2. Different examples given in this paper also show the centeredness of the curve-skeleton.

A curve-skeleton should remain stable against small changes in the shape. In particular, small changes introduced by noise should not affect the curve-skeleton significantly. Although the medial axis based on which we extract the curve-skeleton is not stable under shape perturbations [Wol92, ACK02, AJDE04], the curve-skeleton remains stable. The reason is that the unstable parts of the medial axis do not contribute to our curve-skeleton. Figure 6(a) and (b) show a noisy Hand (we generate it by perturbing the points of a smooth Hand shown in Figure 13) and its medial axis respectively. As we can see the medial axis has a number of spikes because of noise. However, these spikes are close to the boundary and hence have small MGF value as the coloring of the medial axis shows. The erosion process erodes these spikes before reaching the "middle" of the medial axis, where the curve-skeleton is. Figure 6(c) shows the curve-skeleton of noisy Hand, which is almost the same as the one of smooth Hand in Figure 8 though it has more wiggles.

The curve-skeleton computed by our algorithm remains stable under certain deformations where the topology of the shape does not change and the geodesic distance between any two points on the surface does not change much. Defor-

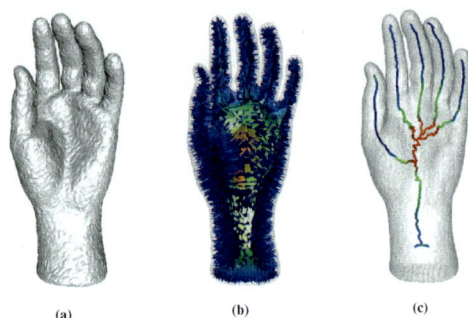

Figure 6: *(a) Noisy Hand, (b) the medial axis of the noisy Hand colored with MGF values, (c) the curve-skeleton.*

mations in animated characters mostly belong to this category. Figure 7 shows the curve-skeletons for a series of deformed horses generated by Sumner and Popovic [SP04]. As we can see, the structure of the curve-skeleton remains unchanged as the horse surface deforms. Although the curve-skeletons are deformed, the length of each component and its eccentricity value (the coloring of the curve-skeletons) almost remain the same.

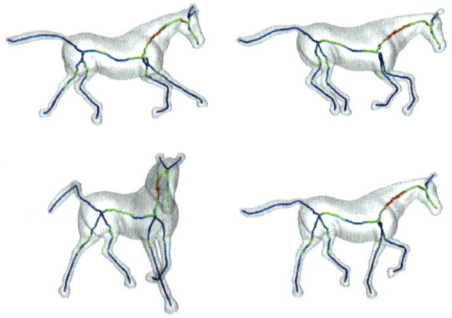

Figure 7: *The curve-skeletons for the animated horses.*

A curve-skeleton is junction detective if it encodes the different logical components of the shape. Although the definition of the logical components of a 3D shape is vague in general, it is often obvious for some classes of shapes. We observe from various examples that the curve-skeleton as computed by our algorithm is junction detective. Since the curve-skeleton is a 1D curve, there are three types of curve-skeleton points: boundary points that have a half interval neighborhood on the curve-skeleton, regular points that have an interval neighborhood and the rest called junction points. The junction points attach different branches of the curve-skeleton. As shown in Figure 8, the logical components of Armadillo, including torso, arms, legs, tail, ears, mouth, fingers and toes, have a one-to-one correspondence with the branches (colored differently) of the curve-skeleton.

The curve-skeletons of other models in this paper such as Female, Boy, Hand and Horse also show their ability to detect the junctions. Junction detection is a key property of a curve-skeleton which many applications depend on such as mesh decomposition and shape matching.

4.1. Shape eccentricity

In addition to those mentioned in [CSM05], the curve-skeleton extracted by our algorithm can be associated with two other attributes, which make them encode shape information more comprehensively. Our definition and algorithm allow easy computation of these two quantities for each skeleton edge. Consider the hand in Figure 8. We approximate the medial axis using a subset of the Voronoi diagram and hence each skeleton edge E is a Voronoi edge, whose dual Delaunay triangle t has three vertices on the surface (red points). The geodesic paths between each pair of them together form a 'circle' (red circles), called the geodesic circle of E. The length of this circle, denoted $g(E)$, captures the local size of $O_E \subset O$ where O_E corresponds to the skeleton edge E. Let $c(E)$ denote the length of the circumcircle of the dual triangle t (blue circles). The ratio $\frac{g(E)}{c(E)}$ essentially tells how much O_E deviates from a tubular shape. We call this ratio the *eccentricity* of E, denoted by $e(E)$. In Figure 8, the fingers have small eccentricity value and are close to tubular shape while the palm has a big eccentricity value and is more flat. In Section 5.6 we detail the algorithm for identifying tubular/flat regions of a shape using eccentricity.

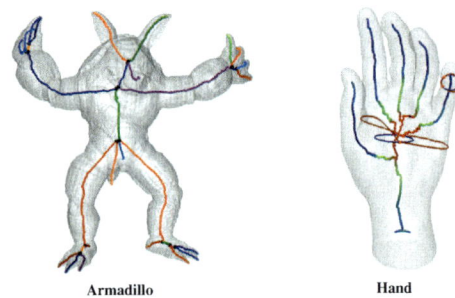

Armadillo   Hand

Figure 8: *Armadillo: the branches of the curve-skeleton correspond to logical components of the shape. Hand: geodesic circles, colors of the curve-skeleton indicate the eccentricity values.*

5. Algorithm details

In this section we give the detailed description of each step of the extraction algorithm described in section 3. We assume that the input surface is a connected triangulated surface, denoted by T.

5.1. Medial axis approximation

In this work we follow the Voronoi diagram filtration of Dey and Zhao [DZ03] to approximate the medial axis. This approach filters the Voronoi diagram of a set of vertices on the surface and retains a set of Voronoi facets to approximate the medial axis. The main problem with this medial axis computation is that the filtration, often guided by some input parameters, leave some unwanted spikes or holes in the approximate medial axis. However our algorithm is not affected by this shortcoming as we avoid the filtration and instead take all the Voronoi facets inside the space bounded by T as a preliminary approximation of the medial axis, denoted by M_T. Eventually this superset of the medial axis gets eroded by our algorithm. Figure 6(b) and Figure 9(a),(b) show the approximated medial axes of Hand and T-shape respectively.

Note that the medial axis of a triangulated surface can be computed exactly [CKM04]. However, this exact computation is expensive. Fortunately, the medial axis approximated by the Voronoi diagram serves our purpose as long as the vertices of the trianguled surface form a dense sampling of the original object.

5.2. MGF Approximation

For a Voronoi facet F on the approximated medial axis, let a_F and b_F be two endpoints of the dual Delaunay edge of F, as in Figure 9(c). We compute the shortest geodesic distance $f(F)$ between a_F and b_F by using the algorithm presented by V. Surazhsky et. al [SSK*05]. The black curve in Figure 9(a) shows the shortest geodesic path between a_F and b_F computed by their algorithm. We approximate the medial geodesic function for any point inside F with this quantity $f(F)$. The medial axis in Figure 9(b) is rendered with different colors for different ranges of MGF values.

We approximate the gradient of the MGF for any point inside a Voronoi facet F as follows. First, we compute the tangent directions v_a and v_b of the shortest geodesic path at the two endpoints a_F and b_F respectively. Next, we project these two vectors onto the Voronoi facet F and approximate the gradient ∇f for any point inside F using the normalized summation $v(F)$ of these two projected vectors. The white arrows in Figure 9(e) show the vector $v(F)$ for the Voronoi facets inside the white circle marked in Figure 9(b).

5.3. Marking

Because of the erosion process, we do not need to mark all skeleton edges in this step. In fact, the actual purpose of the marking step is to find those skeleton edges which form the boundary of the curve-skeleton. Consider a Voronoi edge E and let F be any Voronoi facet incident on it as shown in Figure 9(d). The dot product $d_E(F)$ of $v(F)$ and the inward normal n of the edge E towards F, approximates the flux

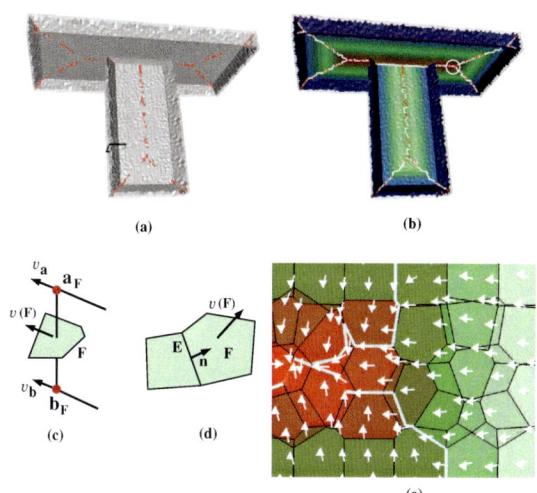

Figure 9: *(a) T-shape: the medial axis, marking step marked the red skeleton edges, (b) the medial axis rendered with the MGF values, cyan skeleton edges collected during erosion, (c) illustration for the MGF gradient computation for each Voronoi facet, (d) illustration for the divergence computation for each Voronoi edge, (e) zoomed area of the white circle in (b).*

flowing into any point on E from the neighborhood in F. We mark the Voronoi edge E as a skeleton edge if $d_E(F) < \theta$ for any incident Voronoi facet F where θ is an input parameter. Actually θ is the only input parameter for the entire curve-skeleton extraction algorithm. We will show its effect later in this section. In addition, to avoid small branches in the final curve-skeleton, at most one edge is allowed to be marked as skeleton edge for a Voronoi facet in this step. Also we do not mark edges on the boundary of M_T. Notice that, whenever an edge is marked, its two endpoints are also marked. Figure 9(a) shows the skeleton edges (red) marked after this step. As we can see they are in the 'middle' of the medial axis but are disconnected. Since MGF has no local minimum in M_2 as Property 2 claims, the skeleton edges marked in this step can not form a loop.

5.4. Erosion

The erosion proceeds by collapsing facets, edges and vertices from M_T gradually. Consider M_T as a cell complex consisting of three types of cells: Voronoi facets, Voronoi edges and Voronoi vertices. A cell τ is a face of another cell σ if τ is on the boundary of σ. We also say σ is a coface of τ. A pair (τ, σ) is a face-coface pair if τ is a face of σ. In our case, there are three types of such pairs: (edge, facet), (vertex,facet) and (vertex,edge). The erosion actually proceeds by collapsing face-coface pairs. One way to think of collapsing a pair (τ, σ) is to push every point on σ and τ onto the

Model	#V	MA	MGF	Erosion	Total
Genus3	6652	0:03	1:13	0:0	1:17
Female	8904	0:08	4:44	0:0	4:52
Horse	15563	0:18	8:31	0:0	8:51
Armadillo	25001	0:18	10:04	0:01	10:25
RockerArm	40171	0:34	36:19	0:01	36:56

Table 1: *Computation times (minutes:seconds) on a 2.8 GHz PC with 1GHz RAM.*

other boundary of σ, see Figure 10. It is the same as σ and τ being eroded. A face-coface pair (τ, σ) is collapsible if σ is the only coface of τ that has not been collapsed so far. If we only collapse the collapsible pairs, the spaces before and after the collapse remain homotopy equivalent.

The MGF value guides the erosion. The assignment of MGF value to every facet in M_T is described in section 5.2. An edge and a vertex is assigned an MGF value equal to the maximal of the MGF values of the facets incident to it. The erosion removes a pair (τ, σ) where τ is not a marked skeleton edge or a vertex in the marking phase and $f(\tau)$ is the least among all such collapsible pairs.

step 1: Initialize a priority queue Q with all the initial collapsible pairs, the one having the smallest $f(\tau)$ being at the top. These pairs are the edges on the boundary of M_T and their cofaces.

step 2: Pop the top element (τ, σ) out of Q and erode σ and τ. The erosion of these two cells may make some adjacent pairs collapsible. If neither of the two cells in a new collapsible pair has been marked in the marking step, push the pair into Q. Repeat step 2 until Q is empty.

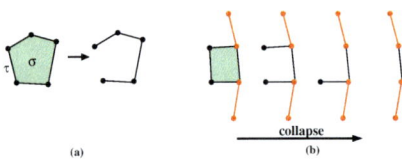

Figure 10: *(a) Collapsing (τ, σ), (b) a series of collapses, red edges and vertices are marked in the marking step.*

Table 5.4 shows the timing of the algorithm for some models.

5.5. Effect of θ

Finally we describe the effect of the *only* user parameter θ, which is used in the marking step to identify the points with certain negative divergence of the gradient field of MGF. As θ decreases, the condition for points being on the curve-skeleton becomes more strict and the resulting curve-skeleton becomes less detailed. Formally, let SK_O^θ be the curve-skeleton of a shape O extracted with parameter θ. We

have $SK_O^{\theta_1} \subseteq SK_O^{\theta_2}$ if $\theta_1 < \theta_2$. Figure 11 shows a series of curve-skeletons of Protein with different θ values. As θ decreases, the curve-skeleton corresponding to the less prominent features go away. When θ reaches a value a little less than -1, no edge and point is marked in the marking step since the condition $d_E(F) < \theta$ is not satisfied for any edge E and any of its incident facet F and the curve-skeleton reaches its simplest form consisting of loops only. Different θ can be chosen for different applications of the curve-skeleton.

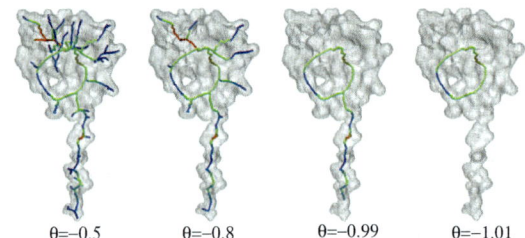

$\theta=-0.5$ $\theta=-0.8$ $\theta=-0.99$ $\theta=-1.01$

Figure 11: *A series of curve-skeletons of Protein with different θs.*

5.6. Computing tubular regions

Many shapes have pronounced features which are perceived to be tubular and flat. This shape information should be readable from the medial axis which encodes the shape compactly. A recent work of Goswami et.al [GDB06] points toward this direction. As a subset of the medial axis, the curve-skeleton together with the eccentricity value can indicate how much the shape differs locally from a tubular one.

First, all skeleton edges are classified into two types, ones with eccentricity values less than a "threshold" and the rest. Second, we compute the geodesic circle for each skeleton edge E and attach the mesh triangles intersected with the geodesic circle with an id number (say 1) if $e(E)$ is less than the threshold and another id number (say 0) otherwise. Figure 12(a) shows the result after this step where the mesh triangles with only one id number are rendered with blue if id number is 1 or white if it is 0, those with no id number are rendered with grey color, and those with two id numbers are rendered with red color. To compute a component we carry out a depth first search starting from a mesh triangle with only one id number and walk to the adjacent triangles until a triangle with a different id number is reached. Figure 12(b) shows the tubular part (in blue) for Genus3 with a threshold 1.41.

Figure 13 shows the identification of tubular parts for two other models.

6. Comparisons

In this section, we make a brief comparison between our algorithm and some existing ones. Our algorithm extracts the

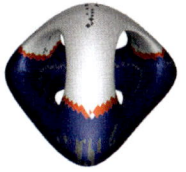 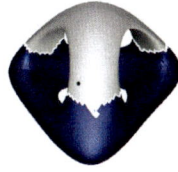

Figure 12: *Genus3: the upper three legs are flat.*

Hand   Mechpart

Figure 13: *The tubular part (in blue) of Hand and Mechpart with threshold* 1.3.

curve-skeleton by eroding the medial axis. Unlike other medial axis based algorithms, the extracted curve-skeleton by our algorithm remains stable against the noise even though the medial axis may not. Distance field based methods works nicely and efficiently for the tubular objects, see [HF05] for recent results. Since the points with the same distances from the boundary of some shapes may form surface patches, the distance field based methods may face difficulty in extracting curve-skeletons for those shapes. The potential field based method is proposed to fix this problem by taking into account the surface area, see for example [CSYB05]. However, in practice, the potential field based method still may fail for the shapes containing thin flat parts. Imagine a very thin flat plate. The variation of the potential field can be very subtle in the middle of this shape since the surface area on the sides (the only place producing different potentials) are relatively small and far away from the middle. Figure 14 shows the comparison results between our method and the potential field based method of [CSYB05]. Notice that the ears of Dog are thin and flat. The surfaces shown in Figure 14(a),(c) are the extracted isosurfaces of the volume data. We choose the parameter 'Potential field strength' to be 8 and 6 and the parameter 'highest divergence points' to be 65 and 20 for (a) and (c) respectively, which give the best results. The curve-skeleton extracted using potential field may not be connected (a) or may not be centered (ears in (c)). One more advantage of our method is that our algorithm needs only *one* user supplied parameter as opposed to many for the existing algorithms.

7. Discussions and Future work

In this paper, we introduce a mathematical definition of curve-skeletons for 3D shapes with connected manifold boundary. We present an algorithm to approximate these curve-skeletons. Extensive experiments show that the approximation algorithm is effective in practice. We also show that the extracted curve-skeletons enjoy many nice properties. It is appropriate to mention that our definition only works for shapes with connected boundary. For otherwise the geodesic distances are undefined between points on different connected components of the boundary.

There are pathological cases where our algorithm fails to extract a curve-skeleton though our definition provides a 1D curve. In these cases, although the boundary of the shape is connected, its medial axis contains a closed surface preventing the erosion process to proceed. One such example can be derived from the famous "house with two rooms" [Hat02], which is a contractible two dimensional subspace of $\mathbb{R}^3$. A small thickening of this "house with two rooms" creates a shape whose medial axis contains a closed surface.

It would be interesting to apply the curve-skeletons extracted by our algorithm in various applications such as shape matching, mesh decomposition, and animation. We plan to address this issue in future work.

8. Acknowledgement

We acknowledge the support of the NSF grants DMS-0310642 and CCR-0430735. We thank AIM@SHAPE, Stanford University, Princeton University, Sumner and Popovic [SP04] and K. Zhou et al [ZHS*05] for providing 3D models, and N. Cornea et al. [CSYB05] for making their code available.

References

[ACK02]   AMENTA N., CHOI S., KOLLURI R.:   The power crust, union of balls, and the medial axis transform. *Comput. Geom.: Theory Applications 19* (2002), 127–153.

[AJDE04]   ATTALI D., J.-D.BOISSONNAT, EDELSBRUNNER H.:   Stability and computation of the medial axis — a state-of-the-art report. In *Mathematical Foundations of Scientific Visualization, Computer Graphics, and Massive Data Exploration*, Möller T., Hamann B., Russell B., (Eds.). Springer-Verlag, 2004.

[AJDL03]   ATTALI D., J.-D.BOISSONNAT, LIEUTIER A.:   Complexity of the delaunay triangulation of points on surfaces : the smooth case. In *Proc. Ann. Sympos. Comput. Geom.* (2003), ACM, pp. 201–210.

[BKS01]   BITTER I., KAUFMAN A., SATO M.: Penalized-distance volumetric skeleton algorithm. *IEEE TVCG 7, 3* (2001).

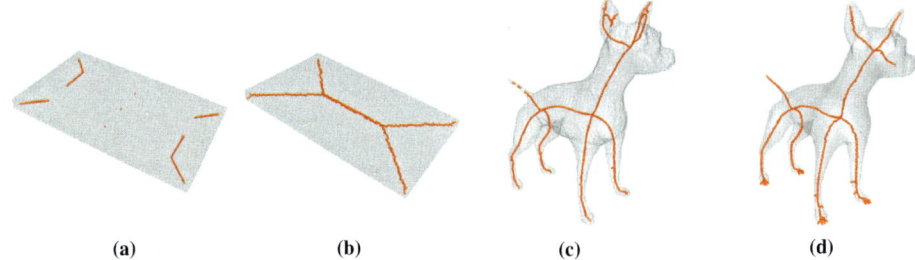

(a)   (b)   (c)   (d)

Figure 14: *(a) and (c) The curve-skeletons extracted by the potential field method. (b) and (d) The curve-skeletons extracted by our algorithm,* $\theta = -0.3$ *for both models.*

[BNB99] BORGEFORS G., NYSTRÈTM I., BAJA G. D.: Computing skeletons in three dimensions. *Pattern Recognition 32*, 7 (1999).

[BST03] BOUIX S., SIDDIQI K., TANNENBAUM A.: Flux driven fly throughs. *Computer Vision and Pattern Recognition* (2003), 449–454.

[CKM04] CULVER T., KEYSER J., MANOCHA D.: Exact computation of the medial axis of a polyhedron. *Computer Aided Geometric Design 21* (2004), 65–98.

[Cos99] COSTA L.: Multidimensional scale space shape analysis. In *IWSNHC3DI'99* (Santorini, Greece, 1999), pp. 214–217.

[CSM05] CORNEA N., SILVER D., MIN P.: Curve-skeleton applications. In *Proc. IEEE Visualization* (2005), pp. 95–102.

[CSYB05] CORNEA N., SILVER D., YUAN X., BALASUBRAMANIAN R.: Computing hierarchical curve-skeletons of 3d objects. *The Visual Computer 21*, 11 (2005), 945–955.

[DZ03] DEY T., ZHAO W.: Approximating the medial axis from the voronoi diagram with a convergence guarantee. *Algorithmica 38* (2003), 179–200.

[Fed96] FEDERER H.: Geometric measure theory. *Classics in Mathematics* (1996), Springer–Verlag.

[GDB06] GOSWAMI S., DEY T. K., BAJAJ C. L.: Identifying flat and tubular regions of a shape by unstable manifolds. In *Proc. 11th ACM Sympos. Solid and Physical Modeling* (2006).

[GK04] GIBLIN P., KIMIA B. B.: A formal classification of 3d medial axis points and their local geometry. *IEEE Transactions on Pattern Analysis and Machine Intelligence 26*, 2 (2004), 238–251.

[Hat02] HATCHER A.: *Algebraic Topology*. Cambridge University Press, 2002.

[HF05] HASSOUNA M. S., FARAG A. A.: Robust centerline extraction framework using level sets. In *IEEE Conf. Computer Vision and Pattern Recognition* (2005).

[OK95] OGNIEWICZ R. L., KÜBLER O.: Hierachic voronoi skeletons. *Pattern Recognition 28*, 3 (1995), 343–359.

[SBTZ99] SIDDIQI K., BOUIX S., TANNENBAUM A., ZUCKER S. W.: The hamilton-jacobi skeleton. In *Proc. ICCV* (Corfu, Greece, September 1999), pp. 828–834.

[SP04] SUMNER R. W., POPOVIC J.: Deformation transfer for triangle meshes. In *Proc. SIGGRAPH 2004* (2004), ACM Press / ACM SIGGRAPH.

[SSK*05] SURAZHSKY V., SURAZHSKY T., KIRSANOV D., GORTLER S., HOPPE H.: Fast exact and approximate geodesics on meshes. In *Proc. SIGGRAPH 2005* (2005), ACM Press / ACM SIGGRAPH, pp. 553–560.

[VL00] VERROUST A., LAZARUS F.: Extracting skeletal curves from 3d scattered data. *The Visual Computer 16* (2000), 15–25.

[Wol92] WOLTER F.-E.: Cut locus and medial axis in global shape interrogation and representation. *MIT Design Laboratory Memorandum 16*, 1 (1992).

[ZHS*05] ZHOU K., HUANG J., SNYDER J., LIU X., BAO H., GUO B., SHUM H.-Y.: Large mesh deformation using the volumetric graph laplacian. In *Proc. SIGGRAPH 2005* (2005), ACM Press / ACM SIGGRAPH, pp. 496–503.

[ZT99] ZHOU Y., TOGA A.: Efficient skeletonization of volumetric objects. *IEEE Trans. Visualization and Comp. Graphics 5*, 3 (1999), 196–209.

Eurographics Symposium on Geometry Processing (2006)
Konrad Polthier, Alla Sheffer (Editors)

Selectively refinable subdivision meshes

Enrico Puppo

Department of Computer and Information Sciences
University of Genova

Abstract

We introduce RGB triangulations, *an extension of red-green triangulations that can support selective refinement over subdivision meshes generated through quadrisection of triangles. Our purpose is to define a mechanism based on local operators that act on subdivision meshes while supporting operations similar to those available in Continuous Level Of Detail models. Our mechanism permits to take an adaptive mesh at intermediate level of subdivision and process it through both refinement and coarsening operations, by remaining consistent with an underlying Loop subdivision scheme. Our method does not require any hierarchical data structure, being based just on color codes and level numbers assigned to elements of a mesh, which can be encoded in a standard topological data structure with a small overhead.*

Categories and Subject Descriptors (according to ACM CCS): I.3.5 [Computer Graphics]: Curve, surface, solid, and object representations

1. Introduction

Subdivision surfaces are used extensively in computer graphics and CAD. Subdivision schemes are based on the recursive refinement of the faces of a geometric mesh and converge to either C^1, or C^2 surfaces. Classical schemes are based on subdivision patterns that are applied to all faces of a mesh at each level. In practice, subdivision is often applied up to a certain level and the resulting mesh is used as an approximation of the limit surface [ZS00].

In some applications, it may be desirable to refine a mesh adaptively. This sort of mechanism is common in Continuous Level Of Detail (CLOD) models developed in the context of free-form mesh modeling and often applied in computer graphics [LRC*02]. In particular, *selective refinement* permits to vary the level of detail (LOD) smoothly across the mesh and dynamically through time. Transition between different LODs should be as smooth as possible and the resulting mesh should always be conforming (i.e., free of cracks). In order to support selective refinement efficiently, it is crucial that a mesh at intermediate LOD can be modified on-line in either way, by refining some parts of it while other parts may be coarsened. To this aim, refinement and coarsening operations must be based on local operators.

Transition between different levels is not easy in classical subdivision schemes, since non conforming situations arise. For instance, the popular Loop [Loo87] and butterfly [DLG90] schemes are based on recursive triangle quadrisection, which gives non-conforming meshes when applied adaptively at different levels of subdivision. Red-green triangulations [BSW83] have been widely used in the literature to obtain adaptive and conforming meshes for such subdivision schemes, but no efficient technique for selective refinement on such meshes has been designed so far.

In this paper, we present RGB triangulations, an extension of red-green triangulations that are built from iterative application of a local operator, namely *edge split*. Our method has the advantage of being progressive and to generate conforming meshes at all intermediate steps, which are consistent with the underlying subdivision scheme. This allows us to design a selective refinement algorithm for subdivision meshes based on triangle quadrisection, which exhibits the same features of the algorithms developed in the context of CLOD models.

We develop our method in the framework of Loop subdivision. We show that all levels of subdivision in the Loop scheme can be obtained and that control points of vertices are computed correctly in our scheme. This implies that our RGB meshes converge to the same surfaces obtained with the classical Loop method.

Differently from classical CLOD models, and from some other models supporting adaptive subdivision, our method does not require any hierarchical data structure. Selective refinement can be performed directly on a RGB mesh, which is maintained in a standard topological data structure, enriched with level numbers and color codes for triangles, edges and vertices, with a small overhead.

RGB triangulations may become a valid substitute or complement for standard subdivision of triangle meshes in solid modelers. Concerning applications in computer graphics, recent advances in reverse subdivision [Sab04] suggest that subdivision surfaces may become a valid alternative to CLOD models for free-form objects. In this view, RGB triangulations provide the tools to manage subdivision surfaces with the same flexibility of CLOD models also in this context.

The rest of the paper is organized as follows. In Section 2 we briefly discuss related work. In Section 3 we introduce the necessary background. In Section 4 we introduce RGB triangulations. In Section 5 we describe the selective refinement algorithm. In Section 6 we describe how to set the position of vertices according to the Loop subdivision scheme. In Section 7 we describe the data structure used to implement selective refinement on RGB triangulations. Finally, in Section 8 we make some concluding remarks.

2. Related work

Subdivision surfaces. The literature on subdivision surfaces is quite extended. The interest reader can refer to [WW02] for a textbook, [ZS00] for a tutorial and [Sab04] for a survey. Here, we will review only those works related to adaptive subdivision of triangle meshes.

Red-green triangulations were introduced in the context of finite element methods [BSW83] as an empirical method to obtain conforming meshes from adaptive subdivision of triangle meshes. Red-green triangulations are usually built through a two-step procedure: first by applying triangle quadrisection adaptively, and then by subdividing some triangles further, through different patterns, to fix non conforming situations. Depending on the underlying subdivision scheme, the geometry of vertices (control points) that lie on the transition between different levels of subdivision may be different form that of the same vertices in a uniformly subdivided mesh. This fact, which is often overlooked, may prevent the correctness of further subdivision or coarsening of a red-green triangulation, unless the subdivision process is repeated from scratch.

Red-green triangulations were used in [ZSS97] to support multiresolution editing of meshes based on the Loop subdivision scheme. Adaptive meshes are computed by reversing subdivision, starting at the finest level and pruning over-refined triangles. Also in this case, a restricted nonconforming mesh is computed first, which is fixed next by

further bisection of some triangles. Relocation of vertices is treated by using a hierarchical data structure that stores the positions of all control points in the uniform subdivision.

In [SHHG01], the quadrisection scheme is decomposed into atomic local operations, called *quarks*, based on the popular *vertex split* operation that is at the basis of Progressive Meshes [Hop96]. A red-green triangulation under the butterfly scheme [DLG90] is obtained through a sequence of quarks. Problems of topological consistency and relocation of vertices are treated by forcing some operations during refinement. The resulting mesh is over-refined with respect to a corresponding red-green triangulation computed with a traditional method. No explicit algorithm for selective refinement is proposed in [SHHG01].

The $\sqrt{3}$ subdivision [Kob00] and the 4-8 subdivision [VZ01] schemes are not based on the classical quadrisection operator. They are naturally adaptive, being both based on local conforming operators.

The $\sqrt{3}$ subdivision alternates triangle trisection (insertion of a new vertex at the center of each triangle) at one level, with edge swap at the next level. This scheme generates triangles that can be regarded as being of *green* and *blue* types in the terminology that we introduce in Section 4. The problem of correct relocation of vertices is addressed in [Kob00]. To this aim, some over-refinement of neighbors of even (green) triangles is imposed. A closed form solution of the subdivision rule permits to compute control points for a vertex at any level on the basis of just its initial position and its limit position. Adaptive refinement is supported, while adaptive coarsening is not investigated explicitly in [Kob00].

The 4-8 subdivision is based on edge split (as in our case) applied to a special case of triangle meshes, called *tri-quad meshes*. An initial tri-quad mesh can be obtained from any triangle mesh by doubling its number of triangles and changing its topology [VZ01]. The correct position of control points is addressed and resolved also in this case with a certain amount of over-refinement of the mesh. Only basic operations are investigated in [VZ01], while no selective refinement algorithm is proposed.

CLOD models. Also the literature on Continuous Level of Detail models is very wide. The interested reader may refer to [LRC*02] for a recent book on this subject. Generally speaking, a CLOD model consists of a base mesh at coarse resolution, plus a set of local modifications that can be applied to the base mesh to refine it. Such modifications are arranged in a hierarchical structure, which consists of a directed acyclic graph (DAG) in the most general case. Meshes at intermediate level of detail correspond to cuts in the DAG, and algorithms for selective refinement work by moving a front through the DAG and doing/undoing modifications that are traversed by this front. This general framework, developed in [Pup98], applies to almost all CLOD models proposed in the literature.

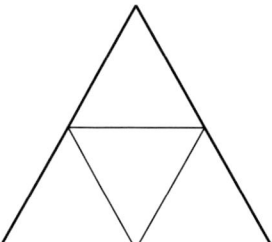

Figure 1: *The triangle quadrisection pattern.*

CLOD models can provide meshes at intermediate LOD, where detail can vary across the mesh and through time, at a virtually continuous scale and with fast procedures that work on-line even for huge meshes. The scheme proposed in [DWS*97] is very popular and most authors refer to it in order to implement their selective refinement algorithms. The outer structure of the algorithm we propose for RGB triangulations is also based on this scheme.

There exist a few CLOD models based on recursive subdivision patterns. The model proposed in [DWS*97] is based on the recursive bisection of right triangles. This rule is also used by several other authors, and may be regarded as a subdivision. It can be applied just to meshes obtained from regular grids (typically representing terrains), while it is not easy to extend it to more general triangle meshes. One generalization is given by 4-k meshes [VG00], which have in fact a strong relation with 4-8 subdivision [VZ01].

3. Background

Triangle meshes. A *triangle mesh* is a triple $\Sigma = (V, E, T)$ where: V is a set of points in 3D space, called *vertices*; T is a set of triangles having their vertices in V and such that any two triangles of T either are disjoint, or share exactly either one vertex or one edge (thus, the mesh is inherently *conforming*); E is the set of edges of the triangles in T, where each edge is taken just once. Standard topological incidence and adjacency relations are defined over the entities of Σ.

We will assume to deal always with *manifold* meshes without boundary, i.e.: each edge of E is incident at exactly two triangles of T; and the *star* of a vertex (i.e., the set of entities incident at it) is homeomorphic to an open disc.

A triangle mesh is said to be *regular* if all its vertices have valence six. In a mesh that is not regular, vertices with a valence different from six are called *extraordinary*.

A *non-conforming mesh* is a structure similar to a mesh, in which triangles may violate the rule of edge sharing: there may exist adjacent triangles t and t' such that one entire edge of t overlaps just a portion of the corresponding edge of t'.

Loop subdivision. The Loop subdivision scheme was introduced in [Loo87] and it is the first and most famous sub-

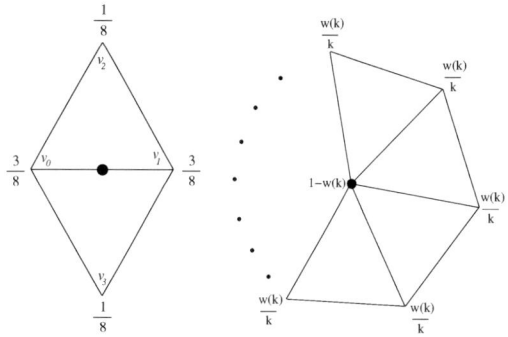

Figure 2: *The Loop subdivision scheme: the stencil used to compute the position of odd vertices (left); the stencil used to compute the position of even vertices (right). Numbers are weights assigned to vertices in the linear combination, k is the valence of the even vertex ($k = 6$ in the regular case) and*
$$w(k) = \frac{5}{8} - \left(\frac{3}{8} + \frac{1}{4} \cos \left[\frac{2\pi}{k} \right] \right)^2.$$

division scheme on triangle meshes. It is an approximating scheme, meaning that the position of control points is changed throughout levels of subdivision, and it converges to a C^2 surface if applied to a regular mesh.

The subdivision pattern is *triangle quadrisection*, as depicted in Figure 1, and it is applied to all triangles of the mesh at each level of subdivision. The position of each new vertex introduced from subdivision (called an *odd* vertex) is computed as weighted sum of vertices from the previous level (called the *even vertices*), as depicted in the stencil on the left of Figure 2. After inserting odd vertices at a given level, all even vertices are relocated according to the stencil on the right of Figure 2. For the sake of brevity we omit here and in the following the scheme for boundary vertices. Our method can be extended easily to treat meshes with boundary too.

Therefore, for a vertex v introduced at level l, there exist an infinite sequence of control points $p^l(v), p^{l+1}(v), \ldots, p^\infty(v)$ that define the positions of v at level l and all successive levels, $p^\infty(v)$ being its position on the limit surface.

For a vertex v of the base mesh position $p^0(v)$ is defined; while for a vertex v introduced at level l position $p^l(v)$ depends on positions p^{l-1} of vertices in its mask. The successive positions in the sequence are computed according to the mask for even vertices, such that for an even vertex v, $p^j(v)$ depends on $p^{j-1}(v)$ as well as on the positions p^j of all its neighbors (which are all odd vertices at level j).

Red-green triangulation. Consider a base mesh and assume the quadrisection scheme is applied adaptively to it. The resulting structure is a non-conforming mesh, as depicted in Figure 3a. This mesh is said to be *restricted* if two

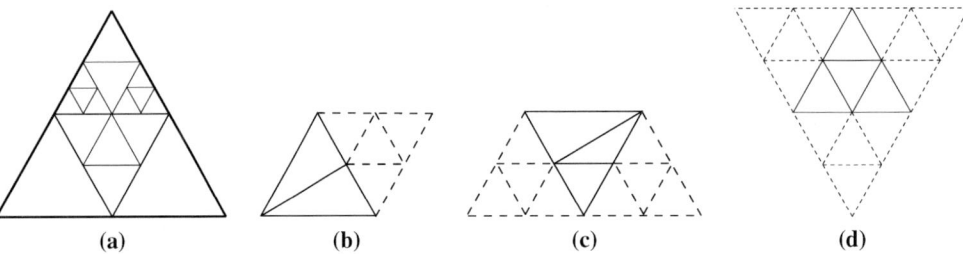

Figure 3: *Red-green triangulation: a non-conforming restricted mesh obtained from adaptive quadrisection (a); bisection is used to fix triangles that have one neighbor at the next level (b); trisection is used to fix triangles that have two neighbors at the next level (c). Triangles that have three neighbors at the next level are subdivided further by quadrisection (d).*

adjacent triangles may differ for no more than one level of subdivision. If a mesh is restricted, it may be made conforming by subdividing some triangles further, by either bisection, trisection, or quadrisection, as depicted in Figures 3b and c. Some authors avoid trisection by forcing further subdivision of triangles in the situation of Figure 3c. This fact, however, may propagate subdivision to adjacent triangles, possibly affecting a large area that will result over-refined.

In case both bisection and trisection are used, a given triangle may be subdivided by ten different patterns: three obtained by rotational simmetry from the pattern depicted in Figure 3b; six obtained by rotational simmetry and mirroring from the pattern depicted in Figure 3c; and one corresponding to the pattern depicted in Figure 3d. Note that, by construction, a triangle t that is subdivided with one of the first nine patterns must necessarily have neighbor(s), at all edges of t that are split, which were subdivided at the next level of subdivision by the thenth pattern.

All triangles that would appear in a standard subdivision are said to be green, while the other triangles that are introduced to make the mesh conforming are said to be red. In the following, we will introduce finer color codes for triangles and edges, in order to develop the details of our method. Green triangles and red triangles generated through bisection, as well as the red triangle generated through trisection, which is depicted as a square triangle in Figure 3c, will maintain the same color codes (green and red, respectively). On the contrary, in our framework, the triangle that is depicted as a skinny isosceles triangle in Figure 3c will be said to be blue. Note that, in both cases, the small equilateral triangle in Figure 3c, as well as all triangles in Figure 3d are green triangles at the next level of subdivision.

4. RGB triangulations

RGB triangulations are defined as all those triangulations that can be built through iterative application of given operators for local subdivision, starting at a base mesh Σ_0. Such operators always produce conforming meshes and can generate all and only those triangles that may appear in a red-green triangulation built from Σ_0, but the possible combina-

tions of such triangles to form a mesh are more numerous that in red-green triangulations. In other words, RGB triangulations form a superset of red-green triangulations, with a higher expressive power. Reverse local operators to coarsen a mesh are also defined that, in combination with the subdivision operators, allows us to support selective refinement.

4.1. Local subdivision operators

Consider a base mesh Σ_0. We assign level zero to all vertices, edges and triangles of Σ_0, and color green to all edges and triangles of Σ_0. In the following, we define local subdivision operators that, when applied iteratively to Σ_0, will generate a conforming mesh where triangles will be colored of green, red and blue; edges will be colored of green and red; and vertices, edges, and triangles will have different levels. Color codes of red-green triangulations are extended here with further codes (blue triangles and colors for edges) in order to control the application of subdivision operators on a local basis.

For the sake of clarity, in our examples we will use regular meshes where all green triangles are equilateral, red triangles are square with one angle of sixty degrees and the other angle of thirty degrees, and blue triangles are isosceles with two angles of thirty degrees. However, our method is not restricted to such constraints, since color codes propagate according to recursive rules that do not depend on either geometry or valence of vertices.

We first define a sub-atomic rule, namely triangle bisection, which can be applied under certain conditions, and produces a certain configuration depending on the color of the triangle to be bisected. Next we define a first atomic rule, namely edge split, as the combination of triangle bisections applied to a pair of adjacent triangles. Finally we add a second atomic rule, namely edge swap, to be applied automatically only in a specific situation.

Triangle bisection can be applied only in the following configurations (see Figure 4):

- **G-bisection:** let t be a green triangle at level l and e be an edge of t (e can be any edge of t and it is always green

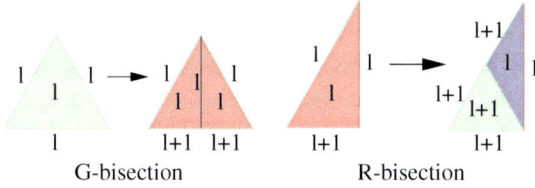

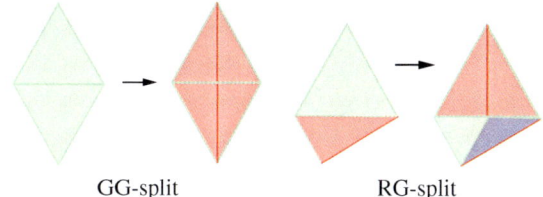

Figure 4: *Triangle bisection: bisection of a green triangle (G-bisection); bisection of a red triangle (R-bisection). Labels denote the level of edges and triangles.*

Figure 5: *Edge split: two green triangles (GG-split); one red and one green triangle (RG-split); two red triangles (RR_1- and RR_2-split). Labels denote the level of vertices and edges.*

and at level l; vertices of t can be at any level $\leq l$). The bisection of t at the midpoint of e generates two red triangles at level l. Each such triangle will have: one green edge at level l (the one common with old triangle t), one green edge at level $l + 1$ (one half of e) and one red edge at level l (the new edge inserted to split t). The new vertex inserted to perform bisection will have level $l + 1$.

- **R-bisection:** let t' be a red triangle at level l and e' its green edge at level l (there is only one such edge). The bisection of t' at the midpoint of e' generates one blue triangle at level l and one green triangle at level $l + 1$. The green triangle is incident at the green edge at level $l + 1$ of old triangle t' and also its other two edges are at level $l + 1$ (the edge inserted to subdivide t', and one half of e'). The blue triangle is incident at the red edge of old triangle t' and has also two green edges at level $l + 1$ (the edge inserted to subdivide t', and the other half of e'). The new vertex inserted to perform bisection will have level $l + 1$.

Edge split consists of the simultaneous bisection of two adjacent triangles t_0 and t_1 by splitting their common edge e, and can occur only if the bisections of both t_0 and t_1 along e are legal according to the rule defined above. It is readily seen that edge split is legal only in the following three cases (see Figure 5):

- **GG-split:** t_0 and t_1 are both green and at the same level;
- **RG-split:** t_0 is green and t_1 is red and they are both at the same level;
- **RR-split:** t_0 and t_1 are both red at level l and e is a green edge at level l. This case may come in two variants (RR_1-split and RR_2-split). Each variant can be recognized by the cycle of colors of edges on the boundary of the diamond formed by t_0 and t_1: this may be either red-green-red-green for RR_1-split, or red-red-green-green for RR_2-split.

BB-swap is applied automatically whenever two blue triangles at level l become adjacent along their red edge at level l (see Figure 6). In this case, such edge is eliminated and the other diagonal of the quadrilateral formed by such two triangles is inserted. The result is a pair of green triangles at level $l + 1$ (all their edges are also green and at level $l + 1$). See Figure 6. In practice, BB-swap will occur immediately after triangle bisection is applied to a red triangle that is already

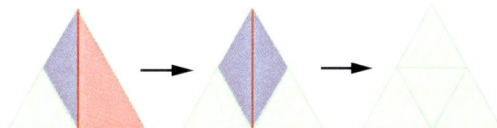

Figure 6: *BB-swap: when a red triangle is bisected, which was already adjacent to a blue triangle, two blue triangles become adjacent along a red edge. Such edge is swapped, thus producing two green triangles at the next level.*

adjacent to a blue triangle along its red edge. Note that, by construction, one of the two new green triangles will have all three vertices at level $l + 1$.

Edge split is the main operator used to perform mesh refinement. It can be applied to legal pairs of adjacent triangles, selected according to rules that drive refinement, while swap is forced whenever two blue triangles become adjacent along a red edge. Note that just green edges can be split, while red edges are only swapped.

4.2. Consistency of RGB triangulations

The family of RGB triangulations from base mesh Σ_0 contains Σ_0 as well as all other meshes can be generated starting at Σ_0 and applying the refinement rules above. We claim that RGB triangulations form a superset of red-green triangulations generated starting at Σ_0 (hence, also of all uniform subdivisions of Σ_0).

We first prove that any red-green triangulation can be built starting at Σ_0 through the local operators defined before. Proof is by induction on the maximum level of subdivision m of a red-green triangulation Σ that subdivides Σ_0. By definition, if $m = 0$ then $\Sigma \equiv \Sigma_0$ and it is obviously a RGB triangulation. Let us suppose now that our claim is true up to level of subdivision $m - 1$. Let Σ' be the mesh obtained from Σ as follows: all green triangles at level m and all red triangles at level $m - 1$ (i.e., red triangles that subdivide green triangles at level $m - 1$ to make the mesh conforming with triangles at level m) are removed, and the holes left are naturally filled with green triangles at level $m - 1$. It is readily seen that Σ' is conforming and its maximum level of subdivision is $m - 1$. Therefore, Σ' can be built through our local operators by inductive hypotesis. Now, in order to obtain back Σ, we must refine all those triangles at level $m - 1$ that were used to fill the holes, each in its proper configuration. Let us call $S = S_r \cup S_g$ this set of triangles, where S_r [S_g] is the subset of S of triangles that will refine into red [green] triangles in Σ. Note that S_r is void if Σ is a uniform subdivision. Consistently with color codes of red-green triangulations, they are all green and at level $m - 1$. We also assign green color to their edges, consistently with our color codes. All remaining triangles of Σ' already belong to Σ.

We refine first all triangles of S_r. Let $t \in S_r$, then t must split according to either bisection or trisection (see Figure 3b and c). Note that, since t is the first triangle that we subdivide, all its neighbors are green at level $m - 1$, and they will eventually be refined into green triangles at level m (by construction rules of red-green triangulations). We may obtain bisection of t by applying GG-split; and trisection of t by applying GG-split followed by RG-split. Note that after this refinement, each triangle adjacent to t along a splitting edge is now subdivided into a pair of red triangles. Let $t' \in S_r$ be another triangle we take for subdivision, and let us consider its neighbors along edges that must be subdivided. If such triangles are green, then t' can be subdivided in the same manner as t. Otherwise, t' might have some red neighbors that have been obtained by subdividing triangles of S_g because of edge splits applied to subdivide some other triangle of S_r processeded before t'. By exhaustive analysis of the possible configurations, it is easy to see that bisection of t' can be obtained by applying either a GG-split or a RG-split, depending on whether the neighbor is green or red; and trisection of t' is obtained by the same operation, followed by either a GR-split or RR2-split, depending on whether the neighbor is green or red. All remaining triangles of S_r can be subdivided as t'.

At this point, we have obtained the correct refinement of all triangles of S_r and partial refinement of all those triangles of S_g that were adjacent to triangles of S_r. Let us consider now $t \in S_g$ and see how it has been refined so far. There are four possibilities:

- All three edges of t have been split. In this case, the triangles that subdivide t are now in the configuration depicted in the center of Figure 6. A BB-swap is sufficient to complete subdivision of t to level m, without affecting its neighbors;
- Two edges of t have been split. In this case, the triangles

that subdivide t are now in the configuration depicted on the left of Figure 6. Note that the neighbor t' of t on its right side (referring to the figure) must necessarily belong to S_g or be a red triangle subdividing a triangle in S_g. Therefore, we first apply either a RG-split or a RR*-split (where * may be 1 or 2 depending on configuration) depending on t' being either green or red. Next we apply edge swap as in the previous case.

- One edge of t has been split. In this case, t has been bisected into two red triangles as in the case of G-bisection. This case is similar to the previous, but the first split must be performed on both sides of t (i.e., on both red triangles refining it).
- No edge of t has been split yet. In this case, all three neighbors of t must either belong to S_g or subdivide triangles of S_g. We first apply either a GG-split or a RG-split, depending on whether the neighbor of t along the splitting edge is green or red. Then we proceed as in the previous case.

By exhaustive analysis of possible configurations, it is easy to see that the neighbors of t affected by the split operations we perform will get to one of the first three configurations. Therefore, the repeated application of these operations will eventually refine all triangles of S_g to level m. At that point, we have obtained mesh Σ.

In order to complete the proof of our claim, we must show that there exist RGB triangulations that are not red-green triangulations. One example is the mesh obtained from a pair of adjacent green triangles by applying a GG-split. The mesh depicted in Figure 9 is a more elaborated example. It is interesting to notice how RGB triangulation may manage fast transitions of LOD better than red-green triangulations. Consider the RGB triangulation on the left of Figure 8. The LOD transition from the base to the apex of the big triangle may be carried out to an arbitrary number of levels by using the same pattern. The equivalent transition using red-green triangulations is depicted on the right side of the same figure and requires about twice the number of triangles.

4.3. Reverse subdivision operators

Since a selective refinement algorithm also needs to reverse subdivision (i.e., to coarsen a mesh), we define also local operators that invert edge split and edge swap.

Triangle merge is the reverse operation of triangle bisection and it is defined as follows:

- **RR-merge:** a pair of red triangles t_0, t_1 of level l that are adjacent at a red edge may merge into one green triangle at level l; the two green edges at level $l + 1$ of t_0 and t_1 are merged to form a new edge e at level l of the new triangle; the red edge and its endpoint at level $l + 1$ disappear.
- **GB-merge:** A pair of adjacent triangles t_0, t_1 such that t_0 is green at level $l + 1$ and t_1 is blue at level l may merge into one red triangle at level l; the common edge of t_0 and t_1 and the endpoint v of such edge that is incident at both

green edges of t_1 disappear; the other two green edges that were incident at v are merged to form one green edge at level l.

Edge merge is the reverse operation of edge split and can be applied to triangles incident at vertices of valence four, such that triangle merge can be applied to pairs of such triangles. The same cases depicted in Figure 5 occur (modifications apply right-to-left in this case):

- **R4-merge** inverts GGsplit;
- **R2GB-merge** invertes GR-split;
- **GBGB-merge** inverts RR_1-split;
- **G2B2-merge** inverts RR_2-split.

A little care must be taken in applying GBGB-merge in order to avoid inconsistencies. Referring to Figure 5, note that the quadrilateral must have two vertices at the same level l and two other vertices at a level lower than l. GBGB-merge must be performed in such a way that the diagonal incident at vertices of level lower than l is maintained.

Reverse edge swap cannot be applied to any pair of adjacent green triangles, otherwise the structure of subdivision would be destroyed. **GG-swap,** which inverts BB-swap, can be applied to a pair of adjacent green triangles t_0 and t_1 at level l if one of them, say t_0, has all three vertices at level $l > 0$. This condition is necessary and sufficient to guarantee that t_0 and t_1 have the same parent triangle t in the subdivision and t_0 is the central triangle obtained by subdividing t. In order to be consistent with refinement rules, reverse edge swap should be applied only if one of the blue triangles generated is removed immediately after through an edge merge operation. We therefore constrain reverse edge swap to occur only in the configurations depicted in Figure 7. These configurations are easy to verify by exploring the stars of vertices common to t_0 and t_1, and checking the colors and levels of their incident triangles.

With these local operators at hand, we can modify any RGB triangulation by either refining or coarsening it locally. This means that we can traverse the whole family of RGB triangulations originated from a given base mesh, reaching any adaptively subdivided mesh that may be needed from an application. This task will be performed through the selective refinement algorithm described in the following section.

5. Selective refinement

Selective refinement is applied to a RGB triangulation Σ and consists of the iterated application of local operators defined in the previous until some user-defined halt condition is verified.

Following [CDM*04, DWS*97], selective refinement is driven by two priority queues:

- A queue Q_r of refinement operations to be performed to

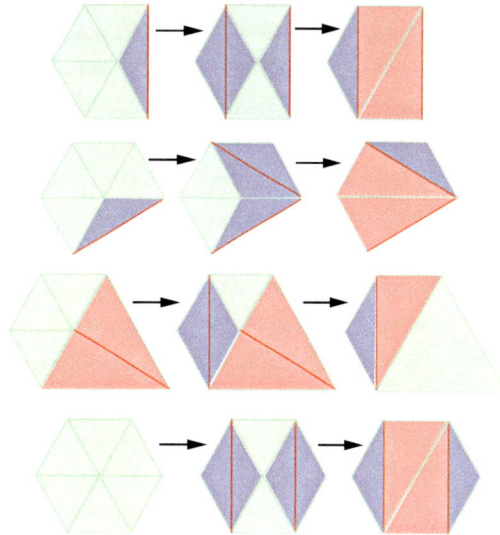

Figure 7: *Legal configurations for applying GG-swap.*

meet LOD requirements. Refinement operations are related to edges to split. Queue Q_r is initialized by considering all green edges of Σ and inserting into Q_r all and only those edges that need refinement. Depending on user needs, priority of an edge may depend, e.g., from its length, or from distance between its midpoint and the position of the vertex to split it at next level of subdivision.

- A queue Q_c of coarsening operations to be performed whenever possible to reduce the size of the mesh without violating LOD requirements. Coarsening operations are related to vertices to be removed through edge merge operations. Queue Q_c is initialized by considering all vertices of Σ that may be removed from an edge merge operation, possibly preceded from suitable edge swap operation(s), and inserting them into Q_c. Priority of a vertex in the queue will be set consistently with the previous case.

After initialization, elements are popped from either Q_r or Q_c, according to higher priority, and related operations are applied to modify Σ, until a halt condition is verified. This may typically depend either on the full satisfaction of LOD requirements, or on the size of the extracted mesh. See [CDM*04] for more details about managing the queues during selective refinement.

The most crucial aspects of the algorithm are related to refinement and coarsening operations to be performed on a RGB triangulation.

Coarsening. Assume a vertex v having level l has been popped from Q_c. If the star of v in Σ has one of the configurations depicted in Figure 7, then the corresponding reverse swap is applied first. At this point, the star of v will have one of the four configurations depicted in Figure 5 (right side). The following operations are performed:

1. Edge merge is applied and Σ is updated accordingly;
2. The four vertices that were adjacent to v are analyzed and those vertices that can be now legally removed are inserted into Q_c; the control point of each vertex is updated (see next section).

Refinement. Refinement is more complex, since edges in Q_r may require further refinement operations on other edges prior to be refined. Assume a green edge e having level l has been popped from Q_r. If the pair of triangles incident at e do not form one of the configurations depicted in Figure 5 (left side) then such triangles are analyzed to trigger recursive edge split operations. Let t be one such triangle:

- if t is a red triangle at level $l - 1$ then its green edge at level $l - 1$ is recursively split;
- if t is a blue triangle at level $l - 1$ then t must be adjacent to a red triangle t' at level $l - 1$ along its red edge; the green edge of t' at level $l - 1$ is split recursively;
- otherwise no action is required.

Performing a single edge split involves the following operations:

1. Compute the control point at level $l + 1$ for the new vertex v (see next section);
2. Update the control point at level $l + 1$ at the four vertices of triangles to be split;
3. Update mesh Σ performing edge split;
4. Test each new green edge generated from split and add it to Q_r if it does not fulfill LOD requirements.

As we will see in the next section, computation of the control point for the new vertex may require some extra refinement of the mesh, necessary to obtain the correct values for the Loop stencil of odd vertices. This is similar to what happens in other adaptive subdivision schemes [Kob00, SHHG01, VZ01]. Since we wish to avoid over-refinement of the result, edge splits performed during computation of control points are tested for LOD requirements. In case one such split is not necessary according to LOD requirements, the new vertex generated from split is marked as temporary and inserted in a queue. A temporary vertex becomes permanent in case one of its incident edges undergoes a standard edge split. At the end of selective refinement, this queue is scanned, and all vertices that are still temporary are removed by performing corresponding edge merge operations.

6. Geometry of control points

So far we have been concerned only with topological changes in a RGB triangulation. We now study the geometry of vertices. As we work in the framework of Loop subdivisions, control points of all vertices are computed at all levels through the proper stencils. However, since RGB triangulations work selectively, it may be not trivial to determine the right vertices to use for a stencil, and their proper control points.

Updates to control points must be done for odd vertices during refinement, and for even vertices both during refinement and during coarsening. We address the three relevant cases in the following.

Control point for an even vertex updated after refinement. When an edge e at level l is split by introducing a new (odd) vertex v, its control point $p^{l+1}(v)$ is computed (see next paragraph) and the control points p^{l+1} of end vertices v_0 and v_1 of e need to be updated with a contribution from $p^{l+1}(v)$. Without loss of generality, let us consider the case of v_0. We adopt a lazy strategy, which generates $p^{l+1}(v_0)$ with a partial value as soon as the first neighbor of v_0 at level $l + 1$ is introduced, and mark it as *restricted*. Then the value of $p^{l+1}(v_0)$ is updated every time another neighbor of v_0 at level $l + 1$ is introduced, by adding its contribution according to the formula used for even vertices in the Loop scheme. After the last neighbor has been introduced, control point $p^{l+1}(v_0)$ gets its correct value by combining the current summation with value $p^l(v_0)$ (that must be already available, by recursion) and is marked *available*.

Control point for an odd vertex introduced via refinement. Let vertex v be introduced at level $l + 1$ of subdivision by splitting an edge e at level l. In order to compute control point $p^{l+1}(v)$, we need to fetch the four vertices in the stencil of v at level l, and their control points p^l (see Figure 2).

In a RGB triangulation Σ, vertices v_0 and v_1 of the stencil of v are the endpoints of edge e to split, so they are found immediately. On the contrary, v_2 and v_3 are not necessarily vertices of the triangles incident at e. In fact, such triangles might have been refined already to higher levels of subdivision. Without loss of generality, let us consider just the case of vertex v_2. Let t be the triangle incident at e on the side of vertex v_2 and let us refer to Figure 8 (left side). If t is green, then vertex v_2 is its vertex opposite to e. Otherwise, t must be red and its neighbor t' on the other side of its red edge e' must be either red or blue. If t' is red, then vertex v_2 is the vertex of t' opposite to t, otherwise the neighbor t'' of t' on the side of 2 is retrieved (this can be decided by looking whether the red edge is incident at v_0 or at v_1). If t'' is green, then vertex v_2 is its vertex opposed to t', otherwise we set $t \leftarrow t''$ and restart the search. This search can be implemented through a simple while cycle and may take $O(m - l)$ time, where m is the maximum level of subdivision, in the worst case (that can occur only in the proximity of an abrupt change of LOD). If Σ were refined uniformly, we will have $m \in O(\log N)$, where N is the number of vertices of Σ. In theory, we may have $m \in O(N)$ in the worst case (e.g., if high level of subdivision is used only in the proximity of a given point and degrades abruptly to level zero in the other parts of the mesh). In practice, since a subdivision mesh is usually built on a small number of levels, and abrupt changes of

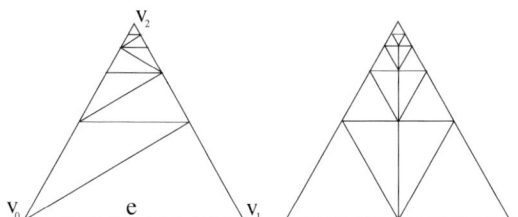

Figure 8: *Left side: finding vertex v_2. The triangle incident at e at level l might have been split through all successive levels of subdivision. Right side: in order to achieve the same transition of LOD from base to apex of the big triangle, almost twice the number of triangles is necessary in red-green triangulations.*

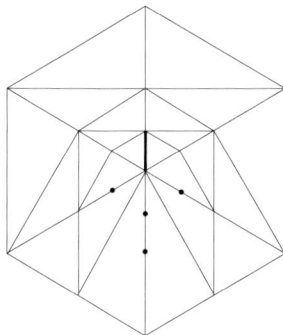

Figure 9: *In order to bisect the edge in bold, vertices marked by bullets must be computed by recursive edge split.*

LOD seldom occur, we may consider this procedure to run in constant time.

Once a vertex $v_i, i = 0, 1, 2, 3$ is fetched, we check whether its control point $p^l(v_i)$ is available or not. If not, we need to compute it recursively by splitting edges in the star of v_i, through the procedure already described in Section 5. All green edges incident at v_i are considered. For each such edge e_j, if it is at level $l' < l$, we split it to generate a new vertex v'_j and its control point $p^{l'}(v'_j)$. Split is repeated until the edge reaches level l. Note that a regular vertex may have green incident edges that differ for at most three levels, thus the number of new vertices to be computed during this operation is usually quite small (see Figure 9).

Control point for an even vertex updated after coarsening. When a vertex v at level $l + 1$ is removed from an edge merge operation, the four vertices that were adjacent to v are checked and, for each such vertex v_i, if its current position in the mesh is at a control point of level higher than l, then it is changed to $p^l(v_i)$. However, the control point $p^l(v_i)$ in the list of control points of v_i is not changed and it is maintained *available* for subsequent processing.

7. Data structure

A RGB triangulation can be maintained in a standard topological data structure for triangle meshes. One possibility is using two dynamic arrays, one for vertices and the other for triangles, with a garbage collection mechanism to manage reuse of locations freed because of coarsening operations. For each triangle, links to its three vertices as well as to its three neighbors are maintained. For each vertex, just a link to one of its incident triangles is maintained (this is sufficient to compute the star of a vertex in optimal time).

This data structure is extended as follows. For each vertex v we store: its level, a link to a linked list of its control points, and a link to the control point in the list corresponding to the position of v in the current mesh. Each control point in a list contains its three coordinates, and a flag to indicate whether it is available or restricted. This flag is actually a counter that is initialized either at six, or at the valence of the vertex if it is an extraordinary vertex, and is decremented each time a contribution to computation of the control point is added. Value zero means that the control point is available.

For each triangle we store its color and its level. We use two different codes for red triangles, depending whether the short (higher level) green edge is followed by the other green edge, or it is followed by the red edge when traversing the triangle counterclockwise. Since two bits are sufficient for the color, and levels in subdivision are usually not many, one byte is sufficient to store both color and level. Edges are addressed as pairs triangle-index, where index can take values 0, 1, 2. Depending on the color of a triangle we store edges in entries 0, 1, 2 in a conventional order (e.g., for a blue triangle 0=red 1=green 2=green). Because we used two different codes for red triangles, this is sufficient to unambiguously determine color and level of each edge.

Since selective refinement is meant to be used dynamically, we set up a caching mechanism to save vertices and their related control points when they are removed from the mesh because of edge merge operations. A caching policy based on least recently used vertex is adopted to manage the cache. Vertices in cache can be restored together with all their control points once they are reinserted in the mesh because of an edge split operation.

8. Concluding remarks

We have introduced RGB triangulations, a mechanism for the subdivision of triangle meshes that can support fully dynamic selective refinement and is compatible with classical subdivision schemes based on the recursive quadrisection of triangles.

We have developed our selective refinement algorithm for the Loop subdivision scheme. However, an analogous algorithm can be developed similarly for the (modified) butterfly subdivision [DLG90, ZSS96]. In that case, subdivision is interpolating, thus each vertex has a fixed position and the data structure becomes simpler. Even vertices do not need any update, while the stencil for odd vertices is larger than in Loop subdivision, and needs a slightly more complex procedure to compute vertex position. In the case of Loop subdivision, it should be also possible to develop a closed form solution for the control points of even vertices, similar to that proposed in [Kob00], which would also allow us to use a simpler data structure. We plan to investigate these issues in our future work.

RGB triangulations are currently under implementation. We expect that our method will outperform red-green triangulations in terms of over-refinement, by maintaining a comparable visual quality. We also plan to compare the performance of RGB triangulations with respect to $\sqrt{3}$-subdivision and 4-8-subdivision, in terms of speed, storage space, visual quality and over-refinement.

Acknowledgments

This work has been partially supported by the European Network of Excellence AIM@SHAPE under contract number 506766, and by Project FIRB-MIUR SHALOM (SHApe modeLing and reasOning: new Methods and tools) funded by the Italian Ministry of Education, University and Research under contract number RBIN04HWR8.

References

[BSW83] BANK R., SHERMAN A., WEISER A.: Refinement algorithms and data structures for regular local mesh refinement. In *Scientific Computing*, Stepleman R., (Ed.). IMACS/North Holland, 1983, pp. 3–17.

[CDM*04] CIGNONI P., DE FLORIANI L., MAGILLO P., PUPPO E., SCOPIGNO R.: Selective refinement queries for volume visualization of unstructured tetrahedral meshes. *IEEE Transactions on Visualization and Computer Graphics 10*, 1 (January/February 2004), 141–159.

[DLG90] DYN N., LEVIN D., GREGORY J.: A butterfly subdivision scheme for surface interpolation with tension control. *ACM Transactions on Graphics 9*, 2 (April 1990), 160–169.

[DWS*97] DUCHAINEAU M., WOLINSKY M., SIGETI D., MILLER M., ALDRICH C., MINEEV-WEINSTEIN M.: ROAMing terrain: Real-time optimally adapting meshes. In *Proceedings IEEE Visualization '97* (Oct. 1997), IEEE, pp. 81–88.

[Hop96] HOPPE H.: Progressive meshes. In *SIGGRAPH 96 Conference Proceedings* (Aug. 1996), Annual Conference Series, ACM SIGGRAPH, Addison Wesley, pp. 99–108.

[Kob00] KOBBELT L.: $\sqrt{3}$ subdivision. In *Proceedings ACM SIGGRAPH 2000* (2000), pp. 103–112.

[Loo87] LOOP C.: Smooth subdivision surfaces based on triangles. Master thesis, University of Utah, Dept. of Mathematics, 1987.

[LRC*02] LÜBKE D., REDDY M., COHEN J., VARSHNEY A., WATSON B., HÜBNER R.: *Level Of Detail for 3D Graphics*. Morgan Kaufmann, 2002.

[Pup98] PUPPO E.: Variable resolution triangulations. *Computational Geometry 11*, 3-4 (1998), 219–238.

[Sab04] SABIN M.: Recent progress in subdivision: a survey. In *Advances in Multiresolution for Geometric Modelling*, Dogdson N., Floater M., Sabin M., (Eds.). Springer-Verlag, 2004, pp. 203–230.

[SHHG01] SEEGER S., HORMANN K., HÄUSLER G., GREINER G.: A sub-atomic subdivision approach. In *Proceedings of Vision, Modeling and Visualization 2001* (Berlin, 2001), Girod B., Niemann H., Seidel H.-P., (Eds.), Akademische Verlag, pp. 77–85.

[VG00] VELHO L., GOMES J.: Variable resolution 4-k meshes: Concepts and applications. *Computer Graphics Forum 19*, 4 (2000), 195–214.

[VZ01] VELHO L., ZORIN D.: 4-8 subdivision. *Computer-Aided Geometric Design 18* (2001), 397–427.

[WW02] WARREN J., WEIMER H.: *Subdivision Methods for Geometric Design*. Morgan Kaufmann, 2002.

[ZS00] ZORIN D., SCHRÖDER P. (Eds.): *Subdivision for Modeling and Animation (SIGGRAPH 2000 Tutorial N.23 - Course notes)*. ACM Press, 2000.

[ZSS96] ZORIN D., SCHRÖDER P., SWELDENS W.: Interpolating subdivision for meshes with arbitrary topology. In *Comp. Graph. Proc., Annual Conf. Series (SIGGRAPH 96)* (1996), ACM Press, pp. 189–192.

[ZSS97] ZORIN D., SCHRÖDER P., SWELDENS W.: Interactive multiresolution mesh editing. In *Comp. Graph. Proc., Annual Conf. Series (SIGGRAPH 97)*, ACM Press (1997). 259-268.

Eurographics Symposium on Geometry Processing (2006)
Konrad Polthier, Alla Sheffer (Editors)

Loop subdivision with curvature control

I. Ginkel[1] and G. Umlauf[1]

[1]Geometric Algorithms Group, Computer Science Department, University of Kaiserslautern, Germany

Abstract

In this paper the problem of curvature behavior around extraordinary points of a Loop subdivision surface is addressed. A variant of Loop's algorithm with small stencils is used that generates surfaces with bounded curvature and prescribed elliptic or hyperbolic behavior. We present two different techniques that avoid the occurrence of hybrid configurations, so that an elliptic or hyperbolic shape can be guaranteed.

The first technique uses a symmetric modification of the initial control-net to avoid hybrid shapes in the vicinity of an extraordinary point. To keep the difference between the original and the modified mesh as small as possible the changes are formulated as correction stencils and spread to a finite number of subdivision steps. The second technique is based on local optimization in the frequency domain. It provides more degrees of freedom and so more control over the global shape.

1. Introduction

Tuning has always been part of developing subdivision algorithms. Already the first publications dealing with subdivision algorithms for surfaces of arbitrary topology use the free parameters of the algorithms to improve the limit shape, e.g. [CC78, Loo87]. Later modifications of the eigenvalues in the frequency domain were used to achieve surfaces with zero or bounded Gauss curvature at the extraordinary points [Sab91, Hol95, PU98b, PU98a, Loo02, Loo03]. Since then, various sufficient conditions on the sub- and subsub-dominant eigenvalues were formulated to minimize polar artifacts or to ensure the ability to generate both elliptic and hyperbolic shapes [SB03, PR04]. Also conditions on the eigenfunctions are known that are necessary to achieve C^k smoothness at extraordinary points [Pra98]. Based on this more sophisticated approaches were developed that modify the eigenvectors to approximate these conditions in order to achieve optimized curvature behavior at extraordinary points [BK04].

When judging the behavior of the curvature near extraordinary points there are two aspects to deal with. The first is to ensure bounded curvature and the second is to avoid generation of so-called hybrid shapes, where neither the elliptic nor hyperbolic components become dominant during the subdivision process. These points with hybrid shape are one reason why subdivision surfaces are not widely used in CAD applications to construct high quality surfaces [KPR04]. Fur-

thermore, there is no simple method for a designer to tell if points with hybrid shape will occur. So it is necessary to detect these points and decide from the control-nets without user-interaction if the hybrid shape should be corrected to an elliptic or hyperbolic shape.

One technique to improve the curvature behavior has recently been presented in [ADS05, ADS06], where a bounded curvature subdivision algorithm was combined with an eigenvalue tuning minimizing the variation of curvature in the so-called shape charts. Though, this reduces the number of hybrid shapes it does not guarantee a prescribed elliptic or hyperbolic shape at the extraordinary points. Our goal is to guarantee that no hybrid shapes are generated. We split the problem into eigenvalue tuning to guarantee bounded curvature behavior and tuning of the eigencoefficients of the given input control-net to ensure purely elliptic or hyperbolic curvature in the vicinity of an extraordinary point.

After recalling the basic principles of analyzing subdivision algorithms and giving sufficient conditions on the modified algorithms in Section 2, we proceed with presenting a bounded curvature algorithm by eigenvalue tuning in Section 3. In Section 4 we propose a method to decide whether to correct the points with hybrid shape to either elliptic or hyperbolic shape. After that two techniques for tuning the eigencoefficients of control-nets are presented in Sections 5 and 6, followed by the conclusion and open problems in Section 7.

2. Analyzing subdivision algorithms

We consider a subdivision surface, which is generated by a stationary, linear and symmetric subdivision algorithm generalizing box- or b-spline subdivision. This allows the use of the standard analysis techniques.

The subdivision surface in the vicinity of an extraordinary point of order n corresponding to an irregularity in the initial mesh of order n can be regarded as the union of the extraordinary point $\mathbf{m}$ and a sequence of spline rings $\mathbf{x}_m$. Each spline ring is represented as a linear combination of real valued functions $\varphi_0, \ldots, \varphi_L$ with control-points $\mathbf{B}_m^0, \ldots, \mathbf{B}_m^L \in \mathbb{R}^3$. Combining the functions in a row vector φ and the control-points in a column vector $\mathbf{B}_m$, the m-th spline ring can be written as $\mathbf{x}_m = \varphi \mathbf{B}_m$. The sequence of control-points $\mathbf{B}_m$ is generated by iterated application of a square subdivision matrix A to the initial data $\mathbf{B}_0$

$$\mathbf{B}_m = A^m \mathbf{B}_0.$$

This yields for the spline rings $\mathbf{x}_m$

$$\mathbf{x}_m = \varphi A^m \mathbf{B}_0.$$

Assume that the subdivision matrix A has eigenvalues $\lambda_0, \ldots, \lambda_L$ with $|\lambda_0| \geq \cdots \geq |\lambda_L|$ corresponding to right eigenvectors $\mathbf{v}_0, \ldots, \mathbf{v}_L$ and linear independent eigenfunctions $\psi_i := \varphi \mathbf{v}_i$ for non-vanishing eigenvalues. Thus, $\mathbf{x}_m$ is represented as

$$\mathbf{x}_m = \sum_{i=0}^{L} \lambda_i^m \psi_i \mathbf{d}_i, \qquad \mathbf{d}_i \in \mathbb{R}^3.$$

For a symmetric subdivision algorithm the block-circulant matrix A can be transformed to a similar block-diagonal matrix $\hat{A}$ with diagonal blocks $\hat{A}_k$ by a discrete block-Fourier transformation F

$$\hat{A} = F^{-1}AF = \operatorname{diag}(\hat{A}_0, \ldots, \hat{A}_{n-1}).$$

The *Fourier index* of an eigenvalue ν of A is defined as

$$\mathcal{F}(\nu) := \{k \in \mathbb{Z}_n : \nu \text{ is eigenvalue of } \hat{A}_k\}.$$

Then, the following conditions are sufficient for the subdivision algorithm to generate regular surfaces with continuous normal and bounded Gauss curvature of arbitrary sign, see [RP06]:

1. All rows of A sum to one, i.e. $\lambda_0 = 1 > |\lambda_1|$.
2. The sub-dominant eigenvalue λ is positive and has algebraic and geometric multiplicity two, i.e.

$$\lambda := \lambda_1 = \lambda_2 > |\lambda_3|,$$

and the characteristic map $\Psi := (\psi_1, \psi_2)$ is injective and regular.
3. The subsub-dominant eigenvalue μ satisfies $\mu = \lambda^2$.
4. The subsub-dominant eigenvalue μ is positive and has algebraic and geometric multiplicity three, i.e.

$$\mu := \lambda_3 = \lambda_4 = \lambda_5 > |\lambda_6|,$$

with Fourier index 0, 2 and $n-2$.

The first condition ensures convergence, the second C^1-regularity, the third bounded curvature and the fourth allows for arbitrary elliptic or hyperbolic shapes. Unfortunately, the standard algorithms do not satisfy these conditions. They have to be modified to fulfill all criteria.

To analyze curvature of subdivision surfaces in more detail we follow [PR04]. Let L be the matrix that orthonormalizes the tangent directions $\mathbf{d}_1$ and $\mathbf{d}_2$. The spline ring $\mathbf{x}_c$ defined by

$$\mathbf{x}_c := (\Psi_c, \psi) \quad \text{with} \quad \Psi_c := \Psi L \quad \text{and} \quad \psi := \sum_{i=3}^{5} \psi_i \langle \mathbf{d}_i, \mathbf{n} \rangle$$

is called the *central surface* of the subdivision surface. It depends on the initial data and provides a tool for judging the behavior of the curvature around extraordinary points a priori. If K_c denotes the Gauss curvature of the central surface $\mathbf{x}_c$, the shape at the extraordinary point $\mathbf{m}$ for generic initial control-nets $\mathbf{B}_0$ can be categorized as

- elliptic in the limit, if $K_c > 0$,
- hyperbolic in the limit, if $K_c < 0$, and
- hybrid, if K_c changes sign.

Even with condition 4. extraordinary points with a hybrid shape can occur, see [KPR04].

3. A bounded curvature variant of Loop's algorithm

Loop's algorithm [Loo87] is a subdivision algorithm for triangular control-nets with vertices of arbitrary valence $n \geq 5$. It is conveniently described by its stencils which are shown in Figures 1(a) and 1(b). This subdivision algorithm satisfies conditions 1. and 2. but not 3. and 4. of Section 2. The relevant eigenvalues for these conditions are

$$\begin{aligned}
\mu_0 &= 5/8 - n\beta \quad \text{and} \\
\mu_i &= 3/8 + c_i/n, \qquad i = 1, \ldots, n-1,
\end{aligned}$$

with $c_i := \cos(2\pi i/n)$ for an n-valent vertex. Note that $\mathcal{F}(\mu_i) = i$ for $i = 0, \ldots, n-1$ and $\mu_1 = \mu_{n-1} > \mu_j$ for $j \neq 1, n-1$ if β is appropriately chosen.

To satisfy condition 4. a triple subsub-dominant eigenvalue $\mu_0 = \mu_2 = \mu_{n-2}$ with Fourier indices 0, 2 and $n-2$ can be achieved by an appropriate choice of the parameter β, see [KPR04]. However, this subdivision algorithm still generates surfaces with unbounded curvature. To satisfy also condition 3. we use the technique in [PU98a, PU98b].

The eigenvalues μ_i for $i = 2, \ldots, n-2$ need to be changed to $\tilde{\mu}_i$ such that

$$\begin{aligned}
\tilde{\mu}_j = \mu_j + \delta_j = \mu_1^2 \quad &\text{for } j = 2, n-2 \quad \text{and} \\
\tilde{\mu}_j = \mu_j + \delta_j < \mu_1^2 \quad &\text{for } j = 3, \ldots, n-3.
\end{aligned}$$

One possible choice for δ_j is

$$\delta_2 := \delta_{n-2} := \mu_1^2 - \mu_2 \qquad \text{and}$$
$$\delta_j := 1/16 - \mu_j \qquad \text{for } j = 3, \ldots, n-3.$$

Using the additional stencil in Figure 1(c) with

$$\gamma_i = f_i + \frac{2}{n} \sum_{j=2}^{n-2} \delta_j c_{ij}, \qquad i = 0, \ldots, n-1,$$

and

$$f_i = \begin{cases} 3/8 & \text{, for } i = 0 \\ 1/8 & \text{, for } i = 1, n-1 \\ 0 & \text{, for } i = 2, \ldots, n-2 \end{cases}$$

condition 3. can be satisfied without affecting conditions 1. and 2. The additional stencil in Figure 1(c) is only used on edges emanating from an irregular vertex. For all the other vertices the usual stencils of Loop's algorithm in Figures 1(a) and 1(b) are used. For the stencil in Figure 1(b) to satisfy condition 4. the parameter β must be set to

$$\beta = \frac{31 - 12c_1 - 4c_1^2}{64n}.$$

Thus, these stencils represent a variant of Loop's algorithm that generates C^1-regular surfaces with bounded Gauss curvature of possibly arbitrary sign.

4. Analyzing and categorizing the initial control-net

To visualize the potential shapes a subdivision algorithm can generate Karciauscas et al. [KPR04] propose the so-called shape charts. If the subdivision matrix has a triple subsubdominant eigenvalue with Fourier index 0, 2 and $n-2$ the third coordinate function ψ of the central surface $\mathbf{x}_c$ can be written as

$$\psi = \sum_{j=3}^{5} \psi_j \langle \mathbf{d}_j, \mathbf{n} \rangle =: a_3 \psi_3 + a_4 \psi_4 + a_5 \psi_5.$$

Here, $\mathbf{n}$ is the surface normal at $\mathbf{m}$ and $a_j := \langle \mathbf{d}_j, \mathbf{n} \rangle$ is the normal component of the eigencoefficient $\mathbf{d}_j$ of the initial control-net $\mathbf{B}_0$. In order to represent all possible shapes categorized by the behavior of K_c the coefficients a_j can be interpreted as barycentric coordinates. The barycentric coordinates $(1,0,0)$ represent the elliptic shape with $K_c > 0$ and the barycentric coordinates $(0,1,0)$ and $(0,0,1)$ represent the hyperbolic shape with $K_c < 0$. Now, for every (a_3, a_4, a_5) the point in the triangle can be colored according to its shape category where red encodes elliptic, green hybrid and blue hyperbolic shapes. This image is the so-called *shape chart*. This concept can easily be transferred to shape charts in polar coordinates [ADS05], which will be used in this paper. Examples of shape charts for the bounded curvature variant of Loop's algorithm of Section 3 are shown in Figure 2.

In order to control the curvature of the subdivision surface at the extraordinary point the hybrid shapes must be corrected to either elliptic or hyperbolic shapes. Therefore,

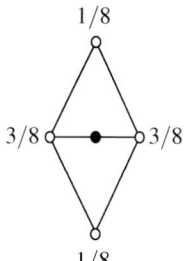

(a) *The stencil of Loop's algorithm for edge-points.*

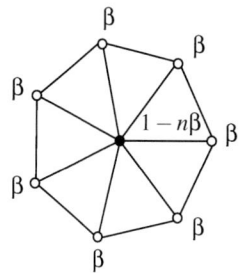

(b) *The stencil of Loop's algorithm for vertex-points of valence n.*

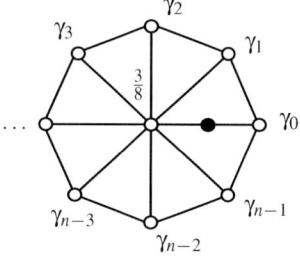

(c) *The additional stencil of the bounded curvature variant of Loop's algorithm for edge-points near a vertex of valence n.*

Figure 1: *Stencils of Loop's algorithm ((a) and (b)) and its bounded curvature variant ((a), (b) and (c)).*

it is necessary to decide what the desired shape is which is represented by the initial control-net $\mathbf{B}_0$. This requires a procedure to find the closest non-hybrid configuration.

If the subdivision surface corresponding to $\mathbf{B}_0$ has a hybrid shape there are several techniques for this decision:

1. Calculate for the pixel in the shape chart corresponding to $\mathbf{B}_0$ the closest non-hybrid pixel.
2. Calculate the curvature of a quadratic least squares fit to the one-ring neighborhood of the irregular vertex in $\mathbf{B}_0$.
3. Calculate the curvature of a quadratic least squares fit to the central surface of $\mathbf{B}_0$.

For the first technique an appropriate 2D distance function on the pixels of the shape chart is necessary. The advantage is that this is computationally simple and fast if the shape chart has been calculated a priori. The disadvantage is that the accuracy is limited by the resolution of the shape chart.

The other two techniques focus on estimating the average quadratic behavior of the surface. Then the sign of the Gaussian curvature of the quadratic fit indicates the desired shape. The second technique focuses on the average global quadratic behavior whereas the third tries to reproduce the

Figure 2: *Shape charts in polar coordinates for the bounded curvature variant of Loop's algorithm of Section 3 for valences 5, 6, 7 and 15. Red encodes elliptic, green hybrid and blue hyperbolic shapes.*

average quadratic behavior in the vicinity of the extraordinary point. Both techniques do not give a specific choice of d_3, d_4, d_5 to guarantee a non-hybrid shape. In combination with the first technique they give a search direction in the shape chart for a non-hybrid pixel.

Remark 1 The tuning method in Section 3 changes the right eigenvectors v_2 and v_{n-2} corresponding to μ_2 and μ_{n-2}. This must be considered for the shape chart analysis.

5. A symmetric technique to manipulate the limit shape

A symmetric approach to control the shape in the vicinity of an extraordinary point is based on the observation that for Loop's algorithm changing the position of the central vertex only affects the limit point d_0 and the coefficient d_3 contributing to the elliptic components of the central surface. The directions d_1 and d_2 spanning the tangent plane as well as d_4 and d_5 defining the hyperbolic components of the central surface are unchanged. This is because of the special structure of the left eigenvectors. In case d_3, d_4 and d_5 represent a hybrid shape, we modify the position of the central vertex such that $\tilde{d}_3$, d_4 and d_5 guarantee a non-hybrid shape.

Assume that the left eigenvectors w_k corresponding to d_k are scaled such that $w_k v_k = 1$ and that the components of w_k corresponding to the one-ring neighborhood $c_1, \ldots, c_n$ around an irregular vertex c_0 are equal to one. For the bounded curvature variant of Loop's algorithm of Section 3 this choice is possible. Then the eigencoefficient d_k is com-

puted as

$$d_k = w_k \cdot [c_0, \ldots, c_n], \qquad k = 0, \ldots, 5.$$

For $k = 3$ the left eigenvector w_3 is of the form of $[n, 1, \ldots, 1]$ which yields for d_3

$$d_3 = \sum_{i=1}^{n} c_i - n c_0.$$

In order to change d_3 to $\tilde{d}_3$ by changing only the irregular vertex c_0 to $\tilde{c}_0$, the corrected eigencoefficient $\tilde{d}_3$ is given by

$$\tilde{d}_3 = \sum_{i=1}^{n} c_i - n \tilde{c}_0.$$

where

$$\tilde{c}_0 = \frac{1}{n} \cdot (d_3 - \tilde{d}_3) + c_0.$$

Restricting the change of d_3 to scaling by α yields

$$\begin{aligned}
\tilde{c}_0 &= \frac{1}{n} \cdot (d_3 - \alpha d_3) + c_0 \\
&= \frac{1 - \alpha}{n} \left(\sum_{i=1}^{n} c_i - n c_0 \right) + c_0 \\
&= \frac{1 - \alpha}{n} \sum_{i=1}^{n} c_i + \alpha c_0.
\end{aligned}$$

This can be written in a stencil where α is computed as in Section 4, shown in Figure 3.

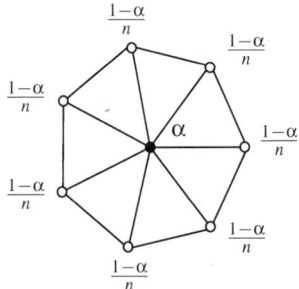

Figure 3: *The stencil for the correction.*

So, the principal of the symmetric technique for manipulating the limit shape can be summarized in the following procedure:

1. Subdivide the initial control-net once.
2. Calculate the eigencoefficients d_3, d_4, d_5 and decide with the shape chart, if a hybrid shape will occur.
3. If a vertex generates a hybrid shape, compute α as in Section 4 and use the correction stencil in Figure 3.
4. Subdivide with the bounded curvature variant of Loop's algorithm of Section 3 without further correction.

Remark 2 The initial subdivision step is necessary, since valid eigencoefficients d_3, d_4, d_5 for Loop's algorithm and its modified variant can only be calculated with a one-ring neighborhood of regular vertices around an irregular vertex.

The fastest possible correction is to set $\alpha = 0$ and thus to completely eliminate all elliptic components by one single correction. So only hyperbolic shapes can be generated.

To decrease the distance between the original and the corrected surface the correction can be slowed down. Therefore, it is possible to choose in the stencil of Figure 3 a fix $\alpha > 1$ for convergence towards an elliptic shape or a fix $\alpha < 1$ for convergence towards a hyperbolic shape. Then a finite number of subdivision and correction steps, i.e. iterating $1. - 3.$ in the above procedure, is used to achieve a non-hybrid shape.

Examples for this are shown in Figures 4, 5, 6 and 7. Two hybrid surfaces and their control-net are shown in Figures 4 and 5. For the visualization of the surface the control-net is subdivided 7 times and shown flat-shaded. A visualization of the Gauss curvature for different corrections applied to these control-nets is shown in Figures 6 and 7. The control-nets are subdivided 10 times and converted to Bézier representation to compute the Gauss curvature K, which is converted to the color $H = 120(1 - \arctan(K)/2\pi)$, $S = 1$ and $V = 1$ in the HSV color model. Here, differences can be clearly observed. The different choices of α show how fast the elliptic component is blended out by the additional scaling of $\mathbf{d}_3$. A choice of α close to one imposes a smaller change to the surface, but requires more subdivision and correction steps to achieve non-hybrid shape. Pictures of the corresponding shaded surfaces show no visible difference.

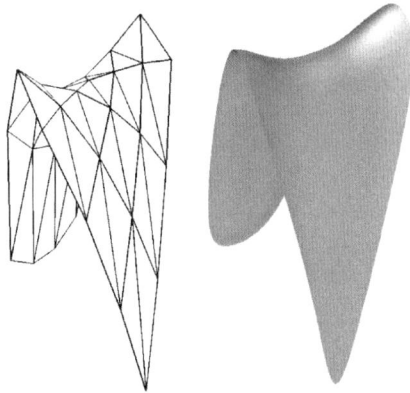

Figure 4: *A control-net with a 5-valent vertex and the corresponding hybrid surfaces generated by the bounded curvature variant of Loop's algorithm.*

Remark 3 The same correction stencil can be applied to the usual algorithm of Loop with β chosen such that $\mu_\beta = \mu_2 = \mu_{n-2}$ resulting in zero Gauss curvature for valence 5 and strictly positive or strictly negative unbounded curvature for valences ≥ 7.

Remark 4 For the Catmull-Clark algorithm $\mathbf{d}_1, \mathbf{d}_2, \mathbf{d}_4$ and $\mathbf{d}_5$ also do not depend on the extraordinary vertex. Therefore

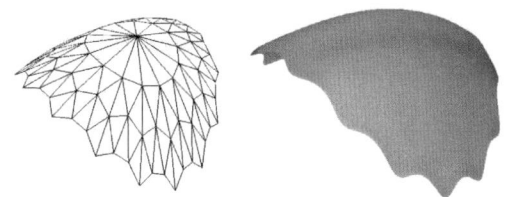

Figure 5: *A control-net with a 15-valent vertex and the corresponding hybrid surfaces generated by the bounded curvature variant of Loop's algorithm.*

a similar correction stencil depending on the one-ring neighborhood around the extraordinary vertex can be derived to control $\mathbf{d}_3$.

This approach only modifies the position of the irregular vertex. This means it is very local and does not give much control over the behavior of the surface away from the extraordinary point. To make the changes less local and gain more control over the global shape it is necessary to control all sub-dominant eigencoefficients and at least to incorporate the one-ring neighborhood of the irregular vertex into the modification.

6. Manipulating the limit shape by local optimization

Recall that the coefficients $\mathbf{d}_k, k = 1, \ldots, 5$, can be calculated with help of the corresponding left eigenvectors $\mathbf{w}_k$ and depend only on the central vertex $\mathbf{c}_0$ and the one-ring neighborhood $\mathbf{c}_1, \ldots, \mathbf{c}_n$. Let $\mathbf{w}_{k,i}$ be the i-th component of $\mathbf{w}_k$ corresponding to the control-point $\mathbf{c}_i$. Then the calculation of $\mathbf{d}_k$ can be written as

$$\mathbf{d}_k = \sum_{i=0}^{n} \mathbf{w}_{k,i} \cdot \mathbf{c}_i,$$

or in a matrix-vector notation

$$L \cdot \mathbf{c} = \mathbf{d}$$

with $L = [\mathbf{w}_{k,i}]_{k=0,\ldots,5}$, $\mathbf{c} = [\mathbf{c}_0 \ldots \mathbf{c}_n]^T$ and $\mathbf{d} = [\mathbf{d}_0 \ldots \mathbf{d}_5]^T$. Regarding this as a system of equations with given matrix L and right hand side $\mathbf{d}$ the system has an exact solution for valence 5 and is under-determined for valences $n > 5$.

In case a given set of coefficients $\mathbf{d}_k$ represents a hybrid shape, we change $\mathbf{d}_3, \mathbf{d}_4$ and $\mathbf{d}_5$ to

$$\mathbf{d}_3' = \mathbf{d}_3 + \kappa_3 \cdot \mathbf{n},$$
$$\mathbf{d}_4' = \mathbf{d}_4 + \kappa_4 \cdot \mathbf{n},$$
$$\mathbf{d}_5' = \mathbf{d}_5 + \kappa_5 \cdot \mathbf{n}$$

such that the new components in normal direction $a_3' = \langle \mathbf{d}_3', \mathbf{n} \rangle$, $a_4' = \langle \mathbf{d}_4', \mathbf{n} \rangle$ and $a_5' = \langle \mathbf{d}_5', \mathbf{n} \rangle$ generate a non-hybrid surface. The scalar factors κ_3, κ_4 and κ_5 are determined as in Section 4. Note that the change of $\mathbf{d}_3, \mathbf{d}_4$ and $\mathbf{d}_5$ is restricted to normal components. Tangential components would not

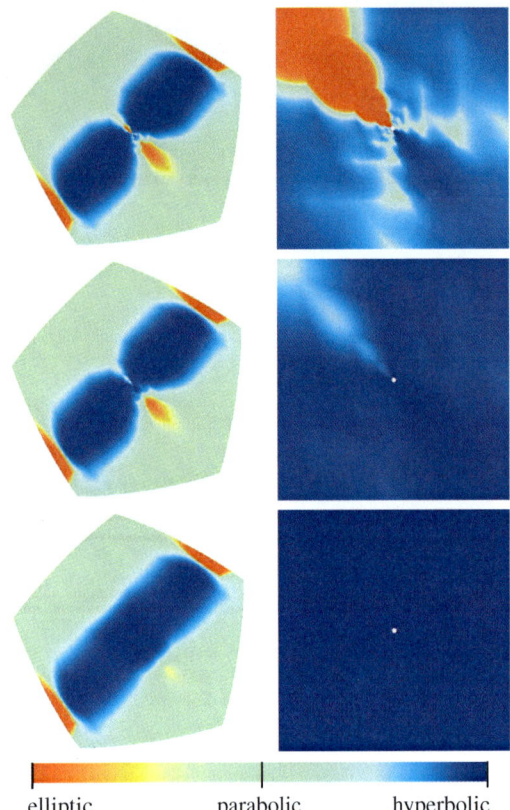

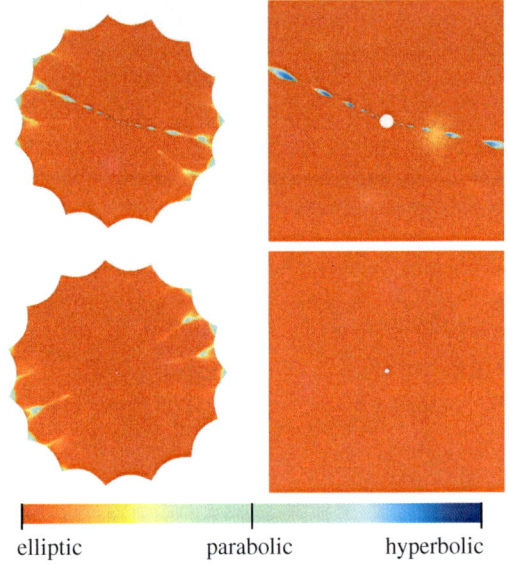

elliptic   parabolic   hyperbolic

Figure 7: *Visualization of the Gauss curvature of the uncorrected hybrid surface (top) of Figure 5 and the surface generated with correction to an elliptic shape with $\alpha = 1.2$ in the first three steps and no correction in the subsequent steps (bottom). The right column shows the corresponding zoom-in at the extraordinary points after 10 subdivision steps.*

elliptic   parabolic   hyperbolic

Figure 6: *Visualization of the Gauss curvature of the uncorrected hybrid surface (top) of Figure 4, the surface generated with slow correction to a hyperbolic shape with $\alpha = 0.95$ in every step (middle) and the surface generated with fast correction to a hyperbolic shape with $\alpha = 0.0$ (bottom). The right column shows the corresponding zoom-in at the extraordinary points after 10 subdivision steps.*

change the corresponding central surface and the scalar factors κ_3, κ_4 and κ_5, but are omitted here for simplicity.

Now, the system of equations is changed to $L \cdot \mathbf{c}' = \mathbf{d}'$ with $\mathbf{d}' = [\mathbf{d}_0, \mathbf{d}_1, \mathbf{d}_2, \mathbf{d}_3', \mathbf{d}_4', \mathbf{d}_5']$. We are now looking for a solution to this system of equations such that the new control-points $\mathbf{c}'$ have minimal distance to the original control-points. This is achieved for the solution $\mathbf{c}'$ that minimizes $\|\mathbf{h}\|$ with $\mathbf{h} := \mathbf{c}' - \mathbf{c}$. Thus, we have to solve

$$L \cdot \mathbf{c}' = L \cdot (\mathbf{h} + \mathbf{c}) = \mathbf{d}'$$

for $\mathbf{h}$. If L^+ is the Moore-Penrose inverse of L, the solution $\mathbf{h}$ has minimal norm if

$$\mathbf{h} = L^+ (\mathbf{d}' - L \cdot \mathbf{c}).$$

This is equivalent to

$$\mathbf{c}' = \mathbf{c} + L^+ (\mathbf{d}' - L \cdot \mathbf{c}).$$

So we get a set of control-points $\mathbf{c}'$ which solve the system $L \cdot \mathbf{c}' = \mathbf{d}'$ and minimize $\sum_i \|\mathbf{c}_i' - \mathbf{c}_i\|^2$.

If we replace the original control-points $\mathbf{c}$ with the new control-points $\mathbf{c}'$ subdivision will result in a surface that has the same limit point $\mathbf{d}_0$, the same directions $\mathbf{d}_1$ and $\mathbf{d}_2$ spanning the tangent plane, but avoids hybrid shapes.

Figures 8 and 9 show the results of applying this optimization technique. The color coding and the shading of the surface is the same as in Section 5.

The eigencoefficients of the new control-net shows a significant change of eigencoefficients of eigenvalues smaller than μ_i, $i = 0, 2, n - 2$. Incorporating also these eigencoefficients to the system of equations avoids this and shifts the change to eigenvalues of magnitude 0 of Fourier index $0, 2, n - 2$. This extends the influence to the two-ring neighborhood but gives more control over the shape of the resulting surface. It might be necessary to subdivide the initial control-net one more time to separate the irregular vertices for this extension. Figure 10 shows the greater impact on the overall shape. It is even visible in the images of the control-nets and the shaded surfaces. Note that the correction κ_4, κ_5 is the same as in Figure 8.

Remark 5 Similar to the change of $\mathbf{d}_3$, $\mathbf{d}_4$ and $\mathbf{d}_5$ a change of $\mathbf{d}_0$ to $\mathbf{d}_0'$ and $\mathbf{d}_1$ to $\mathbf{d}_1' = \alpha_1 \mathbf{d}_1 + \beta_1 \mathbf{d}_2$ and $\mathbf{d}_2$ to $\mathbf{d}_2' = \alpha_2 \mathbf{d}_1 + \beta_2 \mathbf{d}_2$ does not change the normal $\mathbf{n}$, which is essen-

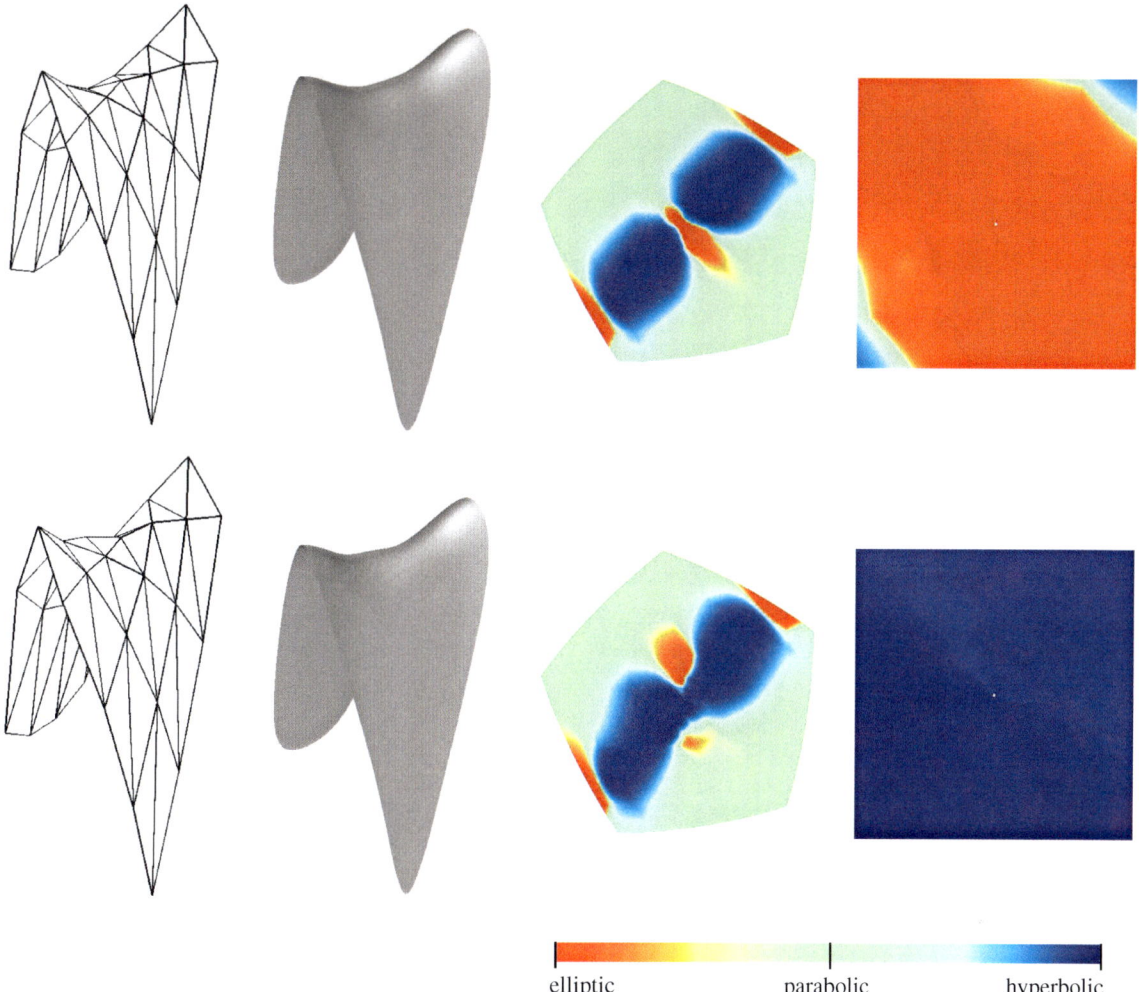

elliptic   parabolic   hyperbolic

Figure 8: *Control-net, flat-shaded surface and visualization of the Gauss curvature of the surface corresponding to the surface in Figure 4 corrected to an elliptic shape using $\kappa_3 = 0.12, \kappa_4 = \kappa_5 = 0$ (top) and to a hyperbolic shape using $\kappa_3 = 0, \kappa_4 = \kappa_5 = 0.12$ (bottom). The right column shows the zoom-in at the extraordinary points after 10 subdivision steps. Compare with the control-net and flat-shaded surface in Figure 4 and the visualization of the Gauss curvature in Figure 6 (top) of the surface with hybrid shape.*

tial for setting up the central surface. These degrees of freedom could also improve $\|\mathbf{h}\|$, but are unused here.

Remark 6 This technique only works if irregular vertices are sufficiently far away from each other, so that changing positions in the one-ring neighborhood only affects the eigencoefficients of one irregular vertex. This can be achieved by subdividing the initial control-net twice.

Remark 7 The proposed technique is similar to the interpolation problem in [HKD93]. The difference is that not only the position of the limit point, but also tangent directions and quadratic behavior are interpolated.

7. Conclusion

We have presented a modified subdivision algorithm that can produce surfaces with arbitrary positive or negative Gauss curvature. Occurrence of hybrid shapes is avoided by modifying the eigencoefficients. The symmetric modification induces only minimal changes to the shape and therefore suits applications in which the modified surface should differ as little as possible from the original surface. If it is necessary to control the position of the limit point corresponding to the extraordinary vertex, for example for an interpolation problem, it is useful to apply the one-ring neighborhood variant of the local optimization. If the shape of the original surface

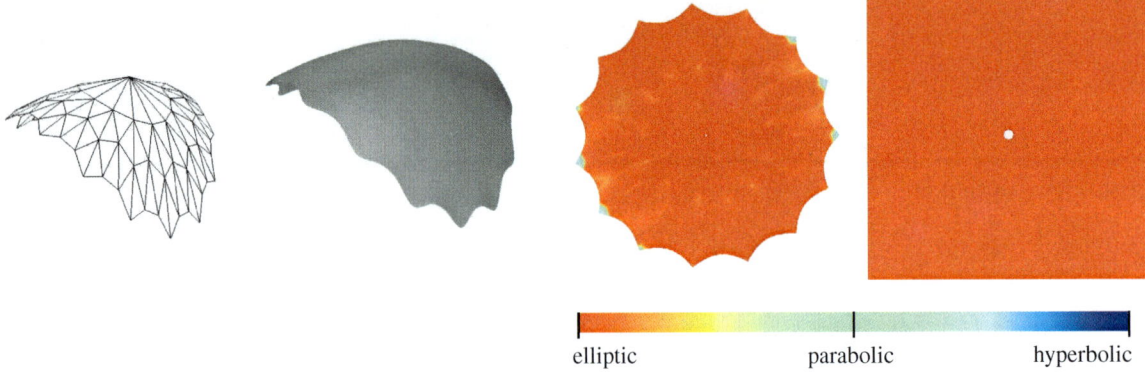

elliptic   parabolic   hyperbolic

Figure 9: *Control-net, flat-shaded surface and visualization of the Gauss curvature of the surface corresponding to the surface in Figure 5 corrected to an elliptic shape using $\kappa_3 = 0.02, \kappa_4 = \kappa_5 = 0$. The right column shows the zoom-in at the extraordinary point after 10 subdivision steps. Compare with the control-net and flat-shaded surface in Figure 5 and the visualization of the Gauss curvature in Figure 7 (top) of the surface with hybrid shape.*

seems unsatisfactory and much control over the shape should be achieved, the extension to the two-ring neighborhood is a good choice.

There are three open questions that we will address in the future. The first is the treatment of special cases for example when two irregular vertices influence each other. The second is to incorporate the unused degrees of freedom for the change of the eigencoefficients in Section 6 into the optimization process and the third is to quantify how much the modified surfaces differ from the surface generated from the unmodified initial control-net.

References

[ADS05] U. H. Augsdörfer, N. A. Dodgson, and M. A. Sabin. A new way to tune subdivision. In M. Desbrun and H. Pottmann, editors, *Eurographics Symposium on Geometry Processing*, 2005.

[ADS06] U. H. Augsdörfer, N. A. Dodgson, and M. A. Sabin. Tuning subdivision by minimising the Gaussian curvature variation near extraordinary vertices. *Computer Graphics Forum*, 25(3), 2006. Proc. Eurographics.

[BK04] L. Barthe and L. Kobbelt. Subdivision scheme tuning around extraordinary vertices. *Computer Aided Geometric Design*, 21(6):561–583, 2004.

[CC78] E. Catmull and J. Clark. Recursive generated b-spline surfaces on arbitrary topological meshes. *Computer-Aided Design*, 10:350–355, 1978.

[HKD93] M. Halstead, M. Kass, and T. DeRose. Efficient, fair interpolation using catmull-clark surfaces. In *Computer Graphics*, ACM SIGGRAPH '93 Proceedings, pages 35–44, 1993.

[Hol95] F. Holt. Towards a curvature continuous stationary subdivision algorithm. *Z.Angew.Math.Mech.*, 76:423–424, 1995.

[KPR04] K. Karciauscas, J. Peters, and U. Reif. Shape characterization of subdivision surfaces - case studies. *Computer Aided Geometric Design*, 21(6):601–614, 2004.

[Loo87] C. Loop. Smooth subdivision surfaces based on thiangles. Master's thesis, Department of Mathematics, University of Utah, 1987.

[Loo02] C. Loop. Bounded curvature triangle mesh subdivision with the convex hull property. *The Visual Computer*, 18(5-6):316–325, 2002.

[Loo03] C. Loop. Smooth ternary subdivision of triangle meshes. In A. Cohen, J.-L. Merrien, and L.L. Schumaker, editors, *Curve and Surface Fitting*, pages 295–302, Saint-Malo, 2003.

[PR04] J. Peters and U. Reif. Shape characterization of subdivision surfaces - basic principles. *Computer Aided Geometric Design*, 21(6):585–599, 2004.

[Pra98] H. Prautzsch. Smoothness of subdivision surfaces at extraordinary points. *Adv. in Comp.Math.*, 9:377–389, 1998.

[PU98a] H. Prautzsch and G. Umlauf. A G^2-subdivision algorithm. In *Geometric Modelling, Computing 13*, pages 217–224, 1998.

[PU98b] H. Prautzsch and G. Umlauf. Improved triangular subdivision schemes. In *Proceedings of the CGI '98*, pages 626–632, 1998.

[RP06] U. Reif and J. Peters. Structural analysis of subdivision surfaces – a summary. In K. Jetter, M. Buhmann, W. Haussmann, and R. Schaback andJ. Stöckler, editors,

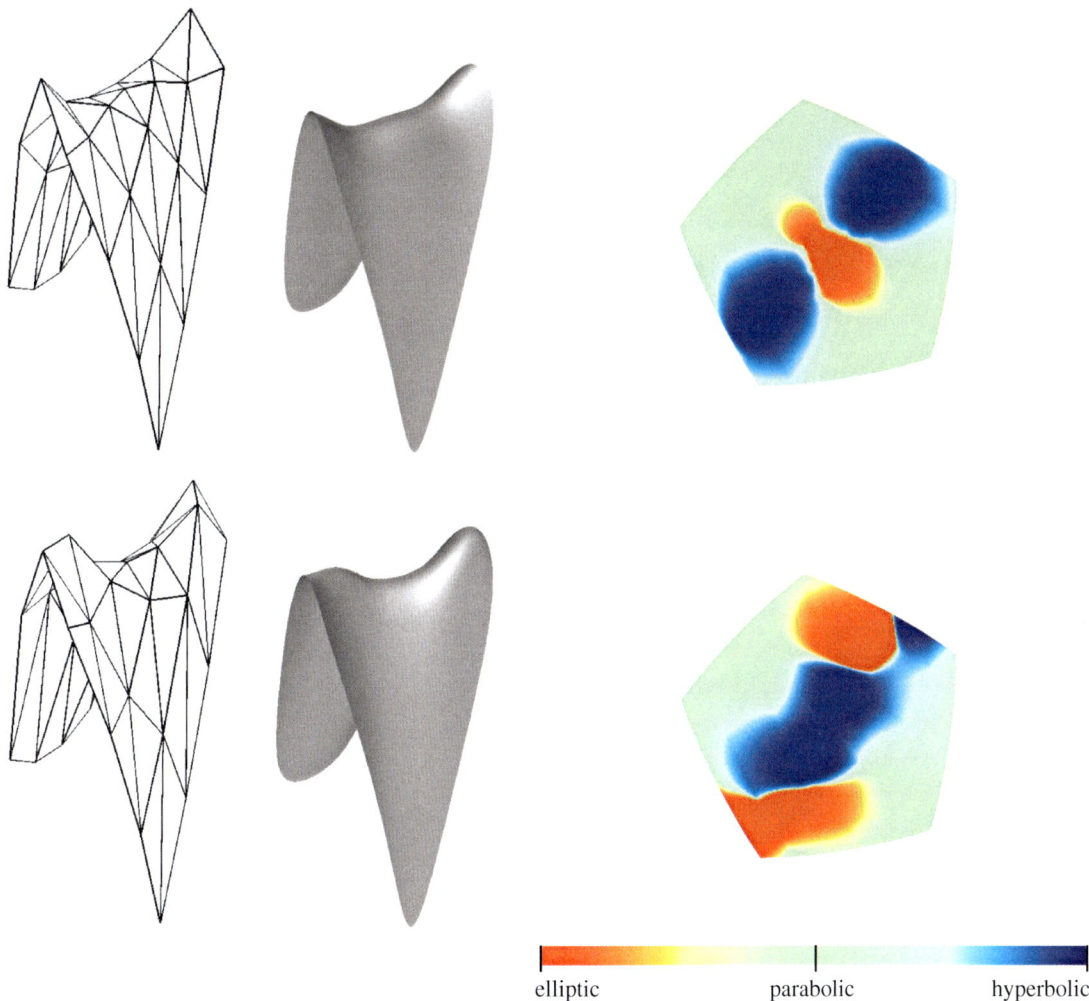

elliptic   parabolic   hyperbolic

Figure 10: *Control-net, flat-shaded surface and visualization of the Gauss curvature of the surface corresponding to the surface in Figure 4 corrected to an elliptic shape using $\kappa_3 = 0.12, \kappa_4 = \kappa_5 = 0$ (top) and to a hyperbolic shape using $\kappa_3 = 0, \kappa_4 = \kappa_5 = 0.12$ (bottom) with the two-ring neighborhood extension. The right column shows the zoom-in at the extraordinary points after 10 subdivision steps. Compare with the control-nets and flat-shaded surfaces in Figures 4 and 8 and the visualization of the Gauss curvature in Figures 6 (top) and 8.*

Topics in multivariate approximation and interpolation. Elsevier, 2006.

[Sab91] M.A. Sabin. Cubic recursive division with bounded curvature. In P.J. Laurent, A. Le Méhauté, and L.L. Schumaker, editors, *Curves and Surfaces*, pages 411–414, 1991.

[SB03] M.A. Sabin and L. Barthe. Artifacts in recursive subdivision surfaces. In A. Cohen, J.-L. Merrien, and L.L. Schumaker, editors, *Curve and Surface Fitting*, pages 353–362, Saint-Malo, 2003.

Eurographics Symposium on Geometry Processing (2006)
Konrad Polthier, Alla Sheffer (Editors)

A C^2 Polar Jet Subdivision

K. Karčiauskas[0] and A. Myles[1] and J. Peters [†1]

[0] University of Vilnius   [1] University of Florida

Abstract

We describe a subdivision scheme that acts on control nodes that each carry a vector of values. Each vector defines partial derivatives, referred to as jets in the following and subdivision computes new jets from old jets. By default, the jets are automatically initialized from a design mesh. While the approach applies more generally, we consider here only a restricted class of design meshes, consisting of extraordinary nodes surrounded by triangles and otherwise quadrilaterals with interior nodes of valence four. This polar mesh structure is appropriate for surfaces with the combinatorial structure of objects of revolution and for high valences.

The resulting surfaces are curvature continuous with good curvature distribution near extraordinary points. Near extraordinary points the surfaces are piecewise polynomial of degree (6,5), away they are standard bicubic splines.

Categories and Subject Descriptors (according to ACM CCS): I.3.5 [Computer Graphics]: Curve, Surface, Solid, and Object Representations

1. Introduction

We present a subdivision algorithm for a restricted class of meshes as depicted in Figure 1. This *polar mesh structure* [KP06b], while very special, is natural for meshes with the combinatorial structure of surfaces of revolution and locally, for many-sided blends and vertices of high valence. Often, for example for reflective surfaces such as inside car headlights, exactly these high-valence blends require good curvature distribution. However, for standard subdivision schemes, high valence leads to visibly poor shape as illustrated in Figure 2.Indeed, standard schemes have been shown to generate saddles even though the initial control net has a convex triangulation [KPR04]. While it is notoriously difficult to argue that a scheme results in high-quality surfaces, or even to define high quality, the proposed scheme does not exhibit the high-valence flaws observed for standard schemes.

Polar jet subdivision has the following properties.
1. Linear, stationary, affine invariant refinement of a control structure.
2. Control nodes of arbitrary valence.
3. Generates *curvature continuous* surfaces that

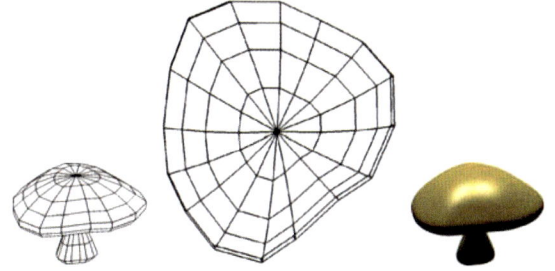

Figure 1: *A polar design mesh: the extraordinary vertex is surrounded by triangles. All quadrilaterals have nodes of valence four.*

4. do not have obvious shape limitations for high valence and
5. can be represented as a sequence of polynomial pieces of degree (6,5).

While the user manipulates a *design mesh*, the scheme refines vectors, called *2-jets*, that are associated with nodes. The approach generalizes bi-cubic spline subdivision to a polar mesh. Polar subdivision schemes can be designed to apply few rules with small footprint directly to the design mesh, as in standard subdivision, but such simplicity results in poorer shape and smoothness when compared with the proposed jet subdivision scheme.

†   supported by NSF DMI-0400214 and CCF-0430891

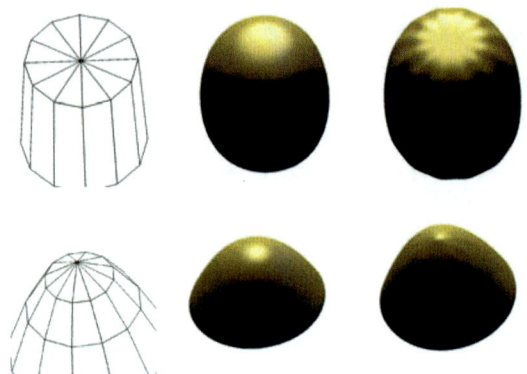

Figure 2: *(left) Polar control net, (middle) polar jet subdivision surface, (right) Catmull-Clark subdivision surface applied to the control net with the central node removed; similar ripples occur near the extraordinary point of the mushroom in Figure 1, also when rendered with Loop's subdivision.*

2. Background

Curvature continuous schemes. A number of piecewise polynomial and even subdivision constructions are known to generate curvature continuous surfaces, for example, as an incomplete listing, [GH89, Pra97, Rei98, Pet02, Loo04]; also a variety of nonpolynomial schemes, based on two Catmull-Clark refinement steps, yield C^2 and even smoother surfaces [CNG00, GH95, YZ04, Lev06]. Due to averaging, shape problems occur in the transition between the 'regular regions' that make up the bulk of the surfaces and the immediate neighborhood of extraordinary points of high valence. Recently, Guided Subdivision [KP06a] suggested a transition by sampling (rather than averaging with) a high-degree spline cap that serves as a shape guide. Polar jet subdivision is inspired by this approach but uses *local jets in place of an explicit guide surface*. This makes the neighborhood of the extraordinary point more responsive to the shape in the adjacent tensor-product region of the surface, and it makes the refinement look more like standard subdivision.

Subdivision surfaces Any of [ZS00, WW02, RP05] give a good introduction to subdivision surfaces. To date no general subdivision algorithm is published that generates everywhere C^2-surfaces without obvious shape deficiencies, restrictions at the extraordinary point or on the overall control net: for example, the scheme [PU98] creates and leverages zero curvature at extraordinary points, [LL03, SW05] join two different subdivision schemes smoothly along edges only and [ZLLT06] creates C^2 subdivision functions with a single extraordinary point and special transitions across edges of the control net. Approaches that associate multiple values and functions with each node, called Hermite or jet subdivision have shown promise in a special case of va-

lence three [XYD06], but do not yet improve the setup at extraordinary points, in general. Tensored circle preserving schemes [MWW01, SD05], double the valence at the poles with each refinement step. In polar subdivision the valence stays fixed.

3. Jet Initialization and Refinement

A *polar design mesh* consists of extraordinary nodes, surrounded by one layer of triangles, and of quadrilaterals with nodes of valence four otherwise (see Figures 1, 2). The extraordinary nodes need only be separated by one layer of nodes of valence four: for example, two pyramids with their polygonal bases joined yield a valid polar design mesh.

Quadrilaterals in the input net will be interpreted as part of a standard uniform bicubic B-spline control net (with adjacent triangles viewed as degenerate quadrilaterals). Section 4 explains how bicubic B-spline subdivision can be interpreted as jet subdivision. Triangular facets are, in each subdivision step, split into a smaller triangle attached to the extraordinary node and a quad. The quads form a ring with jets that, in the mth itera-

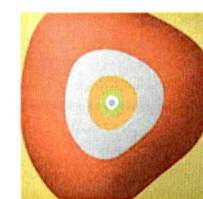

Figure 3: *A nested sequence of surface rings* $\mathbf{x}^1$, $\mathbf{x}^2$, ... *converging towards the extraordinary point.*

tion, define a surface ring $\mathbf{x}^m$ (see Figure 3). The bulk of this paper explains how to define and refine the jets associated with the triangular facets.

The actual refinement of jets is hidden from the end user. A designer manipulates a familiar design control mesh, where the nodes have only position. The jets are automatically generated and refined (see Sections 3.2 and 3.3).

3.1. Refinement per sector

The refinement of a subdivision control net can be *localized* by splitting the control net into overlapping, consistently refined subnets. For example, Catmull-Clark subdivision can be localized to subnets consisting of one quadrilateral and its neighbors. For polar jet subdivision, (one coordinate of) the surface corresponding to triangle i is defined by a subnet of four jets (see Figure 4)

$$\mathbf{c}_i^0 \in \mathbb{R}^6, \ \mathbf{v}_i^0 \in \mathbb{R}^9,$$
$$\mathbf{v}_{i+1}^0 \in \mathbb{R}^9, \ \mathbf{e}_i^0 \in \mathbb{R}^3.$$

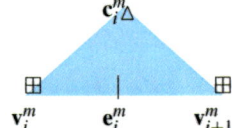

Figure 4: *The four jets forming* $\mathbf{j}_i^0$ *defining sector i. The icon $\boxplus$ stands for the 3×3 grid of entries* $\mathbf{v}_i^0$*, the icon $|$ for jets* $\mathbf{e}_i^0$ *and $\triangle$ for* $\mathbf{c}_i^0$*.*

As shown in Figure 5, the central jet $\mathbf{c}_i^0$ is represented as a 'triangular' Bézier mesh of depth two i.e. six entries. We assign the icon △ to this jet as a mnemonic help (see Figure 4). The vertex jets $\mathbf{v}_i^0$ are tensor-degree 2-jets in B-spline form (icon ⊞, compare with Figure 5), i.e. 3×3 entries and the interspersed edge-jets $\mathbf{e}_i^0$ are 2-jets in a single variable in B-spline form, 3 entries, with icon |. That is, we use B-spline representation where possible since Bézier representation is particular to one patch and in that sense unsymmetric. We group the jets into the column vector

$$\mathbf{j}_i^0 := [\mathbf{c}_i^0; \mathbf{v}_i^0; \mathbf{v}_{i+1}^0; \mathbf{e}_i^0] \in \mathbb{R}^{27}.$$

There is one such vector for each x, y, z- coordinate. The sharing of the vertex jets $\mathbf{v}_i^0$ and the relation $\mathbf{c}_{i+1}^m = R\mathbf{c}_i^m$ (R listed in Appendix 6.1), guarantee consistency of the localized computation.

3.2. Initialization of jets from a design mesh

To allow a designer to work with only positional information, we automatically generate the jets $\mathbf{j}_i^0$ from a *design control mesh* $\mathbf{d}$. Let $\mathbf{d}_0$ be the position of the extraordinary node, $\mathbf{d}_1, \ldots, \mathbf{d}_N$ the 1-link (1-ring) of direct neighbors and $\mathbf{d}_{N+1}, \ldots, \mathbf{d}_{2N}$ the 2-link as shown in Figure 6. The six entries $\mathbf{c}_i^0(1), \ldots, \mathbf{c}_i^0(6)$ of the central jets, defining position, tangent and curvature at the extraordinary point, are initialized as

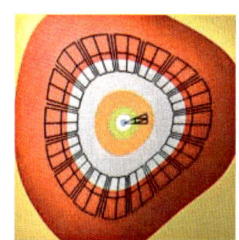

Figure 5: *The jets of $\mathbf{j}^0$ embedded in 3-space (see also Figure 13).*

$$\mathbf{c}_0^0(1) := \frac{5}{9}\mathbf{d}_0 + \frac{4}{9N}\sum_{j=1}^{N}\mathbf{d}_j, \qquad (1)$$

For $j = 2, \ldots, 6$, and $\quad \ell := [10, 10, 9, 10 - \cos(\frac{2\pi}{N}), 9]$,

$$\mathbf{c}_0^0(j) := \frac{\ell(j)}{18}\mathbf{d}_0 + \frac{1}{N}\sum_{k=0}^{N-1}L_k(j,1)\mathbf{d}_k + \frac{1}{N}\sum_{k=0}^{N-1}L_k(j,2)\mathbf{d}_{N+k}.$$

For $i = 1, \ldots, N-1$, $\quad \mathbf{c}_i^0 := R^i\mathbf{c}_0^0$.

The matrices $R \in \mathbb{R}^{6 \times 6}$ and $L_k \in \mathbb{R}^{5 \times 2}$ and the initialization of

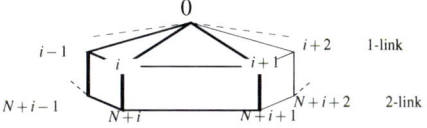

Figure 6: *Indexing the design mesh* $\mathbf{d}$.

the jets $\mathbf{v}_i^0$ and $\mathbf{e}_i^0$ that define the shape at the opposite edge of the triangle are listed in Appendix 6.1. If $N \in \{3, 4\}$, the 1-link and the 2-link of the design mesh are each interpreted as a periodic uniform cubic B-spline curve control net. Each

is uniformly subdivided to double the valence. This is a standard trick for polar meshes (the adjacent regular mesh need not be refined!) to improve shape without resorting to special rules for low valences [KP06b].

3.3. Local Subdivision Matrix

The local jet control net of the sector at level $m + 1$ depends only on the sector at level m via the local subdivision matrix A that does not change with m or sector i (see Figures 7,8):

$$\mathbf{j}_i^{m+1} = A\,\mathbf{j}_i^m, \qquad A \in \mathbb{R}^{27 \times 27}.$$

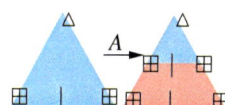

Figure 7: *Local jet refinement conceptually splits each triangle into a smaller triangle and a quadrilateral.*

We choose A so that (i) the refined center jet is determined entirely by the center jet; (ii) the center jet is preserved under binary refinement (by $A_{\triangle,\triangle}$, listed in the Appendix 6.2); (iii) each refined jet $\mathbf{v}_i^{m+1}$ depends only on the central jet (via $A_{\triangle,\boxplus}$) and $\mathbf{v}_i^m$ (via $A_{\boxplus,\boxplus}$); (iv) each refined jet $\mathbf{e}_i^{m+1}$ depends only on the central jet (via $A_{\triangle,|}$) and $\mathbf{e}_i^m$ (via $A_|$). The iconic subscripts hint at the role in mapping old jets to new jets. This yields the simple structure (submatrices listed in Appendix 6.2):

$$A := \begin{bmatrix} A_{\triangle,\triangle} & 0 & 0 & 0 \\ A_{\triangle,\boxplus} & A_{\boxplus,\boxplus} & 0 & 0 \\ A_{\triangle,\boxplus}R & 0 & A_{\boxplus,\boxplus} & 0 \\ A_{\triangle,|} & 0 & 0 & A_| \end{bmatrix}.$$

Choices (i) and (ii) imply that we know the limit jet at the extraordinary point. Two goals determine A in detail: C^2 continuity and changing the surface shape gradually from the boundary of the triangular facets (represented by $\mathbf{v}_i^m$, $\mathbf{v}_{i+1}^m$ and $\mathbf{e}_i^m$) to the central jet $\mathbf{c}_i^m$. The dominant eigenvalues of A are associated with scaling the central jet; the lower eigenvalues with the fading contributions from the boundary.

Theorem 1 The matrix A has a full complement of linearly independent eigenvectors with eigenvalues

$$1, \frac{1}{2}, \frac{1}{2}, \frac{1}{4}, \frac{1}{4}, \frac{1}{4}, \underbrace{\frac{1}{8}, \ldots, \frac{1}{8}}_{7-\text{fold}}, \underbrace{\frac{1}{16}, \ldots, \frac{1}{16}}_{7-\text{fold}}, \underbrace{\frac{1}{32}, \ldots, \frac{1}{32}}_{7-\text{fold}}.$$

Proof By (ii) (cf. Appendix 6.2), $A_{\triangle,\triangle}$ has the dominant eigenvalues $1, \frac{1}{2}, \frac{1}{2}, \frac{1}{4}, \frac{1}{4}, \frac{1}{4}$. The eigenvalues of $A_|$ are $\frac{1}{8}, \frac{1}{16}, \frac{1}{32}$ and $A_{\boxplus,\boxplus} = \text{diag}(A_|, A_|, A_|)$. Independence of eigenvectors is checked by explicit eigendecomposition.

4. Representation in Bézier form

To evaluate, we now associate functions with the jets, just as Catmull-Clark subdivision associates one bicubic B-spline with each quadrilateral facet vertex of valence

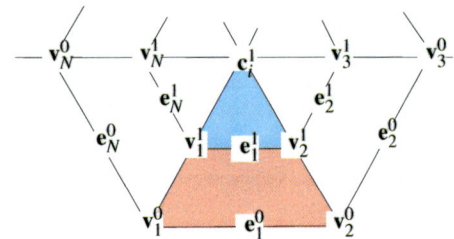

Figure 8: *Jets after one refinement step: the input jets $\mathbf{v}_i^0$, $\mathbf{c}^0$ and $\mathbf{e}_i^0$ generate a new layer of jets $\mathbf{v}_i^1$, $\mathbf{e}_i^1$ and new representations $\mathbf{c}_i^1$ of the central jet. Triangular facets are split into a smaller triangular and a quadrilateral facet. The jets associated with the quadrilateral facet define a segment $\mathbf{x}_i^m$ of a surface ring $\mathbf{x}^m$ (cf. Figure 10).*

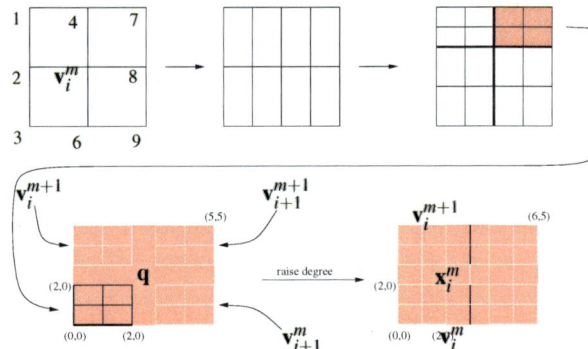

Figure 12: *(top) Conversion of $\mathbf{v}_i^m$ by knot insertion to four abutting 3×3 blocks of Bézier coefficients (right). (bottom) The upper right block fills as quarter of a 6×6 control net of a Bézier patch of degree $(5,5)$.*

four. In fact, the latter has an alternative and equivalent interpretation as jet subdivision: each node and its eight neighbors form a tensor 2-jet $\bar{\mathbf{v}}$ (see Figure 9). Blending the information from four such vertex jets, 4 times 9 degrees of freedom, the Bézier patch $\mathbf{q}$ is in general of degree $(5,5)$. But since the jets overlap and hence contain redundant information, the resulting patch is only of degree $(3,3)$, as expected. In the same spirit, near the extraordinary point, each sector $\mathbf{x}_i^m$, $i \in 0,\dots,N-1$, of a surface ring is defined by the four vertex jets $\mathbf{v} := [\mathbf{v}_i^{m+1}, \mathbf{v}_{i+1}^{m+1}, \mathbf{v}_i^m, \mathbf{v}_{i+1}^m]$, plus a small perturbation from $\mathbf{e} := [\mathbf{e}_i^m, \mathbf{e}_i^{m+1}]$ as follows (cf. Figure 10) By knot

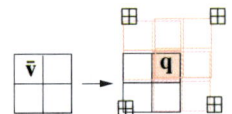

Figure 9: *Bicubic B-spline control net as overlay of four jets of type $\mathbf{v}$.*

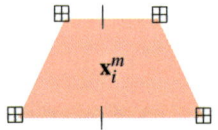

Figure 10: *Jets defining a patch $\mathbf{x}_i^m$.*

and, skipping to a surface ring by eigendecomposition of A [Sta98], yields fast evaluation at arbitrary points.

5. Discussion and Continuity

Implementing jet subdivision is more complex than implementing standard subdivision schemes: compared to standard subdivision there is an additional jet initialization, i.e. one matrix multiplication per design net node (Appendix 6.1); the local subdivision matrix A has about nine times as many entries as standard schemes that work directly on the mesh (Appendix 6.2) and evaluation amounts to applying a more complex linear transformation (Appendix 6.3) than the usual change of basis from B-spline to Bézier representation. Also, because it subdivides polar meshes, the proposed

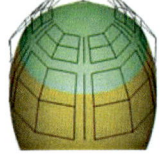  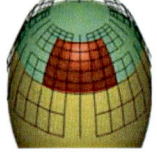

Figure 11: *Conversion of Figure 12 illustrated in 3-space.*

insertion, the four jets $\mathbf{v}$ define the 36 Bézier coefficients of an auxiliary polynomial patch $\mathbf{q}$ (Figures 11, 12). Raising the degree of $\mathbf{q}$ to six in the circular direction and replacing the central column of coefficients by coefficients obtained from jets $\mathbf{e}$ results in a Bézier patch $\mathbf{x}_i^m$ of degree $(6,5)$. The Appendix summarizes the process as a matrix multiplication. The patch can be evaluated by standard algorithms

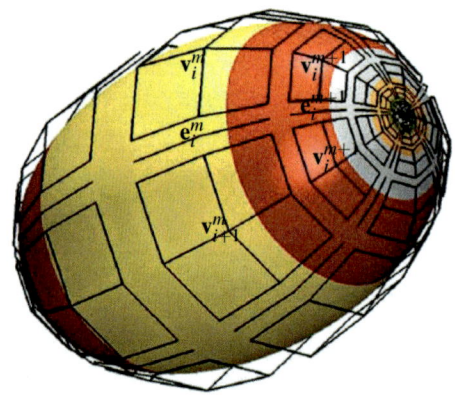

Figure 13: *Summary of the construction: Vertex jets $\mathbf{v}$ and edge jets $\mathbf{e}$ whose icons ⊞ and | mimic the representation of the jets as mesh fragments. (Typically, the user will not see or manipulate the jets directly.) The middle, gold surface is of degree $(3,3)$, the other rings $\mathbf{x}^m$ are of degree $(6,5)$.*

scheme has a different flavor on quadrilateral and triangular input facets.

However, from the user's point of view, the extra effort and different structure are hidden by the design mesh and the automatic initialization of the jets; and the extra computational and storage effort is negligible since it is confined to extraordinary nodes and, due to sparsity, all formulas reduce to computing products of small matrices and vectors. Therefore, for some applications, the hidden complexity may be an acceptable price to obtain good continuity and shape properties that cannot be obtained by applying few rules with small footprint directly to the design mesh as in standard subdivision. (We derived and optimized the parameters of a number of such direct polar, almost everywhere C^2 subdivision algorithms for comparison.)

To *verify curvature continuity*, we note that adjacent patches $\mathbf{x}_i^m$ and $\mathbf{x}_{i+1}^m$ join C^2 since their transversal derivatives are defined by the same two jets $\mathbf{v}_i^{m-1}$ and $\mathbf{v}_i^m$ by the process just detailed in Section 4. Therefore each ring is C^2. Also due to shared jets, the surface rings $\mathbf{x}^m$ and $\mathbf{x}^{m+1}$ join C^2. In each sector, separately, the eigenvectors of A with eigenvalue 2^{-k} define, by the construction of Section 4, (eigen)functions f_{kj} so that each sector of the surface ring has the expansion

$$\mathbf{x}_i^m = a_0 f_{00} + \frac{1}{2^m}(a_{10} f_{10} + a_{11} f_{11}) + \frac{1}{4^m}(a_{20} f_{20} + a_{21} f_{21} + a_{22} f_{22}) + \sum_{k=3}^{5} \sum_{j=0}^{6} 2^{-km} a_{kj} f_{kj}. \tag{2}$$

In particular, the eigenvectors allow us to choose

$$f_{00} = 1, \quad (f_{10}, f_{11}) = \rho_i,$$
$$\{f_{20}, f_{21}, f_{22}\} = \{f_{10}^2, f_{10} f_{11}, f_{11}^2\}, \tag{3}$$

where ρ_i is one sector of the map defined in Appendix 6.4. Abbreviating $(u,v) := 2^{-m} \rho_i(s,t)$ and $g(u,v) := x_i^m(s,t)$, (2) becomes

$$g(u,v) = a_0 + a_{10}u + a_{11}v + a_{20}u^2 + a_{21}uv + a_{22}v^2 + g_3(u,v) + g_4(u,v) + g_5(u,v), \tag{4}$$

where the term g_k corresponds to $\sum_{j=0}^{6} 2^{-km} a_{kj} f_{kj}$, i.e. satisfies $2^{-k} g_k(u,v) = g_k(u/2, v/2)$. Differentiating, we get

$$\partial_u^i \partial_v^j g_k(u/2, v/2) = 2^{-k+i+j} \partial_u^i \partial_v^j g_k(u,v), \quad i+j < 3. \tag{5}$$

Therefore the derivatives of $g(u,v)$ have well-defined limits at $(0,0)$. Since the jets $\mathbf{c}_i^0$ consistently define a single unique polynomial of degree two at the extraordinary point, these limits are the same up to second order for all sectors; that is, the surface is C^2.

Note that the third equation of (3) implies that $\mathbf{x}^m$ reproduces ρ^2, a polynomial of degree six in the circular direction. So, while interpolating the jets $\mathbf{v}_i^m$ with patches of degree $(5,5)$ suffices to create a sequence of C^2 joined C^2 rings, we needed to add edge jets in the circular direction for C^2 continuity at the extraordinary point.

Compared to schemes like [Pra97, Rei98], we attribute the observed overall good curvature distribution and shape retention (Figures 16, 17) to the fact that we do not impose a fixed low degree polynomial shape but work with jets. This yields a gradual transition between the main bicubic tensor-product spline body and the extraordinary point, especially for high valences. (Large polar meshes inherit the well-known macroscopic shape of bicubic splines and their pole regions are either less challenging or have less predictable desirable shape than the scenarios in Figure 17.) Compared to Catmull-Clark subdivision, C^2 continuity at the extraordinary point helps the curvature distribution (see Figures 1 and 2). The attenuation of the influence of the outer jets $\mathbf{v}_i^m$ and $\mathbf{e}_i^m$ and dominance of the central jet $\mathbf{c}_i^m$ is controlled by the local subdivision matrix A. This matrix is essentially determined by the requirements (i) through (iv) of Section 3.3 and by the constraints (3). The few remaining degrees of freedom are set to even out the attenuation. Figure 16 illustrates that the twin goals of high smoothness and shape preservation can be achieved simultaneously.

Focus on polar nets made it possible to completely state in print, a specific scheme, the first of its kind. Clearly, there are different choices of local refinement matrices A to be explored, different mesh layouts, patch representations, levels of continuity or number of variables. We must, however, expect that initialization, refinement and change of basis matrices in these scenarios will be more complex.

References

[CNG00] COTRINA NAVAU J., GARCIA N. P.: Modelling surfaces from planar irregular meshes. *Comput. Aided Geom. Design 17*, 1 (2000), 1–15.

[GH89] GREGORY J. A., HAHN J. M.: A C^2 polygonal surface patch. *Comp Aided Geom Design 6*, 1 (1989), 69–75.

[GH95] GRIMM C. M., HUGHES J. F.: Modeling surfaces of arbitrary topology using manifolds. *Computer Graphics 29*, Annual Conference Series (1995), 359–368.

[KP06a] KARČIAUSKAS K., PETERS J.: Guided Subdivision. *Comp Aided Geom Design* (2006). accepted subject to revision.

[KP06b] KARČIAUSKAS K., PETERS J.: Surfaces with polar structure. *Computing* (2006), 1–8. to appear, http://www.cise.ufl.edu/research/SurfLab/papers.

[KPR04] KARČIAUSKAS K., PETERS J., REIF U.: Shape characterization of subdivision surfaces – Case studies. *Comp. Aided Geom. Design 21*, 6 (2004), 601–614.

[Lev06] LEVIN A.: Modified subdivision surfaces with continuous curvature. In *SIGGRAPH, ACM Transactions On Graphics* (2006), p. to appear.

[LL03] LEVIN A., LEVIN D.: Analysis of quasi-uniform subdivision. *Appl Comp Harm Anal*, 15 (2003), 18–32.

[Loo04] LOOP C.: Second order smoothness over extraordinary vertices. In *Symp Geom Processing* (2004), pp. 169–178.

[MWW01] MORIN G., WARREN J. D., WEIMER H.: A subdivision scheme for surfaces of revolution. *Comp Aided Geom Design 18*, 5 (2001), 483–502.

[Pet02] PETERS J.: C^2 free-form surfaces of degree (3,5). *Comp Aided Geom Design 19*, 2 (2002), 113–126.

[Pra97] PRAUTZSCH H.: Freeform splines. *Comput. Aided Geom. Design 14*, 3 (1997), 201–206.

[PU98] PRAUTZSCH H., UMLAUF G.: Improved triangular subdivision schemes. In *Computer Graphics International* (1998), pp. 626–632.

[Rei98] REIF U.: TURBS—topologically unrestricted rational B-splines. *Constr. Appr. 14*, 1 (1998), 57–77.

[RP05] REIF U., PETERS J.: Topics in multivariate approximation and interpolation. In *Structural Analysis of Subdivision Surfaces – A Summary* (2005), et al. K. J., (Ed.), pp. 149–190.

[SD05] SABIN M., DODGSON N.: A circle-preserving variant of the four-point subdivision scheme. In *Mathematical Methods for Curves and Surfaces: Tromso 2004* (2005), Daehlen M., Schumaker, (Eds.), Nashboro Press, pp. 275–286.

[Sta98] STAM J.: Exact evaluation of Catmull-Clark subdivision surfaces at arbitrary parameter values. In *SIGGRAPH* (1998), pp. 395–404.

[SW05] SCHAEFER S., WARREN J. D.: On C^2 triangle/quad subdivision. *ACM Trans. Graph 24*, 1 (2005), 28–36.

[WW02] WARREN J., WEIMER H.: *Subdivision Methods for Geometric Design*. Morgan Kaufmann Publishers, 2002.

[XYD06] XUE Y., YU T. P.-Y., DUCHAMP T.: Jet subdivision schemes on the k-regular complex. *Comp Aided Geom Design* (2006). to appear.

[YZ04] YING L., ZORIN D.: A simple manifold-based construction of surfaces of arbitrary smoothness. *ACM TOG 23*, 3 (Aug. 2004), 271–275.

[ZLLT06] ZULTI A., LEVIN A., LEVIN D., TEICHER M.: C^2 subdivision over triangulations with one extraordinary point. *Comp Aided Geom Design 23*, 2 (feb 2006), 157–178.

[ZS00] ZORIN D., SCHRÖDER P. (Eds.):. *Subdivision for Modeling and Animation* (2000), Course Notes, ACM SIGGRAPH.

6. Appendix

This appendix contains three sparse matrices, broken into small pieces to allow easy implementation of the algorithm.

The first matrix (Section 6.1) initializes the jets and has pieces L_k, L_e, L_v and R. The second matrix, A refines the local jets at each subdivision step (Section 6.2). The third matrix, K, converts jets to polynomial pieces in Bézier form (Section 6.3).

Throughout, the sector indices $i = 1, \ldots, N$, are counted modulo the valence N. We abbreviate $c := \cos(\frac{2\pi}{N})$. diag(vector) is the diagonal matrix with diagonal 'vector', and $A(:, i : j)$ denotes columns i through j of a matrix A.

6.1. Jet Initialization matrix

With the indexing of the design mesh $\mathbf{d}$ as in Figure 6,

$$\mathbf{v}_i^0 := L_v \mathbf{d}_v, \qquad \mathbf{e}_i^0 := L_e \mathbf{d}_e,$$

where

$$\mathbf{d}_v := [\mathbf{d}_0, \mathbf{d}_{i-1}, \mathbf{d}_{N+i-1}, \mathbf{d}_i, \mathbf{d}_{N+i}, \mathbf{d}_{i+1}, \mathbf{d}_{N+i+1}],$$
$$\mathbf{d}_e := [\mathbf{d}_v, \mathbf{d}_{i+2}, \mathbf{d}_{N+i+2}].$$

$$L_v := \frac{1}{14400} \begin{bmatrix} 4020 & 3850 & 475 & 5236 & 646 & 154 & 19 \\ 1320 & 4300 & 1150 & 5848 & 1564 & 172 & 46 \\ 240 & 3400 & 2500 & 4624 & 3400 & 136 & 100 \\ 4020 & 1078 & 133 & 7084 & 874 & 1078 & 133 \\ 1320 & 1204 & 322 & 7912 & 2116 & 1204 & 322 \\ 240 & 952 & 700 & 6256 & 4600 & 952 & 700 \\ 4020 & 154 & 19 & 5236 & 646 & 3850 & 475 \\ 1320 & 172 & 46 & 5848 & 1564 & 4300 & 1150 \\ 240 & 136 & 100 & 4624 & 3400 & 3400 & 2500 \end{bmatrix},$$

$$L_e := \frac{1}{28800} \begin{bmatrix} 8040 & 154 & 19 & 9086 & 1121 & 9086 & 1121 & 154 & 19 \\ 2640 & 172 & 46 & 10148 & 2714 & 10148 & 2714 & 172 & 46 \\ 480 & 136 & 100 & 8024 & 5900 & 8024 & 5900 & 136 & 100 \end{bmatrix}.$$

The matrix L_k for initializing the central jet $\mathbf{c}_0^0$,

$$L_k := \begin{bmatrix} w_1 c_k + \frac{4}{9} & w_9 c_k \\ w_1 c_{k-1} + \frac{4}{9} & w_9 c_{k-1} \\ w_2 c_k^2 + 2 w_1 c_k - w_3 & 4 w_5 c_k^2 + 2 w_9 c_k + w_6 \\ w_2 c_k c_{k-1} + w_1(c_k + c_{k-1}) - w_4 & 4 w_5 c_k c_{k-1} + w_9(c_k + c_{k-1}) + w_6 c \\ w_2 c_k c_{k-2} + 2 w_1 c_{k-1} + w_7 & 4 w_5 c_k c_{k-2} + 2 w_9 c_{k-1} + w_8 \end{bmatrix},$$

depends on k via $c_k := \cos(\frac{2\pi k}{N})$ and

$$w_1 := \frac{11}{81}(2+c), \quad w_9 := \frac{-w_1}{110}, \quad w_2 := \frac{88}{81}c^2 - \frac{56}{135}c + \frac{136}{405},$$

$$w_3 := \frac{44}{81}c^2 - \frac{28}{135}c - \frac{659}{2025}, \quad w_4 := w_3 c + \frac{4}{9}(c-1),$$

$$w_5 := -\frac{1}{405}c^2 + \frac{1}{135}c + \frac{1}{405}, \quad w_6 := \frac{2}{405}c^2 - \frac{2}{135}c + \frac{7}{4050},$$

$$w_7 := -2 w_3 c^2 - \frac{898}{2025}c^2 - \frac{28}{135}c + \frac{1339}{2025},$$

$$w_8 := \frac{4}{405}c^4 - \frac{4}{135}c^3 - \frac{2}{135}c^2 + \frac{2}{135}c + \frac{47}{4050}.$$

The matrix

$$R := \begin{bmatrix} 1 & 0 & 0 & 0 & 0 & 0 \\ 0 & 0 & 1 & 0 & 0 & 0 \\ 2(1-c) & -1 & 2c & 0 & 0 & 0 \\ 0 & 0 & 0 & 0 & 0 & 1 \\ 0 & 0 & 2(1-c) & 0 & -1 & 2c \\ 4(1-c)^2 & -4(1-c) & 8(1-c)c & 1 & -4c & 4c^2 \end{bmatrix} \qquad (6)$$

encodes C^2 constraints between the central jets $\mathbf{c}_i^0$.

6.2. The local subdivision matrix A

The submatrixes of the local subdivision matrix $A \in \mathbb{R}^{27 \times 27}$ that maps $\mathbf{j}_i^{m+1} = A\,\mathbf{j}_i^m$ are $A_{\triangle,\triangle}, A_|, A_{\boxplus,\boxplus}, A_{\triangle,\boxplus}, A_{\triangle,|}$ where the iconic subscripts hint at the role in mapping old jets to new jets.

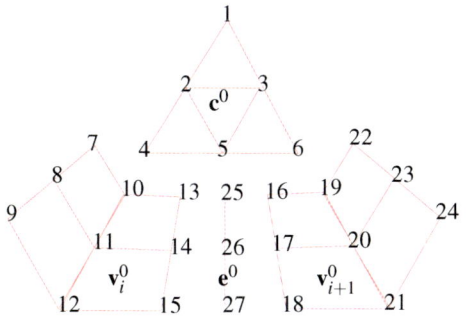

Figure 14: *Indexing of local jet representations in vector* $\mathbf{j}$. *For example,* $\mathbf{j}_{i-1}^m(16:24) = \mathbf{v}_i^m = \mathbf{j}_i^m(7:15)$.

$$A_{\triangle,\triangle} := \frac{1}{4}\begin{bmatrix}4&0&0&0&0&0\\2&2&0&0&0&0\\2&0&2&0&0&0\\1&2&0&1&0&0\\1&1&1&0&1&0\\1&0&2&0&0&1\end{bmatrix}, \qquad A_| := \frac{1}{576}\begin{bmatrix}278&-109&11\\488&-160&14\\608&-112&8\end{bmatrix},$$

$$A_{\boxplus,\boxplus} := \operatorname{diag}(A_|, A_|, A_|),$$

$$A_{\triangle,\boxplus}(:,1) := 1 - \sum_{j=2}^{6} A_{\triangle,\boxplus}(:,j) - \sum_{j=7}^{16} A_{\boxplus,\boxplus}(:,j),$$

$$k_0 := 28 - 2\mathsf{c} + \mathsf{c}^2\,, \quad k_1 := 8 + \mathsf{c}\,, \quad k_2 := 8 + 7\mathsf{c}\,,$$

$$k_3 := 28 + 94\mathsf{c} + 49\mathsf{c}^2\,, \quad k_4 := 52 + 34\mathsf{c} - 5\mathsf{c}^2\,,$$

$$A_{\triangle,\boxplus}(:,2:6) := \frac{1}{1280(2+\mathsf{c})^2}\operatorname{diag}(1,4,16,1,4,16,1,4,16)\cdots$$

$$\cdots \begin{bmatrix}3(1294+553\mathsf{c}-461\mathsf{c}^2)&-108(15-4\mathsf{c})&46k_3&-276k_2&414\\27(26+3\mathsf{c}-15\mathsf{c}^2)&-36(7-4\mathsf{c})&10k_3&-60k_2&90\\12(1-\mathsf{c})(5+4\mathsf{c})&-18(1-\mathsf{c})&k_3&-6k_2&9\\3(322+223\mathsf{c}+445\mathsf{c}^2)&828(1-\mathsf{c})&46k_4&828\mathsf{c}&-414\\9(2-\mathsf{c}+29\mathsf{c}^2)&180(1-\mathsf{c})&10k_4&180\mathsf{c}&-90\\-12(1-\mathsf{c})(1+2\mathsf{c})&18(1-\mathsf{c})&k_4&18\mathsf{c}&-9\\-3(178-761\mathsf{c}-11\mathsf{c}^2)&-36(1-34\mathsf{c})&46k_0&276k_1&414\\-3(86-139\mathsf{c}-\mathsf{c}^2)&-108(1-2\mathsf{c})&10k_0&60k_1&90\\-36(1-\mathsf{c})&-18(1-\mathsf{c})&k_0&6k_1&9\end{bmatrix}.$$

With $\ell_0 := -190 + 1169\mathsf{c} + 11\mathsf{c}^2$,

$$\ell_1 := -122 + 211\mathsf{c} + \mathsf{c}^2\,, \quad \ell_2 := 49 + 4\mathsf{c} + \mathsf{c}^2\,,$$

$$A_{\triangle,|}(:,1) := 1 - \sum_{j=2}^{6} A_{\triangle,|}(:,j) - \sum_{j=1}^{3} A_|(:,j)$$

$$A_{\triangle,|}(:,2:6) := \frac{1}{2560(2+\mathsf{c})^2}\operatorname{diag}(1,4,32)\cdots$$

$$\cdots \begin{bmatrix}3\ell_0 & 3\ell_0 & 1656 & 92\ell_2 & 1656\\3\ell_1 & 3\ell_1 & 360 & 20\ell_2 & 360\\-27(1-\mathsf{c}) & -27(1-\mathsf{c}) & 18 & \ell_2 & 18\end{bmatrix}.$$

6.3. The change of basis operator

The lower left nine Bézier coefficients of the auxiliary patch $\mathbf{q}$ of degree $(5,5)$ representing four vertex jets $\mathbf{v}_i^m$ are, by knot insertion,

$$\begin{bmatrix}\mathbf{q}(0,2) & \mathbf{q}(1,2) & \mathbf{q}(2,2)\\\mathbf{q}(0,1) & \mathbf{q}(1,1) & \mathbf{q}(2,1)\\\mathbf{q}(0,0) & \mathbf{q}(1,0) & \mathbf{q}(2,0)\end{bmatrix}$$

$$= \mathbf{q}(0:2,0:2) := K\mathbf{v}_i^m$$

$$K := \frac{1}{36}\begin{bmatrix}4&4&1&8&8&8&4&4&1\\0&0&0&8&8&2&8&8&2\\0&0&0&0&0&0&16&16&4\\6&3&0&12&6&0&6&3&0\\0&0&0&12&6&0&12&6&0\\0&0&0&0&0&0&24&12&0\\9&0&0&18&0&0&9&0&0\\0&0&0&18&0&0&18&0&0\\0&0&0&0&0&0&36&0&0\end{bmatrix}.$$

Here, the entries of $\mathbf{v}_i^m$ are in their order as in Figure 12, *left*, i.e. $\mathbf{v}_i^m\!\left(\begin{smallmatrix}1&4&7\\2&5&8\\3&6&9\end{smallmatrix}\right)$. The matrix K can also be used to generate the remaining entries of $\mathbf{q}$ by applying the straightforward symmetries of the indices of $\mathbf{q}$ and the jets:

$$\mathbf{q}(0:2,5:3) \leftarrow \mathbf{v}_i^{m+1}\!\left(\begin{smallmatrix}3&6&9\\2&5&8\\1&4&7\end{smallmatrix}\right), \quad \mathbf{q}(5:3,5:3) \leftarrow \mathbf{v}_{i+1}^{m+1}\!\left(\begin{smallmatrix}9&6&3\\8&5&2\\7&4&1\end{smallmatrix}\right),$$

$$\mathbf{q}(0:2,0:2) \leftarrow \mathbf{v}_i^m\!\left(\begin{smallmatrix}1&4&7\\2&5&8\\3&6&9\end{smallmatrix}\right), \quad \mathbf{q}(5:3,0:2) \leftarrow \mathbf{v}_{i+1}^m\!\left(\begin{smallmatrix}7&4&1\\8&5&2\\9&6&3\end{smallmatrix}\right).$$

After raising the degree of $\mathbf{q}$ to six in the circular direction and renaming it $\mathbf{x}_i^m$, we replace the central column of coefficients by

$$\mathbf{x}_i^m(3,0:2) = \frac{1}{9}\begin{bmatrix}4&4&1\\6&3&0\\9&0&0\end{bmatrix}\mathbf{e}_i^m, \quad \mathbf{x}_i^m(3,3:5) = \frac{1}{9}\begin{bmatrix}0&0&9\\0&6&3\\4&4&1\end{bmatrix}\mathbf{e}_i^{m+1},$$

where the three entries of $\mathbf{e}_i^m$ are indexed, in the natural order, i.e. as $\begin{bmatrix}1\\2\\3\end{bmatrix}$.

6.4. The map ρ

The map

$$\rho : [0,1]^2 \times \{1,\ldots,N\} \to \mathbb{R}^2$$

is a C^2 spline ring of piecewise degree $(3,1)$. Its outer and inner layers differ by a scaling factor $1/2$, $\rho_{1i} := 2\rho_{0i}$ where

$$\rho_{0i} := \begin{bmatrix}\cos(i\alpha)\\\sin(i\alpha)\end{bmatrix}, \quad \alpha := \frac{2\pi}{N}.$$

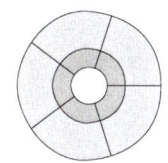

Figure 15: *The maps* ρ *and* $\rho/2$ *for* $N = 5$.

and each represents a periodic planar uniform cubic spline. We note that ρ is regular and injective and ρ and $\rho/2$ join C^2 (see Figure 15), tesselating the plane with common center.

Acknowledgements The presentation benefited from the helpful feedback of the reviewers. The work was supported by NSF Grants DMI-0400214 and CCF-0430891.

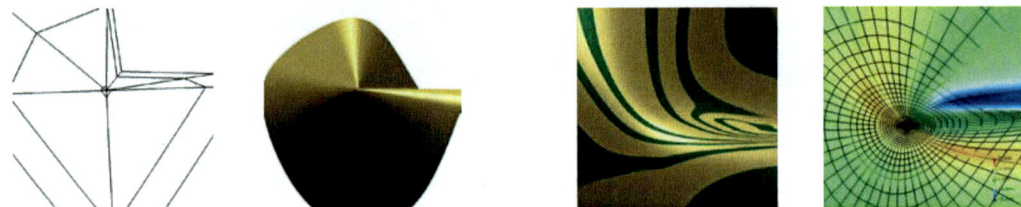

Figure 16: *A semisharp curvature continuous blend attests to the ability of polar guided subdivision to both preserve* macro-scopic shape *and* local *curvature continuity. Note the small design mesh triangles at the center, intended to force a rapid transition near the extraordinary node. The right figures show reflection lines and mean curvature at the enlarged center.*

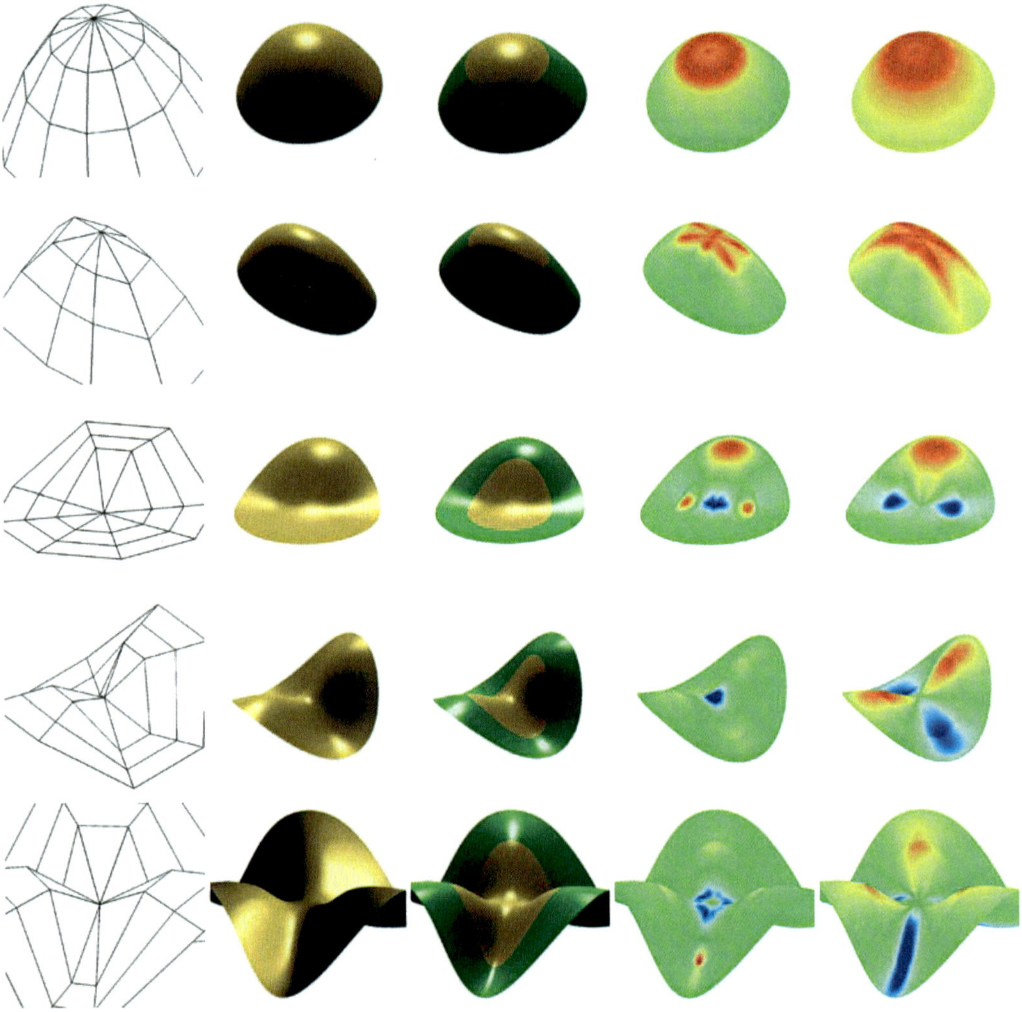

Figure 17: *Shape and smoothness analysis of the neighborhood of an extraordinary point. (top to bottom) basic cone, elongated cone, saddle, single ridge, monkey saddle, (left to right) control net, shape, patches of degree (3,3) shaded green, Gauss curvature and mean curvature texture centered at 0=green.*

Eurographics Symposium on Geometry Processing (2006)
Konrad Polthier, Alla Sheffer (Editors)

.

Rectangular Multi-Chart Geometry Images

Nathan A. Carr[1], Jared Hoberock[2], Keenan Crane[2] and John C. Hart[2]

[1] Adobe Systems Inc.   [2] University of Illinois, Urbana-Champaign

Abstract

Many mesh parameterization algorithms have focused on minimizing distortion and utilizing texture area, but few have addressed issues related to processing a signal on the mesh surface. We present an algorithm which partitions a mesh into rectangular charts while preserving a one-to-one texel correspondence across chart boundaries. This mapping permits any computation on the mesh surface which is typically carried out on a regular grid, and prevents seams by ensuring resolution continuity along the boundary. These features are also useful for traditional texture applications such as surface painting where continuity is important. Distortion is comparable to other parameterization schemes, and the rectangular charts yield efficient packing into a texture atlas. We apply this parameterization to texture synthesis, fluid simulation, mesh processing and storage, and locating geodesics.

Categories and Subject Descriptors (according to ACM CCS): I.3.5 [Computer Graphics]: Geometric algorithms

Keywords: Mesh parameterization, face clustering, texture atlas, geometry images.

1. Introduction

Many powerful surface processing operations can be expressed as the solution of partial differential equations (PDEs) on a surface, including feature-sensitive smoothing, reaction-diffusion texturing, texture synthesis, mesh editing, fluid flow, and geodesic tracing. Traditional approaches for solving surface PDEs rely on an irregular mesh over the surface or in space. Recently, geometry images [GGH02] have provided an efficient square domain that allows surface PDEs to be solved through simpler image operations.

The main problem with using geometry images to generate a domain for surface PDEs is that mapping the surface to a single rectangular chart incurs a large amount of distortion. High parametric distortion reduces the precision and efficiency of surface processing applications. Multi-chart geometry images [SWG*03] reduce distortion by cutting the mesh into multiple irregular pieces, each retessellated with a regular triangle mesh and packed into a single atlas. The added flexibility of multiple cuts reduces distortion, but irregular charts do not pack efficiently and require additional processing of the "topological sideband" to determine pixel neighbors across charts.

We present a new surface parameterization scheme, demonstrated in Fig. 1, that decomposes a triangle mesh into

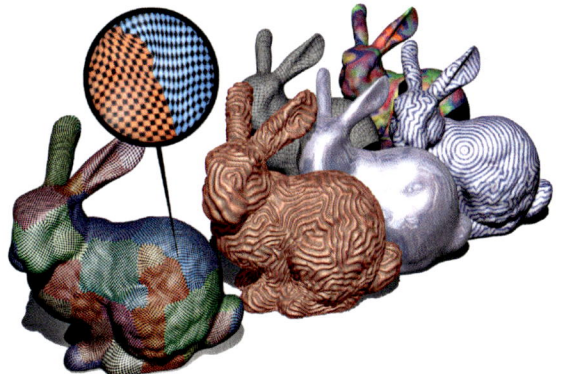

Figure 1: *We describe a new mesh clustering method that creates rectangular patches whose texels align across boundaries to conveniently support the implementation of various surface processing applications.*

four-cornered quasi-rectangular clusters which map to rectangular charts in parameter space, and whose texels align across shared boundaries. This representation facilitates simple, efficient surface PDE solutions using image processing operations over a set of images with well-defined neighbors at their boundaries. Such an organization eases the GPU-accelerated implementation of surface mesh processing.

A key feature of our approach is the use of *parameterization distance* to control cluster shape. This distance approximates geodesic distance within a cluster-centric coordinate system. We gain flexibility in chart selection by permitting T-junctions and self-neighboring charts, which helps minimize parametric distortion, as discussed in Sec. 6.

Section 2 compares our approach to other methods for parameterizing meshes and solving PDEs on a surface. Section 3 describes our algorithm, which extends iterative clustering to find developable charts that map naturally into rectangular regions. Section 7 demonstrates the utility of this new data structure with a variety of surface processing applications. Section 6 compares the distortion incurred by this representation to other options, establishing it as one of the most attractive choices for surface PDE processing. Section 8 discusses the limitations of this method, and ideas for its further improvement.

2. Previous Work

2.1. Surface Parameterization

Recent work has demonstrated the utility of decomposing meshes into a regular domain. Geometry images [GGH02] use a regular domain to implicitly encode the connectivity of a mesh, allowing standard image compression to be applied to mesh geometry. Carr and Hart [CH02] and Purnomo *et al.* [PCK04] show how square charts can be efficiently packed (and repacked [CH04]) to eliminate wasted texture space.

Polycube-maps warp and project geometry on the quadrilateral faces of a manual cuberille exoskeleton [THCM04], whereas our approach constructs its charts automatically on meshes, even those whose small intricate features would significantly challenge Polycube construction.

Purnomo *et al.* [PCK04] merged triangles based on coplanarity into clusters, straightened them into polygons [SSGH01], and used barycentric subdivision to divide them into quadrilaterals. Boier-Martin *et al.* [BMRJ04] constructed a centroidal Voronoi diagram on planar clusters to yield a hexagon-dominant mesh subdivided into a quad mesh. Dong *et al.* [DBG*06] construct a quad mesh from a smoothed, subdivided Morse-Smale complex of a shape harmonic. These techniques yield semi-regular quadrangulations with no T-junctions, whereas we generate rectangular clusters on developable (not necessarily planar) surface segments with T-junctions.

Our cluster growth seeks to minimize distortion while constraining growth to a rectangular shape. It is an extension of Sorkine *et al.* [SCOGL02], which constructs intermediate parameterizations during cluster growth to prevent clusters from exceeding a distortion bound. We also parameterize

during cluster growth, but with additional rules governing cluster shape.

Iso-charts uses iso-map's geodesic metric extension to multidimensional scaling to grow large contiguous charts within a distortion bound [ZSGS04]. Alternatively such near-developable charts may be formed by iterative clustering based on fitting to a union of conics [JKS05]. Because we parameterize during cluster growth, we use parameterization distance as an approximation of geodesic distance to constrain chart growth to rectangular forms.

Other considerations of cluster shape include forcing their boundaries through regions of high curvature [LPRM02], or (additionally) lower average visibility [SH02]. Sander *et al.* [SSGH01] constrained greedy cluster growth that avoided crossing base domain boundaries, and straightened the boundaries of clusters as a post-process, whereas our greedy cluster growth straightens the sides of its rectangular clusters through the cluster growth rules.

Our proposed method is only guarantees C^0 continuity across charts, whereas techniques such as globally smooth parameterization [KLS03] and global conformal parameterization [GY03] produce smoother parameterizations. More recent work has extended this method to form globally smooth parameterizations containing nearly uniform sized square charts [RLL*05].

We use a modified chessboard metric to coerce k-means clustering to form rectangles, whereas Hausner [Hau01] used a modified manhattan metric to construct squarish clusters used to artistically tile planar image regions.

2.2. Surface PDEs.

Our goal of establishing a continuous mapping between a surface and an array of rectangular images is a novel contribution among a recent flurry of similar methods that use a parameterization as a basis for solving surface partial differential equations on a surface.

Bertalmio *et al.* [BSCO01] derive formulas for diffusion on a volumetric isosurface by lifting the dynamics to a voxel grid in the embedding space, and applied the results to texture filtering, reaction diffusion texturing and the visualization of the surface's principal curvature flow. Such Eulerian space-grid formulations are more accurate but consume too much space and work at a fixed resolution that ignores surface features.

Sibley and Taubin [ST04] implement diffusion across the irregular chart boundaries of an atlas, whereas the texels of our RMCGIs are pre-aligned across chart boundaries by design to overcome these concerns.

Bajaj and Xu [BX03] implemented flow on a loop-subdivided triangulated surface to perform anisotropic (feature-sensitive) smoothing, and Stam [Sta03] implement flows on Catmull-Clark surfaces. They both use the metric tensor to reduce the effects of parametric distortion on flow, and overcome the effects of irregular valence through repeated subdivision about extraordinary points. While the metric tensor is designed to accommodate the distortion of length due to differences in element size, it does not completely eliminate these effects [Sta03]. We too depend on subdivision around extraordinary points, but our method automatically forms an evenly distributed base mesh for the subdivision.

Shi and Yu [SY04] also implemented flow on a mesh by representing fluid fields directly on the mesh topology. This couples fluid detail to tessellation. However, since their implicit Lagrangian technique traces velocity vectors on the mesh surface, it proves robust in the face of extraordinary vertices. Our method handles these points as a special case.

Lui *et al.* [LWC05] solved PDEs on a global conformal parameterization of a surface using the metric tensor to reduce the effects of distortion, with applications in fluid flow, segmentation, denoising and inpainting applied to the surface signal. Conformal parameterization produces quads that meet at right angles, but with wide variances in element size that focuses computation in arbitrary regions unrelated to shape or signal. In contrast, our proposed algorithm parameterizes surfaces in a curvature sensitive manner into uniformly-sized quadrilaterals that reduce the effects of element-size variation on the underlying PDE solution, and distribute computation evenly across the surface.

3. Algorithm Overview

Our algorithm is inspired by the greedy flattening bounded-distortion parameterization algorithm of Sorkine *et al.* [SCOGL02] and the iterative k-means mesh clustering algorithm used by Sander *et al.* [SWG*03], Schlafman *et al.* [STK02], Cohen-Steiner *et al.* [CSAD04], and more recently [JKS05]. An overview of our approach is as follows:

Phase I: Iterative Face Clustering (Sec. 4)
  1. Select an initial set of seed faces to form clusters
  2. Grow clusters outward from seed faces
  3. While large gaps of unassigned triangles remain
    Re-orient each cluster along its local parameter axes
    Regrow clusters outward from seed faces
    Insert a new seed face
  4. Add remaining faces to nearest cluster
Phase II: Chart Parameterization (Sec. 5)
  5. Determine chart boundaries
  6. Solve each chart's local parameterization

Figure 5 illustrates this process. Phase I uses iterative clustering and greedy flattening to determine clusters of faces that parameterize with low distortion into roughly rectangular regions. Phase II forces each cluster into a rectangular region of texture space such that edges shared across chart boundaries sample the same number of texels in both

charts. The resulting parameterization provides a seamless set of rectangular domains useful for rendering, storage, and computation.

4. Rectangular Cluster Formation

Standard methods for iterative mesh clustering begin with a collection of disjoint seed faces and attach remaining faces to the cluster whose seed face is *closest*, according to a given metric (often a combination of Euclidean distance and orientation). Triangles containing the centroids of these clusters become seed faces in the next iteration, and the process repeats until cluster centers stabilize.

Our approach differs from this standard algorithm in three ways. First, we parameterize new triangles during cluster growth to predict and track chart distortion, with additional rules that avoid jagged boundaries and holes in the cluster. Second, the "distance" metric uses a per-cluster coordinate frame, and is approximated by distance in the parameter domain. Third, we adjust our reference frame at the end of each iteration to prevent charts extending away from the original axes (which can occur even thorugh we are using a chessboard metric for constraints).

4.1. Parameterization Distance

To constrain chart growth to a rectangular shape, we must estimate geodesic distances on the meshed surface. We can take the same parameterization used during cluster growth to bound chart distortion, and use it to approximate geodesic distance. Distances in this domain correspond to geodesic distance altered by the path integral of the parametric distortion.

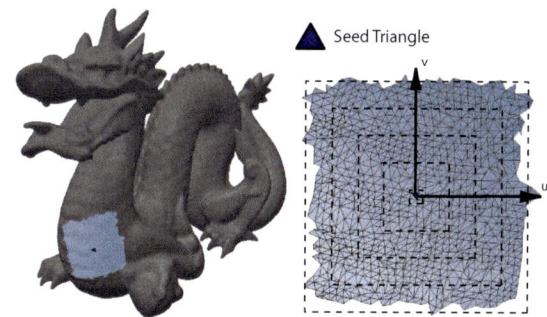

Figure 2: *An isolated cluster grown under the chessboard metric forms a square in the parameterization domain.*

We encourage rectangular clusters by changing our metric from the usual Euclidean L^2 distance that forms round clusters to a chessboard L^∞ distance whose equidistant "circles" are squares, shown in Fig. 2. We further generalize to *rotated* rectangular clusters by formulating the oriented, anisotropic metric $L_{a,\theta}^{\infty}$ such that magnitude is measured

$$||u,v||_{a,\theta}^{\infty} = \max \left(\frac{|u\cos\theta - v\sin\theta|}{a_u}, \frac{|u\sin\theta + v\cos\theta|}{a_v} \right) \tag{1}$$

for aspect ratio a and orientation θ.

Since each cluster has its own unique distance metric, the notion of distance does not extend globally across the mesh. This certainly defies the definition of a distance metric and somewhat confounds the direct assignment of triangles to their "nearest" cluster, but still assigns triangles to the chart which is in some sense the best.

4.2. Frontier Face Parameterization

Frontier faces lie just outside of a cluster, sharing at least one edge with the cluster's boundary. Before we can include a frontier face to a cluster, it must be flattened into the parameterization domain so we can evaluate the four inclusion criteria: (1) its parametric distance to the cluster center, (2) the amount of distortion it would contribute to the cluster, (3) its potential for fold-over, and (4) its effect on boundary smoothness. Face parameterization depends on the number of edges shared with the cluster boundary:

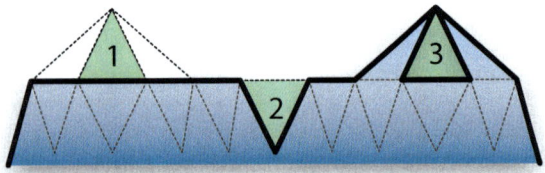

Figure 3: *Three cases of triangles adjacent to the frontier on the corresponding number of its edges.*

Case 1: One shared edge. The two vertices of the shared edge inherit the cluster's parameterization coordinates, but the third vertex is assigned new coordinates. Coordinates can be assigned by a rigid unfolding of the triangle about its shared-edge hinge, but a simple look ahead during flattening can reduce distortion and lead to larger, more efficient clusters. Following Sorkine *et al.* [SCOGL02], we rigidly unfold any frontier faces that share the same free vertex, and assign to the free vertex the average of the resulting parameterization coordinates for it.

Case 2: Two shared edges. All three vertices inherit the parameterization coordinates from the cluster boundary. If the frontier face fails the inclusion criterion, then it is re-evaluated as a Case-1 frontier face with respect to the edge it shares with the most recently added cluster face, while its second shared edge potentially forms a seam.

Case 3: Three shared edges. An annulus has formed and the frontier face is filling it. The triangle is evaluated as in Case-2 with respect to the two edges it shares with the most recently added contour faces, and the third edge shared with an uncooperative contour face becomes a seam.

The frequency of these Case-3 situations is reduced by biasing the order of cluster growth. When a Case-2 frontier face satisfies the cluster candidacy requirements, it is immediately added to the cluster, bypassing the priority queue usually used to include the best candidates first. This order bias also helps to smooth the cluster boundary.

4.3. Candidacy Criteria

Once a frontier face has been parameterized, the following criteria are evaluated.

Parameterization Distance to Seed Face. We measure distance to the cluster center by evaluating (1) using parameters a, θ of the cluster's coordinate frame. Note that the same triangle may be a frontier face of multiple clusters, but should be added to the closest one.

Incurred Distortion. Numerous parameterization distortion metrics exist, e.g. [SSGH01, SdS00, ZMT05], and nothing in our algorithm precludes the use of any of these. We use the metric of [SCOGL02] which penalizes both stretch and shrinkage for parameterized triangles, and is minimized when the parameterization is perfectly isometric.

Foldover Prevention. To test the parameterization of the frontier face for intersection with the existing cluster parameterization, it is sufficient to test the frontier face's edges against the parameterization of the cluster boundary. We maintain a quad-tree of parameterized boundary edges to hasten intersection queries.

Boundary Smoothness. Since clusters are mapped to rectangular charts, boundary smoothness is very important. We ensure smoothness through the use of boundary constraints. The *final boundary* of a cluster is the subset of its boundary edges which border either another cluster or a frontier face which failed the candidacy test. Given a midpoint m of a final boundary edge, let n denote the nearest position along the cluster's coordinate axes (which are oriented and stretched by θ, a) for which (1) yields the same distance. (For the midpoint m, the vector n is an outward "normal" quantized to be perpendicular to the desired cluster boundary shape.) We exclude from consideration any frontier face whose centroid makes an incidence angle less than $\alpha \approx 80°$ with respect to any final boundary edge midpoint and its associated outward quantized normal.

Since our distance metric is evaluated at triangle centroids, saw-toothed boundaries between patches are prevalent. We prevent sawtooth boundaries by ignoring distance for Case-2 frontier faces, instead relying on the other three criteria for cluster inclusion. In practice this virtually eliminates boundary irregularities.

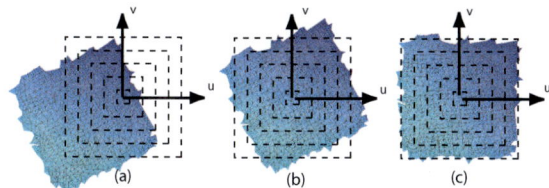

Figure 4: *Patch centering and reorienting process: (a) before, (b) re-centering, (c) reorientation.*

4.4. Cluster Re-alignment

After each patch growth iteration, clusters may not be rectangular in shape or even centered in their local parameter domain. To guide the patch shape toward its rectangular goal, we perform two refinement operations. Fig. 4 shows the operations of the centering and re-orientation process.

Patch Centering. To center a patch, we offset its coordinates by its (negated) center of mass, computed as the area-weighted average of triangle barycenters.

Patch Re-orientation and Size Estimation. We reorient the patch to better align it with the target rectangular shape. Similar to Gottschalk [GLM96], we do principal component analysis over the triangles. We express the texture coordinates of a triangle t using barycentric coordinates $\mathbf{u}^t(\alpha, \beta) = \mathbf{u}_0^t \alpha + \mathbf{u}_1^t \beta + \mathbf{u}_2^t (1 - \alpha - \beta)$, where $\mathbf{u}_0^t$, $\mathbf{u}_1^t$, and $\mathbf{u}_1^t$ are texture coordinates of triangle t. The symmetric 2×2 covariance matrix C is given by

$$C_{ij} = \sum_{t \in T} \int_0^1 \int_0^1 (\mathbf{u}_i^t - \mu_i)(\mathbf{u}_j^t - \mu_j) A_t \, d\alpha \, d\beta \qquad (2)$$

(where $\mathbf{u}_i$ denotes the ith coordinate of $\mathbf{u}$) and has a closed-form expression. The eigenvectors of C define the new coordinate frame. Additionally, by comparing the ratio of the eigenvalues of C, we can compute the target size a for the of the patch to guide the L_a^∞ metric in the next iteration.

4.5. Seed Faces

The clustering process involves the introduction of new seed faces both at the start of algorithm and between growth process iterations. Careful choice of seed face locations can improve the resulting cluster quality and convergence properties of our algorithm. Ideally a new seed face should be placed as far way from mesh boundaries, high curvature regions, and existing cluster boundaries as possible, to maximize the size of its potential cluster.

To do this, we grow an advancing front starting with faces that are adjacent to feature regions (i.e. mesh boundaries, high curvature, or an existing cluster). The front moves over only the faces currently unassigned to any cluster. The last face reached by this front becomes the new seed face to start

a new cluster. We follow [SWG*03] for finding new cluster centers, where distance between faces is measure by the Euclidean distance between their centroids.

Once a new seed face has been found, we parameterize it with its centroid sitting on the origin of its own local parameter domain. Initially we do not know the appropriate aspect scale for this new cluster, so we cannot immediately introduce it into the iterative clustering process. To solve this problem, we perform an outward growth around this seed while leaving existing clusters in place. This growth is performed with a cluster target size of $a = (1, 1)$. Following this growth the cluster is re-centered and re-oriented, and its aspect scale a is updated. At this point the new cluster is ready to take part in the iterative growth process.

4.6. Termination

Though we provide no proof of the convergence, the method does tend to settle and approach a good solution once enough seed faces have been added. Our current implementation adds seed faces until the gaps between patches becomes small. This process could be easily automated, however, we currently do this by visual inspection.

Due to the strictness of the cluster smoothness criteria, some faces remain unassigned to any cluster as shown in the second figure of Fig. 5. A final outward cluster growth phase assigns these remaining faces to their nearest clusters, ignoring distortion, smoothness and even the fold-over criteria.

5. Chart Parameterization

Once clusters are determined, we map each patch to a rectangular chart while aligning texels along adjacent patches. This mapping depends on a particular choice of texture resolution, but can be used with any texture size which maintains the ratio of edge lengths.

5.1. Chart Boundaries and Fixed Vertices

To achieve texel continuity among patches, attention must be paid to the shared boundary segments occurring within and between charts. Each shared segment is assigned a discrete texel length, and the *fixed* end-points of the boundary segments are assigned texel coordinates which lie on texel boundaries.

Two different types of vertices may start or end a boundary segment. A *fixed* vertex is any vertex which resides on the start of a seam where a number of patches come together. A *corner* vertex is any vertex which has been chosen as one of the four corners of a rectangular chart, which may also be at a fixed vertex location. At the end of the clustering phase, fixed vertices are determined by the patch layout, and corner vertices may be freely chosen. Figure 6 shows the possible cases for fixed and corner vertices and the resulting boundary segments.

Figure 5: *Dragon model undergoing the iterative clustering process followed by final mapping.*

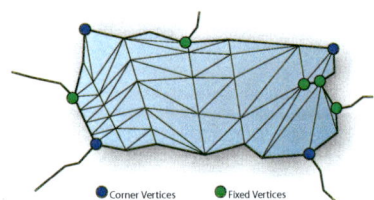

Figure 6: *Fixed and corner vertices for a patch.*

5.2. Patch Size and Corner Determination

For each patch, we walk the boundary in counter-clockwise order and collect fixed and corner vertices into an ordered list ℓ. Corner vertices at this point arise from previously visited adjacent patches whose corners were already assigned.

Discrete texel lengths are assigned to each of the boundary segments. The texel length for any boundary segment shared with a previously visited patch is set to the length of the corresponding segment in the adjacent patch. The remaining boundary segment lengths are directly computed by multiplying the 3D arc length of the segment by a user specified parameter γ to convert to texel length. This value is then rounded up or down to the closest even, positive texel length. The last patch visited on a closed mesh will have no free boundary segment lengths. Requiring an even texel length for all segments between fixed vertices ensures that the perimeter length p of every patch is even and can therefore be mapped to a rectangular texel grid. If γ is too low, it is possible to have a patch with even boundary length that cannot be mapped to a rectangle. In such cases, we repeat the process with an increased γ so that all patches have the necessary resolution.

Corner vertices for the patch are chosen by first selecting a vertex v_{ll} on the boundary of the patch to be the lower left corner of the parameterization. We select v_{ll} by computing the bounding box for the already flattened patch, and choose the vertex closest to any of the bounding box corners. The ordered list ℓ is traversed to find the segment which contains v_{ll}. If v_{ll} is not already in ℓ, then we insert the vertex, which

divides a boundary segment into two pieces. To compute the texel distance of v_{ll} along the divided segment, we use compute v_{ll}'s fractional arc length along the segments and snap it to the closest integer texel location.

We insert the remaining three corner vertices into the ordered list by computing an appropriate texel width and height for the patch. The bounding box aspect ratio α and the texel perimeter p allows us to compute the texel width w of our rectangular patch: $w = \lfloor p/(2.0 + 2.0 * \alpha) \rfloor$. We can then search for a vertex v_{lr} to assign to the lower right corner. This vertex must be placed at w units along the boundary in counter clock-wise order from vertex v_{ll}. We search the ℓ for a vertex that is w units away from v_{ll}. If one is not found, we search the containing segment for a vertex that closest matches this distance and insert it into ℓ, making it exactly w units away from v_{ll}. The remaining two corner vertices are found in a similar manner. In rare cases, no free vertex exists within a segment needing to be subdivided. In such cases we insert a new vertex into the mesh, splitting the two adjacent triangles into four.

5.3. Patch Parameterization and Crossing Edges

After computing the texel size, boundary segment lengths, and corner positions for every patch we can parameterize each patch into a rectangle. The boundary segment lengths and corner vertices determine the location of the boundary of each patch in its local parameter domain. Vertices interior to the patch are solved using a linear spring model [Flo97] which guarantees a valid embedding of the mesh.

Forcing patches into square regions requires special handling to avoid degenerate triangles. We refer to any edge which is not on the boundary of the patch but whose end vertices sit on the boundary as a *crossing edge*. Crossing edges form degeneracies if the vertices of this edge are mapped to the same edge of the rectangular parameter boundary.

Assigning patch boundary vertices in such a way that crossing edges do not result in degeneracies is not always possible. Furthermore, the location of crossing edges may require choosing an edge assignment which results in high

Param.	Quality	Models			
Method	Metric	Gargoyle	Horse	Dragon	Feline
	SE		80.0%		
TMPM	PE		70.0%		
	TE		56.0%		
GI	TE	67.8%	32.4%	42.4%	33.3%
	SE	98.7%	99.2%	92.7%	99.1%
MCGI	PE	72.7%	75.6%	73.1%	75.6%
	TE	71.8%	75.0%	67.8%	74.9%
	SE	72.7%	75.0%	64.8%	67.6%
RMCGI	PE	83.6%	81.7%	82.9%	80.1%
	TE	60.9%	61.3%	53.7%	54.4%

Table 1: *Comparisons of efficiency metrics: stretch (SE), packing (PE) and texture (TE = SE×PE), for Texture Mapping Progressive Meshes (TMPM), Geometry Images (GI), Multi-chart Geometry Images (MCGI) and our Rectangular Multi-chart Geometry Images (RMCGI). PE for RMCGI was measured for packing into a single texture, whereas for per-chart texture maps TE = SE.*

distortion, negating the benefit of rectangular patches. To avoid this problem altogether, we perform local remeshing of any crossing edge that leads to a degeneracy by inserting a new vertex along the edge.

5.4. Parameterization Optimization

To improve the parameterization, we apply non-linear optimization using the stretch metric from [SSGH01]. Our current strategy uses the relaxation scheme proposed by [YBS04]. Stretch error is further minimized by visiting vertices one at a time, optimizing their location in their one-ring by performing a conjugate gradient descent. Non-fixed vertices on patch boundaries may be optimized with one degree of freedom by transforming the two pieces of the one-ring straddling the boundary into a single coordinate system, applying the optimization, and transforming each piece back to its own local coordinate system. Our current implementation guarantees C^0 continuity; higher continuity requires additional constraints [KSG03].

6. Results and Discussion

Distortion Forcing each patch into a perfect rectangle requires additional distortion when compared with algorithms which permit natural chart boundaries. However, increased distortion is balanced well by the high packing efficiency of rectangular charts.

Table 1 measures and compares the distortion of our rectangular multi-chart geometry images (RMCGI) using the efficiency metrics of Sander *et al.* [SSGH01] (including *stretch efficiency:* the area weighted average of the inverted stretch of each surface patch). RMCGI's mapping of each

Model	Tris Charts	Cluster Time/Rate	Param. Time/Rate	Packing Time/Rate
Horse	97K	1m:22s	2m:15s	4.23s
	68	1190 Δ/s	720.2 Δ/s	16.08 □/s
Gargoyle	200K	6m:35s	5m:30s	5.76s
	128	505.7 Δ/s	606.3 Δ/s	22.22 □/s
Feline	100K	3m:58s	1m:39s	4.78s
	120	419.5 Δ/s	1011 Δ/s	25.10 □/s
Dragon	150K	6m:30s	3m:46s	7.98s
	164	384.7 Δ/s	663.8 Δ/s	20.55 □/s
Jerry	94860	1m:30s	1m:50s	3.11s
	64	1060 Δ/s	863.5 Δ/s	20.58 □/s
Bunny	69451	1m:31s	1m:24s	3.69s
	64	765.6 Δ/s	822.7 Δ/s	17.34 □/s

Table 2: *Clustering, parameterization, and packing performance of our algorithm for a variety of models. Note that we use a custom conjugate gradient solver.*

Param. Method	CPU	Param. Rate	Packing Rate
BDPMP (fast)	1.0 GHz P3	7833 Δ/s	N/A
BDPMP (relax)	1.0 GHz P3	657 Δ/s	N/A
FBSPTM	2.4 GHz PC	17 Δ/s	N/A
GI	N/A (2002)	19 Δ/s	N/A
Iso-Charts	3.0 GHz Xeon	1576 Δ/s	1897 c/s
LSCM	1.3 GHz Pentium	643 Δ/s	96,968 c/s
MCGI	N/A (2003)	58 Δ/s	19 c/s
RMCGI	2.0 GHz Athlon	297 Δ/s	22,923 c/s

Table 3: *Comparisons of parameterization and packing performance (in elements per second), for Bounded-distortion Piecewise Mesh Parameterization (fast and relaxation methods), Feature-based Surface Parameterization and Texture Mapping, Geometry Images, Iso-Charts, Least Squares Conformal Mapping, Multi-chart Geometry Images and our Rectangular Multi-chart Geometry Images.*

quasi-rectangular cluster into a perfectly rectangular parameterization incurs additional distortion when compared to algorithms which allow natural chart boundaries (such as multi-chart geometry images). On the other hand, the packing efficiency of these rectangular charts is quite high, and PE=100% if each chart is assigned its own texture map, which results in competitive overall texture efficiency.

We grow surface clusters using a stretch bound of 2.0. Fig. 7 shows that the distortion of RMCGI patches peaks near their boundaries, due in part to the final addition of straggler triangles to the rectangular clusters.

Performance Table 2 decomposes the RMCGI execution time per task and per element. The performance of the shape-constrained *k*-means clustering algorithm is influenced by a complicated combination of mesh size and shape. Parameterization is more clearly a function of the number of charts, modulated by the complexity of the cluster bound-

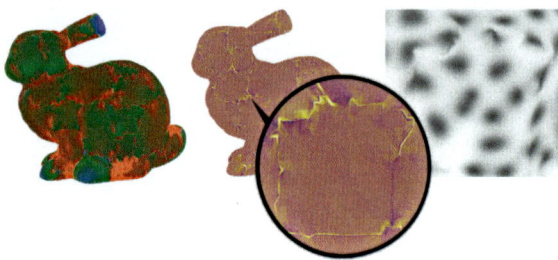

Figure 7: *Parametric stretch is plotted over the bunny (upper-left); red/blue indicate over/under-sampling. Its gradient (right) reveals distortion discontinuities at patch boundaries that can interfere with some PDE processes if special care is not taken to account for distortion.*

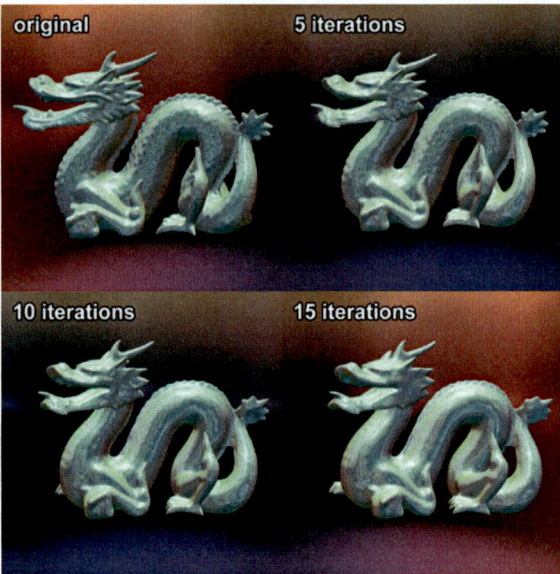

Figure 8: *The Dragon after various iterations of feature-preserving smoothing implemented as a bilateral chart image filter. Smoothing for a 1.4M triangle Dragon was performed at about 7 iterations per second in graphics hardware (GeForce 7800).*

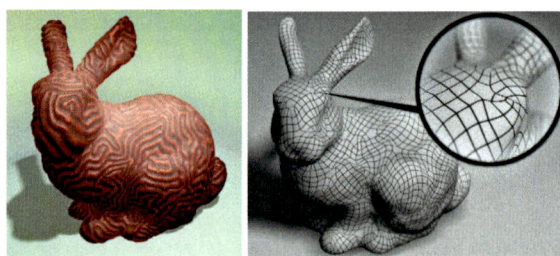

Figure 9: *Reaction diffusion texture synthesis (left) and Catmull-Clark subdivision (right) over the surface of the Stanford bunny.*

aries which in turn is affected by curvature distribution and tessellation frequency. We pack charts into a single texture atlas using the CompaSS rectangular packing software [CM04]. Packing rectangles is significantly easier and more efficient than for other chart shapes.

Table 3 compares RMCGI performance to the median per-triangle parameterization and per-chart packing rates reported for other approaches. While such averages ignore algorithmic complexity, they show that RMCGI runs slower than pure parameterization methods but faster than other regular retessellations which produce geometry-image structures.

Limitations On low-polygon meshes with irregular tessellation, our method produces jagged boundaries, due to a limited number of available decompositions. One might implement a triangle splitting scheme to assist smoothing of chart boundaries, which could also reduce parameteric distortion. Continuity near boundaries could also be improved by a computing per-texel metric from a smooth surface corresponding to the mesh. Figure 7 illustrates minor artifacts near a stretch discontinuity while simulating reaction-diffusion, though these kinds of artifacts are not universally apparent (e.g., 9). Metric discontinuity along chart boundaries is not unique to our approach.

Because our charts are rectangular, we could achieve higher memory efficiency by storing each chart in its own texture, avoiding the wasted texture memory due to gaps between charts in the texture atlas. Current graphics cards only allow sixteen textures to be bound simultaneously, but future (DirectX 10-based) hardware lifts this restriction.

7. Applications

We have implemented several surface processing applications, using a GPU's fragment shaders to convolve data stored in rectangular regions of a single texture. To ensure well-defined neighbors at patch boundaries, we copy a border of texels from neighboring patches between each iteration. Vector data is reoriented into the local parameteric basis with a simple 2x2 matrix multiply. Handling of texels near extraordinary vertices varies depending on the application.

Real-Time Fluid Flow on Surfaces. Our final application solves the incompressible Navier-Stokes equations for fluid flow over the mesh surface. The implementation is based on Stam's stable fluid method [Sta99] and the GPU implementation detailed in [Har04]. Figure 10 shows our results. We achieved 20 fps on a simulation grid with 206,632 cells. Remaining distortion could be corrected using operators in curvilinear coordinates [Sta03].

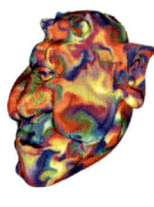

Figure 10: *Results of user interaction with real-time fluid flow evaluated by the GPU using a RMCGI.*

Texture Synthesis. A seamless parameterization is useful for synthesizing textures which are continuous over the surface. We generate wrinkles on the Stanford Bunny (figure 9) using the method of Witkin & Kass [WK91]. While the metric tensor corrects for element area and shape, we find distortion nevertheless affects the result further motivating our efforts to generate low-distortion chart images.

Figure 11: *Isocontours of approximately equal geodesic distance, computed with a regular kernel in a fragment shader using the fast marching method. (Left to right: Feline, Jerry, Dragon, Bunny.)*

Geodesics. Solving the Eikonal PDE $|\nabla T| = 1$ after setting $T = 0$ at a single source vertex yields a distance field over the surface, shown in Figure 11. Geodesic paths from any vertex to this source vertex can then be found by a simple downhill flow on T. The fast marching method [Set96] gives an approximate solution to the Eikonal equation on a regular grid. We apply RMCGIs to a variation on the fast marching method that better preserves the circular shape of the front over the mesh [NK02] using a fragment shader kernel operating on triangles constructed from adjacent fragments. Extraordinary vertices confound this approach, so their distance values are simply averaged from their neighbors. Jagged contours distant from the seed point are an artifact of the approximation method and are unrelated to texture stretch.

Quad Meshes. Geometry images [GGH02, SWG*03] implicitly encode a regularly tessellated mesh in a texture. Figure 9 shows a quad mesh reconstructed from a RMCGI of the bunny, rendered as a Catmull-Clark surface. The closeup shows how geometry is reconstructed near extraordinary vertices in the patch arrangement: each texel adjacent to an extraordinary vertex becomes a vertex of an n-gon, which can be further tessellated into quads and triangles if desired. (This set of quads is unrelated to the arrangement of rectangular charts formed by the atlas, which itself may contain T-junctions.)

Smoothing. We implement feature-preserving smoothing with bilateral filtering [TM98]. Seamless smoothing along chart boundaries is handled by the boundary copy, yielding simpler implementation compared to other parameter-space smoothing methods [ST04]. Extraordinary vertices are handled by averaging among the same texels used to construct the n-gon in quad mesh reconstruction. The 1.4M triangle Dragon in figure 8 was smoothed at a rate of approximately seven smoothing iterations per second.

8. Conclusion

We have demonstrated a novel method for generating low distortion rectangular charts on manifold surfaces. These charts provide good utilization of graphics hardware, and the low distortion mapping assists the fidelity of a signal processed on the surface. We have demonstrated the usefulness of this map by solving various PDEs over the surface domain using regular image operations on rectangular charts. These applications would yield greater error if solved over a single geometry image [GGH02] and would be less efficient and harder to implement with multi-chart geometry images [SWG*03]. Compromises of C^0 continuity and somewhat increased storage are justified by the efficiency and simplicity of rectangular charts, making PDE-based effects more accessible for interactive graphics applications.

Acknowledgments. This research was supported in part by the NSF under ITR OCI-0113968 and SGER SCI-0432257, and by NVIDIA. Result images were rendered in NVIDIA's Gelato. Thanks to David Gu and Pedro Sander for providing mesh data, and to Hui Fang and Yizhou Yu for thoughtful advice.

References

[BMRJ04] BOIER-MARTIN I., RUSHMEIER H., JIN J.: Parameterization of triangle meshes over quadrilateral domains. In *Proc. SGP* (2004), pp. 197–208.

[BSCO01] BERTALMIO M., SAPIRO G., CHENG L.-T., OSHER S.: Variational problems and pdes on implicit surfaces. In *Proc. Workshop on Variational and Level Set Methods in Computer Vision* (2001), pp. 186–193.

[BX03] BAJAJ C. L., XU G.: Anisotropic diffusion of surfaces and functions on surfaces. *ACM TOG 22*, 1 (2003), 4–32.

[CH02] CARR N. A., HART J. C.: Meshed atlases for real-time procedural solid texturing. *ACM TOG 21*, 2 (2002), 106–131.

[CH04] CARR N. A., HART J. C.: Painting detail. *ACM TOG 23*, 3 (2004). (Proc. SIGGRAPH).

[CM04] CHAN H. H., MARKOV I. L.: Practical slicing and non-slicing block-packing without simulated annealing. In *Proc. GLSVLSI* (2004), pp. 282–287.

[CSAD04] COHEN-STEINER D., ALLIEZ P., DESBRUN M.: Variational shape approximation. *ACM TOG 23*, 3 (2004). (Proc. SIGGRAPH).

[DBG*06] DONG S., BREMER P.-T., GARLAND M., PASCUCCI V., HART J. C.: Spectral surface quadrangulation. *ACM TOG 25*, 3 (2006). (Proc. SIGGRAPH).

[Flo97] FLOATER M. S.: Parametrization and smooth approximation of surface triangulations. *CAGD 14*, 4 (1997), 231–250.

[GGH02] GU X., GORTLER S. J., HOPPE H.: Geometry images. In *Proc. SIGGRAPH* (2002), pp. 355–361.

[GLM96] GOTTSCHALK S., LIN M. C., MANOCHA D.: OBBTree: A hierarchical structure for rapid interference detection. In *Proc. SIGGRAPH* (1996), pp. 171–180.

[GY03] GU X., YAU S.-T.: Global conformal surface parameterization. In *Proc. SGP* (2003), pp. 127–137.

[Har04] HARRIS M.: Fast fluid dynamics simulation on the gpu. In *GPUGems*. 2004, pp. 637–665.

[Hau01] HAUSNER A.: Simulating decorative mosaics. In *Proc. SIGGRAPH* (2001), pp. 573–580.

[JKS05] JULIUS D., KRAEVOY V., SHEFFER A.: D-charts: Quasi-developable mesh segmentation. 581–590. (Proc. Eurographics).

[KLS03] KHODAKOVSKY A., LITKE N., SCHROEDER P.: Globally smooth parameterizations with low distortion. *ACM TOG 22*, 3 (2003), 350–357. (Proc. SIGGRAPH).

[KSG03] KRAEVOY V., SHEFFER A., GOTSMAN C.: Matchmaker: constructing constrained texture maps. *ACM TOG 22*, 3 (2003), 326–333. (Proc. SIGGRAPH).

[LPRM02] LEVY B., PETITJEAN S., RAY N., MAILLOT J.: Least squares conformal maps for automatic texture atlas generation. In *Proc. SIGGRAPH* (2002), pp. 362–371.

[LWC05] LUI L., WANG Y., CHAN T. F.: Solving PDEs on manifold using global conformal parameterization. In *Proc. Variational, Geometric, and Level Set Methods in Computer Vision* (2005), pp. 307–319.

[NK02] NOVOTNI M., KLEIN R.: Computing geodesic paths on triangle meshes. In *Proc. WSCG* (2002), pp. 341–348.

[PCK04] PURNOMO B., COHEN J. D., KUMAR S.: Seamless texture atlases. In *Proc. SGP* (2004), pp. 67–76.

[RLL*05] RAY N., LI W. C., LEVY B., SHEFFER A., ALLIEZ P.: Periodic global parameterization. To appear: ACM TOG, pending revision, 2005.

[SCOGL02] SORKINE O., COHEN-OR D., GOLDENTHAL R., LISCHINSKI D.: Bounded-distortion piecewise mesh parameterization. In *Proc. IEEE Vis.* (2002), pp. 355–362.

[SdS00] SHEFFER A., DE STURLER E.: Surface parameterization for meshing by triangulation flattening. *Proc. Meshing Roundtable* (2000), 161–172.

[Set96] SETHIAN J.: A fast marching level set method for monotonically advancing fronts. *Proc. Nat. Acad. Sci 93*, 4 (1996), 1591–1595.

[SH02] SHEFFER A., HART J. C.: Seamster: Inconspicuous low-distortion texture seam layout. *Proc. IEEE Vis.* (2002), 291–298.

[SSGH01] SANDER P. V., SNYDER J., GORTLER S. J., HOPPE H.: Texture mapping progressive meshes. In *Proc. SIGGRAPH* (2001), pp. 409–416.

[ST04] SIBLEY P. G., TAUBIN G.: Atlas-aware laplacian smoothing. In *Poster: IEEE Vis.* (2004), p. 598.27.

[Sta99] STAM J.: Stable fluids. In *Proc. SIGGRAPH* (1999), pp. 121–128.

[Sta03] STAM J.: Flows on surfaces of arbitrary topology. *ACM TOG 22*, 3 (2003), 724–731. (Proc. SIGGRAPH).

[STK02] SHLAFMAN S., TAL A., KATZ S.: Metamorphosis of polyhedral surfaces using decomposition. *CGF 21*, 3 (2002), 219–228.

[SWG*03] SANDER P. V., WOOD Z. J., GORTLER S. J., SNYDER J., HOPPE H.: Multi-chart geometry images. In *Proc. SGP* (2003), pp. 146–155.

[SY04] SHI L., YU Y.: Inviscid and incompressible fluid simulation on triangle meshes. *Comp. Anim. and Vir. Worlds 15*, 3-4 (2004), 173–181.

[THCM04] TARINI M., HORMANN K., CIGNONI P., MONTANI C.: PolyCube-maps. *ACM TOG 23*, 3 (2004), 853–860. Proc. SIGGRAPH.

[TM98] TOMASI C., MANDUCHI R.: Bilateral filtering for gray and color images. In *Proc. ICCV* (1998), p. 839.

[WK91] WITKIN A., KASS M.: Reaction-diffusion textures. *Computer Graphics 25*, 3 (1991). (Proc. SIGGRAPH).

[YBS04] YOSHIZAWA S., BELYAEV A. G., SEIDEL H.-P.: A fast and simple stretch-minimizing mesh parameterization. In *Proc. SMI* (2004), pp. 200–208.

[ZMT05] ZHANG E., MISCHAIKOW K., TURK G.: Feature-based surface parameterization and texture mapping. *ACM TOG 24*, 1 (2005), 1–27.

[ZSGS04] ZHOU K., SNYDER J., GUO B., SHUM H.-Y.: Iso-charts: Stretch-driven mesh parametrization using spectral analysis. In *Proc. SGP* (2004), pp. 45–54.

Eurographics Symposium on Geometry Processing (2006)
Konrad Polthier, Alla Sheffer (Editors)

Automatic and Interactive Mesh to T-Spline Conversion

Wan-Chiu Li   Nicolas Ray   Bruno Lévy [†]

INRIA-ALICE   University Nancy 2   INRIA-ALICE

Abstract

In Geometry Processing, and more specifically in surface approximation, one of the most important issues is the automatic generation of a quad-dominant control mesh from an arbitrary shape (e.g. a scanned mesh). One of the first fully automatic solutions was proposed by Eck and Hoppe in 1996. However, in the industry, designers still use manual tools (see e.g. cyslice). The main difference between a control mesh constructed by an automatic method and the one designed by a human user is that in the second case, the control mesh follows the features of the model. More precisely, it is well known from approximation theory that aligning the edges with the principal directions of curvature improves the smoothness of the reconstructed surface, and this is what designers intuitively do.

In this paper, our goal is to automatically construct a control mesh driven by the anisotropy of the shape, mimicking the mesh that a designer would create manually. The control mesh generated by our method can be used by a wide variety of representations (splines, subdivision surfaces . . .).

We demonstrate our method applied to the automatic conversion from a mesh of arbitrary topology into a T-Spline surface. Our method first extracts an initial mesh from a PGP (Periodic Global Parameterization). To facilitate user-interaction, we extend the PGP method to take into account optional user-defined information. This makes it possible to locally tune the orientation and the density of the control mesh. The user can also interactively remove edges or sketch additional ones. Then, from this initial control mesh, our algorithm generates a valid T-Spline control mesh by enforcing some validity constraints. The valid T-Spline control mesh is finally fitted to the original surface, using a classic regularized optimization procedure. To reduce the L^∞ approximation error below a user-defined threshold, we iteratively use the T-Spline adaptive local refinement.

1. Introduction

With the advent of scanning technology, it is now reasonably easy to obtain a computer representation of an existing object. Initiatives such as the `aim@shape` network of excellence [aim] federate research efforts in this direction. However, the pioneer Henri Gouraud often mentioned in his talks that even if it is no longer necessary to draw a mesh onto the "object" to digitize it (see the image of Gouraud's wife, from his Ph.D.), constructing a "good" quad-based control mesh from a shape is one of the most important issues in geometry processing. This was also identified by Malcolm Sabin [Sab] as one of the major challenges

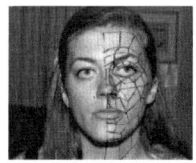

in geometry processing. The one million triangle mesh created by a 3D scanner would be completely different from the polygons carefully chosen and drawn by Gouraud on his wife's face. More precisely, due to convergence properties of the approximation [d'A00], it is well known to both skilled designers and researchers in geometry processing [ASD*03] that an efficient quad-based control mesh needs to adapt the anisotropy of the surface. As explained by d'Azevedo, the edges of the control mesh need to be orthogonal (conformality) and aligned with the principal directions of curvature. The existing automatic solutions [EH96] do not meet this requirement, this is why designers still use interactive tools (e.g. [Rap], [Cyb]). Therefore, converting a scanned mesh into an anisotropy-adapted control-mesh remains a user-intensive process.

In this paper, based on recent advances in Geometry Processing [RLL*06], we present a new method to automatically

---

[†] {wan-chiu.li|ray|levy}@loria.fr

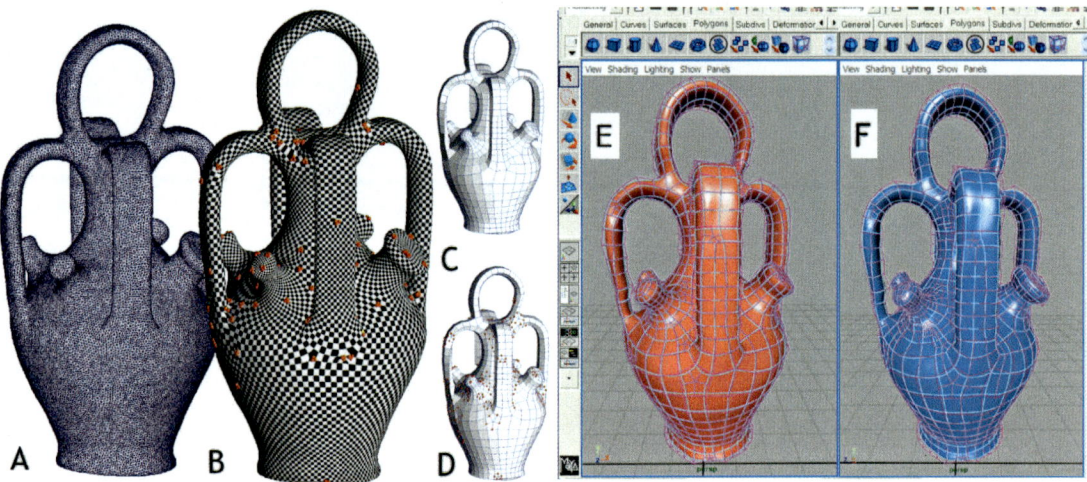

Figure 1: *A: initial triangulated surface; B: PGP (periodic global parameterization). singularities are indicated in red; C: initial control mesh, extracted from the PGP; D: valid T-Mesh; E: T-Spline fitted to the original surface; F: L^∞ fitting with adaptive local refinement.*

construct a quad-dominant control-mesh from a triangulated mesh. In addition, the automatically-constructed control-mesh can be interactively modified by the user. For instance, the user may want to add some control points for further free-form editing. Our method may be used for constructing a wide variety of representations (B-Splines, Nurbs, Catmull-Clark). In the frame of this paper, we demonstrate the automatic conversion of scanned meshes into the T-Spline representation [SZBN03].

Our method is composed of the following steps:

- **extract a quad-dominant control mesh:** We use state-of-the-art methods, such as PGP (Periodic Global Parameterization) shown in Figure 1-B, and extract the control mesh from it (Figure 1-C);

- **manually update the control mesh (optional):** we extend PGP to take into account user-defined information. We also show how the control-mesh may be manually edited, by interactively editing geodesics traced on the original surface;

- **enforce control-mesh validity constraints for T-Splines:** the control mesh of a valid T-Spline is supposed to satisfy some validity conditions. The control mesh obtained at the previous step is transformed to satisfy these conditions (Figure 1-D);

- **fit the control-mesh to the surface:** this is achieved by a classical regularized fitting procedure. The result is fully compatible with readily available industrial software (see Figure 1-E). To minimize the L^∞ norm below a user-defined threshold, we perform adaptive local refinement (see Figure 1-F).

The paper is organized as follows. The remainder of this introduction reviews the previous work. Section 2 shows how to extract an initial control mesh, and presents some tools to interactively modify it. Section 3 explains how to convert this initial control mesh into a valid T-Spline/T-NURCC control mesh (or T-Mesh). The fitting procedure is then explained in Section 4. The paper concludes with some results and suggestions for future work.

Previous work

This section reviews the previous work related with the different steps of our algorithm, control-mesh construction, parameterization and fitting.

1. Quadrilateral control mesh

Manual methods: The trivial solution to obtain a quad-remeshed version of the surface is to draw the boundary curves on the mesh manually as proposed in [KL96] and [MBVW95]. [LLP05] proposed a method with combinatorial data structure which facilitates this kind of curve drawing tasks on meshes. Some commercial softwares, for example, Rapidform [Rap] and Cyslice by Cyberware [Cyb] provide skill designers with tools to patch the objects manually. To provide even more efficient processing, some of them even provide templated-patching for objects with similar shapes and genuses. However, manual patching still requires skilled 3D model designers to obtain a satisfactory effect.

Regular and semi-regular methods: By using a cut-graph that turns a surface into a topological disk, which is then parameterized into a square domain, one can obtain a fully regular control mesh by resampling the geometry image [GGH02]. To produce semi-regular quadrilateral control meshes, [EH96] employed the technique of triangle merging. [BMRJ04] proposed a method based on discrete Lloyd relaxation. Most recently, [DBG*06] have proposed an algorithm that is based on the fact that the Morse-Smale complex induced by any piecewise linear function quadrangulates the surface.

Quad-dominant methods: Some methods consider the anisotropy of the surface since it is optimum from a function approximation point of view. [ASD*03] proposed quad-dominant remeshing method which adapts the anisotropy of the object by explicit integration of the curvature tensor through a parameterization of the surface. Later, [MK04a] improved the method by doing the integration directly on the surface without the need of the parameterization. Nonetheless, explicit integration of stream lines is always plagued by the problem of the seeding and placement of the stream lines. [DKG05] used mixed implicit/explicit schemes with harmonic functions to obtain the control mesh. Finally, [RLL*06] proposed a method that produces the quad-dominant control mesh through the calculation of two periodic functions by implicit integration of a pair of orthogonal vector fields (for example, the two eigenvector fields of the curvature tensor). The quad-dominant control mesh and the parameterization emerge simultaneously from the global numerical optimization process. Seeing the advantages of this method, we have used it to generate our initial quad-dominant control mesh. A more detailed review of this method will be given in Section 2.

2. Global parameterization Fitting a parametric representation to a mesh is much easier if the mesh is parameterized. The first class of methods considers a texture atlas, with multiple charts. Special care needs to be taken considering the smoothness of the parameterization when crossing a chart boundary. In [KLS03], all the charts are simultaneously optimized, with respect to an energy functional taking into account inter-chart transition functions. A similar approach is used in [THCM04]. The second class of methods considers a single "global" parameterization. Gu *et. al*'s approach [GY03] computes the *conformal structure* of a surface. Gortler *et. al* proposed in [GGT04] a general *discrete one-forms* formalism to study this type of methods and prove their validity. Ray *et. al* propose in [RLL*06] the PGP method (Periodic Global Parameterization). We used this latter one, as explained in Section 2.

3. Fitting B-Splines are the most popular representation in CAD/CAM. However, local refinement cannot be done with this representation, since a single control point cannot be inserted without propagating an entire row or column of control point, as done in [EH96]. Surface approximation with subdivision surfaces was also studied [LMH00, LLS01]. These methods minimize some L^∞ errors by subdividing the control mesh *globally*. In order to perform adaptive local insertion of control vertices so as to achieve minimization

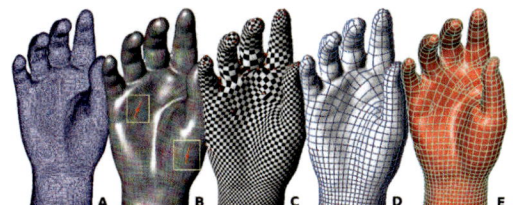

Figure 2: *A: Original surface; B: smoothed control vector field $\vec{K}$ and $\vec{K}^\perp$, interpolating two user-defined directional constraints; C: periodic global parameterization. Singularities are shown in red; D: initial control-mesh; E: A valid T-Mesh.*

of some L^∞ errors, [MK04b] proposed to use Loop subdivision surface with a *local* adaptive refinement procedure. Although this method give satisfactory results, the problem of surface approximation with adaptive local refinement using spline surface is remain unsolved. In the specific case of terrains, T-spline surface fitting to Z-Map models was studied in [ZWS05]. In this paper, we study the T-Spline fitting problem with surfaces of arbitrary topology. Note that other CAD/CAM representations with local refinement capability exist [FB88] (Please refer to [SZBN03] for a summary of the similar methods). Among the possible representations, we chose T-Splines since they can be easily converted to NURBS and subdivision surfaces, and since they offer interesting local refinement capabilities (see further).

2. Creating a Control-Mesh

2.1. Review of Periodic Global Parameterization

To automatically compute an initial quad-dominant control mesh, we use the PGP method described in [RLL*06]. This section gives a quick overview of the method. Given a surface S, two orthogonal unit vector fields $\vec{K}$, $\vec{K}^\perp$ (in practice, $\vec{K}^\perp$ can be determined from $\vec{K}$ and the surface normal), and a user-defined preferred quad edge length ω, PGP generates two periodic functions θ and ϕ defined over S, with their gradients aligned to $\vec{K}$ and $\vec{K}^\perp$ respectively. The control mesh will be then found by extracting certain level sets of θ and ϕ, as explained in the next section.

In our case, since we want to generate an anisotropy-adapted control mesh, the control vector fields $\vec{K}$ and $\vec{K}$ are obtained by smoothing an estimate of the principal directions of curvatures, as in [ASD*03].

The functions θ and ϕ are found by minimizing the following energy functional:

$$F_{PGP} = \int_S \left(\|\nabla\theta - \omega\vec{K}\|^2 + \|\nabla\phi - \omega\vec{K}^\perp\|^2 \right) dS \quad (1)$$

This method is similar to the quad-dominant remeshing in [DKG05], with the two following differences. First, we take into account the geometry of the surface, through the control vector fields $\vec{K}$ and $\vec{K}^{\perp}$. Then, by optimizing F_{PGP} with respect to alternative variables $(u, v) = \cos(\theta), \sin(\theta)$ (resp $(u', v') = \cos(\phi), \sin(\phi)$), we can use implicit integration for both directions, which avoids the uneven spacing of edges and open loops encountered with explicit integration. The resulting energy is a sum of quadratic terms, defined on the edges of the triangulation:

$$F_{PGP} = \sum_{i,j \in E} F_{edge}(i, j)$$

$$F_{edge}(i, j) = \left\| \begin{pmatrix} u_i \\ v_i \end{pmatrix} - \begin{pmatrix} \cos(\beta_{ij}) & -\sin(\beta_{ij}) \\ \sin(\beta_{ij}) & \cos(\beta_{ij}) \end{pmatrix} \begin{pmatrix} u_j \\ v_j \end{pmatrix} \right\|$$

$$\vec{K}_{ij} = 0.5(\vec{K}_i + \vec{K}_j) \quad ; \quad \beta_{ij} = \omega \vec{K}_{ij} \cdot \vec{e}_{ij}$$

$$(2)$$

where, $E = \{\vec{e}_{ij}\}$ is the set of edges of the mesh from the ith to the jth vertex. To minimize F_{PGP}, we fix one of the vertices $u_1 = 1, v_1 = 0, u'_1 = 1, v'_1 = 0$ and minimize F_{PGP} with respect to all the other variables. Since F_{PGP} is a quadratic form, this means solving a sparse symmetric system. We use the conjugate gradient algorithm with Jacobi's preconditioner. The reader is referred to [RLL*06] where some refinements of the method are explained, such as coupling θ and ϕ, modulating the edge length ω, or using an improved solution mechanism for large meshes.

Algorithm 1 Propagation from (θ_i) to (θ_j) along the edge (i, j)

$$\theta_j \leftarrow \arctan(v_i/u_i) \quad ; \quad \vec{n} \leftarrow \vec{e}_{ij}/\|\vec{e}_{ij}\|$$
$$s \leftarrow \operatorname{argmin}_s \left| \theta_i - (\pi/\omega)\vec{n} \cdot (\vec{K}_i + \vec{K}_j) - \theta_j + 2s\pi \right|$$
$$\theta_j \leftarrow \theta_j + 2s\pi$$

Once the alternative variables $(u, v) = (\cos(\theta), \sin(\theta))$ are computed (resp. $(u', v') = (\cos(\phi), \sin(\phi))$), once needs to compute the corresponding variables θ (resp. ϕ). To achieve this, we need to disambiguate the value of θ among all the possible values $\theta + 2k\pi$.

This is done in each triangle individually, by starting at a given vertex of the triangle $\mathbf{p}_1$ with $\theta = \arctan(v_1/u_1)$, and propagating along the three edges, using Algorithm 1. This algorithm chooses among all the possible values of θ_j the one nearest to the optimum value $\theta_i + \beta_{ij}$, where $\beta_{i,j} = \omega \vec{K}_{ij} \cdot \vec{e}_{ij}$ corresponds to the optimum displacement along the edge (i, j).

Once we have reconstructed the parameterization in each individual triangle, we need to check some validity conditions. More precisely, angles around a vertex should sum to 2π, angles around a triangle should sum to π, and vertex neighborhoods should satisfy the wheel compatibility condition (see e.g., [SdS01]). Vertices and triangles that violate these conditions will be referred to as *singular* in what follows.

2.2. Extracting the Control Mesh

Once we have computed the local parameterization in each triangle T, we construct the chart layout. In our setting, the chart boundaries are defined to be the iso-$2k\pi$ lines of θ and ϕ. This defines a set of segments in each triangle. Note that if a triangle T is traversed by an iso-$2k\pi$ line of θ (resp. ϕ), this triangle translated by $2s\pi$ will be traversed at the same location by the iso-$2(k + s)\pi$ line of θ (resp. ϕ). As a consequence, the extremities of these independent segments match along the edges of the triangulation, and the segments form continuous polygonal lines.

Algorithm 2 Control-mesh layout

> **compute chart boundaries:**
> **for each triangle** T
>   **if** T is non-singular
>   **for** $k \in \mathbb{N}$ such that $2k\pi \in [\min_T(\theta), \operatorname{Max}_T(\theta)]$
>   **Line** $l \leftarrow$ line of equation $(\theta = 2k\pi)$
>   **Segment** $S \leftarrow l \cap T$ //in parameter space
>   store S in T
>   store the extremities of S in the corresponding edges of T
>   **end**//for
>   repeat the above procedure for ϕ
>   **end**//if
> **end**//for
> **for each edge** e
>   merge the segment extremities stored in e
>   that share the same geometric location in 3D
> **end**//for
> **for each triangle** T
>   compute the intersections between the edges stored in T
> **end**//for
> remove all dangling segments

Algorithm 2 gives the general algorithm to extract the initial control-mesh layout. Dangling segments, namely, segments with a free extremity, are caused by the singular vertices and triangles.

A key aspect in this algorithm is to have an efficient data structure to represent a set of lines embedded in a surface. We used the embedded simplicial complexes presented in [LLP05] with provably optimum complexity for all the operations. For instance, by storing in each triangle T the set of segments contained by T, it computes all the intersections between the iso$-\theta$ and iso$-\phi$ curves with linear complexity in the total number of triangles. This data structure will also be used in the next section, to facilitate the interactive editing of the control mesh.

2.3. Interactive Editing (*Optional*)

The so-constructed control mesh can be interactively edited by the user. First, it is possible to add constraints in the vector-field smoothing algorithm used to construct the control vector fields $\vec{K}$ and $\vec{K}^{\perp}$. Second, once the initial control-mesh is represented by an embedded simplicial complex [LLP05], all the related manual editing operations and design with geodesic curves can be applied.

Constrained vector field smoothing: To allow the user to interactively edit the control vector field, we use a variant of Hertzmann *et. al*'s method [HZ00]. In this method, the vector field is represented by a set of angles α relative to an initial arbitrary tangent vector field $\vec{t}$. We use a weighted sum of a fitting term and smoothing term.

$$R = (1-\rho)\underbrace{\sum_i \left\|\alpha_i - \alpha_i^0\right\|^2}_{\text{fitting term}} + \rho\underbrace{\sum_{\vec{e}_{ij}} \left\|\alpha_i - \alpha_j - \text{angle}(\vec{t}_i,\vec{t}_j)\right\|^2}_{\text{smoothing term}}$$

(3)

where the angles α_i^0 are computed from the initial values of the vector field $\vec{K}$ at the vertices and $\text{angle}(\vec{t}_i,\vec{t}_j)$ is defined in the range $]-\pi,\pi]$. The user-defined coefficient ρ corresponds to the desired smoothing intensity (in all our examples, $\rho = 0.8$). The smoothing term minimizes the variations of α over an edge $\vec{e}_{ij}$. As before, to make the variable independent of $2k\pi$ translations, we solve for the sines and cosines. User-prescribed directions are simply removed from the degrees of freedom. Figure 2-B shows how two user-defined directional constraints are interpolated.

Manual editing and geodesic design: Using the embedded simplicial complex data structure, all the possible relations between the initial triangulated surface and the control mesh under construction are represented with optimum access time. This makes it easy to implement the following operations (see Figure 3 and the companion video):

- edge straightening: an edge of the control mesh is replaced with a geodesic;

- edge insertion: a new edge is created between two points picked on the surface. All the intersections are kept up-to-date;

- edge deletion.

3. Constructing a T-Spline/T-NURCC

After the automatic and optional interactive steps described in the previous section, we obtain a quad-dominant con-

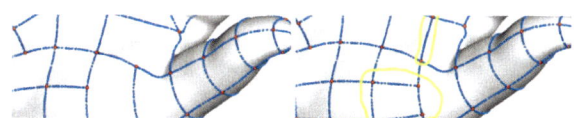

Figure 3: *Manually editing the initial control mesh by adding geodesics (circled). Note that all the intersections are kept up-to-date.*

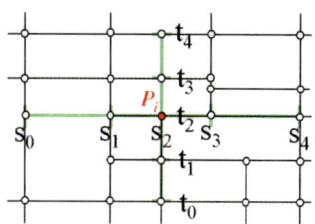

Figure 4: *The pre-image of a T-Mesh*

trol mesh. However, our target T-Splines representation has some requirements characterizing valid control meshes (or T-Meshes) that our control mesh might violate. This section presents an automatic algorithm to enforce these constraints in our control mesh. In addition, we show how to compute the knot-intervals that define the function basis attached to the T-Mesh.

3.1. T-Splines

The main limitation of CAD/CAM representations is their lack of flexibility, and the highly constrained nature of their admissible control meshes. For instance, the control-mesh of a standard B-Spline surface needs to be locally equivalent to a regular grid (a cylinder and a torus are the only possible variations). To overcome this limitation, T-Splines were proposed by Sederberg *et. al* in [SZBN03].

The control points of a T-spline surface are arranged by means of a control grid called a *T-Mesh*, where local refinements can be done. If a T-Mesh forms a rectangular grid, the T-Spline degenerates to a B-spline surface. Figure 4 shows the pre-image of a portion of a T-Mesh. However, despite their support for local refinement and T-junctions, T-Splines cannot have vertices of arbitrary valence in their control mesh. For this reason, Sederberg *et. al* also proposed in [SZBN03] the T-NURCC (Non-Uniform Rational Catmull Clark) representation, that can fill-in the neighborhoods of the extraordinary vertices, while smoothly connecting with the rest of the T-Splines.

In a T-Spline, each edge of the T-Mesh has an associated *knot interval*, that needs to satisfy the following two rules:

Rule 1: The sum of the knot intervals on opposing edges of any face must be equal.
Rule 2: If two T-junctions on opposing edges of a face can be connected without violating the previous rule, that edge must be included in the T-Mesh.

Given a valid T-Mesh, the equation of a T-Spline is defined by:

$$\mathbf{S}(s,t) = \frac{\sum_{i=1}^{n} w_i P_i B_i(s,t)}{\sum_{i=1}^{n} w_i B_i(s,t)}, \quad s,t \in D \quad (4)$$

where: P_i is the ith control point (x_i, y_i, z_i), $i = 1 \cdots n$. The blending function $B_i(s,t) = N[s_{i0}, s_{i1}, s_{i2}, s_{i3}, s_{i4}](s)N[t_{i0}, t_{i1}, t_{i2}, t_{i3}, t_{i4}](t)$, where, $N[s_{i0}, s_{i1}, s_{i2}, s_{i3}, s_{i4}](s)$ is the cubic B-spline basis function associated with the knot vector $s_i = [s_{i0}, s_{i1}, s_{i2}, s_{i3}, s_{i4}]$ and $N[t_{i0}, t_{i1}, t_{i2}, t_{i3}, t_{i4}](t)$ is associated with the knot vector $t_i = [t_{i0}, t_{i1}, t_{i2}, t_{i3}, t_{i4}]$.

The two knot vectors s_i and t_i of the two basis functions of the control point P_i are inferred from the knot information of the T-Mesh as follows (see Figure 4): Let (s_{i2}, t_{i2}) are the knot coordinates of P_i. Consider a ray in parameter space $R(\alpha) = (s_{i2} + \alpha, t_{i2})$. Then s_{i3} and s_{i4} are the s coordinates of the first two s-edges intersected by the ray in the positive α direction. A s-edge is a vertical line segment of constant s. The s_{i0}, s_{i1} and t_i are found similarly.

We first ensure T-Spline validity almost everywhere, and "push" the problems by generating new extraordinary vertices when this cannot be avoided (the neighborhood of these extraordinary vertices will be replaced by T-NURCCes, as explained in the next section):

1. whenever possible, enforce Rule 2 by inserting the missing edges in each loop;

2. for all remaining invalid loops (including N-sided loops), convert them into extraordinary vertices as explained below (see also Figure 5).

- A: In each even N-sided loop, a new vertex is inserted and connected to every two vertex of the loop. This creates N/2 additional quads and one extraordinary vertex;

- B: In each odd N-sided loop, a new vertex is inserted and connected to a new T-vertex in each edge of the loop. This creates N additional quads, N T-vertices and one extraordinary vertex;

- C: similarly, odd N-sided loops with a vertex of valence 3 on the border are remeshed as shown in Figure 5.

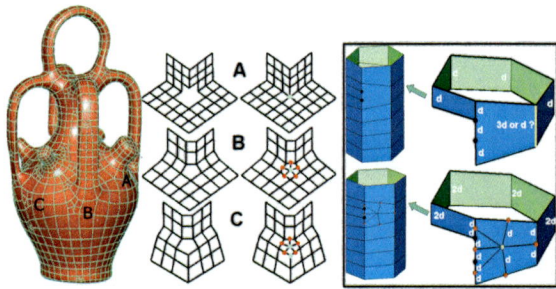

Figure 5: *Left: from N-sided polygon to extraordinary vertices; Right: consistent knot interval assigning through the use of extraordinary vertices.*

Each time an edge is subdivided, its associated knot interval is divided by two. These three operations guarantee that knot intervals remain coherent (see Figure 5-Right).

After this step, the control mesh is a valid T-Spline control mesh everywhere, except in the neighborhood of extraordinary vertices. The concerned quads are replaced with T-NURCCes, as explained in the next section.

3.2. Extraordinary vertices and T-NURCCes

A T-NURCC is a NURCC, which is a modification of cubic NURSSes [SZSS98], with T-junction in the spirit of T-splines. NURCCes is a generalization of both tensor product non-uniform B-spline surfaces and Catmull-Clark surfaces. It enforces the constraint that opposing edges of each four-sided face have the same knot interval. The enforcement of this constraint makes the local subdivision of NURCCes possible. In our implementation, we conceptually apply to each extraordinary vertex two steps of local subdivision (the figure shows a valence-3 vertex). This generates the additional control points marked in red, and expressed by linear combinations of the initial control points. The coefficients of these linear combinations (that depend on the valence of the vertex) are given in Sederberg *et. al*'s paper (and not repeated here for paper length considerations). In practice, we keep a version of the original mesh, and apply local subdivision to a copy. While applying the subdivisions, we store in the newly created control points the list of original control point they depend on together with the coefficients. This representation can be directly used in the subsequent fitting steps, as explained in the next section.

4. Fitting

For surface approximation, in order to measure a defined error metric, one needs to have a correspondence between the approximating and the original surface. Parameterization-free methods [MK04b, DIS03], mostly meant to fit point clouds, geometrically project each sample onto the approximating surface. Parameterization-based methods establish the correspondence by parameterizing the original surface and then identifying the parameter values. Our approach belongs to this latter class of methods.

Given a parameterization $(s,t) \mapsto \mathbf{S}(s,t) \in \mathbb{R}^3$ of the surface, we will minimize the following energy functional, as done in classical regularized fitting methods (see e.g. [Gre94]):

$$E = E_{fit} + \sigma E_{fair}$$

where:
$$\begin{cases} E_{fit} & = & \int \|\mathbf{S}(s,t) - \mathbf{M}(s,t)\|^2 ds dt \\ E_{fair} & = & \int \left((\frac{\partial^2 \mathbf{S}}{\partial s^2})^2 + 2(\frac{\partial^2 \mathbf{S}}{\partial s \partial t})^2 + (\frac{\partial^2 \mathbf{S}}{\partial t^2})^2 \right) ds dt \end{cases}$$
(5)

In this equation, $\mathbf{M}(s,t)$ denotes a parameterization of the original triangulated surface. As often done in Splines fitting, we approximate the fitting term E_{fit} by using a discrete set of m samples:

$$E_{fit} \simeq \sum_{k=1}^{m} \|\mathbf{S}(s_k, t_k) - (x_k, y_k, z_k)\|^2$$

For each sample (x_k, y_k, z_k) of the original surface, (s_k, t_k) denotes its coordinate in parameter space. The natural idea would be to use the original vertices of the surface, but it is better to re-sample it so that the operation is less sensitive to the resolution of the mesh. The re-sampling is done by using a regular grid of samples in each face in the parameter space (we used 10×10 samples per face in our implementation). The corresponding points on the surface is found easily by linear interpolation in the facets.

The thin-plate energy E_{fair} avoids wiggles of the spline surface. However, if the coefficient σ is set to be too large, the final spline surface may fit less to the original surface. In our examples, we used $\sigma = 0.05$.

Note that by construction, our control mesh and associated knot vector defines a standard (or semi-standard) T-Spline. Therefore, the denominators of the T-Spline is identically one, and we can focus on the numerator:

$$\mathbf{S}(s, t) = \sum_{i=1}^{n} \mathbf{P}_i \mathbf{B}_i(s, t)$$

Each coordinate x, y, z can be processed independently. For the x coordinate, the fitting term is given by:

$$E_{fit}^x = \sum_{k=1}^{m} \left(\sum_{i=1}^{n} X_i \mathbf{B}_i(s_k, t_k) - x_k \right)^2 \qquad (6)$$

where X_i (resp Y_i, Z_i) denote the coordinates at the control point $\mathbf{P}_i$. The X_i's that minimize Equation 6 are also the solution of a linear system $A^t A X = A^t b$, where the coefficients of the $m \times n$ matrix A are given by $a_{k,i} = B_i(s_k, t_k)$ and right-hand side by $b_k = x_k$. The unknown vector X corresponds to all the x coordinates of the control points. Adding the fairing term E_{fair}, the linear system becomes:

$$\left(A^t A + \sigma (A_{ss}^t A_{ss} + 2 A_{st}^t A_{st} + A_{tt}^t A_{tt}) \right) X = A^t b \qquad (7)$$

where the coefficients $(\cdot_{k,i})$ of the $m \times n$ matrices A_{ss}, A_{st}, A_{tt} are the second order derivatives $B_{iss}(s_k, t_k), B_{ist}(s_k, t_k)$ and $B_{itt}(s_k, t_k)$ of the basis functions B_i respectively.

To solve the regularized fitting problem, the remaining two difficulties are to determine the parameters (s_k, t_k) associated with the vertices of the original surface, and then accumulating the contributions of all the basis functions to construct the matrices $A, A_{ss}, A_{st}, A_{tt}$ and the right-hand-side b. These two issues will be explained in the next two sections.

4.1. Computing the parameters (s_k, t_k)

The parameterization of the original mesh is obtained as follows. As explained in the previous section, the control mesh

is composed of a set of quadrilaterals. By keeping the relation between the control mesh and the surface, it is possible to retrieve in each loop of the control mesh the set of triangles it contains (using a greedy algorithm). Then, two different cases occur, according to the presence of singularities (see Section 2).

- *none of the triangles and vertices is singular:* a parameterization can be easily reconstructed from the (θ, ϕ) values, by greedily assembling the triangle in parameter-space (see [RLL*06], [SdS01]);

- *the loop contains singular triangles and/or vertices:* we re-parameterize the interior of the loop using [Flo03].

Note that all the loops with edges modified by the user are considered to belong to the second category. After all the loop interiors are parameterized, we need to improve cross-boundary continuity. The charts that did not contain singularities already have a globally smooth parameterization. Therefore, we simply apply a relaxation procedure to the re-parameterized charts and their neighbors, by optimizing the smoothness of transition functions, as explained in [SPPH04].

4.2. Constructing and solving the linear system

To solve our regularized fitting problem (Equation 5), the most natural way would be to proceed on a patch-by-patch basis. This would traverse the matrices $A, A_{ss}, A_{st}, A_{tt}$ row by row, and would make it possible to directly construct the final matrix of the linear system without storing these intermediate matrices.

However, constructing the pre-images of each patch is non-trivial. For this reason, we prefer to iterate on the control nodes. This means we consider one basis functions $\mathbf{B}_i(s, t)$ at a time, with a simpler pre-image. As a consequence, we store the matrices $A, A_{ss}, A_{st}, A_{tt}$ and construct them column by column. After the traversal of all basis functions, the final matrix of the system is finally assembled (see Equation 7).

The basis functions are piecewise defined in a neighborhood around the pre-image of each control point $\mathbf{P}_i$ (see Figure 4). Each basis function $\mathbf{B}_i$ is completely defined by the T-Mesh and associated knot vectors around the control point $\mathbf{P}_i$. To retrieve them, we first fix arbitrary coordinates (s_0, t_0) to P_i and greedily propagate the knot intervals around it until the region of influence $D_i = [s_{i0}, s_{i4}] \times [t_{i0}, t_{i4}]$ is completely determined. The pre-image looks like the one shown in Figure 4.

According to this local parameterization, P_i influences the T-spline patches that correspond to the faces intersected by D_i. The patch of each face is mapped to $[0, 1]^2$ with $(0, 0)$ set at a corner of the face. Therefore, when the influence of P_i is added to the matrices, its pre-image needs to be pre-scaled accordingly. For example, in the pre-image of the T-Mesh, if the size of the rectangle of an influenced face, F, are d and e

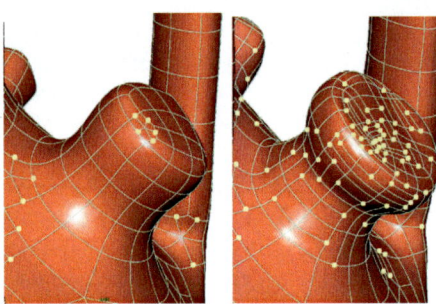

Figure 6: *Adaptive local refinement splits some faces to better capture complex geometry.*

in the s and t directions respectively, the s and t knot vectors of P_i with respect to F should be scaled by $1/d$ and $1/e$ respectively. Once the knot-vectors and region of influence D_i of the basis function B_i are determined, we update the corresponding column in the matrices A, A_{ss}, A_{st} and A_{tt}. After all control points are processed, the final matrix and right-hand side of the system are constructed. All the matrices are represented by column-major sparse data structures (CCS format, for Column Compressed Storage). We use the readily sparse direct solver TAUCS. As a consequence, the inverse of the matrix can be reused to find the X, Y and Z components of the control mesh coordinates. Note that sparse direct solvers perform so well that inversing the matrix is faster than solving a linear system with preconditioned conjugate gradient (see [BBK05], [Lev05] and the timings in the results section).

4.3. Adaptive L^∞ fitting

Global L^2 fitting operates with a fixed number of control points and thus a fixed degree of freedom is sometimes not sufficient to reconstruct the original surface. Therefore, more degree of freedom must be added. Generally, this is done by global refinement of the control mesh, which adds superfluous control points to already low approximation error regions. On the contrary, since we are using T-splines with support for local refinement, new control points can be inserted *locally* in regions of high approximation error (see Figure 6). Thanks to the local support of T-splines, there is no need to carry out the global L^2 fitting every time a new control point is added. Only a smaller linear system needs to be solved involving only the patches of the control mesh affected by the local refinement operation. The L^∞ metric is defined as follows:

$$L^\infty(\mathbf{S}, \mathbf{M}) = max_S \|(\mathbf{S}(s,t) - \mathbf{M}(s,t))\|^2 \qquad (8)$$

where $\mathbf{M}(s,t)$ denotes a parameterization of the original surface. This error metric is evaluated by regularly sampling the parameter space of each face.

We iteratively apply the local refinement procedure described below to the face of worst L^∞ approximation error until it drops below a user-defined threshold.

The local refinement of T-spline is one of its invaluable properties. It is also called local knot insertion (please see [SCF*04] for more details). New control points are inserted into the T-Mesh without changing the geometry of the original T-spline surface. The algorithm recovers the T-Spline validity constraints (Section 3) by iteratively inserting new control points:

1. Insert new control point(s) into the T-Mesh.

2. If any basis function is missing a knot dictated by Rule 1 for the current T-Mesh, perform the necessary knot insertions into that basis function.

3. If any basis function has a knot that is not dictated by Rule 1 for the current T-Mesh, add an appropriate control point into the T-Mesh.

4. Repeat Steps 2 and 3 until there are no more new operations.

The face with highest L^∞ approximation error is splitted into two rectangles. Knot intervals are updated accordingly (i.e. set to 0.5 for the subdivided edges). We compute the refinements in both directions, and choose the one which performs best in reducing the approximation error. Note that the knot-insertion algorithm may introduce a few additional control points by propagation into the T-Mesh (see Figure 6).

The T-spline local refinement algorithm preserves the geometry of the original T-spline surface. However, our goal is to use these newly introduced degrees of freedom to approximate better the original meshed surface. Therefore, a local fitting process is performed after the local refinement. Note that since the T-Splines function have local degrees of freedom, a smaller linear system needs to be solved. The vector X gathering all the X_i coordinates of the control nodes is split into X_f, the set of control points influenced by the new control point (*f*ree to move) and X_l, the set of control points that will remain *l*ocked. The new degrees of freedom and coupling terms on the boundary of the refined patch are determined by the sparsity pattern of the matrix A. The fitting term is given by:

$$F_{fit}(X_f) = \left\| [A_f | A_l] \left[\frac{X_f}{X_l} \right] - b \right\|^2$$

where A is split into A_f and A_l according to X_f and X_l. The new degrees of freedom are then given by the solution of the linear system:

$$A_f^t A_f X_f = A_f^t A_l X_l - A_f^t b$$

The terms introduced by the fairing energy have the same structure. Since we have a small number of coefficients and since the location of the new control points is not far away from the optimum, we use a conjugate gradient algorithm, that converges in a few iterations.

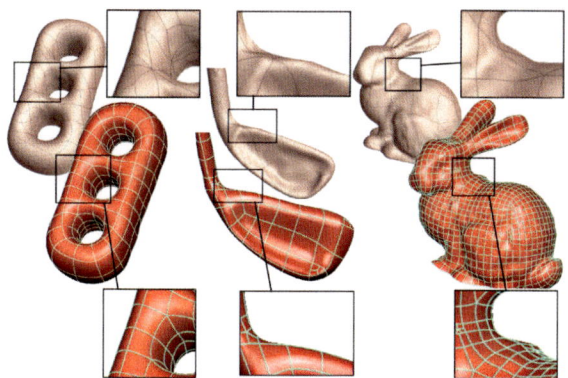

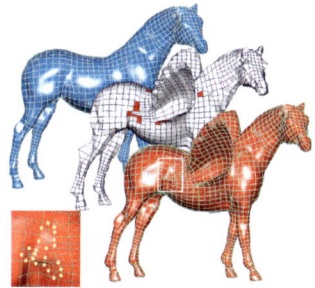

Figure 9: *Conversion from classical mesh models and interactive editing of T-Splines in Maya. The closeup shows how N-sided facets are replaced with T-NURCCes. We also show how this facilitates model editing (pasting the wings of the gargoyle onto the horse).*

Figure 7: *Comparison between the results of [EH96] and ours: note how the symmetry and anisotropy are respected by our approach.*

Figure 8: *Our method applied to a high-genus object. From left to right: initial mesh, fitted control mesh and surface. This example also shows the robustness of our method to mesh with poor quality.*

5. Results and Conclusions

Figure 7 compares Eck and Hoppe's results with ours (note that Eck and Hoppe's images reproduced here only show patch boundaries, each patch has a 4x4 control nodes array, therefore control mesh sizes are comparable). As can be seen, our method better respects the symmetries (see e.g. the three-holes torus) and the anisotropy of the objects, as a designer would do. As a consequence, the resulting surfaces do not have wrinkles (see closeups). We show in Figure 8, 9 and 10 our method applied to data sets of various topologies

and geometries. For all these examples, less than 15 edges were added by the user. Note that the rocker is topologically equivalent to a torus. Therefore, it would be possible to create a control mesh without any singularity. However, we think that the control mesh constructed by our method is a more natural, since it better takes the geometry of the object into account. This is even more obvious in the scanned hand shown in Figure 2. Since the surface is a topological disk, the control mesh could have no extraordinary vertex, but any skilled designer would create one at each finger tip, as our algorithm does, to better balance distortions and adapt the geometry.

Moreover, The *global* energy minimized by PGP overcomes the uneven placement of stream lines obtained with the *local* seeding strategy used in [ASD*03].

Table 1 gives the number of control points obtained for all the models, without and with adaptive local refinement (the L^∞ threshold was set to 0.2% of the bounding box diagonal). As far as timings are concerned, the adaptive fitting algorithm did converge in less than one minute for all these models.

Conclusion: In this paper, we have proposed a method for automatic and interactive mesh to T-spline conversion. Our algorithm proposes an initial solution, that can be manually refined by the user. Using these automatic and manual tools, a complex model can be converted in less than 15 minutes and loaded in industrial software. This is significantly faster than fully manual solutions existing in commercial software. In future work, we think that a better mathematical characterization of the singularities may even further reduce this time, by leading to a fully automatic solution.

	No. of vertices	Control nodes	Control nodes (locally refined)
rocker	23k	2021	3692
botijo	41k	1471	2644
horse	10k	2046	2724
vase	2k	2120	2891

Table 1: *Number of control nodes for various models.*

Acknowledgments

We thank the European NoE aim at shape and the INRIA ARC GEOREP for funding this research.

Figure 10: *Converting a scanned mesh into a T-Spline. From left to right: L^2 fitting, L^∞ fitting with local refinement (with and without the control mesh super-imposed).*

References

[aim]   AIM AT SHAPE:. http://shapes.aim-at-shape.net/index.php.

[ASD*03]   ALLIEZ P., STEINER D. C., DEVILLERS O., LEVY B., DESBRUN M.: Anisotropic Polygonal Remeshing. *ACM TOG (SIGGRAPH)* (2003).

[BBK05]   BOTSCH M., BOMMES D., KOBBELT L.: Efficient linear system solvers for mesh processing. In *IMA Mathematics of Surfaces XI, Lecture Notes in Computer Science* (2005).

[BMRJ04]   BOIER-MARTIN I., RUSHMEIER H., JIN J.: Parameterization of triangle meshes over quadrilateral domains. In *SGP* (2004), Eurographics.

[Cyb]   CYBERWARE:. http://www.cyberware.com/products/cyslice.html.

[d'A00]   D'AZEVEDO E.-F.: Are bilinear quadrilaterals better than linear triangles? *SIAM J. of Scientific Computing 22*, 1 (2000), 198–217.

[DBG*06]   DONG S., BREMER P.-T., GARLAND M., PASCUCCI V., HART J. C.: Spectral surface quadrangulation. In *SIGGRAPH* (2006).

[DIS03]   DODGSON N., IVRISSIMTZIS I., SABIN M.: Curve and surface fitting: Saint malo. In *Nashboro Press, St Malo, France, A. Cohen et al., Eds., vol.2* (2003).

[DKG05]   DONG S., KIRCHER S., GARLAND M.: Harmonic functions for quadrilateral remeshing of arbitrary manifolds. In *Computer Aided Geometry Design, Special Issue on Geometry Processing* (2005).

[EH96]   ECK M., HOPPE H.: Automatic reconstruction of B-spline surfaces of arbitrary topological type. In *SIGGRAPH* (1996).

[FB88]   FORSEY D. R., BARTELS R. H.: Hierarchical b-spline refinement. In *SIGGRAPH conf. proc.* (1988).

[Flo03]   FLOATER M. S.: Mean value coordinates. *CAGD 20* (2003), 19–27.

[GGH02]   GU X., GOTLER S., HOPPE H.: Geometry images. In *Siggraph* (2002).

[GGT04]   GORTLER S., GOTSMAN C., THURSTON D.: *One-Forms on Meshes and Applications to 3D Mesh Parameterization.* Tech. rep., Harvard University, 2004.

[Gre94]   GREINER G.: Variational design and fairing of spline surfaces. In *Computer Graphics Forum Volume 13, Issue 3* (1994), pp. 143–154.

[GY03]   GU X., YAU S.-T.: Global conformal surface parameterization. In *Symposium on Geometry Processing* (2003), ACM.

[HZ00]   HERTZMANN A., ZORIN D.: Illustrating smooth surfaces. In *SIGGRAPH* (2000).

[KL96]   KRISHNAMURTHY V., LEVOY M.: Fitting smooth surfaces to dense polygon meshes. *Computer Graphics 30*, Annual Conference Series (1996), 313–324.

[KLS03]   KHODAKOVSKY A., LITKE N., SCHRODER P.: Globally smooth parameterizations with low distortion. *ACM TOG (SIGGRAPH)* (2003).

[Lev05]   LEVY B.: Numerical methods for digital geometry processing. In *Israel Korea Bi-National Conference* (2005).

[LLP05]   LI W. C., LEVY B., PAUL J.-C.: Mesh editing with an embedded network of curves. In *Shape Modeling International conference proceedings* (2005).

[LLS01]   LITKE N., LEVIN A., SCHRÖDER P.: Fitting subdivision surfaces. In *Proceedings of Scientific Visualization* (2001).

[LMH00]   LEE A., MORETON H., HOPPE H.: Displaced subdivision surfaces. In *SIGGRAPH conf. proc.* (2000).

[MBVW95]   MILROY M. J., BRADLEY C., VICKERS G. W., WEIR D. J.: G1 continuity of b-spline surface patches in reverse engineering. *Computer-Aided Design 27*, 6 (1995), 471–478.

[MK04a]   MARINOV M., KOBBELT L.: Direct anisotropic quad-dominant remeshing. In *Proc. Pacific Graphics* (2004).

[MK04b]   MARINOV M., KOBBELT L.: Optimization techniques for approximation with subdivision surfaces. In *Symposium on Solid Modeling and Applications* (2004), ACM.

[Rap]   RAPIDFROM:. http://www.rapidform.com/.

[RLL*06]   RAY N., LI W. C., LEVY B., SHEFFER A., ALLIEZ P.: Periodic global parameterization, 2006. Accepted pending revisions.

[Sab]   SABIN M.:. http://www.cl.cam.ac.uk/~mas33/challenges.htm.

[SCF*04]   SEDERBERG T. W., CARDON D. L., FINNIGAN G. T., NORTH N. S., ZHENG J., LYCHE T.: T-spline simplification and local refinement. *ACM TOG (SIGGRAPH)* (2004).

[SdS01]   SHEFFER A., DE STURLER E.: Parameterization of faceted surfaces for meshing using angle based flattening. *Engineering with Computers 17* (2001), 326–337.

[SPPH04]   SCHREINER J., PRAKASH A., PRAUN E., HOPPE H.: Inter-surface mapping. *ACM TOG (SIGGRAPH)* (2004).

[SZBN03]   SEDERBERG T. W., ZHENG J., BAKENOV A., NASRI A. H.: T-splines and T-NURCCs. *ACM TOG (SIGGRAPH)* (2003).

[SZSS98]   SEDERBERG T. W., ZHENG J., SEWELL D., SABIN M.: Non-uniform subdivision surfaces. In *SIGGRAPH* (1998).

[THCM04]   TARINI M., HORMANN K., CIGNONI P., MONTANI C.: Polycube-maps. *ACM TOG (SIGGRAPH)* (2004).

[ZWS05]   ZHENG J., WANG Y., SEAH H. S.: Adaptive t-spline surface fitting to z-map models. *GRAPHITE 2005* (2005).

Eurographics Symposium on Geometry Processing (2006)
Konrad Polthier, Alla Sheffer (Editors)

Designing Quadrangulations with Discrete Harmonic Forms

Y. Tong[1]   P. Alliez[2]   D. Cohen-Steiner[2]   M. Desbrun[1]

[1]Caltech   [2]INRIA Sophia-Antipolis, France

Abstract

We introduce a framework for quadrangle meshing of discrete manifolds. Based on discrete differential forms, our method hinges on extending the discrete Laplacian operator (used extensively in modeling and animation) to allow for line singularities and singularities with fractional indices. When assembled into a singularity graph, these line singularities are shown to considerably increase the design flexibility of quad meshing. In particular, control over edge alignments and mesh sizing are unique features of our novel approach. Another appeal of our method is its robustness and scalability from a numerical viewpoint: we simply solve a sparse linear system to generate a pair of piecewise-smooth scalar fields whose isocontours form a pure quadrangle tiling, with no T-junctions.

1. Introduction

Partitioning a surface into quadrilateral regions is a common requirement in computer graphics, computer aided geometric design and reverse engineering. Such quad tilings are amenable to a variety of subsequent applications due to their tensor-product nature, such as B-spline fitting, simulation with finite elements or finite differences, texture atlasing, and addition of highly detailed modulation maps. Quad meshes are particularly useful in modeling as they aptly capture the symmetries of natural or man-made geometry, allowing artists to design simple surfaces using a quite intuitive placement of quad elements. Automatically converting a triangulated surface (issued from a 3D scanner for instance) into a quad mesh is, however, challenging. Stringent topological conditions make quadrangulating a domain or a surface a rather constrained and global problem [Ede00]. Application-dependent meshing requirements such as edge orthogonality, alignment of the elements with the geometry, sizing, and mesh regularity add further hurdles. In this paper, we propose a framework for quadrangle tiling of arbitrary triangulated surfaces that allows for a precise user-guided control over the design of the final pure-quad mesh. We show how to use discrete harmonic forms to solve for two piecewise-smooth scalar fields such that their respective isocontours create a mesh with well-shaped quadrangles at geometrically pertinent edge locations.

1.1. Previous Work

Due to their wide appeal in various communities, quad meshes have been the subject of a large number of papers presenting different algorithms for the generation of isotropic or anisotropic quad elements. Comprehensive reviews, found for instance in [ACSD*03, BMRJ04, AUGA05, DKG05], hint at a need for algorithms offering more *control* on the mesh regularity, as well as on the shape, size and alignment of the mesh elements with geometric or semantic features. A clustering-based method presented in [BMRJ04] manages to limit the number of extraordinary vertices in the final mesh, without guaranteeing the location of singularities. A recent Morse-theoretic approach in [DBG*06] pro-

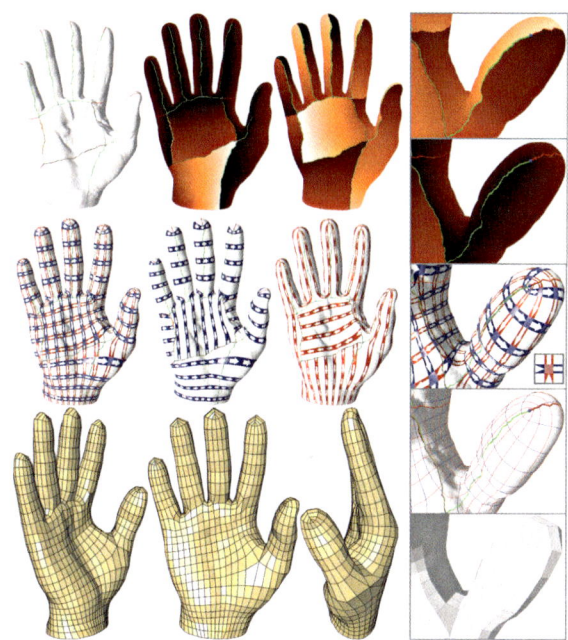

Figure 1: *Scanned Hand. From a triangulated surface and a set of line singularities assembled into a singularity graph, our technique solves a linear, modified Laplace equation to get two potentials (top); The pair of 1-forms associated to the potential differentials is specified as either regular, reverse or switch across singularity lines (center). An isocontouring of these potentials results in a pure-quad mesh with non-integer index singularities capturing the geometry (see close-up, right), and no T-junction (bottom).*

vides improved results, but the modified relaxation [KLS03] involved in this method still allows no control over design, resulting in singularities at conspicuous places and elements of arbitrary shapes. Another technique allows fully regular quadrilateral meshes (except along a seam) through the use of holomorphic discrete 1-forms [GY03]. Unfortunately, this holomorphic requirement leaves little control over the local alignment of the mesh elements and creates potentially large area distortion, even after optimization [JWYG04]. A recent technique proposes a radically different approach to conformal parameterization with *arbitrary* cone singularities [KSS06]: distortion is concentrated at carefully chosen places so as to allow better, global control of area distortion and hence over mesh sizing. Unfortunately, this non-linear method cannot directly control alignment with features, and/or guarantee proper matching of quads through patch boundaries.

A recent trend towards a better control of alignment focuses on vector field topology. A first approach proposed in [ACSD*03, MK04] consists in tracing *curvature lines*, thereby enforcing proper alignment of the mesh edges while creating a natural quad-dominant network. The placement of these lines are based on local decisions, resulting in numerous hanging lines all over the mesh: T-junctions and poor regularity of the mesh result from this greedy line selection. When targeting higher mesh regularity, a better approach defines these lines as isolevels and steepest descents of a *global* potential [DKG05]. As a result of this type of contouring, the lines are either closed curves (so-called isoparametric flow lines), or streamlines (gradient flow lines) obtained by numerical integration, leading to less T-junctions and irregularity on the final quad mesh. This method allows some design control through user-defined selection of a number of local extrema of the potential (be they points, or even polylines). However, each local extremum corresponds to an index 1 in the gradient field of the potential; because of the Hopf-Poincaré index theorem, this means that a number of *other* singularities (most likely saddles, of index -1) will be consequently created too, as the indices of all singularities of the vector field must sum up to the Euler characteristic of the surface. Therefore, design control does not scale nicely as each additional constraint increases shape distortion of the tiles on the rest of the surface.

Finally, Ray *et al.* [RLL*05] recently introduce another contouring technique performing a non-linear optimization of periodic parameters to best align directions along two given orthogonal vector fields, offering more freedom on the type of singularities than any previous approach. In particular, indices of type $1/2$ and $1/4$ can be introduced, allowing a satisfactory balance between area distortion and alignment control. However, even after a curl-correction step modulating the norm of the vector field to minimize the number of point singularities, this method does not provide *direct control* over the placement of singularities (*i.e.*, irregular vertices in the final remesh). Conspicuous imperfections may appear in the final mesh at seemingly random places.

1.2. Approach and Contributions

The recently introduced use of isolines as a basis for remeshing seems particularly appealing from a practical point of view: it naturally privileges regularity of the resulting quad mesh, and is numerically more robust than the use of streamlines as it alleviates the need for numerical integration. Isovalues can eventually be changed in order to adapt spacing between isolines and aspect ratios of the resulting quads. Therefore, we propose to design an algorithm to find

two piecewise continuous scalar functions with harmonicity properties, such that *their respective contouring provide the final pure-quad mesh with no T-junctions.*

Based on the solid foundations of discrete differential forms, our technique improves upon previous methods in several unique respects. In particular, we provide control not only on the position of singularities and their (possibly fractional) indices, but also the way these singularities are interconnected in the final mesh. This information on the topological structure of the output mesh is encoded in what we call a *singularity graph*, specified by the user or semi-automatically computed using the curvature tensor of the surface. This singularity graph is also a convenient way to specify preferred directions for the quads depending on their locations. Moreover, the core of our algorithm relies on novel extensions of the *well-known cotangent formula* that enriches the space of discrete harmonic functions: we thus stay within the framework of linear algebra, avoiding non-linear minimization required in [RLL*05, KSS06] that can impair scalability.

2. Rationale and Theoretical Set-Up

We first describe our approach from a theoretical point of view. We will make use of the language of differential forms [DKT05] as most of the successful meshing techniques so far (using harmonic parameterizations or integral lines of orthogonal vector fields) can be elegantly formulated using the notion of *discrete differential forms*—a fact already noticed in [DKG05]. It is a trivial matter to discretize the equations written in this particular form; in particular the exterior derivative d of a node-based linear function is simply the difference of the node values for each oriented edge. See Appendix A for a brief overview.

2.1. Local Quadrangulation as Contouring

We start by using a "reverse-engineering" argument. Suppose that we *already* have a small surface patch composed of locally "nice" quadrangles, the notion of nice being highly application-dependent. From this mesh, we can first set a local (u, v) coordinate system (with directions e_u and e_v) of the surface to be aligned with the edges of the quads. We can then define a metric $\langle , \rangle$ such that $\langle e_u, e_v \rangle = 0$ everywhere, and such that lengths of each quad edge are unit. Thus, the mesh is locally defined by **integer u- and v-isovalues**. In addition the gradients of the two parameters ∇u and ∇v are *orthogonal* in the prescribed metric. The way we have defined the metric also guarantees that we must 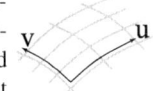 have the magnitudes of the gradients equal to each other. The two conditions together are known as the Cauchy-Riemann equations for the parameters u and v of this patch:

$$\langle \nabla u, \nabla v \rangle = 0 \quad \text{and} \quad \langle \nabla u, \nabla u \rangle = \langle \nabla v, \nabla v \rangle.$$

These two equations can be elegantly formulated using the differentials of u and v, as well as the Hodge star induced by our metric. Indeed, the two 0-forms u and v simply satisfy:

$$du = \star dv$$

Notice that we can deduce (by applying d and $d\star$ to the previous equation) that $d \star dv = d \star du = 0$, hence du and

dv are both coclosed. Since $d \circ d = 0$, both are also closed. Therefore, du and dv must be *harmonic*. In more traditional notation, both gradient fields are curl- and divergence-free. Another consequence of the coclosedness of the two differentials is that both u and v are *also harmonic*, i.e., their Laplacian vanishes. These properties explain the popularity of harmonic functions in Euclidean space, where orthogonality means $\pi/2$ angles, hence leading to well-shaped quads [DKG05, GY03]. We will also stick to the Euclidean metric for now, to keep our explanations simple.

2.2. Towards Global Contouring

To extend the basic principle explained in the previous section from a local quad mesh to a global quad mesh, one needs to overcome a number of issues.

Necessity of discontinuities First, globally continuous harmonic scalar potentials are too restrictive for quad meshing purposes. In fact, for the frequent case of a genus-0 closed manifold , there are *no* globally continuous harmonic potentials other than the constant ones, of little worth. A classical way to deal with this problem is to add pole singularities, which amounts to piercing little holes at various locations on the surface. For example: pierce a sphere once at the top and once at the bottom; what remains is a globally continuous harmonic potential u, with extrema at the two poles, thus with flow lines defining longitudes. However, the corresponding v potential cannot be globally continuous since its derivative dv has closed flow lines, namely latitudes. Therefore, the only hope to truly extend the contouring approach is to allow the potentials to be *piecewise continuous*, i.e., only continuous inside non-overlapping patches of the manifold—akin to the traditional notion of charts [GH95, YZ04, KLS03]. We may find a potential v that is continuous everywhere except on a line joining the two poles, on which the jump of the v-value equals a constant. Note that in this case v is discontinuous but dv is not. However, we will see that continuous harmonic 1-forms are not sufficient in general. The reason is that they can only model singularities with *integer* indices, that is poles and saddles. As Figure 2 depicts, these types of singularities create significant distortion in the quadrangulation. To be able to generate fractional singularities, one needs to allow for certain types of discontinuities on 1-forms.

Compatibility conditions Acknowledging the lack of global continuity, we now assume that the 0-forms (the potentials u and v) can contain *singularities*, i.e., jumps along certain edges. Similarly, their differentials (the 1-forms du and dv, akin to the gradients of each potential) may have singularities at the same locations, i.e., the vector fields representing these 1-forms may jump across patch boundaries. We will denote such a boundary between two continuous patches a ***singularity line***. Fortunately, we will be able to set *simple compatibility conditions* on the jumps of potentials and of their differentials that will guarantee that a global contouring of u and v results in a proper tiling. More precisely, linear constraints of continuity on the two potentials

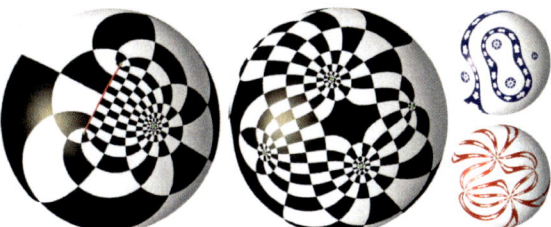

Figure 2: *Undesirable Singularities. Left: a pole (in green) induces too much distortion while a cut (in red) creates T-junctions in the final tiling. Middle and right: more than two poles (of index 1) on a genus-0 surface inevitably create singularities with negative index (saddles), creating large and distorted n-gons.*

can trivially ensure *continuity of the isolines*: indeed, if we trace all isolines with integer values, then the necessary and sufficient compatibility conditions are that the jumps of the potentials should be integer. On the other hand, the *smoothness of the isolines* will be ensured by a (tweaked, yet still linear) condition of "harmonicity" of the two potentials at patch boundaries. This last condition is, in fact, a continuity condition for 1-forms across the singularity line. We thus call this condition *singular continuity* to convey the notion of smoothness *modulo* the presence of a singularity.

Singular Continuity of Discrete Forms As mentioned above, obtaining a quad-dominant tiling on a disk-like patch through contouring two 0-forms u and v is rather easy. However, enforcing a proper tiling throughout the surface requires strong compatibility conditions at each singular line. Fortunately, only three different types of singular continuity across two neighboring patches can happen: *regular* (when both u and v directions individually match between the two patches), *reverse* (when both u and v directions change their orientations across the boundary), and *switch* (when the u and v directions are switched on the shared boundary). Only then can we get a globally consistent tiling of the surface.

2.3. Enforcing Singular Continuity of Forms

We now go over the various cases of continuity. As our technique uses two linear equations per vertex, we describe the different vertex types that we can encounter on a mesh: a vertex can be strictly within a patch, or on a particular type of singularity line.

Free Vertices When a vertex i is within a patch, i.e., not on any singularity line, we simply wish to enforce harmonicity of both 0-forms u and v. Consequently, the celebrated harmonicity condition [PP93] is imposed on this vertex, yielding:

$$\sum_{j \in \mathcal{N}(i)} w_{ij} \begin{pmatrix} u_i - u_j \\ v_i - v_j \end{pmatrix} = 0$$

where the index j goes through all the immediate neighboring vertices of i, u_k (resp., v_k) represents the value of u (resp., v) at vertex indexed k. For the Euclidean metric (as often used in graphics), the weights w_{ij} are the well-known sum of the cotangents of the angles opposite to edge ij.

Vertices with Regular Continuity When a vertex is on a *regular* singularity line between two patches, we assume that the fields u and v are smooth across the patch boundary *mod-*

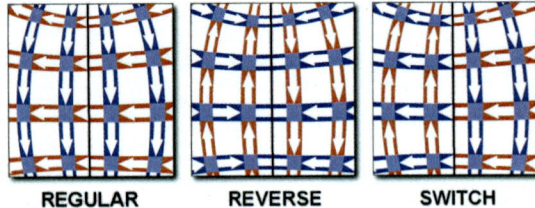

REGULAR   **REVERSE**   **SWITCH**

Figure 3: *Singular continuity. Three different types of continuity through a singularity line. Blue/red arrows are along isolines of u/v.*

ulo a constant offset. That is, if we call u^- (resp., v^-) the potential u of this vertex using its value from one of the patches, and u^+ (resp., v^+) the value at the same vertex but considering its value from the other patch, we wish to have:

$$u^- - u^+ = P_1 \qquad v^- - v^+ = P_2, \qquad (1)$$

where P_1 and P_2 are two arbitrary integer constants associated with this particular patch boundary (we will discuss how to choose their values adequately later on). Obviously, enforcing this "equality modulo offset" will guarantee that integer isolines of u and v do match up at the boundary. Notice also that it corresponds to guaranteeing continuity of the 1-forms du and dv, as $d(u^+ - u^-) = du^+ - du^- = 0$. Finally, to ensure smoothness of these isolines, we enforce harmonicity of the two potentials taking the jump into account (see inset for conventions used):

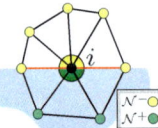

$$\sum_{j \in \mathcal{N}^-(i)} w_{ij} \begin{pmatrix} u_i^- - u_j \\ v_i^- - v_j \end{pmatrix} + \sum_{j \in \mathcal{N}^+(i)} w_{ij} \begin{pmatrix} u_i^+ - u_j \\ v_i^+ - v_j \end{pmatrix} = 0$$

Fortunately, one realizes that the above conditions can be rewritten using only *one* value of u and *one* value of v for our boundary vertex i—therefore alleviating the need for storing two different values, one on each side of the singularity line. Indeed, if we assume $u_i \equiv u_i^-$, and thanks to Eq. (1):

$$\boxed{\sum_{j \in \mathcal{N}(i)} w_{ij} \begin{pmatrix} u_i - u_j \\ v_i - v_j \end{pmatrix} = \sum_{j \in \mathcal{N}^+(i)} w_{ij} \begin{pmatrix} P_1 \\ P_2 \end{pmatrix}.}$$

Notice that this equation is a simple variant of the former case, modifying the right hand side to impose the correct conditions on each side of the boundary.

Vertices with Reverse Continuity This time, we want the 0-forms u and v to change orientation when crossing the patch

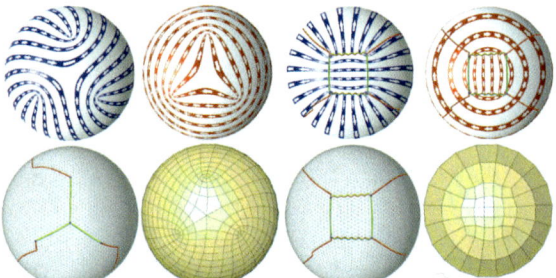

Figure 4: *Line Singularity as Basis of Many Singularities. Trisector (left) and square (right) singularities can be obtained by creating a graph of line singularities.*

boundary. That is, we wish to have $du^+ = -du^-$, and $dv^+ = -dv^-$. These constraints are easily enforced by defining:

$$u^+ + u^- = Q_1 \qquad v^+ + v^- = Q_2,$$

where Q_1 and Q_2 are two integer constants associated to the boundary on which the vertex lies. We now enforce harmonicity of the two potentials at i *modulo* the reversal:

$$\sum_{j \in \mathcal{N}^-(i)} w_{ij} \begin{pmatrix} u_i^- - u_j \\ v_i^- - v_j \end{pmatrix} + \sum_{j \in \mathcal{N}^+(i)} w_{ij} \begin{pmatrix} u_j - u_i^+ \\ v_j - v_i^+ \end{pmatrix} = 0$$

Again, one notices that a simpler expression using only one value for vertex i and a non-zero right-hand side, is:

$$\boxed{\sum_{j \in \mathcal{N}^-(i)} w_{ij} \begin{pmatrix} u_i - u_j \\ v_i - v_j \end{pmatrix} + \sum_{j \in \mathcal{N}^+(i)} w_{ij} \begin{pmatrix} u_i + u_j \\ v_i + v_j \end{pmatrix} = \sum_{j \in \mathcal{N}^+(i)} w_{ij} \begin{pmatrix} Q_1 \\ Q_2 \end{pmatrix}.}$$

This last expression preserves the *symmetric* nature of the Laplacian matrix. This is a particularly nice feature: state-of-the-art linear solvers have been shown to scale very well on such a problem [TCR05, BBK05].

Vertices with Switch Continuity Finally, for vertices on a singularity line on which we want u and v to switch, we simply enforce that $du^+ = dv^-$ and $dv^+ = -du^-$. Notice the extra minus sign, because switching u and v reverses one of the two directions. Again, these conditions are satisfied if:

$$v^- - u^+ = R_1 \qquad v^+ + u^- = R_2,$$

Finally, to ensure smoothness of these isolines, we enforce harmonicity of both potentials given this discontinuity through:

$$\sum_{j \in \mathcal{N}^-(i)} w_{ij} \begin{pmatrix} u_i^- - u_j \\ v_i^- - v_j \end{pmatrix} + \sum_{j \in \mathcal{N}^+(i)} w_{ij} \begin{pmatrix} v_j - v_i^+ \\ u_i^+ - u_j \end{pmatrix} = 0$$

The resulting symmetric expression, using only one value for the vertex i and a non-zero right-hand side, is now:

$$\boxed{\sum_{j \in \mathcal{N}^-(i)} w_{ij} \begin{pmatrix} u_i - u_j \\ v_i - v_j \end{pmatrix} + \sum_{j \in \mathcal{N}^+(i)} w_{ij} \begin{pmatrix} u_i + v_j \\ v_i - u_j \end{pmatrix} = \sum_{j \in \mathcal{N}^+(i)} w_{ij} \begin{pmatrix} R_2 \\ R_1 \end{pmatrix}.}$$

Notice there is an analogous formula for what we could call reverse-switch continuity vertices, namely when we want to switch u and $-v$.

2.4. Properties of Singular Continuity

Although quite simple, the four cases we discussed above provide an already rich repertoire of singularities. In particular, the previously mentioned case of a genus-0 object with two poles can be handled quite simply by linking the two poles with a singularity line: this "virtual" cut on the sphere creates one single patch touching itself along a *regular continuity* boundary. Now, the two potentials u and v can be computed per vertex by solving a modified Laplace equation, with vertices along the singularity line having different coefficients and non-zero right-hand sides.

Independence of Boundary Positions One remarkable property of the previous equations is that the exact position of the various boundaries between patches does *not* affect the final result: any boundary line in the same homology class as the original one will result in the same quad mesh. Although the 0-forms *will* be different (since their

Figure 5: *Line Singularity. From left to right: Piecewise-continuous harmonic potentials u and v (color-shaded); Red and blue arrows depict the direction of the potential gradients; a checkerboard is mapped onto the ellipsoid using (u, v) as texture coordinates; when the singularity line is wiggly, the two potential functions change, but their isolines remain exactly identical to the previous case.*

jumps will be located at distinct locations), their contouring will be exactly the same: only the local sign of their gradients will be affected in the reverse continuity case, while the gradient of u will become the gradient of v in the switch continuity case. In both cases, the union of the isolines of u and v remains the same! Therefore, the only real parameters are the set of constants, chosen for each boundary (that we called P_1, P_2, Q_1, Q_2, R_1, and R_2 previously). This is quite convenient, as no special effort needs to be spent on getting "straight" boundary lines (see Figure 5, right). In other words, only the *topology* of the patches is needed.

Other Typical Singularities Various other singularities can be achieved by designing a proper choice of boundary continuity between various patches. For instance, a trisector singularity, quite typical in direction fields, is obtained by assembling three concurrent lines, all of continuity type "reverse". A square singularity, *i.e.*, four index-1/4 poles forming a square-shaped index-1 singularity, is assembled from four lines in the shape of a square, with type "regular", "switch", "reverse", and "switch" in cyclic order (see Figure 4). Notice that these cases create significantly less distortion, and by design, no T-junctions. We will provide, in the next section, a simple implementation method to handle all these singularities (and more) in a unified manner using a singularity graph; but the equations provided above can already accommodate all these cases.

2.5. Discussion

Harmonic Forms Basis Globally continuous 1-forms are of interest only for objects of genus $g > 0$, the space of harmonic functions having dimension $2g$. Gu *et al.* [GY03, JWYG04] propose algorithms to compute and optimize those functions after having extracted the homology generators. In Appendix B, we explain how these computations can be derived in a simpler manner, simplifying both implementation and algorithmic complexity: we turn their asymmetric $E{\times}E$ linear solve into a symmetric $V{\times}V$ linear system. However, our novel treatment of harmonicity with singularities allows us to enrich these harmonic form bases significantly. One can indeed place a set of canonical singularities on the input mesh, then build a basis of the space of harmonic 1-forms (du and dv) that each singularity (point, line, or square) generates. This procedure requires solving a linear system per basis element, where only the right-hand side of the Laplace equations needs to be changed for every element. Nonetheless, finding a basis of all harmonic forms is

only one step: the next issue is to find a *good* linear combination of these basis elements so as to compute our two potentials, and through contouring, a tiling. A user selection of coefficients for each basis element is highly non-intuitive since each 1-form has global support, making the design delicate at best—and certainly not scalable.

Using Optimization or Guidance Vector Fields An automatic selection of the coefficients could be achieved by providing a (most likely non-linear) optimization in order for the final mesh to best follow a given *guidance vector field* to achieve the same goal as in [RLL*05], or to improve control over sizing [JWYG04]. This approach can be also simplified by finding the best linear combination of basis elements by computing the L^2 *inner product* between each 1-form in the basis and the 1-form representing the guidance field [DKT05]—turning the non-linear optimization into a linear projection. Unfortunately, designing a guidance vector field on a manifold is, in itself, a challenge. In fact, we believe that designing a guidance field would be, in a sense, as complicated as performing a quad remeshing of the manifold: indeed, contouring does not require global orientability (only directions matter), whereas vector fields do.

Instead, we propose a radically different approach that allows an intuitive and scalable solution to quad mesh design. A whole *network of line singularities* is defined and leveraged to truly capture the topology of arbitrary manifolds. Next we present a way to design this network, the *singularity graph*, that is the backbone behind our mesh design.

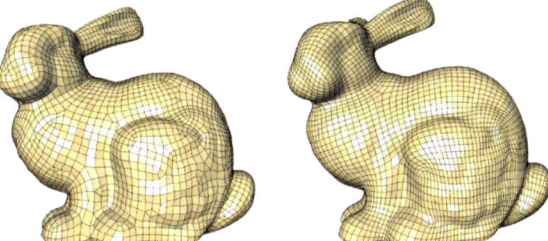

Figure 6: *Bunny: Comparison between a model (top, 6415 faces among which 124 are non-quads, 314 irregular vertices), courtesy of [RLL*05]) and our approach (bottom, 6575 quads). Notice the regularity of our mesh with only 34 irregular vertices (more views of a coarser version can be seen in Fig. 12).*

3. Designing a Singularity Graph

Allowing quad mesh design flexibility requires the use of a potentially large set of patches: the more patches we define, the better we can control the local alignment of edges as well as the local area distortion. Thus, we propose a user-guided (or automatic if needed) way to create this patch layout through the definition of a *singularity graph*, representing a topological "template" linking the singularities.

3.1. Definition of Singularity Graph

We call a *singularity graph* a meta-mesh whose meta-faces are non-overlapping patches of the original mesh, and whose meta-edges are assigned one of the *singularity continuity* conditions described in the previous section (Figure 4). This structure is similar to the notion of *coarse mesh* in subdivision surfaces, as the resulting quad mesh based on this graph only have irregular vertices *at* meta-vertices.

Topological Structure of the Singularity Graph The vertices of the singularity graph (called meta-vertices to avoid ambiguity) are a subset of the vertices of the input triangle mesh. They should be thought of as salient points of the manifold, as they will live at the intersection of several regular patches in the final quad mesh—but additional meta-vertices can also be added virtually anywhere, allowing a *better control* of the output quad mesh as we will discuss later. Connectivity between meta-vertices define the meta-edges of the singularity graph. Each of these meta-edges are made out of two half-edges, oppositely oriented; these half-edges will be useful in explaining further details of our approach. Note that, as pointed out in Section 2.4, the exact positioning of the meta-edges will *not* affect the final mesh as demonstrated in our results (e.g., Figure 10). Finally, every cycle of half meta-edges defines a (meta-)face of the singular graph. Such a face corresponds to a patch in which our 0-forms will be smooth and continuous. In this section, we will call F (resp., E) the number of meta-faces (resp., meta-edges).

Determining Types of Singular Continuity Given a singular graph, we must assign to each meta-edge a particular *singular continuity* type as defined in Section 2. These assignments can be automatically obtained *if* we first tag each meta-halfedge as u, $-u$, v, or $-v$ according to their alignment with increasing or decreasing directions of the parameter (see Figure 7). Indeed, we would ideally like to map each meta-face of the graph to an *orthogonal polygon* in the parameter domain to *guarantee* the existence of a regular quadrangulation inside this patch. That is, the face should conceptually look like a simple polygon in the (u, v) parameter plane with only angles multiple of $\pi/2$ (see 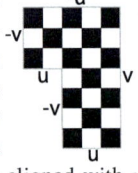 inset)—in fact, the choice of which edge is aligned with u vs. v does *not* affect the final quad mesh, so this polygon can be arbitrarily rotated by multiples of $\pi/2$ too. This condition imposes a constraint on the half-edge assignments, and we will provide an automatic procedure to enforce it on each face. After such an assignment is provided, the corresponding *singular continuity* types for all meta-edges becomes simple, as, for each pair of half-edge assignments, correspond the following continuity types:

- *Regular*: $\{u, -u\}$, $\{-u, u\}$, $\{v, -v\}$, $\{-v, v\}$;
- *Reverse*: $\{u, u\}$, $\{v, v\}$, $\{-u, -u\}$, $\{-v, -v\}$;
- *Switch* : $\{u, v\}$, $\{v, -u\}$, $\{-u, -v\}$, $\{-v, u\}$, (and symmetrically, $\{-u, v\}$, $\{-v, -u\}$, $\{u, -v\}$, $\{v, -u\}$).

3.2. Assisted Singularity Graph Creation

While the user can always either design or modify a given singularity graph and its current assignments interactively, an assisted creation of the graph is important when the manifold to remesh is quite complex. In particular, finding a geometrically-faithful graph is particularly important if the resulting mesh is expected to provide a concise approximation of the original mesh. The following generative procedures have been quite useful in our experiments—specific automatic graph design algorithms can be devised too.

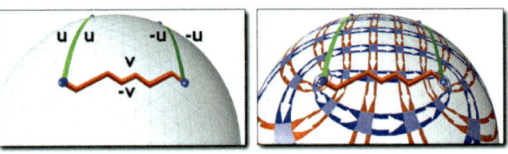

Figure 7: *Tags on half meta-edges. The continuity type of meta-edges is found by first tagging half-metaedges with $u, v, -u,$ or $-v$.*

Placement of Meta-Vertices The *curvature tensor* offers a relevant guidance when it comes to alignment of the quad mesh edges with geometrically relevant directions. This tensor defines two orthogonal principal curvature directions $e_{\min}$ and $e_{\max}$ everywhere, except at the so-called umbilics. As we ideally seek a quad mesh with edges aligned with principal directions, it is natural to place the singularities of the mesh at the singularities of the principal foliations, i.e., the umbilics. To estimate the positions of these umbilics, we estimate the curvature tensor of the surface using the method advocated in [CSM03]. The support of this estimator must be chosen sufficiently large so as to avoid having too many unnecessary singularities in the presence of noise. The user can also define a maximum number of singularities.

Generation and Tagging of Meta-Edges We eventually want the u and v isolines to align with the principal directions of the surface. Since meta-edges represent a *template* of the final quad mesh, it is natural to take as meta-edge paths the ones that are most-aligned with the curvature lines of the input mesh. To find these paths, our technique performs two passes of a region-growing procedure over the input mesh triangles, starting at the one-rings of the meta-vertices. The growth procedure is in essence a multi-source Dijkstra algorithm over the dual graph, where each dual edge is weighted in accordance with its length multiplied by its alignment with $e_{\min}$ for the first pass, and $e_{\max}$ for the second pass. When two region fronts meet, their originating meta-vertices are connected with a meta-edge (if not already connected). The result is a graph connecting the meta-vertices, as desired. The resulting connectivity is a mere product of the geometry itself, as two meta-vertices are connected iff there is a path roughly aligned with a principal direction (see the examples depicted in Figure 8).

Tagging halfedges becomes simple: we can for instance compute the total geodesic curvature of every patch's boundary (i.e., the Gaussian curvature of the patch). We then compute the portion of geodesic curvature around each meta-vertex of a patch (from halfway of the previous meta-edge to halfway of the next meta-edge): this value, divided by the total geodesic curvature of the patch, scaled by 2π, and then rounded to the closest multiple of $\pi/2$ tells us how the tag of the next edge is linked to the previous tag. An initial meta-

Figure 8: *Automatic Generation of Singularity Graph. Left: generation on the ellipsoid. The color map depicts the cumulative distance to the umbilics taking into account the alignment with κ_{min} and κ_{max}. Right: a similar process demonstrated on a cartoon hand.*

edge is tagged u arbitrarily (remember that rotations of the patch's assignments are equivalent). Since the sum of these values will be 2π, we have created a closed face in parameter space, with geometrically-derived angles.

3.3. Designing a Tiling

Designing the final quadrangulation amounts to deciding on an assignment of (u,v) values *at each corner of every meta-face*. These are, indeed, the *only* constraints that need to be fixed for the modified Laplace equation to be solvable: once these "meta-parameters" are known, all the constants defined per patch-boundary in the continuity equations described in Section 2 are determined. Thus, we begin by computing these values by solving a small meta-system. We will then inject the resulting meta-parameters into the system of (modified) Laplace equations for the final solve.

Meta-Parameters The constraints on these parameters depend only on the differences of u and v values on each meta-edge. Let Du (resp., Dv) be such a difference in u (resp., v) value over a meta-edge. In each meta-face, the sum of all the Du's of half-edges tagged as u *must* equal the sum of the Du's for those tagged as $-u$: the meta-face is a *closed* polygon in parameter space. The same argument applies for the sum of Dv of v edges and that of $-v$ edges (in the language of differential forms, this states that du and dv must be closed on the meta-mesh too). Similarly, as we wish to have isolines stitching properly across meta-faces, we must have equal differences for each half-edge of a meta-edge. Thus, these differences are E coefficients that need to be set, and there are $2F$ linear constraints on them (one for u and one for v per meta-face). We now set up in a small linear system these $2F$ constraint equations for E variables. However, the constraints can (and will often) be redundant. We thus use Gauss elimination to find the independent equations. This process is extremely fast since the graph contains typically three orders of magnitude fewer edges than the original mesh, and all the coefficients in this meta linear system are either 1 or -1 (since Du (or Dv) is computed as the simple difference between two parameter values). Additionally, notice that the $2F$ constraint equations sum to zero: we can thus guarantee that there will be *at least* E-$2F$+1 number of independent variables. Since a meta-face is homeomorphic to an orthogonal polygon, each meta-face has at least 4 edges. Therefore, $E \geq 2F$, and there is always at least one degree of freedom. More DOFs can be added by enriching the graph. The *free* meta-parameters now need to be set. Ob-

viously, the user can enter values based on the number of isolines desired on those meta-edges (this is, indeed the geometric meaning of Du, see example inset on a bunny ear).

Automatic assignment can also be done by simply entering the closest integer value to the actual

meta-edge *length*: this provides a good approximation of the most geometrically-relevant number of isolines—as it helps providing a more isometric parameterization, therefore minimizing area distortion.

```
Code to assemble the iᵗʰ line of the linear system AU = B
where A is the modified Laplacian matrix (note: V is the # of unconstrained vertices).
We use the conventions defined in Section 2.
For each half-edge (vᵢ, vⱼ), do:
    c ← wᵢⱼ; swap ← false; εᵤ ← 1; εᵥ ← 1; bᵤ ← 0; bᵥ ← 0
        if (i or j is on singularity line, but both not on the same line)
        switch (type(j))
            case On-Regular:
                getConstantsOfLine(P₁,P₂);  bᵤ−= P₁ * c;  bᵥ−= P₂ * c;
            case OnReverse:
                getConstantsOfLine(Q₁,Q₂);
                εᵤ ← −εᵤ;  εᵥ ← −εᵥ;  bᵤ−= Q₁ * c;  bᵥ−= Q₂ * c;
            case OnSwitch:
                swap = true;  getConstantsOfLine(R₁,R₂);
                εᵥ ← −εᵥ;  bᵤ−= R₁ * cⱼ;  bᵥ−= R₂ * cⱼ;
            default: do nothing;
        switch (type(i))
            case OnRegular:
                getConstantsOfLine(P₁,P₂); bᵤ+= P₁ * c;  bᵥ+= P₂ * c;
            case OnReverse:
                getConstantsOfLine(Q₁,Q₂);
                εᵤ ← −εᵤ;  εᵥ ← −εᵥ;
                bᵤ ← −bᵤ;  bᵥ ← −bᵥ;  bᵤ+= Q₁ * c * εᵤ;  bᵥ+= Q₂ * c * εᵥ;
            case OnSwitch:
                getConstantsOfLine(R₁,R₂);
                εᵥ ← −εᵥ;  swap ←!swap;
                temp ← bᵤ;  bᵤ ← −bᵥ;  bᵥ ← −temp;
                bᵤ+= R₂ * c;  bᵥ+= R₁ * c;
            default: do nothing.
    A[i,i]+= c;  A[i+V,i+V]+= c;  B[i]+= bᵤ;  B[i+V]+= bᵥ;
    if (swap)
        A[i,j+V] = −c * εᵤ;  A[i+V,j] = −c * εᵥ;
    else  A[i,j] = −c * εᵤ;  A[i+V,j+V] = −c * εᵥ;
```

Figure 9: *Pseudocode of the Laplacian Matrix Assembly.*

Final Solve Once the meta-parameters are set, we can finally assemble the global linear system for the 0-forms u and v of the original mesh. The system is created by assembling two linear equations per vertex v_i (depending if it is free or on a line singularity, as discussed in Section 2), and none for the vertices on corners of meta-faces, as they are already determined by the meta-parameters. Whenever one of these linear equations involves a meta-face corner, its value (i.e., one of the meta-parameters) is constrained and therefore substituted and moved to the right-hand side of the system. The matrix of this system is sparse and symmetric. *By construction, a contouring of the u and v potentials will stitch automatically into a pure quad mesh.* We provide pseudocode in Figure 9 for all possible cases of continuity conditions to facilitate this matrix assembly.

4. Implementation Details and Results

In our implementation, we use a half-edge data structure for the input mesh as well as for the singularity graph. The assembly of the modified Laplacian matrix is performed by going through each input mesh half-edge and checking the type of vertices at each end. We used the supernodal multi-frontal Cholesky factorization option of TAUCS [TCR05] as our linear solver since it handles sparse symmetric positive-definite systems very efficiently (the linear systems for the results shown in this paper took between 0.1 to 23s to solve, our running times being consistent with the ones reported in [BBK05]). Generating a singularity graph takes on the order of a minute or two, since only the corners need to be chosen by the user.

Controlling Alignment and Distortion The user can finely control the *alignment* of the mesh with features or semantically relevant directions. Inserting more meta-vertices along a given meta-edge, and tagging the newly-created meta-edge with the exact same tags as the original one will do the trick: the quads created nearby will align with these meta-edges as the isocurves are more constrained by the added meta-vertices. Similarly, while the presence of saddles is sometimes *unavoidable* on certain manifold (as stated by the Hopf-Poincaré theorem), the user can minimize their distortion by altering the singularity graph: adding a meta-face over a saddle will "split" it into four lower-index saddles, reducing the effective distortion quite significantly—at the price of three added irregular vertices.

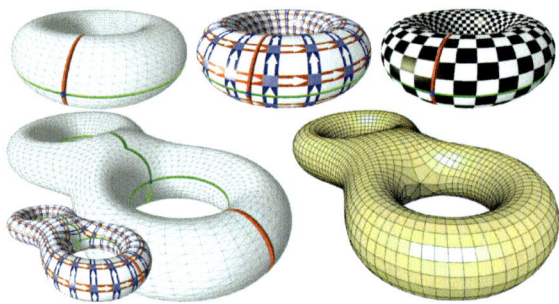

Figure 10: *Non-trivial Topology. Top: genus-1 example; Bottom: genus-2 model—a saddle (imposed by Hopf-Poincaré theorem) is present on a meta-vertex. The final meshes are still pure quad. Both singularity graphs were defined by homology generators.*

Controlling Quad Angles Notice that, unlike previous approaches [GY03], we did not require our pair of 1-forms du and dv to be holomorphic (*i.e.*, related through the Hodge star) to allow for more design flexibility. Therefore, using our method with an unreasonable choice of meta-parameters will result in regions where the quads are grossly non orthogonal. One way to palliate this issue is to find the meta-parameters so that the pair (du, dv) minimizes (in the $\mathcal{L}^2$ sense) the 1-form $du - \star dv$. We can also proceed as explained in Appendix C to truly design a stitched *conformal* parameterization—the slightly-modified singular continuity conditions involved in this variant will be explored in a future paper. An alternative approach is to play with another parameter: the Hodge star $\star$. Although beyond the scope of this paper, our method can indeed be applied for arbitrary metrics. Altering the metric locally amounts to change the coefficient w_{ij} used in the Laplacian operator. Theoretically speaking, the optimal quad shape for a *best* surface approxi-

mation derives from the curvature tensor—a research direction we wish to exploit soon.

Boundaries Boundaries are handled as follows. Once the assignments of (u, v) at each corner of the meta-faces are done, we go through the edges tagged as boundary between two such corners, and force the boundary values to be linearly interpolating the two corner values. This will force the iso-value of our potentials to follow the boundaries (Figure 11).

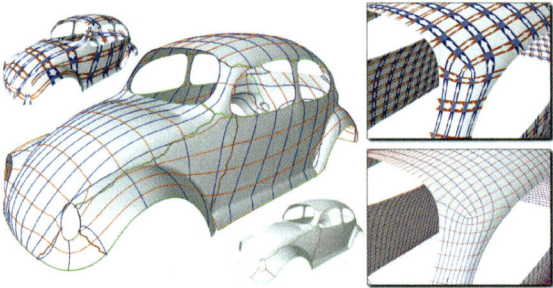

Figure 11: *Quad Remeshing with Boundaries. A triangulated beetle model with boundaries is remeshed with a singularity graph.*

Mesh Extraction The final mesh is easily extracted through integer contouring: we walk in the triangulation from an integer intersection of u and v to the next. Note that we may cross a line singularity; in that case, we account for the singularity type to be able to resume the walk on the other side and find the next intersection. The output is, by construction, a *pure quadrilateral mesh*.

Comparisons & Limitations Our approach may look loosely related to the global linear system proposed in [KLS03] (recently extended to quads in [DBG*06]). However, the difference is significant: while these previous approaches try to make the excess angle of each corner (meta-vertex) be as close to 0 as possible, we instead fully embrace the specificity of the meta-vertices by forcing their Gaussian curvature to multiples of $\frac{\pi}{2}$ (*i.e.*, specific cone singularities). To the best of our knowledge, *no other*

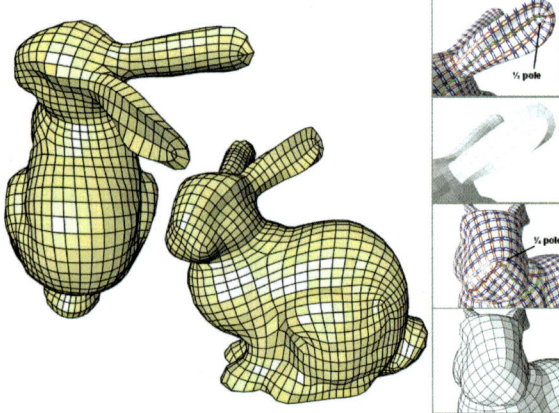

Figure 12: *Remeshing the Stanford Bunny. Left: Pure-quad remeshing. Right: detail of a half-pole, and a quarter-pole (using a switch and a regular line incident to the singularity). The half-pole becomes a degree-2 vertex, incident to two quads, with two nearly collinear edges. The degree-2 vertex can optionally be removed by merging its two incident quads into one.*

approach allows such a unique solution with a simple linear solve. Nonetheless, we acknowledge that if the user decides to significantly override the values on meta-vertices automatically prescribed by our technique, the resulting meshes may have folds and significant stretch. In our experience, except for vastly inappropriate values, the results are always satisfactory. Note that compared to previous work, the main strengths of our method are the simplicity (and scalability) of computations (a simple linear solve followed by isocontouring) and a direct *control* over the singularities. In particular, compared to [RLL*05], we do not resort to a non-linear minimization or to a curl-correction phase, and we allow a precise placement of the resulting (non-integer index) poles.

5. Conclusion

We have presented a theoretical and algorithmic approach for designing quadrangle tilings from arbitrary triangulated surface meshes. Our algorithm computes two piecewise smooth harmonic scalar functions, whose isolines tile the input surface into quadrangles, without any T-junctions. Our main contribution is an extension of the discrete Laplace operator which encompasses several types of line singularities. The resulting two discrete differential 1-forms are either regular, opposite or switched along the singularity graph edges. We show that this modification guarantees the continuity of the isolines across the lines, while the locations of the isolines themselves depend on the global solution to the modified Laplace equation over the whole surface. Design flexibility is provided through specification of the type of each line singularity of the graph, as well as the number of isolines along independent meta-edges to control quad sizes.

Besides the mere interest of creating quad meshes, useful in a slew of applications, we are interested in studying the theoretical consequences of the framework we proposed. In particular, understanding the feasibility and consequences of producing ***arbitrary cone singularities*** using only linear algebra is extremely valuable. This can be easily achieved by defining continuity conditions with arbitrary rotation of a 1-form. The relaxation proposed in [KLS03] could also be used to automatically reduce distortion near irregular vertices if a fully-automatic quad remeshing is sought after.

Acknowledgments The authors wish to thank Peter Schröder for his continuous help, and Alex McKenzie for his helpful comments. Sponsors include NSF (CAREER CCR-0133983, and ITR DMS-0453145), DOE (DE-FG02-04ER25657), and the EU Network of Excellence AIM@SHAPE (IST NoE No 506766).

References

[ACSD*03] Alliez P., Cohen-Steiner D., Devillers O., Lévy B., Desbrun M.: Anisotropic polygonal remeshing. *ACM Trans. Graph. 22*, 3 (2003).

[AUGA05] Alliez P., Ucelli G., Gotsman C., Attene M.: Recent advances in remeshing of surfaces. STAR AIM@SHAPE, January 2005.

[BBK05] Botsch M., Bommes D., Kobbelt L.: Efficient linear system solvers for mesh processing. In *IMA Conf. on Math. of Surfaces* (2005), pp. 62–83.

[BMRJ04] Boier-Martin I., Rushmeier H., Jin J.: Parameterization of triangle meshes over quadrilateral domains. In *Symp. on Geometry processing* (2004), pp. 193–203.

[CSM03] Cohen-Steiner D., Morvan J.-M.: Restricted delaunay triangulations and normal cycle. In *Proceedings of the Symp. on Computational Geometry* (2003), pp. 312–321.

[DBG*06] Dong S., Bremer P.-T., Garland M., Pascucci V., Hart J. C.: Spectral surface quadrangulation. to appear at ACM SIGGRAPH '06, July 2006.

[DKG05] Dong S., Kircher S., Garland M.: Harmonic functions for quadrilateral remeshing of arbitrary manifolds. *Computer Aided Design (Special Issue on Geometry Processing) 22*, 4 (2005), 392–423.

[DKT05] Desbrun M., Kanso E., Tong Y.: Discrete differential forms for computational modeling. In *Discrete Differential Geometry*. ACM SIGGRAPH Course Notes, 2005.

[Ede00] Edelsbrunner H.: Mathematical problems in the reconstruction of shapes, 2000. Talk at MSRI's Workshop on Computational Algebraic Analysis (http://msri.mathnet.or.kr/).

[EW05] Erickson J., Whittlesey K.: Greedy optimal homotopy and homology generators. In *SODA* (2005), pp. 1038–1046.

[GH95] Grimm C., Hughes J.: Modeling surfaces of arbitrary topology. In *Proceedings of ACM SIGGRAPH* (July 1995), pp. 359–369.

[GY03] Gu X., Yau S.-T.: Global conformal parameterization. In *Symposium on Geometry Processing* (2003), pp. 127–137.

[JWYG04] Jin M., Wang Y., Yau S.-T., Gu X.: Optimal global conformal surface parameterization. In *IEEE Visualization* (2004), pp. 267–274.

[KLS03] Khodakovsky A., Litke N., Schröder P.: Globally smooth parameterizations with low distortion. *ACM Trans. Graph. 22*, 3 (2003), 350–357.

[KSS06] Kharevych L., Springborn B., , Schröder P.: Discrete conformal mappings via circle patterns. *ACM Trans. on Graphics 25*, 2 (2006).

[MK04] Marinov M., Kobbelt L.: Direct anisotropic quad-dominant remeshing. In *Proceedings of the Pacific Graphics* (2004), pp. 207–216.

[PP93] Pinkall U., Polthier K.: Computing discrete minimal surfaces and their conjugates. *Experimental Mathematics 2(1)* (1993), 15–36.

[RLL*05] Ray N., Li W. C., Lévy B., Sheffer A., Alliez P.: Periodic global parameterization. *Preprint found at www.loria.fr/~levy/publications/* (2005).

[TCR05] Toledo S., Chen D., Rotkin V.: TAUCS. Available at http://www.tau.ac.il/~stoledo/taucs, 2005.

[YZ04] Ying L., Zorin D.: A simple manifold-based construction of surfaces of arbitrary smoothness. *ACM Trans. on Graphics 23*, 3 (2004), 271–275.

Appendix A: Discrete Forms Glossary

For completeness, we provide a glossary of the terms traditionally involved in the use of discrete differential forms, more details and references can be found in [DKT05].

● *Discrete forms*: A discrete k-form on a piecewise-linear manifold assigns a real number to every oriented k-simplex of the manifold. 0-forms are discrete versions of continuous scalar fields, while 1-forms are discrete versions of vector fields (or of the differential 1-forms they represent).

● *Exterior Derivative*: The exterior derivative operator associates to each k-form ω a particular $(k+1)$-form $d\omega$. If ω is a 0-form (valued at each node), *i.e.*, a function on the vertices, then $d\omega$ evaluated on any oriented edge $v_1 v_2$ is equal to $\omega(v_1) - \omega(v_2)$. If ω is a 1-form, then $d\omega(v_1 v_2 v_3) = \omega(v_1 v_2) + \omega(v_2 v_3) + \omega(v_3 v_1)$ for any triangle $v_1 v_2 v_3$. The exterior derivative is thus similar to the gradient operator for 0-forms, and to the curl operator for 1-forms. It is not difficult to check that $d \circ d = 0$ (the curl of any gradient field is always null).

● *Closed and Exact Forms*: ω is called closed if $d\omega = 0$. For 1-forms, this means that the sum of the values on the directed edges around each face equals zero. Poincaré lemma states that every such closed 1-form has a local potential (a 0-form u with $\omega = du$) inside any disk-like patch. If there is such a potential such that $\omega = du$ on

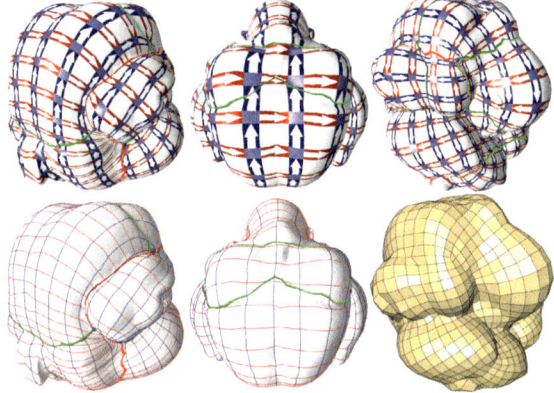

Figure 13: *Omotondo Model. With a simple singularity graph, a quad mesh following relevant geometric directions is created.*

the *whole* space, the 1-form ω is called exact. Since $d \circ d = 0$, exact forms are always closed.

• *Cohomology*: The difference between exactness and closedness of 1-forms on surfaces is determined by the topology of the surface. More precisely, two closed forms are said to belong to the same cohomology class if their difference is *exact*. Cohomology classes of 1-forms form a vector space whose dimension, called the first Betti number, counts the number of loops/holes in the surface.

• *Hodge star operator* $\star$: on 1-forms, it is the discrete analog of applying a rotation of $\frac{\pi}{2}$ to a vector field. Divergence can be written as $d*$ on a 1-form. Note that the Hodge star is metric dependent: if we change the metric, the Hodge star changes accordingly.

• *Harmonic Forms*: for $k > 0$, a k-form ω such that $\star\omega$ is closed is called *coclosed*. A k-form closed *and* coclosed is called harmonic. It can be shown that there is exactly one harmonic form in each cohomology class. Hence, harmonic 1-forms form a vector space of dimension equal to the first Betti number of the surface. A scalar function f is called harmonic if its Laplacian Δf vanishes. The exterior derivative of a scalar field is harmonic iff this field is harmonic.

Appendix B: Harmonic Form Bases

According to the continuous theory, there are $2g$ independent continuous harmonic 1-forms defined on a closed, orientable genus-g surface. Each element of the 1-form basis is associated with a loop, called the homology generator. These homology generators are basically $2g$ loops on the surface that allow to cut open the surface into a disk-like domain. For a detailed discussion, see [GY03]. Often, a surface M has a boundary ∂M, which can have m connected components (i.e., m boundaries). And naturally, if we look for quads orthogonal to the boundaries, we must have one of the 1-forms normal to the boundary, and the other tangential to the boundary. The continuous theory states that normal harmonic 1-forms are associated with the relative homology group $H_1(M, \partial M)$, which includes $m-1$ independent paths connecting the various boundaries, plus the aforementioned $2g$ genus-induced homological generators. Additionally, the space of tangential harmonic forms for a surface with m boundaries has $2g+m-1$ basis elements, associated with $m-1$ closed boundaries in addition to the $2g$ loops (all these loops forming the homology group $H_1(M)$).

We propose a simplified way of computing discrete counterparts of these 1-forms that necessitates *no double-covering*; in fact, it only requires solving a simple Poisson equation. First, any known method (for instance, [EW05]) can be used to find the homology generator loops, which are organized in pairs (each pair corresponds to one circle that will cut a handle and one circle which is around the hole of the handle, as in Figure 10 for the torus).

• For closed surfaces, there is one harmonic form in each cohomology class, so we propose to *pick a closed 1-form ξ from one cohomology class, and solve for a scalar field f such that $d \star df = d \star \xi$. Then $\omega = \xi - df$ is a harmonic 1-form.* Finding the independent ξ's is trivial thanks to *Poincaré duality*, as we now detail. The cohomology class associated with one loop is nothing but the set of all the closed forms that will integrate to 1 around that loop, and to 0 on all other loops. First take one pair of loops (c_1, c_2). Assigning the value 1 to edges *crossing* the loop (say c_1) in one given direction, and 0 to all other edges, defines ξ. Indeed, this is a closed form since the sum of the values will be 0 on all faces (for faces with non-zero edges, *i.e.*, near c_1, they always have a pair of them which cancel out; in other words, the loop always enter, then leave any triangle that it goes through). This 1-form integrates to 1 on c_2, and 0 on all other loops. Thus, ω calculated as above will be a harmonic

form for c_2, and we can get the other $(2g - 1)$ 1-form this way. To get the potential, traverse the surface starting from any seed vertex and integrate along edges (as done in [GY03]). The *gradient* of this potential corresponds to the continuous 1-form that we solved for.

• To compute the normal harmonic forms associated with the $2g$ loops, we use exactly the same method. The only difference is that we need to set the values of the potential at the boundary to 0. Now for the 1-forms associated with the $m - 1$ paths, we put all vertices on the connected component of the boundary incident to the source of the path to 1, and assign 0 to all other boundary vertices (assuming, without loss of generality, that the paths start from the same boundary component). Solving the Laplace equation with Dirichlet boundary conditions gives us a harmonic 0-form whose gradient *is* the 1-form sought after. Although equivalent in result, our method is simpler to implement and *much* faster to compute than the double covering technique proposed in [GY03].

• Finally, tangential forms are easily dealt with based on Hodge and Lefschetz duality between homology and relative homology $H_1(M) \cong H_1(M, \partial M)$. Indeed, the Hodge star of the space of normal forms equals the space of tangential forms. Therefore, we can take the Hodge star of all $2g + m - 1$ normal forms η_i as described above, compute their circulations around the $m - 1$ loops along the boundary and the $2g$ loops for handles. These circulations form a matrix P, the so-called *period matrix*. The tangential 1-form corresponding to the i-th homology class in the basis is then $P^{-1}(e_i)$, with e_i the vector with 1 in the i-th entry and 0 elsewhere.

Notes The singularity graph presented in this paper is another simple alternative to what we just proposed that avoids the need for dual-primal discrete form conversion. Indeed, if we restrict ourselves to the use of "regular" singularity lines, we can get all global harmonic 1-forms directly. In particular, notice that when you use the homology generators as the singularity graph, you get exactly $2g$ free parameters.

Appendix C: Orthogonality & Quantization

To ensure orthogonality, we must have a pair of 1-forms that are Hodge star of each other up to a sign (the pair is then called holomorphic). In order to have quadrangles using integer-contouring, we must have the circulation around $2g + 2(m - 1)$ loops being integer numbers. This means orthogonality is not achievable in many cases (for instance, when there are irrational number entries in the period matrix). But as we can use rational numbers to approximate real numbers to arbitrary precision, we can increase the resolution of the mesh to get closer to orthogonality. More precisely, one can fix a denominator q and round the entries in the inverse period matrix to the closest multiple of $1/q$. Then we use $1/q$ as our contouring step. Better orthogonality is achieved for high values of q, giving a trade-off between orthogonality and coarseness of the resulting quad mesh.

In our approach using the singularity graph, we can obtain holomorphy of our forms by changing the linear system slightly. Instead of the singular continuity types that we used, we have to use conditions such as $du^+ = *du^-$ (and variants), resulting in a modified Laplace equation for u only. From the 0-form u, we can then compute $*du$, and assign v to its integral over each patch. Suppose that u is quantized is described above, all we need to do is to quantize the integral of $*du$ along each edge too. Using the corner values for (u_i, v_i) as the input for the meta linear system will lead to a ***stitched conformal mapping***. In other word, we can relax the constraint that each patch must be mapped to an orthogonal polygon in the parameter domain, and ensure the orthogonality between du and dv directly instead.

Eurographics Symposium on Geometry Processing (2006)
Konrad Polthier, Alla Sheffer (Editors)

Reliable Implicit Surface Polygonization using Visibility Mapping

Gokul Varadhan[1], Shankar Krishnan[2], Liangjun Zhang[1], and Dinesh Manocha[1][†]

[1]Department of Computer Science, University of North Carolina, Chapel Hill, U.S.A
[2]AT&T Labs - Research, Florham Park, New Jersey, U.S.A
http://gamma.cs.unc.edu/VM/

Abstract

We present a new algorithm to compute a topologically and geometrically accurate triangulation of an implicit surface. Our approach uses spatial subdivision techniques to decompose a manifold implicit surface into star-shaped patches and computes a visibilty map for each patch. Based on these maps, we compute a homeomorphic and watertight triangulation as well as a parameterization of the implicit surface. Our algorithm is general and makes no assumption about the smoothness of the implicit surface. It can be easily implemented using linear programming, interval arithmetic, and ray shooting techniques. We highlight its application to many complex implicit models and boundary evaluation of CSG primitives.

Keywords: contouring, Marching Cubes, set operations, implicit modeling, topology

1. Introduction

Implicit surfaces are used to represent shapes of arbitrary topology in computer graphics and geometric modeling. As compared to other surface representations, implicits offer many advantages in terms of performing geometric operations like Boolean operations, blending, warping and offsets. Some early applications of implicit surfaces were the modeling of blobby shapes and of objects with biological or natural appearances. Recently, implicit surfaces have been shown to be useful for surface reconstruction, point-based modeling, and simulation.

In this paper, we address the problem of polygonizing an implicit surface. Our goal is to develop reliable techniques that preserve the topology of the implicit surface and compute a geometrically accurate polygonization. Topology of the implicit surface contains information about its connected components and genus. Many geometric operations like Boolean operations can result in a shape whose topology is very different from that of the primitives. It is important to capture all the topological features for CAD, medical and molecular modeling applications. Implicits are also used to reconstruct topologically accurate continuous surfaces from point clouds. These surfaces are defined as weighted combinations or blends of basis functions and have been applied to datasets consisting of millions of points. However, computing a topologically reliable polygonization of these complex implicit

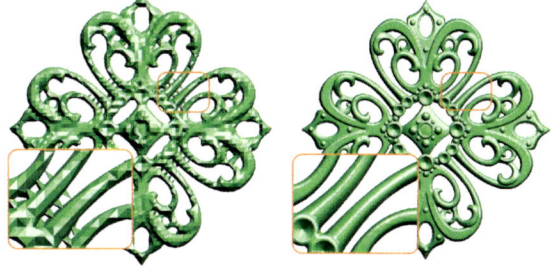

Figure 1: *Polygonization of a complex MPU model:* The right image shows a complex MPU model, Filigree, which has genus 65 and many topological features. This implicit was defined using 514K point samples by blending quadric surfaces using B-splines weights. The left image shows that previous spatial subdivision techniques may not compute an accurate polygonization, when the parameter used to select the grid size is not adequate. Our algorithm, based on Visibility mapping, generates an accurate polygonization of this model, as shown in the image on the right, in about 4 minutes.

surfaces can be a challenge. Other applications of polygonization arise in mathematical visualization, where we want to accurately display the shape of complex surfaces.

The implicit surface polygonization problem has been studied for more than two decades. Prior methods can be classified into spatial subdivision techniques [WMW86, Blo88, HW90], scalar field isosurface reconstruction [LC87, KBSS01, JLSW02], and algorithms based on Morse theory [SH97]. Most applications use some combination of spatial subdivision and isosurface reconstruction. These algorithms are implemented using uniform or adaptive grids, and can handle complex models. In practice, their accuracy varies as a function of the grid resolution and the

[†] Supported in part by ARO Contracts DAAD19-02-1-0390 and W911NF-04-1-0088, NSF awards 0400134 and 0118743, ONR Contract N00014-01-1-0496, DARPA/RDECOM Contract N61339-04-C-0043 and Intel.

coordinate system used for spatial subdivision. As a result, current spatial subdivision techniques may not provide rigorous guarantees on the topology of the reconstructed surface. On the other hand, polygonization algorithms based on Morse theory and critical point analysis can provide topological and geometric guarantees for smooth surfaces. However, these algorithms have been limited to relatively simple surfaces and do not extend to Boolean combinations.

Main Results: We present a new algorithm to polygonize implicit surfaces. Our approach is restricted to compact, manifold surfaces that can be represented as the zero set of a continuous function. We compute a topologically accurate polygonization of the implicit surface and construct a homeomorphism, i.e. a continuous bijection with a continuous inverse, between the implicit surface and the polygonal approximation. This homeomorphism is evaluated by computing ray intersections with the implicit surface.

Our algorithm uses spatial subdivision techniques along with interval arithmetic to decompose the implicit surface into surface patches. We ensure that each patch satisfies the *star-shaped* property, i.e. there exists a guard in space that can 'see' all the points on the patch. We use this property to compute a *visibility map* that projects each point on the patch to the boundary of the grid cell. Our algorithm computes a triangulation of the image of the visibility map on the boundary of the grid cell. Finally, these triangles are back-projected using the inverse of the visibility map to compute a homeomorphic triangulation of the patch. We ensure continuity between adjacent patches and generate a watertight polygonization of the overall surface. Some of the main *features* of our algorithm are:

Generality: Our algorithm is applicable to a broad class of implicits. These include zero sets of elementary functions, blends of locally fit smooth functions such as Multiple Partition-of-Unity (MPUs), Moving Least Squares (MLS), and Boolean combinations.

Reliable Polygonization: Given a manifold surface, our algorithm ensures that the polygonized approximation is topologically equivalent to the implicit surface. We also satisfy a two-sided Hausdorff bound between the original model and the polygonal approximation. Moreover, we do not need to perform any crack patching to handle adaptive grids.

Complex Models: Our approach can handle complex implicit surfaces that arise from point-cloud reconstruction algorithms or Boolean combinations. We use techniques based on spatial subdivision and voxelization to accelerate the computation.

Parameterization: Our algorithm computes a piecewise *star-shaped parameterization* of the implicit surface. Furthermore, the homeomorphism between the implicit surface and our polygonal approximation can be combined with mesh parameterization algorithms to compute a global parameterization of the implicit surface.

Quality of Polygonization: Star-shaped parameterization is a special case of spherical parameterization. This property can

be exploited to generate a triangulation with good aspect ratios using the method proposed by Praun and Hoppe [PH03].

There exist prior implicit surface polygonization algorithms that provide topological guarantees. However, all of these algorithms assume a smooth implicit surface. As a result, they are not applicable to surfaces defined using Boolean operations. To the best of our knowledge, our algorithm is the first topology preserving polygonization algorithm that can handle Boolean operations. We have applied our algorithm to polygonize MPU surfaces generated using point cloud reconstruction algorithms, algebraic surfaces, and boundary evaluation of CSG models.

Organization: The rest of the paper is organized in the following manner. Section 2 presents our polygonization algorithm based on visibility mapping. We show that our algorithm computes a topologically and geometrically accurate polygonization in Section 3. We describe our spatial subdivision algorithm based on star-shaped decomposition in Section 4. We highlight the performance of our algorithm on different benchmarks in Section 5 and compare it with prior approaches in Section 6. Finally, we highlight some of its limitations in Section 7.

2. Implicit Surface Polygonization

Our algorithm uses a divide-and-conquer approach, subdividing the implicit surface into patches. In this section, we present our algorithm to polygonize each patch independently and to generate a watertight triangulation of the implicit surface. We first present our notation and define visibility mapping based on the star-shaped property.

2.1. Notation and Terminology

We use lower case bold letters such as $\mathbf{p}, \mathbf{q}$ to refer to points in $\mathbb{R}^3$. The notation $\mathbf{pq}$ denotes the line segment between the two points $\mathbf{p}$ and $\mathbf{q}$. The notation $(\mathbf{p}_1, \ldots, \mathbf{p}_n)$ denotes the n-sided polygon whose vertices are the points $\mathbf{p}_1, \ldots, \mathbf{p}_n$. Let $\mathscr{I} : \mathbb{R}^3 \to \mathbb{R}$ denote the implicit function; our goal is to compute a triangulation of its zero set. In practice, $\mathscr{I}$ could represent a function of the form $f(x, y, z)$. Furthermore, we allow CSG combinations of simple closed-form functions or other procedural models that can be expressed as elementary functions or weighted combination of simple elementary functions (e.g. blends, MPUs and MLS surfaces). With a slight abuse of notation, we will also use $\mathscr{I}$ to denote the implicit surface; we assume that $\mathscr{I}$ is a manifold. Our algorithm computes a topology-preserving polygonal approximation $\widetilde{\mathscr{I}}$ of the implicit surface $\mathscr{I}$. We construct a homeomorphism $\mathscr{H}$ between $\mathscr{I}$ and $\widetilde{\mathscr{I}}$.

Our algorithm uses a volumetric grid that has convex polyhedral cells. In the rest of the paper, for the sake of simplicity, we will assume that all the cells are cube-shaped. When referring to the cell as a geometric primitive, we will refer to it as a voxel. The boundary of a cell consists of faces, edges, and vertices and each of these is represented as a closed set. We use the symbols ϑ, f, e, and v to refer a voxel, a face, an edge, and a vertex, respectively.

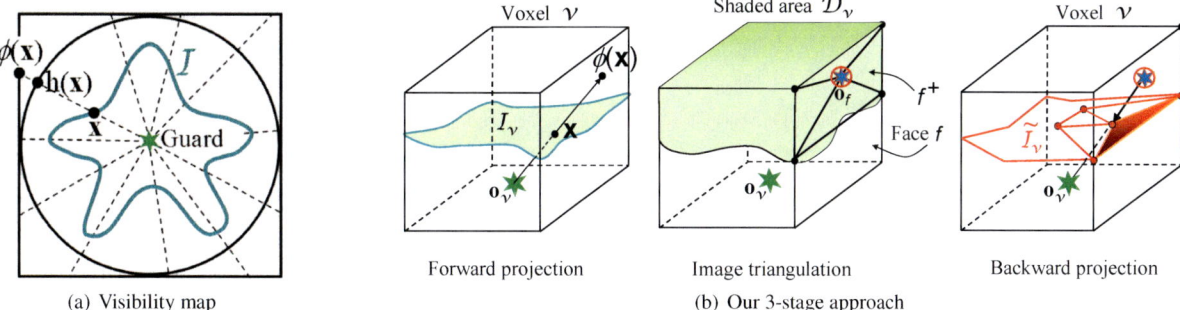

Figure 2: *Fig. (a) shows that each point* **x** *on a star-shaped surface $\mathscr{I}$ can be mapped onto the the unit cube (or the unit sphere) using the visibility map ϕ (or h). Fig. (b) highlights the three steps of our algorithm: Forward projection that projects the star-shaped surface to the voxel boundary, image triangulation which triangulates the image on the voxel boundary, and backward projection, where the triangulation on the boundary is back-projected to compute triangulation of the implicit surface. We only show a subset of the triangles on the surface.*

We use the symbol ∂S to denote the boundary of a set S. By a *restriction* of a set S with respect to another set T, we mean $S \cap T$, which we denote as S_T. Typically, S is a 2D manifold surface and T may correspond to a voxel or a face. The restriction operator has higher precedence over the boundary operator, i.e. $\partial S_T = \partial(S_T)$.

2.2. Visibility Mapping

In this subsection, we introduce visibility mapping and use it to polygonize implicit surfaces. We first present the intuition behind our algorithm. Let us assume that the implicit surface $\mathscr{I}$ is star-shaped. A surface is *star-shaped* if there exists a point in $\mathbb{R}^3$ (called a guard) that can 'see' every point on the surface. In essence, the star-shaped property of a surface captures its visibility. We exploit the fact that a star-shaped surface $\mathscr{I}$ has a star-shaped parametrization, a special case of spherical parameterization. Without loss of generality, we assume that the origin is the guard of $\mathscr{I}$. There exists a one-to-one map $h : \mathscr{I} \to \mathbb{S}^2$ that maps each point on $\mathscr{I}$ to a point on the unit sphere $\mathbb{S}^2$. This function is expressed as: $h(\mathbf{x}) = \frac{\mathbf{x}}{\|\mathbf{x}\|_2}$, where $\|\|_2$ denotes the Euclidean norm. The map $h()$ is the spherical projection operation. See Fig. 2(a). Note that if $\mathscr{I}$ is a closed star-shaped surface then h is a bijection and we can map $\mathscr{I}$ to $\mathbb{S}^2$ and vice-versa. In practice, we consider a mapping function $\phi : \mathscr{I} \to \mathbb{C}^2$ that maps $\mathscr{I}$ to the unit cube $\mathbb{C}^2$: $\phi(\mathbf{x}) = \frac{\mathbf{x}}{\|\mathbf{x}\|_\infty}$ where $\| . \|_\infty$ denotes the max-norm. Like h, ϕ is also a bijective mapping. We refer to ϕ as the *visibility map*. See Fig. 2(a). The visibility map can be thought of as a perspective projection operation onto the unit cube. It provides a simple method for triangulating $\mathscr{I}$: we first triangulate the boundary of $\mathbb{C}^2$ and then map these triangles to $\mathscr{I}$ using ϕ^{-1}. Since ϕ is a bijection, ϕ^{-1} is well-defined. This yields a triangulation of $\mathscr{I}$. Evaluating the visibility map ϕ (or its inverse ϕ^{-1}) is simple: it reduces to shooting a ray from the guard and computing intersection of the ray with the boundary of $\mathbb{C}^2$ (or with $\mathscr{I}$).

We extend the above idea to the case of a general implicit surface by adopting a divide-and-conquer approach. Conceptually, we decompose $\mathscr{I}$ into a set of star-shaped patches and polygonize each star-shaped patch using the visibility maps

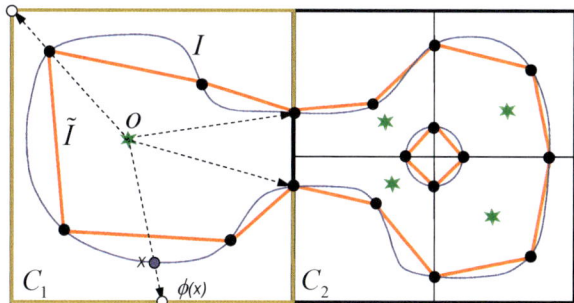

Figure 3: *2D implicit curve approximation using our algorithm:* The figure shows the 2D visibility grid for an implicit curve $\mathscr{I}$. The portion of $\mathscr{I}$ in cell C_1 is star-shaped property with respect to guard O. By projecting this portion onto the boundary of the cell C_1, we compute a one-to-one onto mapping ϕ between any point x on $\mathscr{I}$ and its projection $\phi(x)$ on the boundary of the cell. We use this mapping to compute a homeomorphic polyline approximation of $\mathscr{I}$ within C_1. The cell C_2 is generated after an additional level of quadtree subdivision; the portion of $\mathscr{I}$ in C_2 is also star-shaped.

for that patch. Note that we compute a different map for each patch. In practice, we do not compute such a star-shaped decomposition explicitly but rather use spatial subdivision techniques (see Section 4). We also ensure that there is continuity between the polygonization of neighboring star-shaped patches. This results in a continuous water-tight triangulation of $\mathscr{I}$.

2.3. Polygonization within a Cell

Our algorithm uses spatial subdivision techniques like octree decomposition to generate a grid that satisfies certain visibility properties. We call a volumetric grid $\mathscr{G}$ a **visibility grid** if it satisfies the following conditions:

1. Every voxel ϑ in $\mathscr{G}$, $\mathscr{I} \cap \vartheta$ is star-shaped with respect to (w.r.t) some point $\mathbf{o}_\vartheta$ in ϑ. We call the point $\mathbf{o}_\vartheta$ a guard of ϑ.
2. Every face f in $\mathscr{G}$, $\mathscr{I} \cap f$ is star-shaped w.r.t some point $\mathbf{o}_f$ in f. We call the point $\mathbf{o}_f$ a guard of f.

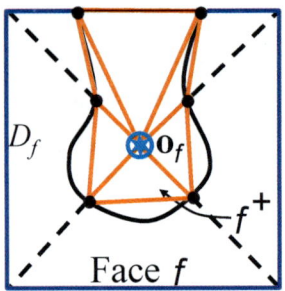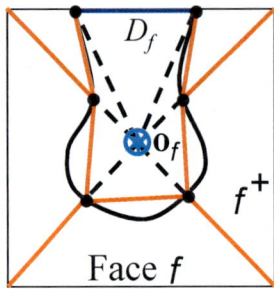

Figure 4: *Image triangulation:* The two figures show two different cases of image triangulation of a face f depending on whether the guard $\mathbf{o}_f$ of face f belongs to f^+ or not. In each case, we compute a triangulation $\widetilde{f^+}$ of f^+.

The visibility grid enforces the star-shaped property within every voxel and face of the grid. Our algorithm requires that that $\mathscr{I}$ intersect the cells in the visibility grid in a non-degenerate manner, i.e., $\mathscr{I}$ should not tangentially intersect any of the voxels, faces, or edges of the grid.

Let us consider $\mathscr{I}_\vartheta = \mathscr{I} \cap \vartheta$, the restriction of $\mathscr{I}$ to the voxel ϑ of a cell. $\mathscr{I}_\vartheta$ is star-shaped and has a guard $\mathbf{o}_\vartheta$ that lies inside ϑ. Without loss of generality, we assume that $\mathbf{o}_\vartheta$ lies inside $\mathscr{I}$. Our algorithm can be easily extended to handle the symmetric case when $\mathbf{o}_\vartheta$ lies outside $\mathscr{I}$. Our overall approach consists of three main steps:

1. **Forward projection:** We define a map $\phi_\vartheta : \mathscr{I}_\vartheta \to \mathscr{D}_\vartheta$ that projects each point on $\mathscr{I}_\vartheta$ to a portion $\mathscr{D}_\vartheta$ of the boundary of the voxel. We refer to ϕ_ϑ as the *forward visibility map* and $\mathscr{D}_\vartheta$ as the *image* of the forward visibility map.
2. **Image triangulation:** We triangulate $\mathscr{D}_\vartheta$ to compute a triangulation $\widetilde{\mathscr{D}_\vartheta}$.
3. **Backward projection:** We backproject the triangles in $\widetilde{\mathscr{D}_\vartheta}$ to obtain a triangulation $\widetilde{\mathscr{I}_\vartheta}$ of $\mathscr{I}$. The backward projection is achieved using a map $\widetilde{\phi}_\vartheta : \widetilde{\mathscr{D}_\vartheta} \to \widetilde{\mathscr{I}_\vartheta}$, which is defined using the inverse of the forward visibility map. We refer to $\widetilde{\phi}_\vartheta$ as the *backward visibility map*.

We now explain each of the above three steps in more detail. Fig. 3 illustrate our algorithm for a 2D planar curve.

2.3.1. Forward Projection

We define a function $\tau_\vartheta : \mathscr{I}_\vartheta \to \partial\vartheta$ that maps $\mathscr{I}_\vartheta$ to the boundary of the voxel (see Fig. 2).

$$\tau_\vartheta(\mathbf{x}) = \mathbf{o}_\vartheta + \parallel \vartheta \parallel \phi(\mathbf{x} - \mathbf{o}_\vartheta),$$

where $\parallel \vartheta \parallel$ denotes half of the length of voxel ϑ. It follows that τ_ϑ is a 1-1 map. In general, if $\mathscr{I}$ intersects the boundary of the voxel, τ_ϑ will not be an onto function. Let $\mathscr{D}_\vartheta = \tau_\vartheta(\mathscr{I}_\vartheta)$ denote the range of τ_ϑ, as shown in Fig. 2(b). In the rest of the paper, we will refer to the bijection $\phi_\vartheta : \mathscr{I}_\vartheta \to \mathscr{D}_\vartheta$, where $\phi_\vartheta(\mathbf{x}) = \tau_\vartheta(\mathbf{x})$. ϕ_ϑ is the *forward visibility map* and $\mathscr{D}_\vartheta$ is its *image*. Furthermore, $\phi_\vartheta(\mathbf{x})$ can be evaluated by shooting a ray from $\mathbf{o}_\vartheta$ in the direction of $\mathbf{x}$ and computing the intersection of the ray with the boundary of the voxel.

2.3.2. Triangulating the Image

Our goal is to compute a triangulation of $\mathscr{D}_\vartheta$. We need to ensure C^0 continuity between the triangulations of $\mathscr{I}$ between adjacent voxels. We utilize the fact that $\mathscr{D}_\vartheta$ consists of points on the boundary of ϑ that are not visible to the guard of ϑ, i.e. the ray from the guard to such a point must intersect $\mathscr{I}$. The star-shaped property ensures that such a point must necessarily lie outside $\mathscr{I}$. We use this property to express $\mathscr{D}_\vartheta$ in terms of faces of ϑ. Let $f_i, i = 1, \ldots, 6$, denote the faces of ϑ. Let f_i^+ be the set of points in f_i that lie outside $\mathscr{I}$, i.e.,

$$f_i^+ = \{\mathbf{x} \mid \mathbf{x} \in f_i, \mathscr{I}(\mathbf{x}) > 0\}.$$

$\mathscr{D}_\vartheta$ is given by: $\mathscr{D}_\vartheta = \bigcup_{i=1}^6 f_i^+$, as shown in Fig. 2. Therefore the problem of triangulating $\mathscr{D}_\vartheta$ reduces to triangulating each f_i^+, which is a portion of face f_i.

It suffices to triangulate just *minimal faces*; a face f is minimal if no face in the grid is a strict subset of f. A non-minimal face is a union of a set of minimal faces. We now present our algorithm to triangulate f^+ for a minimal face f in the visibility grid. We do not compute f^+ explicitly; rather, we take advantage of the star-shaped property, i.e., $\mathscr{I} \cap f$ is star-shaped w.r.t a point $\mathbf{o}_f \in f$. We define a visibility mapping in 2D to compute this triangulation. Our approach again performs three main steps:

1. **Forward projection:** We define a 2D forward visibility map $\phi_f : \mathscr{I}_f \to \mathscr{D}_f$, which projects $\mathscr{I}_f$ onto a portion of ∂f. This visibility map is defined on a 2D domain and is similar to ϕ_ϑ. $\mathscr{D}_f$, the image of ϕ_f, is a polyline.
2. **Image triangulation:** Divide $\mathscr{D}_f$ into a set of line segments as explained below.
3. **Backward projection:** Backproject the line segments to obtain a 'triangulation $\widetilde{f^+}$ of f^+.

We consider two separate cases based on whether or not $\mathbf{o}_f$ belongs to f^+:

Case 1: $\mathbf{o}_f \in f^+$
First we compute $\mathscr{D}_f$, the image of ϕ_f. In this case, $\mathscr{D}_f$ consists of those points in ∂f that lie inside $\mathscr{I}$. Therefore, $\mathscr{D}_f$ can be computed by finding the intersections points between $\mathscr{I}$ and the edges of f. These intersection points together with the vertices of f partition ∂f into two sets of line segments. A subset $\mathscr{L}_1$ of these line segments belong to $\mathscr{D}_f$. Let $\mathscr{L}_2$ denote the remainder of line segments.

1. For each edge, $\mathbf{ab} \in \mathscr{L}_1$, output a triangle $(\mathbf{o}_f, \phi_f^{-1}(\mathbf{a}), \phi_f^{-1}(\mathbf{b}))$.
2. For each edge $\mathbf{ab} \in \mathscr{L}_2$, output a triangle $(\mathbf{o}_f, \mathbf{a}, \mathbf{b})$.

This is illustrated in Fig. 4. The union of all the triangles generated by these steps is the triangulation $\widetilde{f^+}$ of f^+. We evaluate ϕ_f^{-1} by shooting a ray from $\mathbf{o}_f$ and computing the intersection of the ray with $\mathscr{I}$.

Case 2: $\mathbf{o}_f \notin f^+$
As in Case 1, we first compute a set of line segments, $\mathscr{L}_1$, which belong to $\mathscr{D}_f$. In this case, $\mathscr{D}_f$ consists of those points

in ∂f that lie outside $\mathcal{I}$. For each edge $\mathbf{ab} \in \mathcal{L}_1$, we output a quadrilateral $(\mathbf{a}, \mathbf{b}, \phi_f^{-1}(\mathbf{b}), \phi_f^{-1}(\mathbf{a}))$. We triangulate each quadrilateral using one of its diagonals. This quadrilateral may degenerate into a triangle if either $\mathbf{a}$ or $\mathbf{b}$ corresponds to an intersection point between $\mathcal{I}$ and ∂f (Fig. 4). The union of all the triangles generated above is the triangulation $\widetilde{f^+}$ of f^+.

Image Retriangulation: A finer triangulation $\widetilde{f^+}$ can be obtained by refining the initial triangulation of f^+. This can be computed in two ways: (a) by subdividing the line segments in $\mathcal{L}_1$ and $\mathcal{L}_2$ or (b) by subdividing the triangles in $\widetilde{f^+}$.

2.3.3. Backward Projection

We use the method presented above to triangulate f^+ for each face f of the cell. We represent this triangulation of $\mathcal{D}_\vartheta$ as $\widetilde{\mathcal{D}}_\vartheta : \widetilde{\mathcal{D}}_\vartheta = \bigcup_{i=1}^{6} \widetilde{f_i^+}$. Each triangle in $\widetilde{\mathcal{D}}_\vartheta$ has the property that all its vertices belong to $\mathcal{D}_\vartheta$. Therefore any point $\mathbf{x} \in \widetilde{\mathcal{D}}_\vartheta$ can be expressed in terms of barycentric coordinates:

$$\mathbf{x} = u\mathbf{a} + v\mathbf{b} + w\mathbf{c},$$
$$u, v, w \geq 0, \quad u + v + w = 1, \quad \mathbf{a}, \mathbf{b}, \mathbf{c} \in \mathcal{D}_\vartheta.$$

We define a map $\widetilde{\phi}_\vartheta : \widetilde{\mathcal{D}}_\vartheta \to \mathbb{R}^3$:

$$\widetilde{\phi}_\vartheta(\mathbf{x}) = u\phi_\vartheta^{-1}(\mathbf{a}) + v\phi_\vartheta^{-1}(\mathbf{b}) + w\phi_\vartheta^{-1}(\mathbf{c}).$$

Since $\mathbf{a}, \mathbf{b}$, and $\mathbf{c}$ belong to $\mathcal{D}_\vartheta$, $\phi_\vartheta^{-1}(\mathbf{a}), \phi_\vartheta^{-1}(\mathbf{b})$, and $\phi_\vartheta^{-1}(\mathbf{c})$ are well-defined. We refer to $\widetilde{\phi}_\vartheta$ as the *backward visibility map*. In essence, the backward map back-projects the triangles in $\widetilde{\mathcal{D}}_\vartheta$ by using the inverse of the forward visibility map ϕ_ϑ and barycentric coordinates. The backward projection produces the approximation $\widetilde{\mathcal{I}}_\vartheta = \widetilde{\phi}_\vartheta(\widetilde{\mathcal{D}}_\vartheta)$ in the voxel. The overall approximation $\widetilde{\mathcal{I}}$ corresponds to the union of the approximations within the individual voxels:

$$\widetilde{\mathcal{I}} = \cup_\vartheta \widetilde{\mathcal{I}}_\vartheta$$

The resulting approximation $\widetilde{\mathcal{I}}$ is a continuous, watertight polygonal surface.

2.4. Continuity

The approximation $\widetilde{\mathcal{I}}$ has C^0 continuity. This follows from the fact that the approximation within two adjacent voxels "match" along the common face. Consider two voxels ϑ_i, ϑ_j that share a face f, i.e., $\vartheta_i \cap \vartheta_j = f$. Then the images of the forward visibility maps of ϑ_i and ϑ_j are identical along face f. Our image triangulation procedure also maintains this property. Therefore, we have $\mathcal{D}_{\vartheta_i} \cap f = \mathcal{D}_{\vartheta_j} \cap f$ and $\widetilde{\mathcal{D}}_{\vartheta_i} \cap f = \widetilde{\mathcal{D}}_{\vartheta_j} \cap f$. Since the approximations $\widetilde{\mathcal{I}}_{\vartheta_i}$ and $\widetilde{\mathcal{I}}_{\vartheta_j}$ are obtained by backprojecting $\widetilde{\mathcal{D}}_{\vartheta_i}$ and $\widetilde{\mathcal{D}}_{\vartheta_j}$, these approximations 'match' along f, i.e., $\partial(\widetilde{\mathcal{I}}_{\vartheta_i}) \cap f = \partial(\widetilde{\mathcal{I}}_{\vartheta_j}) \cap f$. The following lemma is used to formally prove the continuity:

LEMMA 1 *Let ϑ_i and ϑ_j be two adjacent voxels that share a face f, i.e., $\vartheta_i \cap \vartheta_j = f$. Then $\partial(\widetilde{\mathcal{I}}_{\vartheta_i}) \cap f = \partial(\widetilde{\mathcal{I}}_{\vartheta_j}) \cap f$.*

We note that the same formulation also applies to adaptive grids.

3. Accuracy of Polygonization

In this section, we analyze topological properties of our polygonization algorithm. We also extend the polygonization algorithm to improve its accuracy and compute a parameterization.

3.1. Topology Preservation

Our goal is to prove that $\widetilde{\mathcal{I}}$ is topologically equivalent to $\mathcal{I}$. In particular, we show that there exists a homeomorphism between $\widetilde{\mathcal{I}}$ and $\mathcal{I}$. Our proof follows the following outline:

1. We define a *local map* $\mathcal{H}_\vartheta : \mathcal{I}_\vartheta \to \widetilde{\mathcal{I}}_\vartheta$ in each voxel ϑ. We show that $\mathcal{H}_\vartheta$ is a bijection. Furthermore, we prove that the local maps of two adjacent voxels match each other along the common face. We combine these local maps to obtain a homeomorphism $\mathcal{H} : \mathcal{I} \to \widetilde{\mathcal{I}}$, and thereby establish topological equivalence between $\mathcal{I}$ and $\widetilde{\mathcal{I}}$.

2. The local map $\mathcal{H}_\vartheta$ is a composition of three bijections:

   a. The forward visibility map ϕ_ϑ.
   b. *The image transfer map*: For each face f, we define a map $\delta_f : f^+ \to \widetilde{f^+}$ that maps the image f^+ on face f to its triangulation $\widetilde{f^+}$. We refer to δ_f as the *image transfer map* of f. By combining the image transfer maps of the faces, we obtain an image transfer map $\delta_\vartheta : \mathcal{D}_\vartheta \to \widetilde{\mathcal{D}}_\vartheta$ for the voxel.
   c. The backward visibility map $\widetilde{\phi}_\vartheta$.

   The local map $\mathcal{H}_\vartheta$ is equal to $\widetilde{\phi}_\vartheta \circ \delta_\vartheta \circ \phi_\vartheta$.

We state the main results of topological equivalence. The proofs of all these lemmas and the theorem are presented in the appendix. We first state the properties of the image transfer map of a face and voxel.

LEMMA 2 *Given a face f in the visibility grid, there exists a bijection $\delta_f : f^+ \to \widetilde{f^+}$.*

LEMMA 3 *Given a voxel ϑ in the visibility grid, there exists a bijection $\delta_\vartheta : \mathcal{D}_\vartheta \to \widetilde{\mathcal{D}}_\vartheta$.*

Since $\mathcal{H}_\vartheta$ is a composition of bijective mappings, it is a bijection. See Appendix. This fact combined with the fact that our approximation generates a C^0 mesh can be used to arrive at the following main result (proof of result is in [VKZM06] and the supplementary material provided with this paper):

THEOREM 1 *$\mathcal{I}$ is homeomorphic to $\widetilde{\mathcal{I}}$ ($\mathcal{I} \approx \widetilde{\mathcal{I}}$).*

Evaluating the Homeomorphism: Our formulation of the homeomorphism is constructive. We can evaluate $\mathcal{H}$ and its inverse using ray-shooting. Consider a point $\mathbf{x} \in \mathcal{I}$. Let $\mathbf{x}$ belong to a voxel ϑ. Then $\mathcal{H}(\mathbf{x}) = \mathcal{H}_\vartheta(\mathbf{x})$. Recall that $\mathcal{H}_\vartheta$ is a composition of three bijections: (a) the forward visibility map ϕ_ϑ, (b) δ_ϑ, and (c) the backward visibility map $\widetilde{\phi}_\vartheta$. Each of the three bijections can be evaluated by ray-shooting. Therefore, computing $\mathcal{H}_\vartheta(\mathbf{x})$ also reduces to ray-shooting. We use this property in Section 3.5 to compute a parameterization of the implicit surfaces.

3.2. Geometric Accuracy

We extend our algorithm to generate a polygonization $\widetilde{\mathscr{I}}$ with a bounded Hausdorff error. Given any $\varepsilon > 0$, our algorithm outputs $\widetilde{\mathscr{I}}$ such that the two-sided Hausdorff distance between $\widetilde{\mathscr{I}}$ and $\mathscr{I}$ is less than ε. One way to bound this error is to require that all the grid cells in the volumetric grid be smaller than ε. However, this can incur a huge penalty in the performance as a large number of star-shaped tests would need to be performed (Section 4). Rather, we bound $\mathscr{I}$ by a set of voxels of size ε and check if $\widetilde{\mathscr{I}}$ is contained by this set of voxels. Such a voxelization is precomputed for the implicit surface using interval arithmetic (see Section 4.3).

When $\mathscr{I}$ is defined in terms of Boolean operations over a set of primitives, we exploit the fact that the boundary of $\mathscr{I}$ is a subset of the union of boundaries of the primitives. Consider a voxel ϑ that is intersected by n primitives, $\mathscr{S}_1, \ldots, \mathscr{S}_n$. In this case, we check if the Hausdorff distance between $\widetilde{\mathscr{I}}_\vartheta$ and $\mathscr{S}_i \cap \vartheta$ is less than ε for each $i = 1, \ldots, n$. This provides a sufficient condition under which the Hausdorff distance between $\widetilde{\mathscr{I}}$ and $\mathscr{I}$ is also bounded by ε. A different technique for bounding Hausdorff error is presented in [AAB02].

3.3. Quality of Polygonization

We can also generate a triangulation with good aspect ratios by using the remeshing techniques proposed by [PH03, THCM04]. Each patch in our case has a *star-shaped parameterization*, which is a special case of spherical parameterization – a property that is utilized by these remeshing techniques. We can use these techniques to guide the image triangulation step of our algorithm (Section 2.3.2). Since the image triangulation is done on the faces of the grid, we automatically ensure continuity across adjacent voxels.

3.4. Sharp Features

We use a technique similar to Dual contouring [JLSW02] to obtain an accurate polygonization near sharp features. For each triangle in $\widetilde{\mathscr{I}}$, we use the normals at the three vertices to estimate a *minimizing vertex* as described in [JLSW02]. Then we output a dual polygonal mesh by combining the minimizing vertices of adjacent triangles: each triangle in $\widetilde{\mathscr{I}}$ corresponds to a vertex in the dual mesh, an edge in $\widetilde{\mathscr{I}}$ corresponds to a dual edge, and a vertex in $\widetilde{\mathscr{I}}$ corresponds to a polygon in the dual mesh. By triangulating these polygons, we obtain a triangulation that can capture sharp features.

3.5. Parameterization

A star-shaped surface can be parameterized using spherical coordinates, i.e. latitude and longitude. Both the original implicit surface $\mathscr{I}$ and our approximation $\widetilde{\mathscr{I}}$ consist of star-shaped patches, denoted as $\mathscr{I}_\vartheta$ and $\widetilde{\mathscr{I}}_\vartheta$. As a result, our algorithm automatically computes a piecewise star-shaped parameterization of $\mathscr{I}$.

Our algorithm can also compute a global parameterization of implicit surfaces. In general, computing a good parameterization of implicit surfaces is a difficult problem [Ped95],

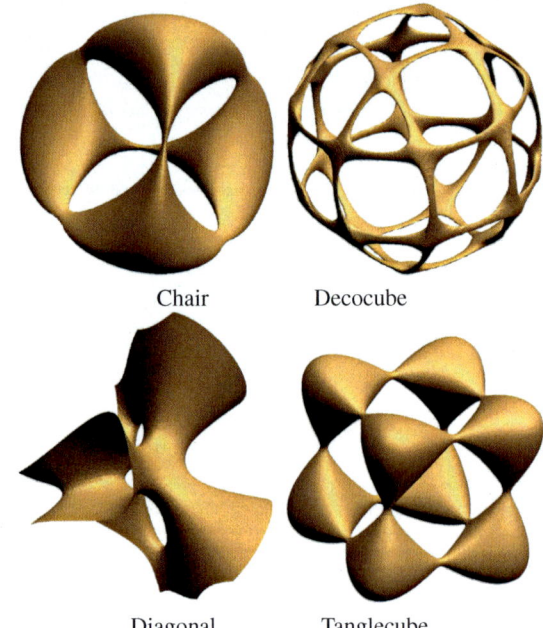

Chair   Decocube

Diagonal   Tanglecube

Figure 5: *Polygonization of algebraic surfaces:* Our algorithm is able to compute reliable polygonization of these high genus algebraic surfaces. The running time is about 10 sec for these surface, except for Decocube that takes about 52 seconds.

In contrast, there exist a host of techniques to parameterize polygonal meshes [FH05]. Our algorithm reduces the problem of implicit surface parameterization to a mesh parameterization problem. We exploit the fact that our algorithm constructs a homeomorphism $\mathscr{H} : \mathscr{I} \rightarrow \widetilde{\mathscr{I}}$ between the implicit surface and our polygonal approximation. Furthermore, $\mathscr{H}$ can be evaluated by ray-shooting. Therefore, by computing a mesh parameterization $\mu : \mathscr{X} \rightarrow \widetilde{\mathscr{I}}$, we obtain an implicit surface parameterization $\lambda : \mathscr{X} \rightarrow \mathscr{I}$, where $\lambda = \mathscr{H}^{-1} \circ \mu$.

4. Spatial Subdivision

Our polygonization algorithm assumes that each patch of the implicit surface restricted to a voxel/face of a cell is star-shaped. In this section, we present our spatial subdivision algorithm to decompose the space into a visibility grid with respect to $\mathscr{I}$. We perform an octree subdivision and check whether the portion of $\mathscr{I}$ restricted to a voxel ϑ, i.e. $\mathscr{I}_\vartheta$, is star-shaped. A similar test is also performed on the faces of each cell.

In general, a star-shaped surface does not have a unique guard. The set of all guards is called the *kernel*. Thus the surface is star shaped if its kernel is non-empty. For polygonal meshes, testing whether the kernel is non-empty can be performed efficiently using linear programming. However, it is hard to perform the kernel test exactly for non-linear primitives. We perform a conservative test using linear programming and interval arithmetic for many types of implicit surfaces, by extending the approach described by Varadhan *et al.* [VKSM04]. Our algorithm is based on two oracles:

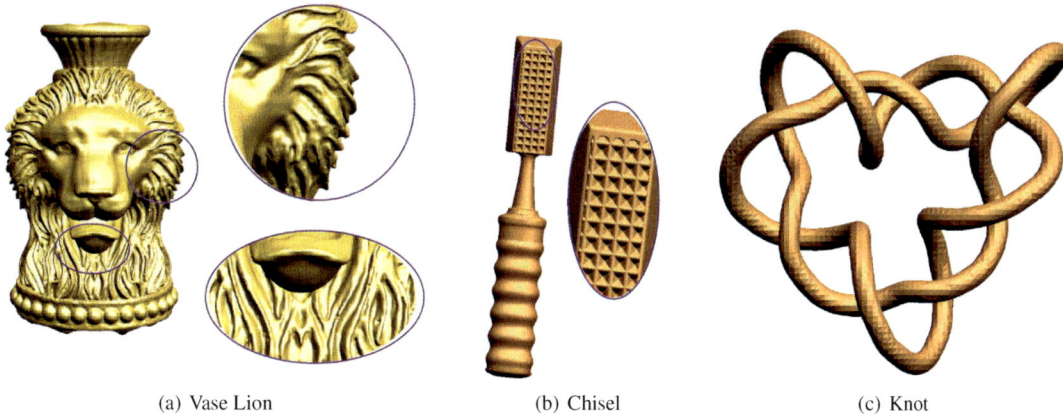

(a) Vase Lion   (b) Chisel   (c) Knot

Figure 6: *Polygonization of complex MPU models: The figure shows three MPU models:* Vase Lion *with* 200K *points,* Chisel *with* 142K *points and and* Knot *with* 28K *points.*

- **O1:** Given a voxel ϑ and a parameter n, this oracle returns n points inside ϑ, which lie on $\mathscr{I}$. It returns an error code if $\mathscr{I}_{\vartheta} = \emptyset$. As long as the surface passes through a cell, this oracle can be implemented by generating line segments uniformly inside the cell and look for intersections with the surface. The generation of line segments is done by 3 sets of two-plane parameterization of the rays as described in Levoy *et al.* [LH96]. As the density of samples in each plane is increased, we get more line segments to intersect the surface.

  Computing the intersection of the line segments with the surface reduces to a ray shooting problem. Implicit surfaces like MPU implicits [OBA*03] are defined as the zero set of a convex combination of quadric surfaces. Dealing with quadric surfaces is fairly well studied in the literature. Further, techniques to compute roots of high-degree univariate polynomials have matured over the last decade and many efficient and reliable numerical algorithms exist [Bin96].

- **O2:** Given a point $\mathbf{p}$ on $\mathscr{I}$, this oracle returns the unit normal vector to $\mathscr{I}$ at $\mathbf{p}$. This is basically gradient computation.

Our method initially estimates a point inside the cell that is likely to be a point in the exact kernel (if one exists). Once we have a witness point in the kernel, we use interval arithmetic to determine if the point obtained is indeed a witness to the exact kernel. If the interval arithmetic test is positive, we can guarantee that the surface is star-shaped. Otherwise, we subdivide the voxel until the condition is satisfied.

4.1. Kernel Witness Computation

We start by invoking oracle **O1** to generate a set of points inside the voxel. We use a simple heuristic to decide the number of points. If V is the volume of the voxel, we generate $m = cV^{2/3}$ points (roughly proportional to the surface area) with an appropriate constant c. The value of c depends on the curvature bounds of the surface inside the voxel and the lower bound of the voxel size during subdivision. Let the point set be $\mathscr{Q} = \{\mathbf{q}_1, \mathbf{q}_2, \ldots, \mathbf{q}_m\}$. We then invoke oracle **O2** to compute the unit normal vector on each of the points obtained from the first step. Let the normal vector at point $\mathbf{q_i}$ be $\mathbf{n_i}$. We define

m linear constraints corresponding to halfspaces supported by the tangent planes $((\mathbf{q_i} - \mathbf{x})^T \mathbf{n_i} > 0, \mathbf{x} \in \mathbb{R}^3)$. We also include the six linear constraints defining the faces of the cell. We solve for the feasibility of the linear program to determine a witness.

We choose a point that is roughly in the center of the approximate kernel. To find this point, we augment the linear program by adding a slack variable δ to each of the constraints - $(\mathbf{q_i} - \mathbf{x})^T \mathbf{n_i} + \delta > 0$. We also set the objective function to maximize δ. This heuristic relies on the fact that as m increases, the kernel computed with linear programming converges to the true kernel. It relies on the following fact: Consider an ε-sampling of the implicit surface using the isophotic metric (Euclidean distance + arc length on Gaussian sphere) [PSH*04]. Then the kernel boundary computed using the samples will approximate the true kernel boundary (in terms of Hausdorff error) to within a monotonic function of ε.

It can be seen that the kernel monotonically shrinks (in terms of set containment) with increasing number of sampled points. The point most interior in the kernel is also the most stable to perturbations in the linear constraints. Hence, if the true surface is star-shaped, this point is likely to be (heuristically) the best candidate for being a guard. Let the point computed using linear programming be $\mathbf{p}$. We need to verify if $\mathbf{p}$ is actually a witness to the exact kernel. Such a witness would satisfy $0 \notin (\mathbf{x} - \mathbf{p})^T \mathbf{n}(\mathbf{x})$, for all points $\mathbf{x}$ on the primitive inside the voxel where $\mathbf{n}(\mathbf{x})$ is the normal to the surface at $\mathbf{x}$.

4.2. Application to Implicit Surfaces

The algorithm described above is applicable to all surfaces, as long as the oracles are available. As a result, our approach is general. However, if surfaces have a certain structure in terms of their description, we take advantage of that to provide more efficient interval tests. Many implicits are represented as zero set of an *elementary function*, i.e. built from a finite combination of constant functions, field operations algebraic, exponential, and logarithmic functions and their inverses under repeated compositions. For such surfaces, the normal vector at a point can be defined by the gradient function which are el-

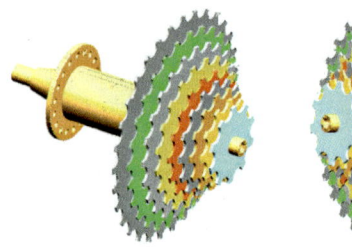
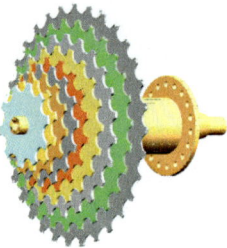
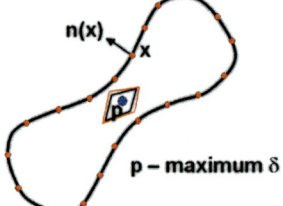
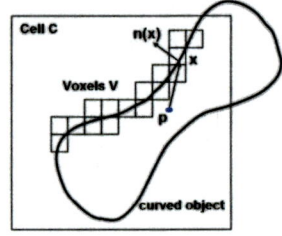

Figure 7: CAD model: This CAD model consists of 14 solids designed using a total of 84 Boolean operations on cones and cylinders. Our algorithm took on an average of 15 secs to compute the boundary of each solid.

Figure 8: *Star-shaped test on implicit surfaces: The left figure shows the candidate point selection. The red points are a result of implementing oracle **O1**. The right figure shows the kernel membership test. Given a cell C, we compute an initial voxelization V of the primitive and use the intervals in V during the interval arithmetic step.*

ementary functions themselves. This allows us to define our star-shaped condition.

MPU Implicits: Many surface reconstruction algorithms use MPU implicits to define an approximate implicit surface for a given set of points with normals [OBA*03]. Inside localized regions of space, the set of points inside an octree voxel define a local shape function (a quadric function). The MPU implicit is defined as

$$\mathscr{I}(\mathbf{x}) = \frac{\sum_i w_i(\mathbf{x}) Q_i(\mathbf{x})}{\sum_i w_i(\mathbf{x})}, \qquad (1)$$

where the weight functions $w_i(\mathbf{x})$ are defined as quadratic B-splines for each octree voxel and $Q_i(\mathbf{x})$ is a local quadric surface based on the points inside the voxel. Based on this definition, we express the gradient function of $f(\mathbf{x})$ as

$$\nabla \mathscr{I}(\mathbf{x}) = \frac{\sum_i (w_i(\mathbf{x}) \nabla \mathbf{Q_i}(\mathbf{x}) + Q_i(\mathbf{x}) \nabla \mathbf{w_i}(\mathbf{x}))}{\sum_j w_j(\mathbf{x})} \qquad (2)$$

This expression is plugged into the interval test given a candidate kernel point. Since the interval test is a sign test, we eliminate the denominator since it is a positive weight function.

Moving Least Squares (MLS): Moving Least Squares are commonly used to render point-set surfaces and can be used to generate a topology preserving reconstruction of point samples, under certain conditions [Ame04, Kol05].

We use the formulation of Kolluri [Kol05] in the following discussion. Let S be the input point set. Let the unit normal at point $\mathbf{s_i} \in S$ be $\mathbf{n_i}$. According to Kolluri [Kol05], the MLS is the zero set of

$$\mathscr{I}(x) = \frac{\sum_{\mathbf{s_i} \in S} W_i(\mathbf{x})((\mathbf{x} - \mathbf{s_i})^T \mathbf{n_i})}{\sum_{\mathbf{s_j} \in S} W_j(\mathbf{x})}, \qquad (3)$$

where $W_i(\mathbf{x}) = e^{-\|\mathbf{x} - \mathbf{s_i}\|^2 / \varepsilon^2} / A_i$. A_i is the number of samples inside a ball of radius ε centered at $\mathbf{s_i}$.

The implicit surface, defined above, is star-shaped with respect to a point $\mathbf{p}$ if $(\mathbf{x} - \mathbf{p})^T \nabla \mathbf{I}(\mathbf{x}) > 0$. Let the cardinality of set S be k. Let N be the $3 \times k$ matrix whose i^{th} column is $\mathbf{n_i}$, W be the $k \times 1$ vector whose i^{th} element is $W_i(\mathbf{x})$ and S be the $k \times 3$ matrix whose i^{th} row is $(\mathbf{x} - \mathbf{s_i})^T$. Then the star-shaped

test boils down to evaluating the interval expression

$$tr(NW(\mathbf{x} - \mathbf{p})^T) - 2tr(S^T NW(\mathbf{x} - \mathbf{p})^T S)/\varepsilon^2, \qquad (4)$$

$tr()$ is the matrix trace function. The first term of the expression is a 3×3 matrix and the 2nd term is a $k \times k$ matrix.

4.3. Interval Arithmetic and Termination

The main advantage of using interval arithmetic is that all our computations are validated and once it terminates we can be sure that the surface is divided into star-shaped pieces. In this section, we will discuss issues of termination with our interval arithmetic approach. Our subdivision algorithm terminates when one of the following interval conditions are satisfied:

1. The star-shaped condition is satisfied, i.e. $0 \notin (\mathbf{x} - \mathbf{p})^T \nabla \mathscr{I}(\mathbf{x})$
2. The cell does not contain the implicit surface, i.e. $0 \notin \mathscr{I}(x)$

For manifolds, cells far away from the surface will satisfy condition 2 and we terminate the subdivision. For smooth surfaces (C^1 or C^2), the gradient function will be away from zero sufficiently close to the surface and hence condition 1 will be satisfied after finite number of subdivisions.

There are two conditions under which our interval test, as described, will not be satisfied.

- If the candidate kernel witness point $\mathbf{p}$ lies inside the cell, $(\mathbf{x} - \mathbf{p})$ vanishes.
- If the medial axis of the surface lies inside the cell, the gradient of the implicit surface vanishes.

We alleviate these problems by precomputing a dense voxelization of the implicit surface as shown in Fig. 8(b). The motivation to perform the voxelization is to generate a set of intervals that are as close to the implicit surface as possible. While computing the kernel witness point, we ensure that it lies outside the intervals in the voxelization. Given a grid cell, we find the intervals that intersect the cell and perform the interval test on each of the intervals. If the test on each of the intervals does not contain zero, the candidate point is a valid witness to the kernel and we terminate the subdivision. For smooth surfaces, the medial axis is separated from the surface (local feature size is strictly positive) and the sufficiently dense voxelization will not include the medial axis inside the intervals.

The only case that is left is when the cell contains a part of the surface that has a singularity (condition 1 above is not satisfied because gradient function is zero on the surface). In this case, interval arithmetic will not terminate. In our implementation, once the cells are roughly the size of the intervals we stop the subdivision process and resort only to linear programming to test for the star-shaped property. While this does not provide a rigorous guarantee, this heuristic will work unless we are dealing with highly pathological cases.

The performance of interval arithmetic depends on the number of false negatives which results in unnecessary subdivision. The precomputed voxelization serves the purpose of significantly improving the performance.

4.4. Boolean Combinations

We use the star-shaped property to perform the test on Boolean combinations of primitives. We use a conservative test based on the following property: if $\mathscr{S}_1$ and $\mathscr{S}_2$ are two star-shaped primitives with a common guard, then $\mathscr{S}_1 \odot \mathscr{S}_2$ is also star-shaped where $\odot$ denotes a Boolean operation such as union and intersection. This is because

$$Kernel(\mathscr{S}_1) \cap Kernel(\mathscr{S}_2) \subseteq Kernel(\mathscr{S}_1 \odot \mathscr{S}_2)$$

The above test reduces the star-shaped test for a Boolean combination to star-shaped tests on the individual primitives. We can test whether the individual primitives are star-shaped w.r.t a common guard by using linear programming and interval arithmetic techniques as discussed earlier in the section. There are two important advantages of the above test. First, this test can be performed without an explicit representation of the final solid corresponding to the Boolean operation. Second, this test does not require the final surface to be smooth; this is crucial because Boolean operations typically generate non-smooth surfaces with sharp features. Fig. 7 shows application of our algorithm to Boolean operations.

5. Implementation and Performance

Our algorithm uses three basic components: (a) linear programming to verify if a a polyhedral primitive or a set of points samples satisfy the star-shaped test, (b) interval arithmetic to check whether a curved primitive intersects a cell and if it satisfies the star-shaped test, and (c) ray shooting to evaluate the forward and backward visibility maps, image transfer map, and the homeomorphism. Each of these three components are relatively simple to implement. Moreover, there exist public domain packages for many of these components, e.g., *QSOPT* for linear programming.

We have tested our algorithms on three kinds of benchmarks: MPU-based reconstruction [OBA*03] of point cloud, polygonization of algebraic surfaces, and CSG models. Figs. 1, 5, 6, and 7 highlight our results on these benchmarks. Table 1 shows a performance comparison with Bloomenthal's algorithm [Blo88] and the algorithm by Boissonnat & Oudot [BO05]. While our algorithm is somewhat slower than Bloomenthal's algorithm, it has the advantage that it provides guar-

Model	Bloomenthal	Boissonnat [BO05]	Our Method
Filigree	380	519	241
Chisel	47	79	58
Vase-lion	205	13,582	234
Chair	1.0	?	10.6
Tanglecube	0.35	17.6	11
Decocube	1.89	?	52
Diagonal	0.93	9.0	9.1

Table 1: Performance comparison: This table compares the performance of our algorithm with Bloomenthal's algorithm and the algorithm by Boissonnat & Oudot for various models. All the timings are in seconds. The question mark indicates cases for which we were not able to successfully run the algorithm.

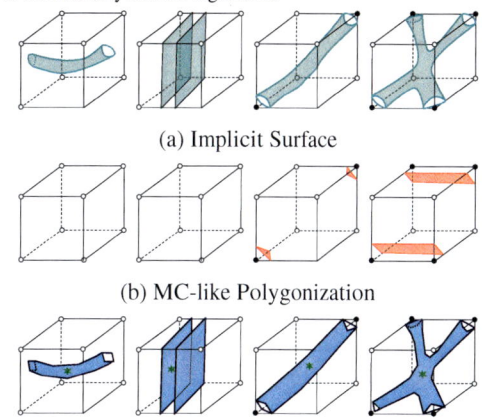

(a) Implicit Surface

(b) MC-like Polygonization

(c) Visibility Mapping Based Polygonization

Figure 9: This figure compares our algorithm with MC-like algorithms for a set of cell configurations. The top row of figures show different cell configurations; the middle row shows polygonization obtained using an MC-like algorithm; the bottom row shows polygonization produced by our algorithm. In some cases, MC-like algorithms generate no surface output.

antees on the accuracy of the output. Compared to Boissonnat & Oudot's algorithm, which also provides guarantees, our algorithm is significantly faster and more scalable to complex models.

6. Comparison with Prior Methods

In this section, we compare and contrast features of our algorithm with prior techniques to polygonize implicit surfaces.

6.1. Isosurface Extraction Methods

One of the commonly used approaches for isosurface extraction is Marching Cubes (MC) and its extensions [LC87, KBSS01, JLSW02]. We refer to these algorithms collectively as MC-like algorithms. The input to MC-like algorithms is a sampled scalar field obtained by discretizing the continuous implicit function upto a certain resolution. However, the main issue in these approaches is generating the scalar field at an appropriate resolution such that the reconstructed surface is accurate. An inadequate grid resolution may result in an inaccurate output when the implicit surface has *small components*, *thin sheets*, or *needle-shaped features*, as shown in Fig. 9.

Assume that the subdivision algorithm has computed a guard within each cell in Fig. 9. It is possible that the subdivision algorithm may need to generate sub-intervals within the cell to reliably compute the guard. Once such a guard has been computed, our polygonization algorithm can reliably handle all these cases as shown in Fig. 9.

Topological ambiguity: Many solutions have been proposed to fix topological ambiguities that can arise in some cases of the original MC algorithm [WG90, CGMS00]. In some case, these algorithms can provide topological guarantees by assuming some type of interpolation, e.g., trilinar interpolation, on the sampled scalar field. These topological ambiguity handling algorithms do not address the issues of generating a grid with adequate resolution to provide global topological guarantees with respect to the original implicit surface.

Topology preserving sampling: Varadhan et al. [VKSM04] presented an adaptive grid generation algorithm such that the final surface reconstructed using MC has the same topology as the implicit surface. However, this algorithm is rather conservative and uses much stronger criteria for sample generation. Moreover, our polygonization algorithm can explicitly compute the homeomorphism and compute a parameterization of the implicit surface.

6.2. Spatial Subdivision Methods

Spatial subdivision techniques have been widely used to polygonize implicit surfaces [Blo88, Vel90, HW90, SFYC96, JLSW02, WGG99, AG01]. These algorithms either generate a uniform grid, octree or a kd-tree, or perform tetrahedral-based polygonization. These algorithms can also handle non-manifold implicit surfaces [BF95]. However, to the best of our knowledge, these spatial subdivision algorithms do not provide topological guarantees on the final polygonization. Few algorithms such as [KB89] choose to visualize implicit surfaces by ray-tracing rather than polygonization, but these are complementary to our work.

6.3. Topology Preserving Polygonization

Some algorithms can guarantee a topology preserving polygonization of implicit surfaces [BNO96, SH97, BCSV04, vOW04]. These algorithms assumes a C^2 continuous implicit surface. Under this assumption, these algorithms attempt to isolate the critical points. They do not apply to Boolean operations wherein the critical points corresponding to the final solid are unknown.

We have demonstrated the application of the meshing algorithm described in [BO05] in Table 1. It can only handle smooth surfaces and can not deal with Boolean operations. It is significantly slower than our algorithm and can not deal with very complex inputs.

6.4. Interval Arithmetic

Synder [Sny92] presented an adaptive subdivision method for computing a topologically accurate polygonal approximation of an implicit curve/surface. This method keeps subdividing cells until a parameterizability criterion is satisfied This parameterizability criterion would not be satisfied if the implicit surface has a sharp feature, e.g., in the case of a surface derived from a Boolean operation. Furthermore, the computational techniques presented in Snyder's paper to verify the parameterizability criterion are not applicable to Boolean operations.

Plantinga and Vegter [PV04] present a similar method for implicit surface polygonization. The condition of Plantinga and Vegter imposes the constraint that the gradient of the implicit function should not vary by more than $\pi/2$. As a result, it is applicable to only smooth surfaces and cannot handle objects defined by Boolean operations.

7. Limitations

Our approach has a few limitations. We assume that the implicit surface is a manifold. Our interval arithmetic based star-shaped decomposition algorithm can be conservative in practice. Furthermore, the star-shaped test can generate a high number of sub-intervals in order to compute a guard inside a voxel. Our algorithm cannot handle cases where the implicit surface has singularities such as self-intersections or tangential intersections.

8. Conclusion and Future Work

We present a new algorithm to polygonizing implicit surfaces based on star-shaped decomposition. We compute a visibility mapping for each star-shaped patch within the cell and use these maps to compute a polygonization that is homeomorphic to the original surface. Our algorithm is relatively simple to implement and not prone to robustness problems or crack patching that arise in other techniques based on adaptive subdivision. We demonstrate the performance of our algorithm on many complex benchmarks, including non-smooth surfaces, and compute a watertight polygonization. In terms of performance, our algorithm is considerably faster than previous methods that can provide rigorous guarantees on the topology and can handle complex models. However, our current implementation is not fast enough for interactive applications.

There are many avenues for future work. We would like to extend our approach to handle non-manifold surfaces. We would like to compute a global parameterization of implicit surface that can give bounds on distortion. It may be useful to develop point-reconstruction algorithms for non-smooth surfaces based on the MLS formulation [Kol05] and generate a topology preserving polygonization using our approach. Another challenge is to faithfully reconstruct all the sharp features in the input. Finally, we would like to improve the performance of our algorithm for real-time implicit modeling.

Acknowledgements

We would like to acknowledge SensAble Technologies Inc. for providing the Filigree and Vase-lion models (`http://shapes.aim-at-shape.net`).

References

[AAB02] ANDUJAR C., AYALA D., BRUNET P.: Topology simplification trough discrete models. pp. 88–105.

[AG01] AKKOUCHE S., GALIN E.: Adaptive implicit surface polygonization using marching triangles. *Computer Graphics Forum 20*, 2 (2001), 67 –80. ISSN 1067-7055.

[Ame04] AMENTA N.: Defining point-set surfaces. *ACM Trans. on Graphics (Proc. of ACM SIGGRAPH) 23* (2004), 264–270.

[BCSV04] BOISSONNAT J., COHEN-STEINER D., VEGTER G.: Isotopic implicit surface meshing. *ACM Symposium on Theory of Computing* (2004), 301–309.

[BF95] BLOOMENTHAL J., FERGUSON K.: Polygonization of Non-Manifold implicit surfaces. In *SIGGRAPH 95 Conference Proceedings* (1995), pp. 309–316.

[Bin96] BINI D.: Numerical computation of polynomial zeros by means of aberth's method. *Numerical Mathematics 13* (1996), 179–200.

[Blo88] BLOOMENTHAL J.: Polygonization of implicit surfaces. *Comput. Aided Geom. Design 5*, 4 (1988), 341–355.

[BNO96] BOTTION A., NUIJ W., OVERVELD K. V.: How to shrinkwrap through a critical point: an algorithm for the adaptive triangulation of surfaces with arbitrary topology. *Implicit Surfaces* (1996), 53–72.

[BO05] BOISSONNAT J., OUDOT S.: Provably good sampling and meshing of surfaces. *Graphical Models* (2005).

[CGMS00] CIGNONI P., GANOVELLI F., MONTANI C., SCOPIGNO R.: Reconstruction of topologically correct and adaptive trilinear isosurfaces. *Computers & Graphics 24*, 3 (2000), 399–418.

[FH05] FLOATER M., HORMANN K.: *Surface parameterization:: A tutorial and survey.* Tech. rep., 2005.

[HW90] HALL M., WARREN J.: Adaptive polygonalization of implicitly defined surfaces. *IEEE Computer Graphics and Applications 10*, 6 (Nov. 1990), 33–42.

[JLSW02] JU T., LOSASSO F., SCHAEFER S., WARREN J.: Dual contouring of hermite data. *ACM Trans. on Graphics (Proc. SIGGRAPH) 21*, 3 (2002).

[KB89] KALRA D., BARR A. H.: Guaranteed ray intersections with implicit surfaces. In *Computer Graphics (SIGGRAPH '89 Proceedings)* (1989), vol. 23, pp. 297–306.

[KBSS01] KOBBELT L., BOTSCH M., SCHWANECKE U., SEIDEL H. P.: Feature-sensitive surface extraction from volume data. In *Proc. of ACM SIGGRAPH* (2001), pp. 57–66.

[Kol05] KOLLURI R.: Provably good moving least squares. *ACM-SIAM Symposium on Discrete Algorithms* (2005).

[LC87] LORENSEN W. E., CLINE H. E.: Marching cubes: A high resolution 3D surface construction algorithm. In *Computer Graphics (SIGGRAPH '87 Proceedings)* (1987), vol. 21, pp. 163–169.

[LH96] LEVOY M., HANRAHAN P.: Light field rendering. In *SIGGRAPH '96: Proceedings of the 23rd annual conference on Computer graphics and interactive techniques* (New York, NY, USA, 1996), ACM Press, pp. 31–42.

[OBA*03] OHTAKE Y., BELYAEV A., ALEXA M., TURK G., SEIDEL H.: Multi-level partition of unity implicits. *ACM Trans. on Graphics (Proc. of ACM SIGGRAPH) 22* (2003), 463–470.

[Ped95] PEDERSEN H. K.: Decorating implicit surfaces. In *SIGGRAPH 95 Conference Proceedings* (Aug. 1995), Cook R., (Ed.), Annual Conference Series, ACM SIGGRAPH, Addison Wesley, pp. 291–300. held in Los Angeles, California, 06-11 August 1995.

[PH03] PRAUN E., HOPPE H.: Spherical parameterization and remeshing. *ACM Trans. on Graphics (Proc. of ACM SIGGRAPH) 22* (2003).

[PSH*04] POTTMANN H., STEINER T., HOFER M., HAIDER C., HANBURY A.: The isophotic metric and its application to feature sensitive morphology on surfaces. 560–572.

[PV04] PLANTINGA S., VEGTER G.: Isotopic approximation of implicit curves and surfaces. *Symposium on Geometry Processing* (2004).

[SFYC96] SHEKHAR R., FAYYAD E., YAGEL R., CORNHILL F.: Octree-based decimation of marching cubes surfaces. *Proc. of IEEE Visualization* (1996), 335–342.

[SH97] STANDER B. T., HART J. C.: Guaranteeing the topology of an implicit surface polygonization for interactive modeling. In *Proc. of ACM SIGGRAPH* (1997), pp. 279–286.

[Sny92] SNYDER J. M.: Interval analysis for computer graphics. In *Computer Graphics (SIGGRAPH '92 Proceedings)* (July 1992), Catmull E. E., (Ed.), vol. 26, pp. 121–130.

[THCM04] TARINI M., HORMANN K., CIGNONI P., MONTANI C.: Polycube-maps. *ACM Trans. Graph. 23*, 3 (2004), 853–860.

[Vel90] VELHO L.: Adaptive polygonization of implicit surfaces using simplicial decomposition and boundary constraints. In *Eurographics '90* (Sept. 1990), Vandoni C. E., Duce D. A., (Eds.), North-Holland, pp. 125–136.

[VKSM04] VARADHAN G., KRISHNAN S., SRIRAM T. V. N., MANOCHA D.: Topology preserving surface extraction using adaptive subdivision. In *Eurographics Symposium on Geometry Processing* (2004).

[VKZM06] VARADHAN G., KRISHNAN S., ZHANG L., MANOCHA D.: Reliable implicit surface polygonization using visibility mapping. In *http://gamma.cs.unc.edu/VM/supplem.pdf* (2006).

[vOW04] VAN OVERVELD K., WYVILL B.: Shrinkwrap: An efficient adaptive algorithm for triangulating an iso-surface . *The Visual Computer 20*, 6 (2004), 362– 369.

[WG90] WILHELMS J., GELDER A. V.: Topological considerations in isosurface generation extended abstract. *Computer Graphics 24*, 5 (1990), 79–86.

[WGG99] WYVILL B., GALIN E., GUY A.: Extending The CSG Tree. Warping, Blending and Boolean Operations in an Implicit Surface Modeling System. *Computer Graphics Forum 18*, 2 (1999), 149–158.

[WMW86] WYVILL B., MCPHEETERS C., WYVILL G.: Data structure for soft objects. *The Visual Computer 2*, 4 (1986), 227–234.

Eurographics Symposium on Geometry Processing (2006)
Konrad Polthier, Alla Sheffer (Editors)

Robust Principal Curvatures on Multiple Scales

Yong-Liang Yang[1]   Yu-Kun Lai[1]   Shi-Min Hu[1]   Helmut Pottmann[2]

[1]Tsinghua University, Beijing   [2]Vienna University of Technology.

Abstract

Geometry processing algorithms often require the robust extraction of curvature information. We propose to achieve this with principal component analysis (PCA) of local neighborhoods, defined via spherical kernels centered on the given surface Φ. Intersection of a kernel ball B_r or its boundary sphere S_r with the volume bounded by Φ leads to the so-called ball and sphere neighborhoods. Information obtained by PCA of these neighborhoods turns out to be more robust than PCA of the patch neighborhood $B_r \cap \Phi$ previously used. The relation of the quantities computed by PCA with the principal curvatures of Φ is revealed by an asymptotic analysis as the kernel radius r tends to zero. This also allows us to define principal curvatures "at scale r" in a way which is consistent with the classical setting. The advantages of the new approach are discussed in a comparison with results obtained by normal cycles and local fitting; whereas the former method somewhat lacks in robustness, the latter does not achieve a consistent behavior at features on coarse scales. As to applications, we address computing principal curves and feature extraction on multiple scales.

1. Introduction

Differential geometry plays a central role in the analysis of curves and surfaces. Local investigations frequently use differential invariants such as curvatures, but also the global understanding of shapes can benefit from differential geometric entities. Since differentiation is very sensitive to noise, the use of differential invariants requires data smoothing and de-noising prior to computation. This can be done in a global way via appropriate geometric flows [CRT04] or locally, using smooth approximations of the data in an appropriate neighborhood [CP03, GI04, Tau95, TT05]. In both cases, the preservation of features which may not be considered as noise is not an easy task and requires especially adapted algorithms. A related difficulty is the suppression of those details in the geometry which lie below the scale one is interested in. Whereas classical differential geometry cannot be used directly for such important objects as meshes, *discrete differential geometry* is extending the theory to the discrete setting [CSM03, DGS05, HP04]. Some discrete operators work on larger local neighborhoods as well and thus are defined on multiple scales. Though handling noisy data is possible [HP04], this is not the main intent and strength of discrete differential geometry.

The present paper pursues an approach via *integral invariants* obtained by integration over local neighborhoods.

Figure 1: *Feature extraction on multiple scales using PCA on ball neighborhoods of different radii. Darker regions are classified as features on all scales, lighter shaded regions correspond to features extracted at only one or two scales.*

These neighborhoods are constructed by means of balls, whose radius r defines the scale on which one is working. The origin of this method is work on molecular shape analysis [Con86], on 2D shape matching [MHYS04], and feature extraction [CRT04]. Consider a domain D, its boundary surface Φ, a point $\mathbf{p} \in \Phi$. the ball $B_r(\mathbf{p})$ with radius r and center $\mathbf{p}$, and the boundary sphere $S_r(\mathbf{p}) = \partial B_r(\mathbf{p})$. We define *ball and sphere neighborhoods* as $N_b^r(\mathbf{p}) := D \cap B_r(\mathbf{p})$ and $N_s^r := D \cap S_r(\mathbf{p})$, resp. Clarenz et al. [CRT04] and Pauly et al. [PKG03] consider the *surface patch neighborhood* $N_p^r(\mathbf{p}) = \Phi \cap B^r(\mathbf{p})$ and perform PCA on this patch. It turns out that PCA of the patch neighbourhood is less robust against noise than PCA of the ball and sphere neighborhoods.

Contributions and overview. The main contributions of our short paper are: (i) We present the results of a thorough study of PCA on all three neighborhoods described above. (ii) Our analysis also yields definitions of principal curvatures at scale r which are consistent with the classical theory. However, note that PCA yields an integrated quantity, which is not the same as a curvature estimate at a single point. (iii) In Sec. 3, we compare our approach with local fitting [CP03] and normal cycles [CSM03]. (iv) As applications, we study principal curves and feature extraction on multiple scales (Sec. 4). Details and further applications will be provided in a forthcoming paper.

2. Principal component analysis of local neighborhoods

Principal component analysis of a point set A means to compute its barycenter $\mathbf{s} := (\int_A \mathbf{x}\, d\mathbf{x})/(\int_A d\mathbf{x})$ and covariance matrix

$$J(A) := \int_A (\mathbf{x} - \mathbf{s}) \cdot (\mathbf{x} - \mathbf{s})^T d\mathbf{x}. \qquad (1)$$

If A is not full-dimensional but contained in a smooth surface, (1) is understood as a surface integral. The principal directions and principal components of A are given by the eigenvectors $\mathbf{e}_i$ and the corresponding eigenvalues λ_i ($i = 1, 2, 3$) of $J(A)$.

In order to derive results on the principal components of the neighborhoods defined above, we use the principal frame of the surface Φ at $\mathbf{p}$ as coordinate frame and approximate Φ up to second order by a paraboloid Π,

$$\Pi: \; z = \frac{1}{2}(\kappa_1 x^2 + \kappa_2 y^2). \qquad (2)$$

Here, κ_1, κ_2 denote the principal curvatures at $\mathbf{p}$. In order to compute the first two terms in the Taylor expansion of moments with respect to the radius r, it is sufficient to work with Π instead of Φ. By the symmetry of the paraboloid Π, the barycenters of all neighbhorhoods lie on the z-axis, and the eigenvectors $\mathbf{e}_i$ of the corresponding covariance matrices are parallel to the coordinate axes. In the following, we present some results of [PHYK05].

PCA of the ball and sphere neighborhoods. Consider the barycenters $\mathbf{s}_b^r$ and $\mathbf{s}_s^r$ of the ball and sphere neighborhoods $N_b^r = B_r(\mathbf{p}) \cap D$ and $N_s^r = S_r(\mathbf{p}) \cap D$, respectively. Their signed distances d_b^r and d_s^r from the surface are related to the surface's *mean curvature H* via

$$d_b^r = \frac{3}{8}r + \frac{9H}{64}r^2 + O(r^3), \; d_s^r = \frac{1}{2}r + \frac{H}{4}r^2 + O(r^3). \quad (3)$$

Two eigenvectors of the covariance matrix are approximately tangent to the surface, the corresponding eigenvalues M_{b1}, M_{b2} (ball) and M_{s1}, M_{s2} (sphere) read

$$M_{bi}^r = \frac{2\pi}{15}r^5 - \frac{\pi}{48}(2\kappa_i + \kappa_1 + \kappa_2)r^6 + O(r^7), \quad (4)$$

$$M_{si}^r = \frac{2\pi}{3}r^4 - \frac{\pi}{8}(2\kappa_i + \kappa_1 + \kappa_2)r^5 + O(r^6). \quad (5)$$

An important consequence of equations (4) and (5) is that we can *define* principal curvatures on scale r which, as r tends to zero, converge to the actual principal curvatures in the classical sense. We may use the ball neighborhood and compute curvatures $\kappa_{b1}^r, \kappa_{b2}^r$ from M_{b1}^r and M_{b2}^r via (4), or use the sphere neighborhood and compute $\kappa_{s1}^r, \kappa_{s2}^r$ from M_{s1}^r and M_{s2}^r via (5):

$$\kappa_{b1}^r := \frac{6}{\pi r^6}(M_{b2}^r - 3M_{b1}^r) + \frac{8}{5r}, \qquad (6)$$

$$\kappa_{s1}^r := \frac{1}{\pi r^5}(M_{s2}^r - 3M_{s1}^r) + \frac{4}{3r}, \qquad (7)$$

and analogous expressions for $\kappa_{b2}^r, \kappa_{s2}^r$. Our tests showed a very similar and useful behavior of both types of principal curvatures.

PCA of the patch neighborhood. Similar results hold for the patch neighborhood N_p^r, but unfortunately here we have higher sensitivity to noise (see [PHYK05] and the discussion below). Properties of N_p^r not contained in Clarenz et al. [CRT04] are the following: The surface area PA^r of N_p^r is related to the principal curvatures κ_1, κ_2 of Φ at $\mathbf{p}$ by

$$PA^r = \pi r^2 + \frac{\pi}{32}(\kappa_1 - \kappa_2)^2 r^4 + O(r^5). \qquad (8)$$

The two eigenvalues of $J(N_p^r)$ which correspond to the approximate principal directions satisfy

$$M_{pi}^r = \frac{\pi}{4}r^4 + \frac{\pi}{192}(\kappa_1^2 + \kappa_2^2 - 4\kappa_i^2 - 6\kappa_1\kappa_2)r^6 + O(r^7).$$

3. Comparison of principal component analysis with local fitting and normal cycles

The computation of integral invariants based on the ball neighborhood can efficiently be done by means of the fast Fourier transform, whereas for PCA of the sphere neighborhood, we use a geometric method based on an almost uniform multilevel discretization of the sphere S_r [PHYK05]. We have run extensive tests in order to compare PCA of different neighborhoods with the main methods used in geometry processing: normal cycles [CSM03] and local fitting. As representative of the many available methods for local fitting we employ osculating jets [CP03].

Robustness. In order to avoid effects caused by a complicated geometry, we illustrate sensitivity to noise by means of a rather simple surface (Fig. 2). A summary of results concerning the principal curvature directions is depicted in Fig. 3. PCA of the patch neighborhood [CRT04, PKG03] is quite sensitive to noise, normal cycles less so. The effect of

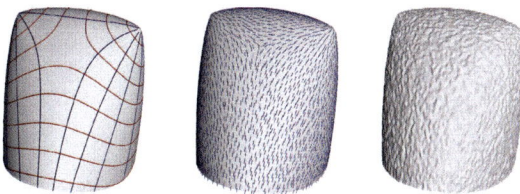

Figure 2: *Principal curvature lines (left) and directions of minimum principal curvature (middle) for a smooth surface ('pillow'); noisy surface (right) used for the robustness test.*

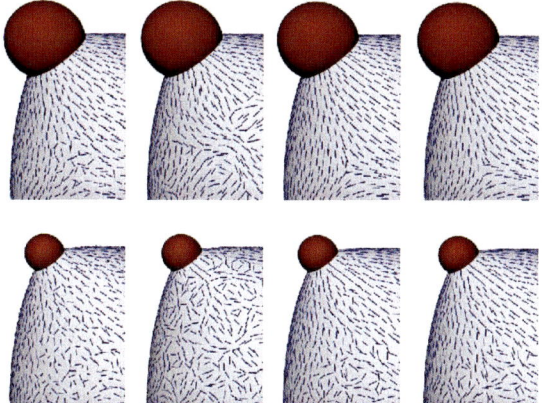

Figure 3: *Directions of minimum principal curvature computed from the noisy surface of Fig. 2, right, but shown over the smooth surface for clarity. From left: normal cycles, PCA (patch), osculating jet, PCA (ball).*

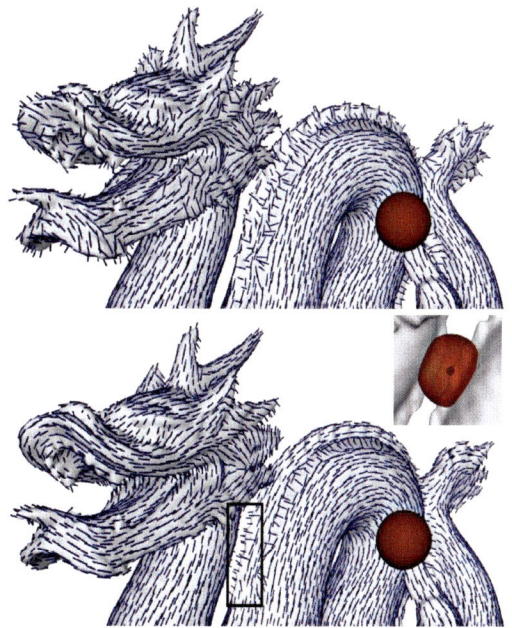

Figure 4: *Directions of minimum principal curvature obtained with osculating jets (top) and PCA/ball (bottom). Larger kernels may have extraneous intersections (see detail) which – if not detected – cause unwanted effects.*

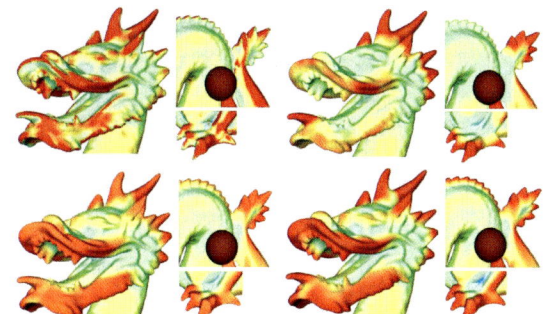

Figure 5: *Maximum principal curvature obtained with osculating jets (top left), normal cycles (top right), PCA/sphere (bottom left), and PCA/ball (bottom right). The jet method exhibits inconsistencies, normal cycles miss some features.*

noise removal by averaging over a greater domain is apparently much more pronounced for PCA of ball and sphere neighborhoods than for the patch neighborhood. The normal cycle method performs in between. For simple surfaces, fitting methods behave like PCA of ball/sphere neighborhoods.

Multi-scale behavior. At a low noise level all PCA based methods respond very well to an increase of the kernel radius and exhibit the expected smoothing and simplification effect. This is also true for normal cycles, although their feature detection capability appears to be somewhat weaker. Local fitting methods have defects when it comes to judging curvature at coarse scales (see Figs. 4 and 5). While using a larger neighborhood has some smoothing effect on the local fitting surface Φ_l, one evaluates curvature of Φ_l at the center of the neighborhood. Here, the sensitivity of classical curvatures to small changes becomes crucial and explains why the overall picture of the extracted quantities for a large neighborhood is by far less smooth than for PCA based methods. Apparently the ball and sphere PCA are most robust to noise and exhibit the desired scaling behavior.

4. Principal curves and feature extraction on multiple scales

The eigenvectors $\mathbf{e}_1^r$, $\mathbf{e}_2^r$ of the covariance matrices of local neighborhoods serve as principal directions of the given surface at the chosen scale r. They are not exactly tangential to the given surface. In fact they should not follow details which are small compared to r. In order to define principal curves on Φ we project $\mathbf{e}_1^r$ $\mathbf{e}_2^r$ onto Φ and integrate the resulting tangential vector fields. The direction of this projection shall be given by the eigenvector $\mathbf{e}_3^r$, which estimates the surface normal. Note that the projected directions are usually not orthogonal to each other, which is actually a numerical advantage for integration. By guiding the whole projection procedure with the directions arising from PCA, principal curves become less sensitive to local surface deviations. Figure 6 illustrates their behavior for different scales. The more we increase r, the better these curves follow the

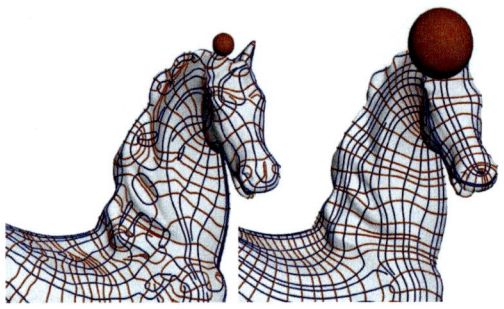

Figure 6: *Principal curves for larger kernel radius (right) better follow the global shape.*

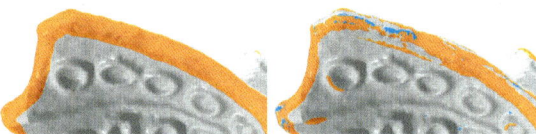

Figure 7: *Feature extraction on a coarse scale with PCA/ball (left) and osculating jet (right).*

Figure 8: *Feature extraction using principal curvatures obtained with PCA on the sphere neighborhood.*

global geometry of the surface. We believe that this tool will be very valuable for global shape understanding (see e.g. [LPW*06]).

Feature regions on 3D models are characterized by at least one large principal curvature. We have shown how to define and compute principal curvatures κ_1^r, κ_2^r on multiple scales, e.g. based on formulae (6) or (7). Hence, we also have a simple tool for robust multi-scale feature extraction: On a given scale r, feature regions are first filtered out by $\max(|\kappa_1^r|, |\kappa_2^r|) > T_c$, where T_c is a threshold on curvatures; T_c can be derived from a few user-specified feature and non-feature samples based on a simple statistical model. Using the sign of the dominant principal curvature, the extracted feature regions are further classified into ridges and valleys.

Feature regions extracted at different scales have been combined into the single image of Fig.1: The dark shaded regions are the persistent features, and lighter shaded regions correspond to features extracted at only one or two scales. See also Figures 7 and 8.

Acknowledgements. This research was supported by the Austrian Science Fund (FWF) under Grant No. P16002-N05 and by the National Science Foundation of China under Grant No. 60225016. We gratefully acknowledge fruitful discussions with Johannes Wallner and his help with the preparation of the final version.

References

[Con86] CONNOLLY M.: Measurement of protein surface shape by solid angles. *J. Mol. Graphics 4* (1986).

[CP03] CAZALS F., POUGET M.: Estimating differential quantities using polynomial fitting of osculating jets. In *Symp. Geometry Processing* (2003), pp. 177–178.

[CRT04] CLARENZ U., RUMPF M., TELEA A.: Robust feature detection and local classification for surfaces based on moment analysis. *IEEE TVCG* (2004).

[CSM03] COHEN-STEINER D., MORVAN J. M.: Restricted Delaunay triangulations and normal cycle. In *ACM Symp. Comp. Graphics* (2003), pp. 312–321.

[DGS05] DESBRUN M., GRINSPUN E., SCHRÖDER P.: *Discrete Differential Geometry: An Applied Introduction.* SIGGRAPH Course Notes, 2005.

[GI04] GOLDFEATHER J., INTERRANTE V.: A novel cubic-order algorithm for approximating principal direction vectors. *ACM TOG 23*, 1 (2004), 45–63.

[HP04] HILDEBRANDT K., POLTHIER K.: Anisotropic filtering of non-linear surface features. *Computer Graphics Forum 23*, 3 (2004), 391–400.

[LPW*06] LIU Y., POTTMANN H., WALLNER J., WANG W., YANG Y.: Geometric modeling with conical meshes and developable surfaces. *ACM TOG 25*, 3 (2006).

[MHYS04] MANAY S., HONG B.-W., YEZZI A. J., SOATTO S.: Integral invariant signatures. In *Proc. European Conf. Computer Vision* (2004), pp. 87–99.

[PHYK05] POTTMANN H., HUANG Q.-X., YANG Y.-L., KÖLPL S.: Integral invariants for robust geometry processing. Geometry Preprint 146, TU Wien, 2005.

[PKG03] PAULY M., KEISER R., GROSS M.: Multi-scale feature extraction on point-sampled geometry. *Computer Graphics Forum 22*, 3 (2003), 281–289.

[Tau95] TAUBIN G.: Estimating the tensor of curvature of a surface from a polyhedral approximation. In *Proc. Int. Conf. Computer Vision* (1995).

[TT05] TONG W.-S., TANG C.-K.: Robust estimation of adaptive tensors of curvature by tensor voting. *IEEE Pattern Anal. Machine Intell. 27*, 3 (2005), 434–449.

Eurographics Symposium on Geometry Processing (2006)
Konrad Polthier, Alla Sheffer (Editors)

A Quadratic Bending Model for Inextensible Surfaces

Miklos Bergou   Max Wardetzky   David Harmon   Denis Zorin   Eitan Grinspun
Columbia University   Freie Universität Berlin   Columbia University   New York University   Columbia University

Abstract

Relating the intrinsic Laplacian to the mean curvature normal, we arrive at a model for bending of inextensible surfaces. Due to its constant Hessian, our isometric bending model reduces cloth simulation times up to three-fold.

Categories and Subject Descriptors (according to ACM CCS): I.3.5 [Computer Graphics]: Computational Geometry and Object Modeling

1. Introduction

Computation of curvature-based energies and their derivatives is a costly component of many physical simulation and geometric modeling applications. Typically the energy density is expressed in terms of elementary symmetric functions of the principal curvatures of the mesh [Cia00, YB02, CDD*04,BS05,TW06]). In general the resulting expressions are nonlinear in the positions of mesh vertices, and the attendant numerics involve costly evaluations of energy gradients and Hessians. Our contribution is to consider the class of isometric surface deformations, arriving at an expression for bending energy which is quadratic in positions. Such quasi-isometric deformations are typical, *e.g.*, for inextensible plates and shells where membrane (stretching) stiffness is greater than bending stiffness by four or more orders of magnitude, hence we focus on cloth simulation as a primary application area.

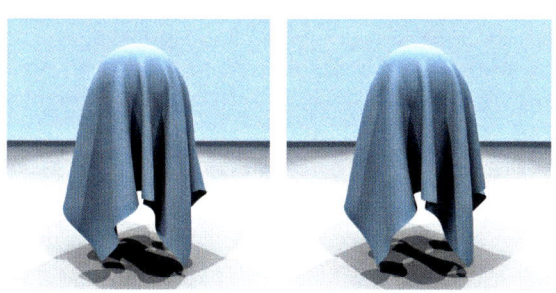

Figure 1: *Final rest state of a cloth draped over a sphere, for* (left) *the proposed isometric bending model and* (right) *the widely-adopted nonlinear hinge model.*

Continuous setting. Consider the bending energy of a deformable surface S:

$$E_b(S) = \frac{1}{2} \int_S H^2 \mathrm{d}A , \qquad (1)$$

where H is mean curvature and $\mathrm{d}A$ is the differential area. $E_b(S)$ is closely related to the Willmore energy of a surface, and the Canham-Helfrich energy of thin bilipid membranes. Note the invariance of $E_b(S)$ under (i) rigid motions and (ii) uniform scaling of the surface: (i) is required for conservation of linear and angular momenta (Nöther's theorem); (ii) affects the characteristic size of folds and wrinkles.

We may rewrite (1) by the following argument. If $\mathbf{x} : S \to \mathbb{R}^3$ denotes the embedding of the surface, the mean curvature normal $\mathbf{H}$ of S can be written as the Laplace-Beltrami, Δ, induced by the Riemannian metric of S, applied to the embedding of the surface: $\mathbf{H} = \Delta\mathbf{x}$. Thus we write (1) as

$$E_b(S) = \frac{1}{2} \int_S \langle \Delta\mathbf{x}, \Delta\mathbf{x} \rangle_{\mathbb{R}^3} \mathrm{d}A, \qquad (2)$$

where $\langle \cdot, \cdot \rangle_{\mathbb{R}^3}$ denotes the standard inner product of $\mathbb{R}^3$.

Central observation. The Laplace-Beltrami, Δ, remains unchanged under isometric deformations of the surface—therefore, for inextensible surfaces, $E_b(S)$ is *quadratic* in positions. Equation (2) together with the assumption of isometric deformation is henceforth called the *isometric bending model* (IBM). Our contribution is to present an analogous *discrete* IBM that is quadratic in positions. Its linear gradient and constant Hessian present an economic model for computing bending forces and their derivatives, enabling fast time-integration of cloth dynamics.

2. Discrete IBM

Our guiding principle for discretization is to maintain the core properties of the smooth isometric bending model. We say that a triangulated surface *deforms isometrically* if its inner metric does not change, *i.e.*, if all edge lengths remain invariant. Denoting the surface's vertex position vector by $\mathbf{x} = (x_0, x_1, ..., x_{n-1})^T \in \mathbb{R}^{3n}$, we require our discrete IBM to provide a bending energy $E_b(\mathbf{x})$ which is (i) quadratic in $\mathbf{x}$ under isometric deformations, (ii) invariant under rigid motions of the mesh, and (iii) invariant under uniform scaling. Observing conditions (i) and (ii) we write E_b as

$$E_b(\mathbf{x}) = \frac{1}{2}\mathbf{x}^T Q \mathbf{x} = \frac{1}{2}\sum_{i,j} Q_{ij} \langle x_i, x_j \rangle_{\mathbb{R}^3},$$

with quadratic form Q, invariant under isometric deformations, hence depending only on intrinsic mesh properties such as connectivity, edge lengths, inner angles, or area. Since E_b is an energy, Q must be positive semi-definite. In [BWH*06] we show that condition (ii) is satisfied if and only if $\sum_i Q_{ij} = \sum_j Q_{ij} = 0$. Condition (iii) says that Q must scale with $1/s^2$ if the whole mesh is scaled by a global factor s. We can then write $Q = L^T M^{-1} L$, where L is invariant under scaling and $\sum_j L_{ij} = 0$, whereas M is symmetric positive definite and scales with s^2. Like Q itself, both L and M are assumed to only depend on intrinsic mesh properties. A *discrete IBM* is then any energy of the form

$$E_b(\mathbf{x}) = \frac{1}{2}\mathbf{x}^T (L^T M^{-1} L)\mathbf{x} = \frac{1}{2}\mathbf{x}^T Q \mathbf{x}. \qquad (3)$$

The quadratic form $Q = L^T M^{-1} L$ corresponds to the *constant energy Hessian*.

One way to obtain a suitable M and L is to discretize the smooth Laplacian, Δ, using the finite element (FE) method,

$$L_{ij} = \int_S \langle \nabla \Phi_i, \nabla \Phi_j \rangle dA, \quad M_{ij} = \int_S \Phi_i \cdot \Phi_j dA,$$

where $\{\Phi_i\}$ is some FE basis, the FE stiffness matrix L is the discrete Laplacian, the FE mass matrix inverse M^{-1} simplifies to division by area in a lumped mass matrix approximation [ZT00], and $L\mathbf{x}$ is the discrete analogue of the smooth mean curvature vector, $\Delta \mathbf{x}$.

One possible FE basis is induced by the usual linear Lagrange elements [PP93, CDD*04]. For cloth simulation, where isometric deformations are associated with bending about edges, we prefer non-conforming (edge-based) Crouzeix-Raviart [CR73] elements. As observed by Hildebrandt and Polthier [HP04], in this setting the mean curvature vector, $L\mathbf{x}$, is associated to edges by construction; in [BWH*06] we show that this version of $L\mathbf{x}$ corresponds to the *linearization* about the flat (planar) rest state of common mean curvature models used in computer graphics: The models of Bridson *et al.* [BMF03], Cohen-Steiner and Morvan [CSM03], Grinspun *et al.* [GHDS03], and Bobenko and Schröder [BS05] fall into this category.

Implementation. We present the implementation tools for the Crouzeix-Raviart IBM; for the derivation see [BWH*06]. In a one-time precomputation step, the constant Hessian, Q, is assembled in the usual manner [ZT00], *i.e.*, by considering contributions from each *local* matrix, $Q(e_i)$, centered about interior edge e_i with stencil consisting of the triangles, t_0, t_1, incident to e_i and their vertices, x_0, x_1, x_2, x_3. With reference to the illustrated labeling convention, we build $Q(e_0)$ by

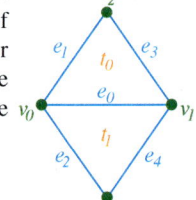

$$Q(e_0) = \frac{3}{2(A_0 + A_1)} K_0^T K_0,$$

where A_i is the area of triangles t_i, and K_0 is the row vector

$$K_0 = (c_{03} + c_{04}, c_{01} + c_{02}, c_{01} + c_{03}, c_{02} + c_{04}),$$

where $c_{jk} = \cot \angle e_j, e_k$. The *local* energy is obtained by $E_b(e_i) = (x_0, x_1, x_2, x_3)Q(e_i)(x_0, x_1, x_2, x_3)^T$. The *global* (total) energy of the system is obtained by summing over all local contributions corresponding to interior edges.

3. Application to the bending of cloth and plates

We apply the discrete IBM to efficient simulation of clothing and thin plates, which is important for feature film production, fashion design, and manufacturing (see the recent surveys [CK05, MTV05, ZJFP04, NG96]). While resistance to bending is much weaker than to stretching, it is the interplay between the two modes that gives cloth its characteristic folds and wrinkles [TW06, Cia00]. Most models of cloth consider separately the bending and in-plane energies:

$$E(S) = E_b(S) + E_p(S).$$

Since any good model of in-plane response (*e.g.*, [TPBF87, BW98]) satisfies the assumption of small in-plane strains, we view the in-plane model as a mechanism (in the spirit of penalty forces) that enforces isometry. Henceforth, we assume that some in-plane model has been chosen, and we focus on presenting our bending model in the context of a complete cloth solver.

Bending models have been extensively studied in graphics; see Thomaszewski's recent survey [TW06]. To our knowledge, all popular models are inherently nonlinear in positions, *e.g.*, they may involve expressions in terms of edge lengths or dihedral angles. Some models are linearized every time step, resulting in the lack of Euclidean-motion invariance. The discrete IBM overcomes both of these disadvantages. As a comparison to our IBM, we implemented a "nonlinear hinge" model similar to [BW98, BMF03, GHDS03].

Elastic and damping forces. Our elastic and dissipative forces depend on the energy gradient and Hessian, respectively. Elastic behavior is governed by the conservative force

$$\mathbf{f}_e(\mathbf{x}(t)) = -\nabla_\mathbf{x} E(S). \qquad (4)$$

Figure 2: *Snapshots from our simulation of a billowing flag. The accompanying movie demonstrates that despite its economy of cost, the proposed bending model achieves qualitatively the same dynamics as popular nonlinear models.*

We model dissipation using the Rayleigh model,

$$\mathbf{f}_d(\mathbf{V}(t)) = -\left(\alpha_1 M + \alpha_2 \Delta E(S)\right)\mathbf{V}(t) , \qquad (5)$$

where $\mathbf{V}(t) = \dot{\mathbf{x}}(t)$ is the velocity, and the damping coefficients α_1 and α_2 govern the decay of low and high frequencies, respectively [ZT00, Hug87]. In implementing IBM, the matrix $-\alpha_2 \nabla^2 E_b(S) = -\alpha_2 Q$ in (5), corresponding to bending, is precomputed once. Similarly, for (4) we compute bending forces via the matrix-vector product $-\nabla_{\mathbf{x}} E_b(S) = -Q\mathbf{x}$. In contrast, the nonlinear hinge requires a relatively expensive computation for both equations.

Dynamics. The time evolution of the cloth is governed by a a coupled system of first order initial value problems (IVPs):

$$\underbrace{\begin{pmatrix} \dot{\mathbf{x}}(t) \\ \dot{\mathbf{V}}(t) \end{pmatrix}}_{\dot{\mathbf{q}}(t)} = \underbrace{\begin{pmatrix} Id & 0 \\ 0 & M^{-1} \end{pmatrix}}_{A} \underbrace{\begin{pmatrix} \mathbf{V}(t) \\ \mathbf{f}_e(\mathbf{x}(t)) + \mathbf{f}_d(\mathbf{V}(t)) \end{pmatrix}}_{\mathbf{b}(\mathbf{q}(t))} ,$$

given $\mathbf{q}(0)$ and the physical mass matrix M.

Numerical treatment. Time discretization of the above system is a well-studied problem (see [Hau04, BWH*06] and references therein). *Explicit* methods adopt the form $\mathbf{q}_{k+1} = \mathbf{q}_k + h A \mathbf{b}_k$; here k is discrete time and h is the time step, *i.e.*, $t = hk$, and $\mathbf{b}_k = \mathbf{b}(\mathbf{q}_k)$. *Implicit* methods search for the root of

$$\mathbf{g}(\mathbf{q}_{k+1}) = \mathbf{q}_{k+1} - \mathbf{q}_k - h A \mathbf{b}(\mathbf{q}_{k+1}) = 0 .$$

We solve this nonlinear system in $\mathbf{q}_{k+1}$ with a Newton solver. Letting $\mathbf{q}_{k+1}^{(0)}$ be the initial guess for $\mathbf{q}_{k+1}$, each Newton iteration improves on the guess,

$$\mathbf{q}_{k+1}^{(i+1)} = \mathbf{q}_{k+1}^{(i)} - (\nabla_{\mathbf{q}}\mathbf{g})^{-1}\mathbf{g}(\mathbf{q}_{k+1}^{(i)}) ,$$

until convergence. This method requires evaluation of both the energy gradient (to compute $\mathbf{g}$) and energy Hessian (to compute $\nabla_{\mathbf{q}}\mathbf{g}$), and the IBM provides an efficient way to do so. Finally, it is often desirable to treat some forces explicitly and others implicitly, using IMEX methods [Hau04].

Results. In an evaluation of two solvers, two problem scenarios, two mesh types, and resolutions ranging from 400 to 25600 vertices, we observe a typical two- to three-fold speedup in simulation times compared to the nonlinear hinge. Figures 1-2 and the accompanying movie provide a

visual point of comparison, and Table 1 summarizes our performance measurements. We observe a seven- to eleven-fold speedup in bending force computation. Since IBM's Hessian is precomputed, we can report only the negligible time required to add it to $\nabla_{\mathbf{q}}\mathbf{g}$; in contrast, the repeated computation of the nonlinear hinge Hessian is costly. Overall speedup will depend on the fraction of total computation associated to bending; to estimate this we conducted several experiments:

Experimental setup. We implemented the implicit solver framework of [BW98] as well as the explicit Euler method. The test framework incorporates (i) the constant strain linear finite element for in-plane response [ZT00, Hug87]; (ii) collision detection using k-DOP trees [KHM*98] and response using Bridson's framework [BFA02]; (iii) the PETSc solver library [BBG*01].

Draping cloth. We simulated heavily-damped draping of a square sheet over a sphere (see Figures 1, 3 and the movie). The draped cloths are qualitatively similar in their final configuration and distribution of folds. Only the final draped shape is important, therefore we used large Rayleigh coefficients thus allowing larger time steps [Hug87].

Billowing flag. We simulated the dynamics of a flag under wind (refer to the movie and Figure 2). The billowing motion of the IBM and nonlinear flag are qualitatively similar. We found no need to readjust material parameters when switching from the nonlinear to the IBM model; as discussed in [BWH*06] the carryover of parameters is expected. We modeled wind by a constant homogeneous velocity field, with force proportional to the projection of the wind velocity onto the area-weighted surface normal.

Conclusion. By restricting our attention to isometric deformations—the natural family of deformations for inextensible thin plates—we obtained a bending energy that is quadratic in positions. We demonstrated the consequent performance benefits and the simplicity of implementation in the context of cloth simulation.

In [BWH*06] we discuss observed benefits in the context of surface-fairing Willmore flows (see Figure 4). While in this setting deformations are not quasi-isometric, the adoption of IBM in an inexact-Newton solver yields a significant speedup.

Draping problem		regular mesh (resolution, in no. vertices)				irregular mesh (resolution, in no. vertices)			
		400	1600	6400	25600	450	2100	6500	22500
Gradient cost (ms)	nonlinear hinge	0.937	3.45	16.4	66.6	1.10	5.43	17.6	67.8
	quadratic IBM	0.081	0.338	2.19	9.15	0.098	0.494	2.32	9.68
Hessian cost (ms)	nonlinear hinge	12.8	54.2	218	890.	15.2	77.2	246	888
	quadratic IBM	0.237	0.963	3.87	15.7	0.266	1.28	3.99	13.6
Explicit step cost (ms)	nonlinear hinge	3.81	6.64	27.5	112.	2.16	9.53	31.4	140.
	quadratic IBM	2.63	2.90	11.9	48.8	0.964	4.35	15.2	76.5
Implicit step cost (ms)	nonlinear hinge	28.6	138	470.	1730	33.9	219	557	1880
	quadratic IBM	11.0	62.7	168	505	13.6	103	219	612

Flag problem		regular mesh (resolution, in no. vertices)				irregular mesh (resolution, in no. vertices)			
		400	1600	6400	25600	450	2100	6500	22500
Gradient cost (ms)	nonlinear hinge	0.975	3.99	16.0	64.0	1.10	5.43	17.8	68.7
	quadratic IBM	0.085	0.341	2.14	8.75	0.099	0.490	2.31	9.28
Hessian cost (ms)	nonlinear hinge	13.4	54.8	212	849	15.2	77.4	247	887
	quadratic IBM	0.251	0.974	3.79	14.99	0.267	1.30	3.96	13.7
Explicit step cost (ms)	nonlinear hinge	1.73	7.05	27.7	112.	1.97	9.80	32.7	134
	quadratic IBM	0.780	3.26	13.3	53.4	0.900	4.54	16.1	70.0
Implicit step cost (ms)	nonlinear hinge	27.6	106	420.	1680	33.5	155	513	1880
	quadratic IBM	9.53	32.9	127	490	12.5	50.4	166	608

Table 1: *Computational cost per time step for two solvers, regular- and irregular-meshes, and multiple resolutions, comparing IBM to the nonlinear hinge, as measured on a Pentium D 3.4GHz, 2GB RAM. Time step cost includes collision handling.*

Figure 3: *The quadratic bending model is valid over the full range of bending to in-plane stiffness ratios, e.g., (left to right) $10^{-5} : 1$, $10^{-3} : 1$, and $10^{-2} : 1$.*

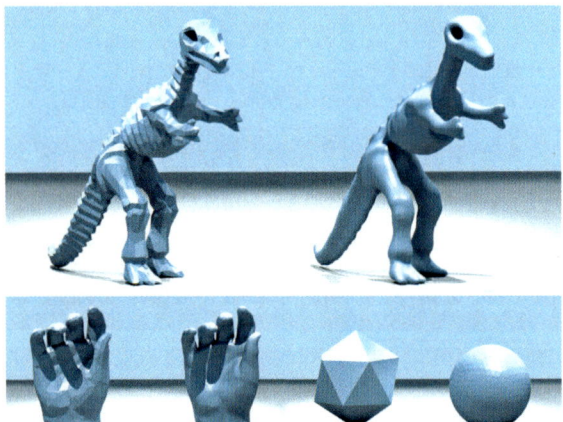

Figure 4: *Willmore flow smoothes a dino, a hand and an icosahedron (44928, 24192, and 5120 triangles, respectively). Smoothing requires 7.47s, 4.42s and 120ms, after one-time Hessian factorization costing 8.77s, 5.31s and 200ms, respectively. Flat shaded.*

Acknowledgements. This work was supported in part by the DFG Research Center Matheon "Mathematics for key technologies" in Berlin, the NSF (MSPA-MCS 0528402), Elsevier, and nVidia. Special thanks to Konrad Polthier for facilitating the meeting leading to this work, and to Charles Han for his tireless production assistance including lighting and rendering.

References

[BBG*01]   BALAY S., BUSCHELMAN K., GROPP W. D., KAUSHIK D., KNEPLEY M., MCINNES L. C., SMITH B. F., ZHANG H.: PETSc homepage. http://www.mcs.anl.gov/petsc, 2001.

[BFA02]   BRIDSON R., FEDKIW R., ANDERSON J.: Robust treatment of collisions, contact and friction for cloth animation. *ACM TOG 21*, 3 (2002), 594–603.

[BMF03]   BRIDSON R., MARINO S., FEDKIW R.: Simulation of clothing with folds and wrinkles. *SCA '03* (2003), 28–36.

[BS05]   BOBENKO A. I., SCHRÖDER P.: Discrete Willmore flow. *Europgraphics Symposium on Geometry Processing* (2005), 101–110.

[BW98]   BARAFF D., WITKIN A.: Large steps in cloth simulation. In *Proceedings of SIGGRAPH* (1998), pp. 43–54.

[BWH*06]   BERGOU M., WARDETZKY M., HARMON D., ZORIN D., GRINSPUN E.: Discrete quadratic curvature energies. TR, Columbia University, 2006.

[CDD*04]   CLARENZ U., DIEWALD U., DZIUK G., RUMPF M., RUSU R.: A finite element method for surface restoration with smooth boundary conditions. *CAGD* (2004), 427–445.

[Cia00]   CIARLET P.: *Mathematical Elasticity, Vol III*. North-Holland, 2000.

[CK05]   CHOI K.-J., KO H.-S.: Research problems in clothing simulation. *Computer Aided Design 37*, 6 (2005), 585–592.

[CR73]   CROUZEIX M., RAVIART P. A.: Conforming and non-conforming finite elements for solving stationary Stokes equations. *RAIRO Anal. Numer. 7* (1973), 33–76.

[CSM03]   COHEN-STEINER D., MORVAN J.-M.: Restricted Delaunay triangulations and normal cycle. *SoCG 2003* (2003), 312–321.

[GHDS03]   GRINSPUN E., HIRANI A. N., DESBRUN M., SCHRÖDER P.: Discrete shells. *SCA '03* (2003), 62–67.

[Hau04]   HAUTH M.: *Visual Simulation of Deformable Models*. PhD thesis, University of Tübingen, 2004.

[HP04]   HILDEBRANDT K., POLTHIER K.: Anisotropic filtering of non-linear surface features. *Comput. Graph. Forum 23*, 3 (2004), 391–400.

[Hug87]   HUGHES T. J. R.: *Finite Element Method - Linear Static and Dynamic Finite Element Analysis*. Prentice-Hall, Englewood Cliffs, 1987.

[KHM*98]   KLOSOWSKI J. T., HELD M., MITCHELL J. S. B., SOWIZRAL H., ZIKAN K.: Efficient collision detection using bounding volume hierarchies of k-dops. *IEEE TVCG 4*, 1 (January-March 1998), 21–36.

[MTV05]   MAGNENAT-THALMANN N., VOLINO P.: From early draping to haute couture models: 20 years of research. *The Visual Computer 21*, 8-10 (2005), 506–519.

[NG96]   NG H. N., GRIMSDALE R. L.: Computer graphics techniques for modeling cloth. *IEEE Computer Graphics & Applications 16*, 5 (Sept. 1996), 28–41.

[PP93]   PINKALL U., POLTHIER K.: Computing discrete minimal surfaces and their conjugates. *Experim. Math. 2* (1993), 15–36.

[TPBF87]   TERZOPOULOS D., PLATT J., BARR A., FLEISCHER K.: Elastically deformable models. In *Proceedings of SIGGRAPH* (1987), pp. 205–214.

[TW06]   THOMASZEWSKI B., WACKER M.: Bending Models for Thin Flexible Objects. In *WSCG Short Communication proceedings* (2006).

[YB02]   YOSHIZAWA S., BELYAEV A.: Fair triangle mesh generation via discrete elastica. In *GMP2002* (Wako, Saitama, Japan, July 2002), Suzuki H., Martin R., (Eds.), RIKEN (The Institute of Physical and Chemical Research), Japa, IEEE, pp. 119–123.

[ZJFP04]   ZHU H., JIN X., FENG J., PENG Q.: Survey on cloth animation. *Journal of Computer Aided Design & Computer Graphics 16*, 5 (2004), 613–618.

[ZT00]   ZIENKIEWICZ O. C., TAYLOR R. L.: *The finite element method: The basis*, 5th ed., vol. 1. Butterworth and Heinemann, 2000.

Eurographics Symposium on Geometry Processing (2006)
Konrad Polthier, Alla Sheffer (Editors)

Overfitting Control for Surface Reconstruction

Yunjin Lee[1]   Seungyong Lee[1]   Ioannis Ivrissimtzis[2]   Hans-Peter Seidel[3]

[1]POSTECH   [2]Coventry University   [3]MPI Informatik

Abstract

This paper proposes a general framework for overfitting control in surface reconstruction from noisy point data. The problem we deal with is how to create a model that will capture as much detail as possible and simultaneously avoid reproducing the noise of the input points. The proposed framework is based on extra-sample validation. *It is fully automatic and can work in conjunction with any surface reconstruction algorithm. We test the framework with a Radial Basis Function algorithm, Multi-level Partition of Unity implicits, and the Power Crust algorithm.*

Categories and Subject Descriptors (according to ACM CCS): I.3.5 [Computer Graphics]: Computational Geometry and Object Modeling; I.6.5 [Simulation and modeling]: Model Development

1. Introduction

In this paper, we deal with the problem of data *overfitting* in surface reconstruction. Overfitting appears when we model a given sample so faithfully that we capture not only information about the underlying surface but also the idiosyncrasies of the sample, that is, the noise in the points. Fig. 1 shows an example of overfitting.

Instead of modifying any of the existing algorithms, we propose a general framework for handling overfitting, which can be used in conjunction with any surface reconstruction technique. Our framework is based on a simple and accurate method for error estimation, *extra-sample validation* [HTF01]. The initial data set is randomly subdivided into two distinct subsets, the *training* set and the *validation* set. Data from the training set are used for trials of surface reconstruction, while the quality of reconstruction is assessed using the validation set. To have trials of reconstruction with increasing surface complexity, we use a hierarchical partition of the training data, based on an octree. We compute a representative training sample for each octree cell and a surface is created by applying a reconstruction algorithm to the training samples from the leaf cells of the octree.

1.1. Related work

In the area of surface reconstruction, [HDD*92, TL94, BBX95, CL96] are some of the earlier algorithms that influenced the field. More recently, implicit techniques emerged

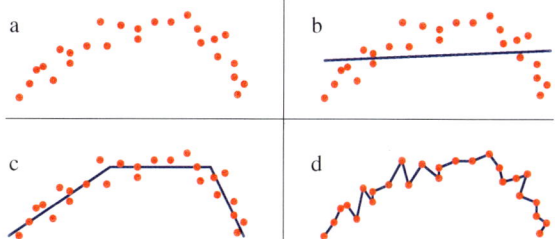

Figure 1: *Curve reconstruction: (a) sample points; (b) underfitted model; (c) correct model; (d) overfitted model.*

as the fastest and more stable techniques. The most common choices of implicits are the radial basis functions (RBFs) [CBC*01, OBS03] and quadrics [OBA*03]. Delaunay tetrahedrization has also been successfully used for surface reconstruction [ACK01, DG03]. In this paper, we experiment with the the techniques in [OBS03] and [OBA*03] and the Power Crust algorithm [ACK01].

In the literature of surface reconstruction, relatively little attention has been paid to the problem of overfitting. Ohtake *et al.* [OBS04] proposed an algorithm which penalizes overfitting by adding a regularization term to the usual distance error metric between the model and a sample. However, they did not present an automatic method to control the regularization term. Steinke *et al.* [SSB05] use Support Vector

Machines for surface reconstruction. They avoid overfitting with a regularization term which is determined with extra-sample validation. However, as in [OBS04], they treat overfitting as a global phenomenon, with one regularization term applied for the whole model. In many cases, this is not a realistic assumption.

2. Overfitting Control Framework

Following the standard terminology, the *training error* of a given model is the error measured against the training data. The *prediction error* is the expected error between the model and any sample coming from the same source as the training data. Overfitting usually arises when we try to minimize the training error instead of the prediction error. Indeed, the training error typically decreases monotonically with the model complexity, and eventually becomes zero when we interpolate the training data. In contrast, in a typical behavior shown in Fig. 2, the prediction error first decreases with the model complexity and then increases as we start overfitting.

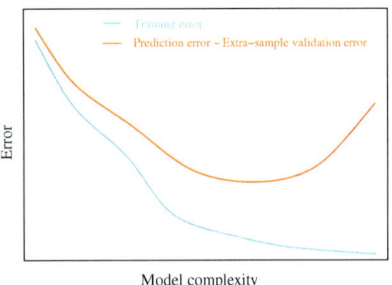

Figure 2: *Typical training and prediction error curves.*

2.1. Simple approach

In a simple approach to overfitting control, we can first split the data into the training and validation sets and then use the training data to reconstruct a series of different surfaces, indexed by one or more algorithmic parameters. We can select the surface with the least error against the validation data as the final output. However, this simple approach has certain limitations. First, the error is computed as an average over the whole surface. This error assumes a uniformly distributed noise, which is rarely the case with scan data. Second, the implementation depends on the specific parameters of a surface reconstruction algorithm and as a result, we need to devise a new strategy for each different algorithm.

2.2. Hierarchical framework

To overcome these limitations, we propose a hierarchical framework based on adaptive spatial subdivision of the input data. We first randomly select a half of the input points to be the training data. The other half of the points are the validation data. The data structure used for spatial subdivision is

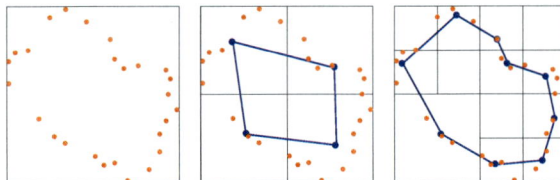

Figure 3: *The red points are training data. The blue points are cell representatives. Piecewise linear interpolating reconstructions are shown in blue.*

a progressively refined octree. At the beginning, the octree consists of a single cell corresponding to the bounding box of the input data. As an initialization step, we recursively subdivide the bounding box up to a few levels.

For a given octree O_l at level l, we split all the leaf cells that have not been marked as "completed" and create a new tree O_{l+1} at the next level $l + 1$. Then, we determine a single representative point for the training points in each leaf cell c_{l+1} of O_{l+1} by weighted averaging. We feed these representative points to the surface reconstruction algorithm of our choice to produce a surface S_{l+1}. For the leaf cells of the previous and current octrees, O_l and O_{l+1}, we compute the validation errors against the surfaces S_l and S_{l+1}. A cell c_l passes the overfitting test if the validation errors are decreasing with the subdivision of c_l itself *or* the majority of the child cells c_{l+1}. In this case, we keep the children of c_l in O_{l+1} and mark them as "uncompleted". Otherwise, overfitting is detected and we mark c_l as completed and remove its children from O_{l+1}. For cells which do not have enough validation points for a reliable error estimate, we also use validation points from neighbor cells. When an octree cell contains only one training point, it is marked as completed.

We repeat this process until all leaf cells of the adaptively refined octree are marked as completed. After completing the adaptive spatial subdivision with overfitting control, we collect the representative points from the leaf cells of the final octree. The final surface is obtained by applying the surface reconstruction algorithm to these points. Fig. 3 illustrates a 2D example.

Our overfitting control framework can work with any surface reconstruction technique, which can be considered as a black box. The input of the black box is two point sets, training data and validation data, and the output consists of the reconstructed surface and the validation error. During the overfitting control process, we do not need the actual surface reconstructions but only the error measurements for the validation data. For an implicit-based technique, such as [OBA*03] and [OBS03], we can directly estimate the distance of a validation point from the surface using the Taubin distance [Tau91], as used in [OBA*03]. In the case of the Power Crust, we measure the error from a surface by the

Metro tool [CRS98]. The following pseudocode summarizes our hierarchical framework.

Overfitting control framework

Input: training and validation data, $\mathcal{P}_{tr}$ and $\mathcal{P}_{vl}$.
Output: octree partitioning of the bounding box of $\mathcal{P}_{tr} \cup \mathcal{P}_{vl}$.

$l = 0$;
O_0 = subdivision of the bounding box of $\mathcal{P}_{tr} \cup \mathcal{P}_{vl}$
        up to a few levels;
$S_0 = MakeSurface(\mathcal{P}_{tr}, O_0)$;
while (O_l has uncompleted cells) **do** {
        subdivide uncompleted cells of O_l to create O_{l+1};
        $S_{l+1} = MakeSurface(\mathcal{P}_{tr}, O_{l+1})$;
        for (each uncompleted cell c_i of O_l) {
                if (test with $\mathcal{P}_{vl}$ detects an overfitting at level $l + 1$) {
                        mark c_i as completed;
                        remove children of c_i from O_{l+1};
                } **else**
                        keep and mark children of c_i in O_{l+1}
                                as uncompleted;
        }
        l++;
}
return O_l;

MakeSurface$(\mathcal{P}, O)$: for each cell c of O, compute a representative for the points of $\mathcal{P}$ contained in c and use a surface reconstruction algorithm to return a surface S.

Finally, if we are not sufficiently rich in data and want the reconstruction to involve all the available data, we can perform a 2-fold cross validation [HTF01]. That is, we repeat the whole process after swapping the training and validation data, and merge the two results.

3. Experimental Results

We tested the proposed framework with the RBF interpolation [OBS03] and the Power Crust [ACK01], chosen as examples of interpolating techniques with and without normals. We also used the MPU implicits [OBA*03] as an approximating technique with the use of a small error bound.

Fig. 5 shows reconstructions by these three algorithms. We used a point set sampled from the *tangle cube*,

$$x^4 - 5x^2 + y^4 - 5y^2 + z^4 - 5z^2 + 11.8 = 0, \quad (1)$$

with added noise. Overfitting control successfully reduced the noise in the data set and the results describe the underlying shape more faithfully. Table 1 compares the reconstruction errors. In all cases, overfitting control reduces the maximum error. In the cases of RBF and MPU, overfitting control also reduces the RMS error. For Power Crust, the reconstruction with overfitting control has a smaller number of polygons and we have an increased RMS error.

		RBF	MPU	Power Crust
Max. Error	single app.	0.0208	0.0260	0.0280
	overf. control	0.0117	0.0114	0.0246
RMS Error	single app.	0.0241	0.0203	0.0351
	overf. control	0.0147	0.0148	0.0407

Table 1: *Reconstruction errors measured by the Metro tool.*

Figure 4: *Left: single RBF. Right: overfitting control.*

Fig. 4 shows the RBF reconstructions of the bunny model without and with overfitting control. The initial point set from the Stanford Digital Model repository contains the original scanning noise. We used the confidence values supplied with the points as weights in the computation of the validation errors.

4. Discussion and Future Work

We propose a framework for the systematic control of overfitting in surface reconstruction. It is fully automatic and can be used in conjunction with any surface reconstruction technique, which can be treated as a black box. The levels of detail are determined by the quality of data, which means that some parts of the reconstruction can have more details than others. On the other hand, as one would expect, there is a computational overhead compared to the corresponding single reconstruction algorithms. We think that this extra computational time is justified as it allows a more informed model selection based on the analysis of the original data.

The effect of overfitting control on the model may be considered similar to that of a postprocessing smoothing step. However, the main difference between these two techniques is that smoothing needs a user-controlled parameter, while overfitting control is based on data analysis. Consequently, in overfitting control, the amount of smoothing is *locally* and *adaptively* determined by the data. In contrast, in a smoothing technique, a user-controlled parameter is *globally* applied to control the smoothing effects because it is tedious or impossible to assign a different parameter value for each specific region.

Recently there has been considerable research on algorithms and representations of point based geometry, such as

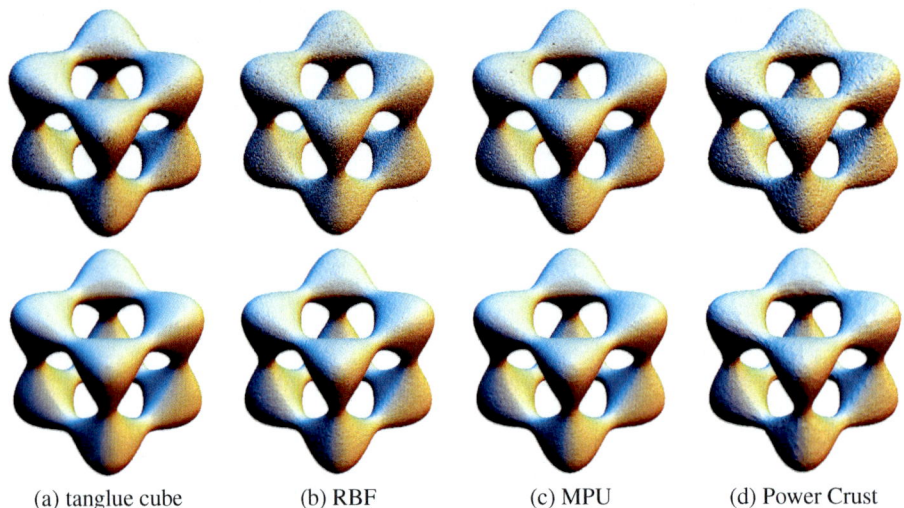

(a) tanglue cube   (b) RBF   (c) MPU   (d) Power Crust

Figure 5: *Bottom left: a point set sample. Top left: the same sample with added noise. The other three figures in the top row show single reconstructions from the noisy sample, while those in the bottom row show the results of overfitting control.*

point set surfaces [ABCO*03]. These techniques are related to this paper in that they also consider processing a given point set to obtain a better one in representing the underlying surface. We hope the main idea of this paper can also be applied to those kinds of techniques, although the details would differ.

Acknowledgements

The bunny model is courtesy of the Stanford Computer Graphics Laboratory. This research was supported in part by the BK21 program, the ITRC support program, and KOSEF (F01-2005-000-10377-0).

References

[ABCO*03] ALEXA M., BEHR J., COHEN-OR D., LEVIN D., FLEISHMAN S., SILVA C. T.: Computing and rendering point set surfaces. *IEEE Transactions on Visualization and Computer Graphics 9*, 1 (2003), 3–15.

[ACK01] AMENTA N., CHOI S., KOLLURI R. K.: The power crust, unions of balls, and the medial axis transform. *Computational Geometry: Theory and Applications 19*, 2-3 (2001), 127–153.

[BBX95] BAJAJ C. L., BERNARDINI F., XU G.: Automatic reconstruction of surfaces and scalar fields from 3D scans. *In Proc. ACM SIGGRAPH 1995* (1995), 109–118.

[CBC*01] CARR J. C., BEATSON R. K., CHERRIE J. B., MITCHELL T. J., FRIGHT W. R., MCCALLUM B. C., EVANS T. R.: Reconstruction and representation of 3D objects with radial basis functions. *In Proc. ACM SIGGRAPH 2001* (2001), 67–76.

[CL96] CURLESS B., LEVOY M.: A volumetric method for building complex models from range images. *In Proc. ACM SIGGRAPH 1996* (1996), 303 – 312.

[CRS98] CIGNONI P., ROCCHINI C., SCOPIGNO R.: Metro: measuring error on simplified surfaces. *Computer Graphics Forum 17*, 2 (1998), 167–174.

[DG03] DEY T. K., GOSWAMI S.: Tight cocone: A water-tight surface reconstructor. *Journal of Computing and Information Science in Engineering 3*, 4 (2003), 302–307.

[HDD*92] HOPPE H., DEROSE T., DUCHAMP T., MCDONALD J., STUETZLE W.: Surface reconstruction from unorganized points. *In Computer Graphics (Proc. ACM SIGGRAPH 1992)* (1992), 71–78.

[HTF01] HASTIE T., TIBSHIRANI R., FRIEDMAN J.: *The Elements of Statistical Learning: Data Mining, Inference, and Prediction*. Springer, New York, 2001.

[OBA*03] OHTAKE Y., BELYAEV A., ALEXA M., TURK G., SEIDEL H.-P.: Multi-level partition of unity implicits. *ACM Transactions on Graphics 22*, 3 (2003), 463–470.

[OBS03] OHTAKE Y., BELYAEV A., SEIDEL H.-P.: A multi-scale approach to 3D scattered data interpolation with compactly supported basis functions. In *Proc. Shape Modeling International* (2003), pp. 153– 161.

[OBS04] OHTAKE Y., BELYAEV A., SEIDEL H.-P.: 3D scattered data approximation with adaptive compactly supported radial basis functions. In *Proc. Shape Modeling International* (2004), pp. 31–39.

[SSB05] STEINKE F., SCHÖLKOPF B., BLANZ V.: Support vector machines for 3D shape processing. *Computer Graphics Forum (Proc. Eurographics 2005) 24*, 3 (2005), 285–294.

[Tau91] TAUBIN G.: Estimation of planar curves, surfaces, and nonplanar space curves defined by implicit equations with applications to edge and range image segmentation. *IEEE Trans. Pattern Analysis and Machine Intelligence 13*, 11 (1991), 1115–1138.

[TL94] TURK G., LEVOY M.: Zippered polygon meshes from range images. *In Proc. ACM SIGGRAPH 1994* (1994), 311–318.

Eurographics Symposium on Geometry Processing (2006)
Konrad Polthier, Alla Sheffer (Editors)

Nonobtuse Remeshing and Mesh Decimation

J. Y. S. Li and H. Zhang

GrUVi Lab, School of Computing Science, Simon Fraser University, Burnaby, BC, Canada

Abstract

Quality meshing in 2D and 3D domains is an important problem in geometric modeling and scientific computing. We are concerned with triangle meshes having only nonobtuse angles. Specifically, we propose a solution for guaranteed nonobtuse remeshing and nonobtuse mesh decimation. Our strategy for the remeshing problem is to first convert an input mesh, using a modified Marching Cubes algorithm, into a rough approximate mesh that is guaranteed to be nonobtuse. We then apply iterative "deform-to-fit" via constrained optimization to obtain a high-quality approximation, where the search space is restricted to be the set of nonobtuse meshes having a fixed connectivity. With a detailed nonobtuse mesh in hand, we apply constrained optimization again, driven by a quadric-based error, to obtain a hierarchy of nonobtuse meshes via mesh decimation.

Categories and Subject Descriptors (according to ACM CCS): I.3.5 [Computer Graphics]: Computational Geometry and Object Modeling

1. Introduction

A *nonobtuse triangle mesh* is composed of a set of nonobtuse triangles, in which every angle is less than or equal to $90°$. The problem of nonobtuse remeshing can be formulated as follows. Given an input mesh M, construct a nonobtuse mesh $\hat{M}$, with *any connectivity*, which smoothly and accurately approximates M. To make effective use of nonobtuse meshes in an interactive setting, it is often desirable to construct a hierarchy of nonobtuse meshes, via decimation. We refer to this problem as that of nonobtuse mesh decimation.

Nonobtuse meshes are of interest in several contexts. First, a nonobtuse triangulation is necessarily a Delaunay one, both in 2D [BMR94] and for surface meshing [BS06]. Secondly, nonobtuse meshes are shown to possess more desirable numerical properties for finite element methods, e.g., in producing Stieltjes matrices [BGR88] or in ensuring faster convergence when physical properties over the discretized domain vary enormously [BMR94]. Also, nonobtuse meshes result in more efficient and more accurate geodesic estimations via fast marching [KS98]. Finally, nonobtusity ensures the validity of planar mesh embedding via discrete Harmonic maps [EDD*95]; it also implies that certain key properties of the well-known discrete Laplacian-Beltrami operator would be analogous to those of the classical Laplacian-Beltrami operator on a surface with Riemannian metric [BS06].

To the best of our knowledge, no known algorithms are guaranteed to produce a nonobtuse mesh which either interpolates or accurately and smoothly approximates a point cloud. The same holds for the remeshing problem. However, there have been studies on nonobtuse 2D triangulation and a great deal of work on remeshing of curved surfaces with angle or other quality criteria [AUGA05]. Bern et al. [BMR94] give an $O(n \log^2 n)$ algorithm for nonobtusely triangulating a n-sided polygon using $O(n)$ triangles. Chew [Che93] develops a refinement scheme based on constrained Delaunay triangulation (DT) to ensure an angle bound of $[30°, 120°]$. Cheng and Shi [CS05] use restricted union of balls to generate an ε-sampling of a surface and extract a mesh from the DT of the sample points. A lower bound on the minimum angle can be as high as $30°$ with proper choice of parameters, but results in a high triangle count. Neither algorithm provides a nonobtuse guarantee; this has been identified as an open problem by Gu and Yau [GY03], and so far only some simple heuristics [GY03] have been suggested.

In this paper, we present a solution for guaranteed nonobtuse remeshing and mesh decimation. To make our presentation concise, we shall only deal with closed, 2-manifold meshes. Handling of mesh boundaries is discussed in an extended version of this paper [LZ06]. The framework we develop is quite general and flexible. It allows us to generate nonobtuse meshes from point clouds with slight modifica-

    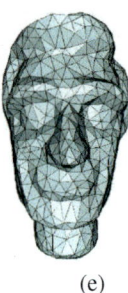

(a)   (b)   (c)   (d)   (e)

Figure 1: *Major components of our approach. (a) Original mesh (34,712 faces). (b) Result of our midpoint-based Marching Cubes algorithm (15,380 faces) shows visible artifacts. (c) After constrained optimization, the approximation error from Metro [CRS98] is $\varepsilon = 0.78\%$, which is the Hausdorff distance as a percentage of the bounding box diagonal. (d) After decimation of 50% of the vertices nonobtusely. Error against (c): $\varepsilon = 0.107\%$. (e) 90% decimated. Error against (c): $\varepsilon = 0.61\%$.*

tions to our existing algorithm. Angle bounds may also be enforced, although the cost of the constrained optimization will increase. Both of these issues are elaborated in [LZ06].

Our strategy for remeshing is to construct an initial nonobtuse mesh and iteratively refine it in a constrained manner to obtain a high-quality approximation of the input mesh M. Utilizing a modified Marching Cubes algorithm, we build a close but rough approximation of M. An added advantage of using Marching Cubes is that we can bound the mesh size by adjusting the grid resolution. Next, we iteratively deform the mesh, in a greedy fashion, to reduce approximation error. The deformation is carried out via quadratic programming with the constraint that each vertex must move within its *feasible region*. Once a final, fine-detailed nonobtuse mesh is generated, we can perform nonobtuse decimation via edge-collapse under similar nonobtuse constraints. See Figure 1.

2. Initial nonobtuse mesh via modified Marching Cubes

To construct the initial nonobtuse mesh $\hat{M}_0$, we rely on a *midpoint-based* Marching Cubes algorithm. Originally appeared in the seminal paper of Lorensen and Cline [LC87], the Marching Cubes algorithm produces a triangle mesh which tessellates the zero level-set of a given 3D scalar field. Given an input mesh M, we generate a signed distance field sampled at the vertices of a regular cubical grid. Distance field generation for open meshes is more difficult, but it can be achieved with extra care [LZ06]. For our purpose, only the signs of the distance values matter. Specifically, instead of using linear interpolation to compute vertex locations [LC87], we insist that each mesh vertex along a cube edge be at the edge's midpoint. With a few additional modifications to mesh connectivity, e.g., with new vertices at cube centers created in some cases, the resulting triangle mesh is provably nonobtuse. Closer examination reveals that the resulting mesh also has no angle smaller than $30°$. Thus it is possible to ensure an angle bound by placing tighter constraints. All these issues are discussed in length in [LZ06].

3. Deform-to-fit via constrained optimization

Given the initial nonobtuse mesh $\hat{M}_0$, we wish to

$$\min. \sum_{v \in V(\hat{M})} \mathcal{L}(v, M), \quad \text{subject to} \quad \hat{M} \in \mathcal{NO}(\hat{M}_0), \quad (1)$$

where $V(\hat{M})$ is the set of vertices in $\hat{M}$, $\mathcal{NO}(\hat{M}_0)$ is the set of nonobtuse meshes with the same connectivity as $\hat{M}_0$, and

$$\mathcal{L}(v, M) = \alpha \cdot \mathcal{D}(v, M) + (1 - \alpha) \cdot \mathcal{S}(v). \quad (2)$$

The point-to-surface distance $\mathcal{D}(v, M)$, where M is the original input mesh, is a *quadric error* [GH97], $\mathcal{S}(v)$ is a smoothness (regularization) term, and $0 \leq \alpha \leq 1$ is a free parameter for the trade-off between error reduction and smoothness. We define $\mathcal{D}(v, M) = Q_v(v)/k$ where in general, $Q_v(x)$, the quadric error [GH97], is the sum of squared distances from x to a set of planes associated with v. In our case, this set is composed of k supporting planes, including that of the triangle T of M that is closest to v and those of the r-ring neighbor triangles of T in M. In our experiments, we set $r = 1$. At vertex v, the smoothness term $\mathcal{S}(v) = ||v - \mathcal{C}(v)||^2$ measures the squared distance between v and the centroid $\mathcal{C}(v)$ of its one-ring neighbors. Solution to the optimization (1) is likely intractable, we thus resort to heuristics. Specifically, we move vertices one at a time in a greedy fashion, under a set of linearized and convexified nonobtuse constraints.

Given a vertex v, let its one-ring vertices be $v_0, v_1, \ldots,$ and v_{k-1}, in order. Consider the one-ring edge $e = (v_i, v_{i+1})$. Let U and V be the two planes orthogonal to e, passing through v_i and v_{i+1}, respectively. Let $sphere(e)$ be the sphere centered at the midpoint q of e with diameter $|e|$ and let p be the intersection between $\overline{vq}$ and $sphere(e)$; refer to Figure 2. Finally, let R be the tangent plane of $sphere(e)$ through p. We define the *feasible region* $\mathcal{F}(e, v)$ of edge e with respect to v as the intersection of the front half spaces defined by U, V, and R, with their normals shown in the figure. The feasible region $\mathcal{F}(v)$ of v is then given by $\mathcal{F}(v) = \bigcap_{i=0}^{k-1} \mathcal{F}[(v_i, v_{i+1}), v]$, which is linear and convex. Note that by construction, $\mathcal{F}(v)$ is always nonempty, since $v \in \mathcal{F}(v)$.

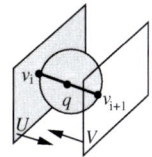

Figure 2: *Left: feasible region $\mathcal{F}[(v_i, v_{i+1}), v]$ is the space delimited by planes U, V, and R. Right: any point between U and V and outside the sphere makes a nonobtuse triangle with v_i and v_{i+1}. Thus the feasible region is a linearized and convexified subset of the actual "nonobtuse region".*

To solve (1), we iteratively deform the mesh to obtain progressively better approximation to M. For each vertex v in the current mesh $\hat{M}$, we compute an optimal position v^* which minimizes $\mathcal{L}(v, M)$ subject to $v \in \mathcal{F}(v)$. To solve this convex quadratic programming problem, we rely on the OOQP solver of Gertz and Wright [GW03]. Then a priority $\mathcal{H}(v)$, given by the *improvement* made by moving v to v^*: $\mathcal{H}(v) = \mathcal{L}(v, M) - \mathcal{L}(v^*, M)$, is computed. The vertex with the highest priority is moved to its optimal position and all vertices influenced by the move will have their priorities recomputed. To update the quadric for a vertex v that has just moved, we need to find a triangle T' in M that is closest to v. In our implementation, we execute a local search within the 3-ring neighborhood of the previous closest triangle T. The greedy optimization stops when $\mathcal{H}(v) \leq 0, \forall v \in V(\hat{M})$.

Experimentally, the nonobtuse meshes produced by optimization alone may be slightly rough. This can be attributed to the fact that our optimization is highly localized. To this end, we suggest to alternate between mesh optimization and constrained Laplacian smoothing, with the former being the first and the last steps of our "deform-to-fit" procedure and the latter encouraging *tangential* movement of the vertices to smooth out any roughness. In the smoothing step, vertices are processed one at a time with no particular ordering. The objective function is simply the smoothness term $\mathcal{S}(v)$.

4. Nonobtuse mesh decimation

Analogous to nonobtuse remeshing, nonobtuse mesh decimation can also be formulated via constrained optimization and solved in an iterative greedy fashion via edge collapse. For each candidate edge $e = (u, v)$, we compute the optimal position $w^* = \operatorname{argmin}_{w \in \mathcal{F}(e)}[Q_u(w) + Q_v(w)]$, for the unified vertex w, where $Q_u(w)$ and $Q_v(w)$ are quadric errors defined as before and the feasible region $\mathcal{F}(e)$ for edge e is determined similarly as in the vertex case, but by the one-ring neighbors of both u and v. The cost for collapsing edge (u, v) is $Q_u(w^*) + Q_v(w^*)$; this is set to ∞ if $\mathcal{F}(e) = \emptyset$. We iteratively collapse the edge with the lowest cost, and update the costs for the new edges as well as those edges that are influenced due to changes to their feasible regions.

5. Experimental results

Now we demonstrate several characteristics of our approach and the quality of the nonobtuse meshes produced. Unless otherwise specified, we choose $\alpha = 0.5$ for mesh optimization and only one step of Laplacian smoothing is applied. The approximation error ε, as defined for Figure 1, measures against the original mesh for remeshing and against the full-resolution nonobtuse mesh in decimation.

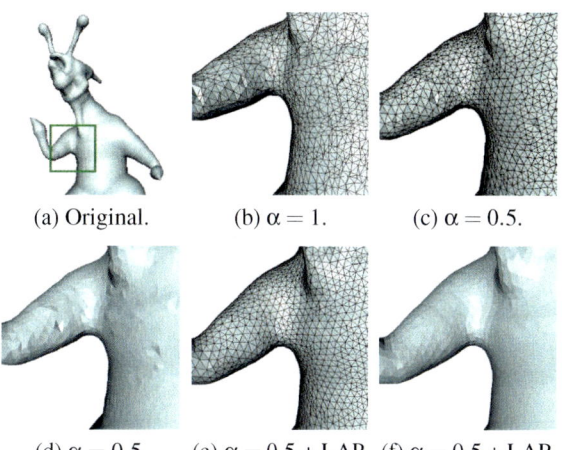

(a) Original. (b) $\alpha = 1$. (c) $\alpha = 0.5$.

(d) $\alpha = 0.5$. (e) $\alpha = 0.5 +$ LAP. (f) $\alpha = 0.5 +$ LAP.

Figure 3: *Importance of smoothness term and smoothing: Without the smoothness term in the objective function, we suffer from poor angle quality, shown in (b). Percentage of small angles (angles less than $30°$) is $\gamma = 7.74\%$. (c) and (d): After weighing in the smoothness term: $\gamma = 0.12\%$. But geometric roughness is quite visible. (e) and (f): With constrained Laplacian smoothing added: $\gamma = 0.03\%$. And the rough geometric features are smoothed out.*

In Figure 3, we show the importance of the smoothness term and smoothing. By comparing (b) and (c), it is apparent that including the smoothness term in the objective function dramatically improves the overall angle quality of the mesh. An additional smoothing step further improves the mesh quality as the roughness is smoothed out, as shown in (f). In Table 1, we report mesh quality and performance statistics from nonobtuse remeshing of six models. As it can be observed in conjunction with Figure 4, our algorithm is capable of producing good angle distributions and approximation quality. In particular, smoothing does help reduce the number of $90°$ angles as shown in the middle column of Figure 4. In our current implementation of nonobtuse decimation, no smoothness term is incorporated into the objective function. Hence, we see some spikes in the $90°$ bins.

6. Discussion and future work

In this paper, we propose a solution for guaranteed nonobtuse remeshing and mesh decimation. We design a modified, midpoint-based Marching Cubes algorithm to con-

Input mesh (#F)	#F	#V	Min. $\angle$	% small $\angle$'s	Metro ε	1: %VM	2: %VM	3: %VM
Armhand (50K)	18798	9401	26.14	0.05	1.00	310.6	397.9	487.3
Bigsmile (34.7K)	15380	7692	19.49	0.04	0.78	308.3	312.9	N/A
Baldhead (15.8K)	18704	9356	20.68	0.06	1.86	407.1	399.8	393.0
Horse (40K)	19880	9944	15.03	0.18	1.31	258.3	266.4	N/A
Man (29K)	68252	34122	10.52	0.14	0.55	328.7	285.1	N/A
Monster (32.5K)	48226	24115	19.79	0.03	0.64	294.8	374.3	N/A

Table 1: *This table shows quality measures and performance statistics for nonobtuse remeshing. We report, from left to right: input mesh and its face count (#F); face (#F) and vertex (#V) counts of the output nonobtuse mesh; minimum angle (Min. $\angle$) in the nonobtuse mesh; percentage of angles less than* $30°$ *(% small $\angle$); approximation error (Metro ε); the number of vertices moved in each mesh optimization step (1, 2, and 3) as a percentage ("1:%VM", "2:%VM", and "3:%VM") of (#V). Note that in many cases, only two optimization steps are needed; in that case, the "3:%VM" column will be marked by "N/A".*

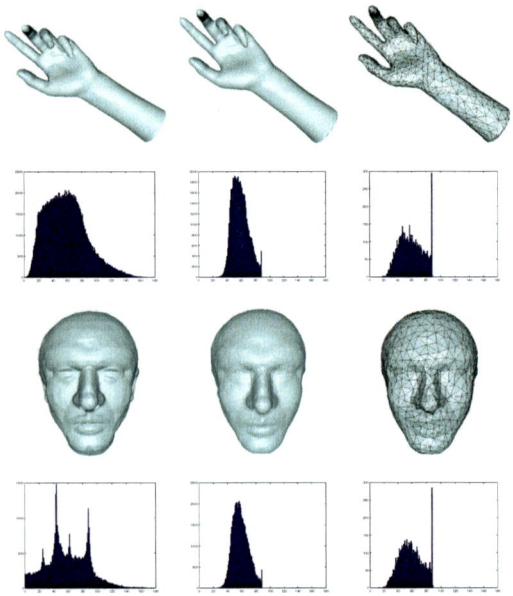

Figure 4: *Visual results and histogram plots of angle distributions. From left to right: Original model; remeshed model; model after 90% of vertices decimated. The histogram plot associated with a particular model is given right below the model. Top: Armhand model. Bottom: Baldhead.*

struct an initial nonobtuse mesh from which a series of local constrained optimizations are performed. Our framework is quite general and flexible and it provides several directions for future work. In particular, we wish to achieve efficient nonobtuse remeshing with angle bounds, nonobtuse or acute mesh generation from point clouds, and effective handling of mesh boundaries. In addition, lazy evaluations, early stopping, and adaptive smoothing and error reduction procedures can be used to improve efficiency. Other more challenging improvements, such as the use of adaptive grids in Marching Cubes and the preservation of sharp features in remeshing, are also worth investigating.

References

[AUGA05] ALLIEZ P., UCELLI G., GOTSMAN C., ATTENE M.: Recent advances in remeshing of surfaces, 2005. Part of the state-of-the-art report of the AIM@SHAPE EU network.

[BGR88] BAKER B. S., GROSSE E., RAFFERY C. S.: Nonobtuse triangulation of polygons. *Disc. & Comp. Geom. 3* (1988), 147–168.

[BMR94] BERN M., MITCHELL S., RUPPERT J.: Linear-size nonobtuse triangulation of polygons. In *ACM Symp. on Comp. Geom. (SoCG)* (1994), pp. 221–230.

[BS06] BOBENKO A. I., SPRINGBORN B. A.: A discrete lapplacian-beltrami operator for simplicial surfaces, February 2006. preprint: arXiv:math.DG/0503219.

[Che93] CHEW P.: Guaranteed-quality mesh generation for curved surfaces. In *ACM SoCG* (1993), pp. 274–280.

[CRS98] CIGNONI P., ROCCHINI C., SCOPIGNO R.: Metro: Measuring error on simplified surfaces. *Computer Graphics Forum 17*, 2 (1998), 167–174.

[CS05] CHENG H.-L., SHI X.: Quality mesh generation for molecular skin surfaces using restricted union of balls. In *IEEE Visualization* (2005), pp. 399–405.

[EDD*95] ECK M., DEROSE T., DUCHAMP T., HOPPE H., LOUNSBERY M., STUETZLE W.: Multiresolution analysis of arbitrary meshes. In *SIGGRAPH* (1995), pp. 173–182.

[GH97] GARLAND M., HECKBERT P. S.: Surface simplification using quadric error metrics. In *SIGGRAPH* (1997), pp. 209–216.

[GW03] GERTZ E. M., WRIGHT S. J.: Object-oriented software for quadratic programming. *ACM Trans. on Math. Software 29* (2003), 58–81.

[GY03] GU X., YAU S.-T.: Global conformal surface parameterization. In *Proc. of SGP* (2003), pp. 127–137.

[KS98] KIMMEL R., SETHIAN J. A.: Computing geodesic paths on manifolds. *Proc. of National Academy of Science USA 95* (1998), 8431–8435.

[LC87] LORENSEN W. E., CLINE H. E.: Marching cubes: A high resolution 3d surface construction algorithm. In *SIGGRAPH* (1987), pp. 163–169.

[LZ06] LI J. Y. S., ZHANG H.: Guaranteed nonobtuse meshes via constrained optimization, 2006. Technical Report TR 2006-13, School of Computing Science, Simon Fraser University.

Eurographics Symposium on Geometry Processing (2006)
Konrad Polthier, Alla Sheffer (Editors)

Size functions for 3D shape retrieval

S. Biasotti, D. Giorgi, M. Spagnuolo and B. Falcidieno

CNR - IMATI - Genova, Italy

Abstract

This paper sketches a technique for 3D model retrieval built on size functions, a mathematical tool to compare shapes. Size functions are introduced for the first time to discriminate among 3D objects, through the proposal of an innovative method to construct size graphs independently of the underlying triangulation. We demonstrate the potential of our approach in a series of comparative experiments with respect to existing techniques.

Categories and Subject Descriptors (according to ACM CCS): I.3.6 [Computer Graphics]: Methodoloy and Techniques

1. Introduction

3D shape classification and retrieval is a very lively research topic. In this paper, we propose a framework to extend to the 3D domain the use of size functions, a shape descriptor which has been extensively applied to content-based image retrieval (see e.g. [VUFF93, CFG05]).

The theory of size functions has been developed since the beginning of the 90's in order to get a new geometric-topological approach to shape discrimination [VUFF93]. The idea is to analyze the growth of a topological space S, according to the increasing values of a real function φ defined on it. In particular, size functions code the topological evolution of S counting the number of connected components which remain disconnected passing from a lower level set of S to another. Since the growth of S is driven by the real function φ, size functions encode geometrical properties in the topological history. Hence they take into account both local and global properties of a shape. A similar approach has been introduced in [ELZ02].

The main contribution of this paper is to exploit and fruitfully enhance their potential for 3D shape comparison. The result is the definition of a technique for 3D shape description and retrieval, which is able to interpret the knowledge embedded in the shape, taking into account structure, topology and geometry.

2. Related work

The majority of the methods proposed in the literature for 3D shape retrieval mainly focuses on the low-level geometry

of shapes, in the sense of considering its spatial distribution or extent in the 3D space [BKS*06, KFR03]. Nevertheless, there is a growing consensus towards high-level descriptors which merge a global topological analysis with local geometric attributes [CZCG04]. For example, the method presented in [HSKK01] addresses 3D shape similarity by using the Reeb graph in a multi-resolution fashion and performs retrieval by means of graph-matching techniques. Similarly, the importance of structural descriptions for shape matching has been recently pointed out in [BMM*03, ZS*05]. Exhaustive surveys on 3D shape searching techniques can be found for example in [TV04, IJL*05].

3. Approach

The aim of this paper is to provide a high-level technique based on size functions to fully reveal the topological information on the shape which is encoded in the representation model. The attractive feature of size functions is that they provide a high-level description which can be readily used to establish a similarity measure between shapes, formalizing qualitative aspects of shapes in a quantitative way.

3.1. Size functions

Given a *size pair* (S, φ), where S is a topological space and $\varphi : S \to \mathbf{R}$ is a continuous function called a *measuring function*, the size function $\ell_{(S,\varphi)} : \{(x,y) \in \mathbf{R}^2 : x < y\} \to \mathbf{N}$ is defined by setting $\ell_{(S,\varphi)}(x,y)$ equal to the number of connected components of the lower level set $S_y = \{P \in S : \varphi(P) \le y\}$, containing at least one point of S_x [dFL05].

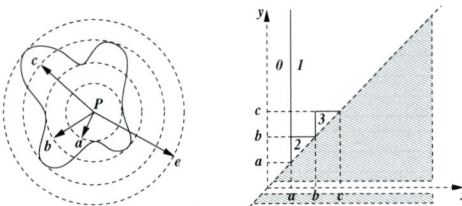

Figure 1: *A size pair and the corresponding size function.*

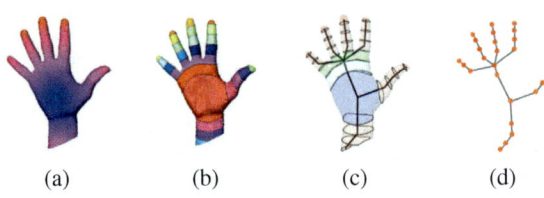

Figure 2: *(a) Evaluation of the distance from the barycenter on the hand model in [aim]. Red and blue colors respectively represent maximum and minimum values. (b) The mesh partition. (c-d) The centerline skeleton.*

In the example in Figure 1 we consider the size pair (S, φ), where S is the curve represented by a continuous line in Figure 1(left), and φ is the function "distance from the point P". The size function associated with (S, φ) is shown in Figure 1(right). The value displayed in each region is the value taken by the size function in that region.

An important property of size functions is that they can always be seen as linear combinations of characteristic functions of triangles (possibly unbounded triangles with vertices at infinity). Hence, by taking the formal series of vertices associated with their right angles (called *cornerpoints* for the bounded triangles and *cornerlines* for the unbounded ones) we get a simple and compact representation [FL01].

The discrete counterpart of a size pair is a *size graph* (G, φ), where $G = (V(G), E(G))$ is a finite graph, with $V(G)$ and $E(G)$ the set of vertices and edges respectively, and $\varphi : V(G) \to \mathbf{R}$ is a measuring function labelling the nodes of the graph [d'A00].

3.2. Building size graphs for 3D shapes

Our idea is to associate with a 3D object a size graph (G^f, φ), where G^f is a centerline skeleton representing S, f is a real continuous function driving the centerline extraction, and φ is a measuring function labelling each node of the graph with local geometrical properties of the original model. This model signature, which combines the structural information provided by the mapping function f with the different information provided by the measuring function φ, produces informative size functions. Beside the obvious improvement in computational efficiency, the skeletal structure reduces the dimensionality of the problem, meanwhile storing sufficient information about the original object. Further details about our method can be found in [BGSF06].

The construction of the centerline skeleton G^f relies on the discretization of the Reeb graph theory defined in [Bia04]. Given a shape represented by a regular triangle mesh M, we subdivide the co-domain $[f_{min}, f_{max}]$ of f : $M \to \mathbf{R}$ considering nv regular values of f, $f_i \in [f_{min}, f_{max}]$, $i = 1, \ldots, nv$. The level sets of f that correspond to these values partition the mesh M into regions, see Figure 2(b). Hence all points belonging to a region or a contour are identified and represented as nodes and arcs of a traditional graph, see Figure 2(c,d).

Four different mapping functions f are considered in our framework, namely the distance from the barycenter, the distance from the center of the bounding sphere, the integral geodesic distance in [HSKK01] and the topological distance from curvature extrema in [MP02].

Once the centerline G^f has been extracted, the *size graph* (G^f, φ) is obtained by defining the measuring function φ : $V(G^f) \to \mathbf{R}$ on the nodes of G^f. For each node $v_R \in V(G^f)$ corresponding to a region R, the value of $\varphi(v_R)$ is defined as a property characterizing the region R or its boundary $B_M(R)$. In particular, we are proposing to use:

- the *area* of the region R;
- the *minimum, maximum and average distance* of the barycenter of R from the region vertices;
- the *length of* $B_M^+(R)$ (resp. $B_M^-(R)$), where $B_M^+(R)$ (resp. $B_M^-(R)$) is the set of connected components of $B_M(R)$ such that the outgoing directions for the mapping function f are ascending (resp. descending);
- the sum of the *pseudo-cone lateral areas* computed for each component of R in $B_M^+(R)$ (resp. $B_M^-(R)$);
- a set of distance functions from selected points on the minimal bounding box of the model.

3.3. 3D model comparison

Once that a size graph (G^f, φ) has been obtained, the definition of the size functions follows the classical one. Denoting by G_y^f the subgraph of G^f obtained by erasing all vertices of G^f at which φ takes a value strictly greater than y, and all edges that connect those vertices to other vertices, the size function of (G^f, φ) is defined by setting $\ell_{(G^f, \varphi)}(x, y)$ equal to the number of connected components of G_y^f, containing at least a vertex of G_x^f.

In order to compute size functions, we have followed the algorithm introduced in [d'A00]. To compare two models we use the matching distance between their size functions, whose suitability for shape comparison has been discussed from the theoretical point of view in [dFL05].

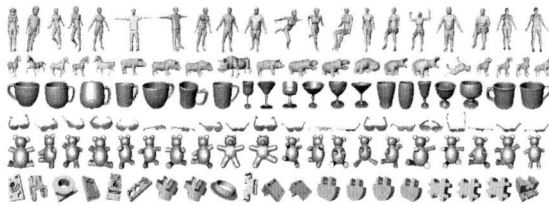

Figure 3: *Our testing models.*

4. Experimental results

We have constructed a database of regular triangle meshes consisting of 5 classes of 20 elements, plus 20 unclassified manufactured models, see Figure 3. The original models of our database were collected from several web repositories ([dre], [aim], [SMKF04], [cae], [mcg]). To validate our results, we have considered the spherical harmonic descriptor [KFR03], the view-based approach in [COTS03] and the Multiresolution Reeb graph described in [HSKK01].

As a first performance parameter, we have considered the *percentage recall*. The recall histogram in Figure 4(a) has been obtained computing, for the rank thresholds $N = 10, 20, \ldots, 120$, the percentage of models in the same class of the query retrieved within the first N items. Results are averaged over the whole database, and indicate that almost 80% of relevant items are retrieved within top 25% of the database (i. e. within the first 30 models; remember that each class contains 20 elements). Figure 4(b) compares the *average rank* for the whole database obtained using size functions with the values obtained by the other techniques. The value obtained with size functions is the lowest one; recall that for this indicator lower values indicate better performance. A further measure we are using to assess the retrieval performance is the *last place ranking* defined in [EBG98], whose values are reported in Figure 4(c). High values within the interval $[0,1]$ indicate good results.

One of the attractive features of our approach is its flexibility. In fact, the core idea of our method is the analysis of properties of real functions describing the shape under study. The role of the real functions is to take into account only the shape properties of the object which are relevant to the problem at hand, as well as to impose the desired invariance properties. When changing the functions, the resulting configurations can give insights on the shape from different perspectives, see Figure 5. These results suggest that our approach could also be used as a finer tool, after a rough filter has been used, or as an instrument to refine queries. Using our technique would allow the user to readily indicate the shape idea he has in mind, through the selection of a set of a features (i.e. mapping and measuring functions) which have a clear and intuitive geometric (and perceptual) significance.

5. Concluding remarks

The proposed shape descriptor presents many desirable properties. Indeed it is:

1. quick to compute: the computation of 120 size functions for the 120 models in the database requires 1.53 second on a 1.73GHz laptop PC-M; the off-line step of computing the size graphs requires 1 minute and 12 seconds;
2. concise to store: on average it requires less than $1K$;
3. easy and quick to compare: evaluating 120×120 matching distances requires 8.55 seconds;
4. invariant under similarity transformations: imposing the desired invariance simply means requiring the invariance for the mapping and measuring functions, without any change in the mathematical model;
5. robust against noise and small extra features;
6. able to discriminate among shapes at many scales, conveying information about their global and local properties.

By summarizing, we have proposed an original framework to extend the use of size functions in the 3D context. We have derived a signature to be extracted from 3D models, which guarantees the topological coding and the geometrical description, and is computationally efficient. Such a representation has been used as a size graph for computing discrete size functions. The experimental results have shown that this approach is promising, and goes into the direction of developing tools to automatically annotate the shape semantic, and to encapsulate it in a digital shape representation.

Acknowledgements

The authors thank Prof. M. Ferri and the Vision Mathematics Group at the Univ. of Bologna for fruitful discussions. This work has been developed in the CNR research activity (ICT-P03) and partially supported by the EU NoE "AIM@SHAPE" (http://www.aimatshape.net).

References

[aim]   http://shapes.aim-at-shape.net.

[BGSF06]   BIASOTTI S., GIORGI D., SPAGNUOLO M., FALCIDIENO B.: *Computing size functions for 3D models*. Tech. Rep. 03, CNR-IMATI-GE, 2006.

[Bia04]   BIASOTTI S.: *Computational topology methods for shape modelling applications*. PhD thesis, Univ. of Genoa, 2004.

[BKS*06]   BUSTOS B., KRIM D., SAUPE D., SCHRECK T., VRANIĆ D.: An experimental effectiveness comparison of methods for 3D similarity search. *IJDL 6*, 1 (2006), 39–54.

[BMM*03]   BIASOTTI S., MARINI S., MORTARA M., PATANÉ G., SPAGNUOLO M., FALCIDIENO B.: 3D shape matching through topological structures. In *DGCI* (2003), vol. 2886 of *LNCS*, pp. 194–203.

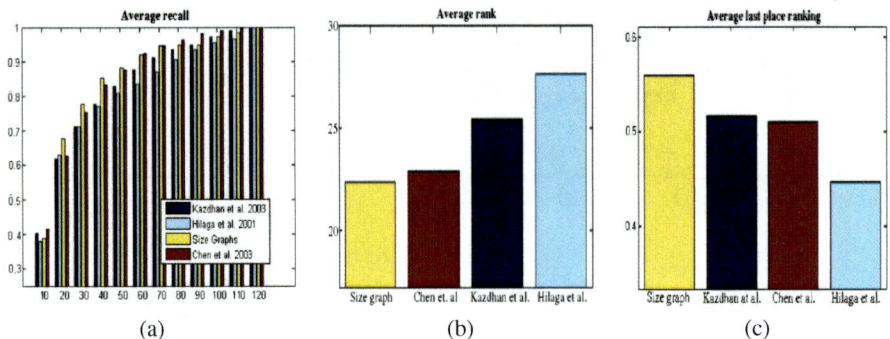

Figure 4: *Comparison with existing methods. (a) Recall, (b) average rank and (c) last place ranking histograms.*

[cae] http://www.hec.afrl.af.mil/HECP/Card1b.shtml #caesarsamples.

[CFG05] CERRI A., FERRI M., GIORGI D.: A new framework for trademark retrieval based on size functions. In *VVG'05* (2005), pp. 167–172.

[COTS03] CHEN D., OUHYOUNG M., TIAN X., SHEN Y.: On visual similarity based 3D model retrieval. *Computer Graphics Forum 22* (2003), 223–232.

[CZCG04] CARLSSON G., ZOMORODIAN A., COLLINS A., GUIBAS A.: Persistence barcodes for shapes. In *SGP'04* (2004), pp. 127–138.

[d'A00] D'AMICO M.: A new optimal algorithm for computing size functions of shapes. In *CVPRIP Alg.s III* (2000), pp. 107–110.

[dFL05] D'AMICO M., FROSINI P., LANDI C.: Using matching distance in size theory. *Int. J. Imaging Systems and Technology 25(6C)* (2005), 4577–4582.

[dre] http://www.designrepository.org.

[EBG98] EAKINS J., BOARDMAN J., GRAHAM M.: Similarity retrieval of trademark images. *Multimedia 5*, 2 (1998), 53–63.

[ELZ02] EDELSBRUNNER H., LETSCHER D., ZOMORODIAN A.: Topological persistence and simplification. *Discrete Comput. Geom. 28* (2002), 511–533.

[FL01] FROSINI P., LANDI C.: Size functions and formal series. *Appl. Algebra Eng. Comm. Comput. 12* (2001), 327–349.

[HSKK01] HILAGA M., SHINAGAWA Y., KOHMURA T., KUNII T. L.: Topology matching for fully automatic similarity estimation of 3D shapes. In *SIGGRAPH* (2001), pp. 203–212.

[IJL*05] IYER N., JAYANTI S., LOU K., KALYANARAMAN Y., RAMANI K.: Three-dimensional shape searching: state-of-the-art review and future trends. *CAD 37*, 5 (2005), 509–530.

[KFR03] KAZHDAN M., FUNKHOUSER T.,

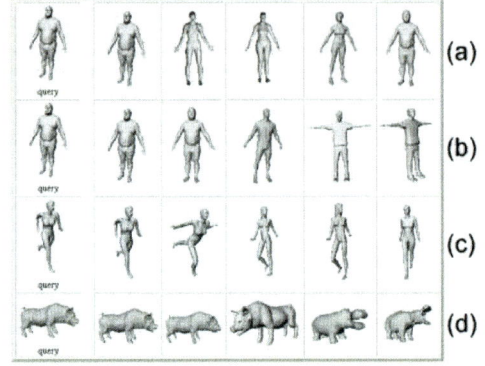

Figure 5: *Query results choosing different mapping and measuring functions. (a) Emphasis on spatial position, (b) robustness, (c) human poses; (d) fine results on animals.*

RUSINKIEWICZ S.: Rotation invariant spherical harmonic representation of 3D shape descriptors. In *SGP'03* (2003), pp. 156–165.

[mcg] http://www.cim.mcgill.ca/ shape/benchMark/.

[MP02] MORTARA M., PATANÉ. G.: Shape-covering for skeleton extraction. *IJSM 8*, 2 (2002), 245–252.

[SMKF04] SHILANE P., MIN P., KAZHDAN M., FUNKHOUSER T.: The Princeton Shape Benchmark. In *SMI'04* (2004), pp. 167–178.

[TV04] TANGELDER J., VELTKAMP R.: A survey of content based 3D shape retrieval methods. In *SMI'04* (2004), pp. 145–156.

[VUFF93] VERRI A., URAS C., FROSINI P., FERRI M.: On the use of size functions for shape analysis. *Biol. Cybernetics 70* (1993), 99–107.

[ZS*05] ZHANG J., , SIDDIQI K., MACRINI D., SHOKOUFANDEH A., DICKINSON S.: Retrieving articulated 3-D models using medial surfaces and their graph spectra. In *EMMCVPR'05* (2005), pp. 285–300.

International Programme Committee

External Reviewers

Author Index